STAR WARE

The Amateur Astronomer's Ultimate Guide to Choosing, Buying, and Using Telescopes and Accessories

THIRD EDITION

Philip S. Harrington

John Wiley & Sons, Inc.

For my daughter, Helen, the star of my life

This book is printed on acid-free paper. ♾

Published by John Wiley & Sons, Inc., New York
Published simultaneously in Canada

ISBN: 0-471-41806-4

Printed in the United States of America

10 9 8 7 6 5 4 3 2

Contents

Preface to the Third Edition

> If the pure and elevated pleasure to be derived from the possession and use of a good telescope . . . were generally known, I am certain that no instrument of science would be more commonly found in the homes of intelligent people. There is only one way in which you can be sure of getting a good telescope. First, decide how large a glass you are to have, then go to a maker of established reputation, fix upon the price you are willing to pay—remembering that good work is never cheap—and finally see that the instrument furnished to you answers the proper tests for telescopes of its size. There are telescopes and there are telescopes . . .

With these words of advice, Garrett Serviss opened his classic work *Pleasures of the Telescope*. Upon its publication in 1901, this book inspired many an armchair astronomer to change from merely a spectator to a participant, actively observing the universe instead of just reading about it. In many ways, that book was an inspiration for the volume you hold before you.

The telescope market is radically different than it was in the days of Serviss. Back then, amateur astronomy was an activity of the wealthy. The selection of commercially made telescopes was restricted to only one type of instrument—the refractor—and sold for many times what their modern descendants cost today (after correcting for inflation).

By contrast, we live in an age that thrives on choice. Amateur astronomers must now wade through an ocean of literature and propaganda before being able to select a telescope intelligently. For many a budding astronomer, this chore appears overwhelming.

That is where this book comes in. You and I are going hunting for telescopes. After opening chapters that explain telescope jargon and history, today's astronomical marketplace is dissected and explored. Where is the best place to buy a telescope? Is there one telescope that does everything well? How should a telescope be cared for? What accessories are needed? The list of questions goes on and on.

Happily, so do the answers. Although there is no single set of answers that are right for everybody, all of the available options will be explored so that you can make an educated decision. All of the chapters that detail telescopes, binoculars, eyepieces, and accessories have been fully updated in this third edition to include dozens of new products. Reviews have also been expanded, based on my own experiences from testing equipment for *Astronomy* magazine as well as from hundreds of comments that I have received from readers around the world!

Not all of the best astronomical equipment is available for sale, however; some of it has to be made at home. Eleven new homemade projects are outlined further in the book. The book concludes with a discussion of how to care for and use a telescope.

Yes, the telescope marketplace has certainly changed in the past century (even in the four years since the second edition of *Star Ware* was released), and so has the universe. The amateur astronomer has grown with these changes to explore the depths of space in ways that our ancestors could not have even imagined.

Acknowledgments

Putting together a book of this sort would not have been possible were it not for the support of many other players. I would be an irresponsible author if I relied solely on my own humble opinions about astronomical equipment. To compile the telescope, eyepiece, and accessories reviews, I solicited input from amateur astronomers around the world. The responses I received were very revealing and immensely helpful. Unfortunately, space does not permit me to list the names of the hundreds of amateurs who contributed, but you all have my heartfelt thanks. I want to especially acknowledge the members of the "Talking Telescopes" e-mail discussion group that I established in 1999. A great group that I encourage you to join. I also want to thank the freewheeling spirit of those who participate in the sci.astro.amateur Internet newsgroup. This book would be very different were it not for today's vast electronic communications network.

I also wish to acknowledge the contributions of the companies and dealers who provided me with their latest information, references, and other vital data. Joe O'Neil from O'Neil Photo and Optical in Toronto, Canada, and Frank Mirasola from Astrotec in Oakdale, New York, deserve special recognition for allowing me to borrow and test equipment.

As you will see, chapter 8 is a selection of build-at-home projects for amateur astronomers. All were invented and constructed by amateur astronomers who were looking to enhance their enjoyment of the hobby. They were kind enough to supply me with information, drawings, and photographs so that I could pass their projects along to you. For their invaluable contributions, I wish to thank P. J. Anway, James Crombie, Chris Flynn, Dave Kratz, Ghyslain Loyer, Kurt Maurer, Randall McClelland, Ed Stewart, Dave Trott, and Glen Warchol.

I wish to pass on my sincere appreciation to my proofreaders for this edition: Chris Adamson, Kevin Dixon, Geoff Gaherty, Richard Sanderson, and my wife, Wendy Harrington. I am especially indebted to them for submitting constructive suggestions while massaging my sensitive ego. Many thanks also to Kate Bradford of John Wiley & Sons for her diligent guidance and help.

Finally, my deepest thanks, love, and appreciation go to my ever-patient family. My wife, Wendy, and daughter, Helen, have continually provided me with boundless love and encouragement over the years. Were it not for their understanding my need to go out at three in the morning or drive an hour or more from home just to look at the stars, this book could not exist. I love them both dearly for that.

You, dear reader, have a stake in all this, too. This book is not meant to be written, read, and forgotten about. It is meant to change, just as the hobby of astronomy changes. As you read through this occasionally opinionated book (did I say "occasionally?"), there may be a passage or two that you take exception to. Or maybe you own a telescope or something else astronomical that you are either happy or unhappy with. If so, great! This book is meant to kindle emotion. Drop me a line and tell me about it. I want to know. Please address all correspondence to me in care of John Wiley & Sons, Inc., 111 River Street, Hoboken, NJ 07030. If you prefer, e-mail me at phil@philharrington.net. And please check out additions and addenda in the Star Ware 3.5 section of my web site, http://www.philharrington.net. I shall try to answer all letters, but in case I miss yours, thank you in advance!

1

Parlez-Vous "Telescope"?

Before the telescope, ours was a mysterious universe. Events occurred nightly that struck both awe and dread into the hearts and minds of early stargazers. Was the firmament populated with powerful gods who looked down upon the pitiful Earth? Would the world be destroyed if one of these deities became displeased? Eons passed without an answer.

The invention of the telescope was the key that unlocked the vault of the cosmos. Though it is still rich with intrigue, the universe of today is no longer one to be feared. Instead, we sense that it is our destiny to study, explore, and embrace the heavens. From our backyards we are now able to spot incredibly distant phenomena that could not have been imagined just a generation ago. Such is the marvel of the modern telescope.

Today's amateur astronomers have a wide and varied selection of equipment from which to choose. To the novice stargazer, it all appears very enticing but very complicated. One of the most confusing aspects of amateur astronomy is telescope vernacular—terms whose meanings seem shrouded in mystery. "Do astronomers speak a language all their own?" is the cry frequently echoed by newcomers to the hobby. The answer is yes, but it is a language that, unlike some foreign tongues, is easy to learn. Here is your first lesson.

Many different kinds of telescopes have been developed over the years. Even though their variations in design are great, all fall into one of three broad categories according to how they gather and focus light. *Refractors,* shown in Figure 1.1a, have a large lens (the *objective*) mounted in the front of the tube to perform this task, whereas *reflectors,* shown in Figure 1.1b, use a large mirror (the *primary mirror*) at the tube's bottom. The third class of telescope, called *catadioptrics* (Figure 1.1c), places a lens (here called a *corrector plate*) in front of the primary mirror. In each instance, the telescope's *prime optic* (objective lens or primary mirror) brings the incoming light to a *focus* and then directs

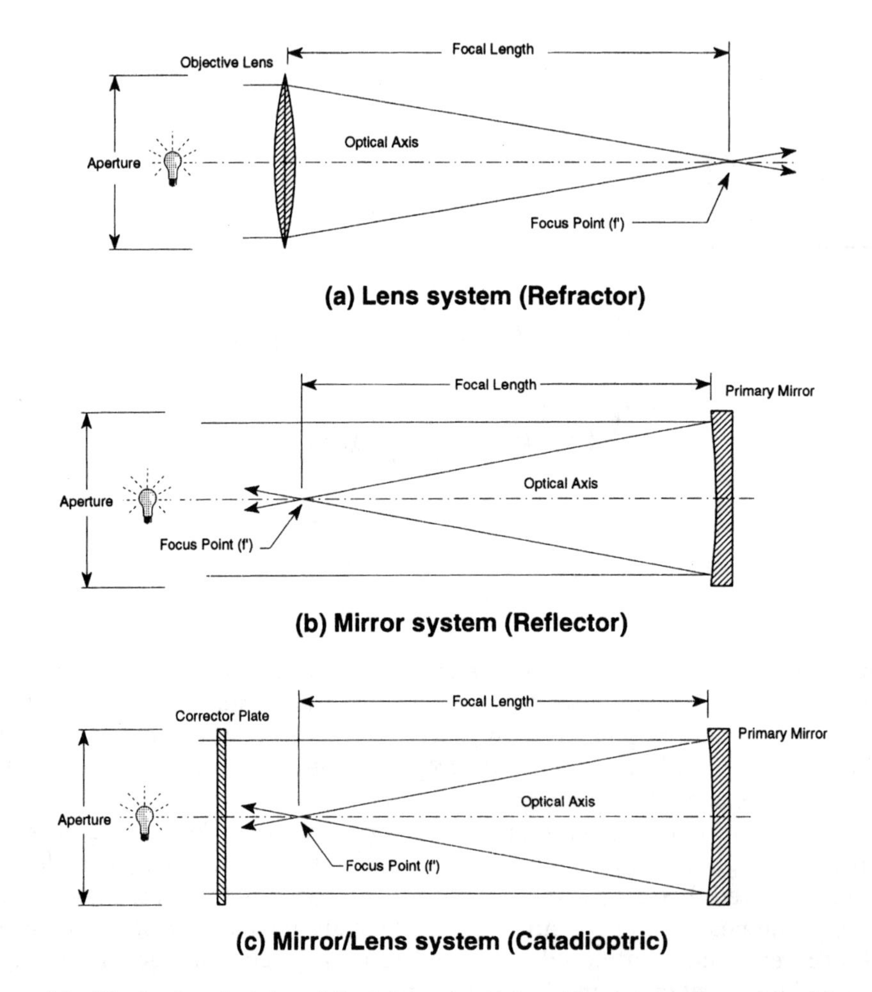

Figure 1.1 *The basic principles of the telescope. Using either a lens (a), a mirror (b), or a combination (c), a telescope bends parallel rays of light to a focus point, or prime focus.*

that light through an *eyepiece* to the observer's waiting eye. Although chapter 2 addresses the history and development of these grand instruments, we will begin here by exploring the many facets and terms that all telescopes share. As you read through the following discussion, be sure to pause and refer to the diagrams found in chapter 2. This way, you can see how individual terms relate to the various types of telescopes.

Aperture

Let's begin with the basics. When we refer to the size of a telescope, we speak of its *aperture*. The aperture is simply the diameter (usually expressed in inches, centimeters, or millimeters) of the instrument's prime optic. In the case of a refractor, the diameter of the objective lens is cited, whereas in reflectors and catadioptric instruments, the diameters of their primary mirrors are specified. For instance, the objective lens in Galileo's first refractor was about 1.5 inches in diameter; it is therefore designated a 1.5-inch refractor. Sir Isaac Newton's first reflecting telescope employed a 1.33-inch mirror and would be referred to today as a 1.33-inch Newtonian reflector.

Many amateur astronomers consider aperture to be the most important criterion when selecting a telescope. In general (and there are exceptions to this, as pointed out in chapter 3), the larger a telescope's aperture, the brighter and clearer the image it will produce. And that is the name of the game: sharp, vivid views of the universe.

Focal Length

The *focal length* is the distance from the objective lens or primary mirror to the *focal point* or *prime focus,* which is where the light rays converge. In reflectors and catadioptrics, this distance depends on the curvature of the telescope's mirrors, with a deeper curve resulting in a shorter focal length. The focal length of a refractor is dictated by the curves of the objective lens as well as by the type of glass used to manufacture the lens.

As with aperture, focal length is commonly expressed in either inches, centimeters, or millimeters.

Focal Ratio

Looking through astronomical books and magazines, it's not unusual to see a telescope specified as, say, an 8-inch f/10 or a 14-inch f/4.5. This f-number is the instrument's *focal ratio,* which is simply the focal length divided by the aperture. Therefore, an 8-inch telescope with a focal length of 56 inches would have a focal ratio (*f-ratio*) of f/7, because $56 \div 8 = 7$. Likewise, by turning the expression around, we know that a 6-inch f/8 telescope has a focal length of 48 inches, because $6 \times 8 = 48$.

Readers familiar with photography may already be used to referring to lenses by their focal ratios. In the case of cameras, a lens with a faster focal ratio (that is, a smaller f-number) will produce brighter images on film, thereby allowing shorter exposures when shooting dimly lit subjects. The same is true for telescopes. Instruments with faster focal ratios will produce brighter images on film, reducing the exposure times needed to record faint objects. However, a telescope with a fast focal ratio will *not* produce brighter images when used visually. The view of a particular object through, say, an 8-inch f/5 and an 8-inch f/10 will be identical when both are used at the same magnification. How bright an object appears to the eye depends only on telescope aperture and magnification.

Magnification

Many people, especially those new to telescopes, are under the false impression that the higher the magnification, the better the telescope. How wrong they are! It's true that as the power of a telescope increases, the apparent size of whatever is in view grows larger, but what most people fail to realize is that at the same time, the images become fainter and fuzzier. Finally, as the magnification climbs even higher, image quality becomes so poor that less detail will be seen than at lower powers.

It's easy to figure out the magnification of a telescope. If you look at the barrel of any eyepiece, you will notice a number followed by *mm*. It might be 26 mm, 12 mm, or 7 mm, among others; this is the focal length of that particular eyepiece expressed in millimeters. Magnification is calculated by dividing the telescope's focal length by the eyepiece's focal length. Remember to first convert the two focal lengths into the same units of measure—that is, both in inches or both in millimeters. A helpful hint: There are 25.4 millimeters in an inch.

For example, let's figure out the magnification of an 8-inch f/10 telescope with a 26-mm eyepiece. The telescope's 80-inch focal length equals 2,032 millimeters ($80 \times 25.4 = 2{,}032$). Dividing 2,032 by the eyepiece's 26-mm focal length tells us that this telescope/eyepiece combination yields a magnification of 78× (read *78 power*), because $2{,}032 \div 26 = 78$.

Most books and articles state that magnification should not exceed 60× per inch of aperture. This is true only under *ideal* conditions, something most observers rarely enjoy. Due to atmospheric turbulence (what astronomers call *poor seeing*), interference from artificial lighting, and other sources, many experienced observers seldom exceed 40× per inch. Some add a caveat to this: Never exceed 300× even if the telescope's aperture permits it. Others insist there is nothing wrong with using more than 60× per inch as long as the sky conditions and optics are good enough. As you can see, the issue of magnification is always a hot topic of debate. My advice for the moment is to use the lowest magnification required to see what you want to see, but we are not done with the subject just yet. Magnification will be spoken of again in chapter 5.

Light-Gathering Ability

The human eye is a wondrous optical device, but its usefulness is severely limited in dim lighting conditions. When fully dilated under the darkest circumstances, the pupils of our eyes expand to about a quarter of an inch, or 7 mm, although this varies from person to person—the older you get, the less your pupils will dilate. In effect,we are born with a pair of quarter-inch refractors.

Telescopes effectively expand our pupils from fractions of an inch to many inches in diameter. The heavens now unfold before us with unexpected glory. A telescope's ability to reveal faint objects depends primarily on the diameter of either its objective lens or primary mirror (in other words, its aperture), not on magnification; quite simply, the larger the aperture, the more light gathered. Doubling a telescope's diameter increases light-gathering power by a factor of four, tripling its aperture expands it by nine times, and so on.

A telescope's *limiting magnitude* is a measure of how faint a star the instrument will show. Table 1.1 lists the faintest stars that can be seen through some popular telescope sizes. Trying to quantify limiting magnitude is any-

Table 1.1 ***Limiting Magnitudes***

Telescope In.	Aperture mm	Faintest Magnitude	Telescope In.	Aperture mm	Faintest Magnitude
2	51	10.3	14	356	14.5
3	76	11.2	16	406	14.8
4	102	11.8	18	457	15.1
6	152	12.7	20	508	15.3
8	203	13.3	24	610	15.7
10	254	13.8	30	762	16.2
12.5	318	14.3			

thing but precise due to a large number of variables. Apart from aperture, other factors affecting this value include the quality of the telescope's optics, meteorological conditions, light pollution, excessive magnification, apparent size of the target, and the observer's vision and experience. These numbers are conservative estimates; experienced observers under dark, crystalline skies can better these by perhaps half a magnitude or more.

Resolving Power

A telescope's *resolving power* is its ability to reveal fine detail in whatever it is aimed at. Though resolving power plays a big part in everything we look at, it is especially important when viewing subtle planetary features, small surface markings on the Moon, or searching for close-set double stars.

A telescope's ability to resolve fine detail is always expressed in *arc-seconds*. You may remember this term from high-school geometry. Recall that in the sky there are 90° from horizon to the overhead point, or zenith, and 360° around the horizon. Each one of those degrees may be broken into 60 equal parts called *arc-minutes*. For example, the apparent diameter of the Moon in our sky may be referred to as either 0.5° or 30 arc-minutes, each one of which may be further broken down into 60 arc-seconds. Therefore, the Moon may also be sized as 1,800 arc-seconds.

Regardless of the size, quality, or location of a telescope, stars will never appear as perfectly sharp points. This is partially due to atmospheric interference and partially due to the fact that light consists of slightly fuzzy waves rather than mathematically straight lines. Even with perfect atmospheric conditions, what we see is a blob, technically called the *Airy disk* (named in honor of its discoverer, Sir George Airy, Britain's Astronomer Royal from 1835 to 1892). Because light is composed of waves, rays from different parts of a telescope's prime optic (be it a mirror or lens) alternately interfere with and enhance each other, producing a series of dark and bright concentric rings around the Airy disk (Figure 1.2a). The whole display is known as a *diffraction pattern*. Ideally, through a telescope without a central obstruction (that is, without a secondary mirror), 84% of the starlight remains concentrated in the

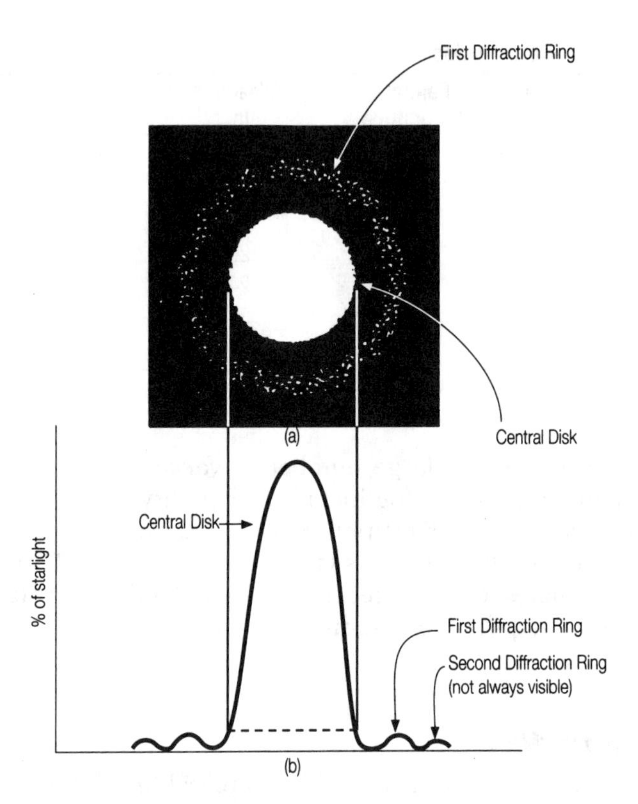

Figure 1.2 *The Airy disk (a) as it appears through a highly-magnified telescope and (b) graphically showing the distribution of light.*

central disk, 7% in the first bright ring, and 3% in the second bright ring, with the rest distributed among progressively fainter rings.

Figure 1.2b graphically presents a typical diffraction pattern. The central peak represents the bright central disk, and the smaller humps show the successively fainter rings.

The apparent diameter of the Airy disk plays a direct role in determining an instrument's resolving power. This becomes especially critical for observations of close-set double stars. How large an Airy disk will a given telescope produce? Table 1.2 summarizes the results for most common amateur-size telescopes.

Although these values would appear to indicate the resolving power of the given apertures, some telescopes can actually exceed these bounds. The nineteenth-century English astronomer William Dawes found experimentally that the closest a pair of 6th-magnitude yellow stars can be to each other and still be distinguishable as two points can be estimated by dividing 4.56 by the telescope's aperture. This is called *Dawes' Limit* (Figure 1.3). Table 1.3 lists Dawes' Limit for some common telescope sizes.

When using telescopes of less than 6-inch aperture, some amateurs can readily exceed Dawes' Limit, while others will never reach it. Does this mean that they are doomed to be failures as observers? Not at all! Remember that

Table 1.2 ***Resolving Power***

Telescope Aperture in.	mm	Diameter of Airy Disk (theoretical) arc-seconds	Telescope Aperture in.	mm	Diameter of Airy Disk (theoretical) arc-seconds
2	51	5.5	14	356	0.78
3	76	3.6	16	406	0.68
4	102	2.8	18	457	0.60
6	152	1.8	20	508	0.54
8	203	1.4	24	610	0.46
10	254	1.1	30	762	0.36
12.5	318	0.88			

Dawes' Limit was developed under very precise conditions that may have been far different than your own. Just as with limiting magnitude, reaching Dawes' Limit can be adversely affected by many factors, such as turbulence in our atmosphere, a great disparity in the test stars' colors and/or magnitudes, misaligned or poor-quality optics, and the observer's visual acuity.

Rarely will a large-aperture telescope—that is, one greater than about ten inches—resolve to its Dawes' Limit. Even the largest backyard instruments can almost never show detail finer than between 0.5 arc-seconds (abbreviated 0.5″)

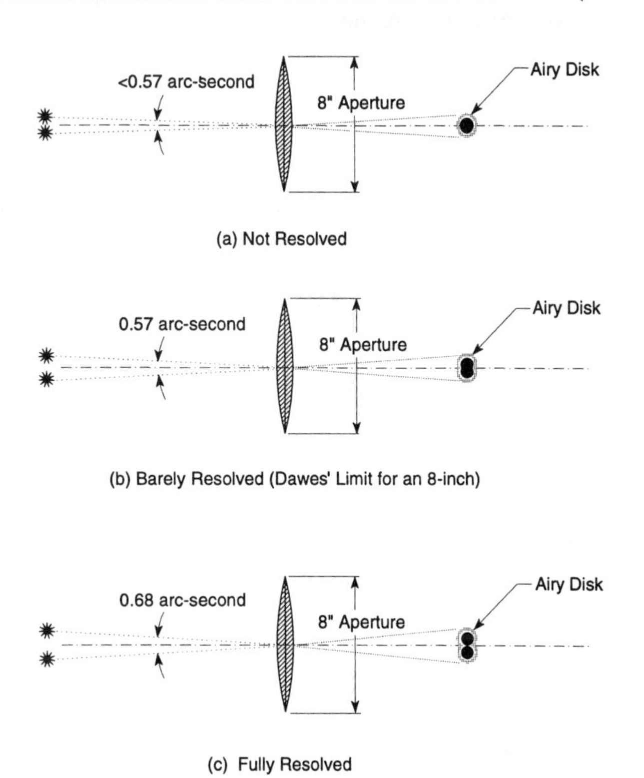

Figure 1.3 *The resolving power of an 8-inch telescope: (a) not resolved; (b) barely resolved, and the Dawes' Limit for the aperture; (c) fully resolved.*

Table 1.3 ***Dawes' Limit***

Telescope in.	Aperture mm	Limit of Resolution arc-seconds	Telescope in.	Aperture mm	Limit of Resolution arc-seconds
2	51	2.3	14	356	0.33
3	76	1.5	16	406	0.29
4	102	1.1	18	457	0.25
6	152	0.76	20	508	0.23
8	203	0.57	24	610	0.19
10	254	0.46	30	762	0.15
12.5	318	0.36			

and 1 arc-second (1″). In other words, a 16- to 18-inch telescope will offer little additional detail compared to an 8- to 10-inch one when used under most observing conditions. Interpret Dawes' Limit as a telescope's equivalent to the projected gas mileage of an automobile: "These are test results only—your actual numbers may vary."

We have just begun to digest a few of the multitude of telescope terms that are out there. Others will be introduced in the succeeding chapters as they come along, but for now, the ones we have learned will provide enough of a foundation for us to begin our journey.

2

In the Beginning . . .

To appreciate the grandeur of the modern telescope, we must first understand its history and development. It is a rich history, indeed. Since its invention, the telescope has captured the curiosity and commanded the respect of princes and paupers, scientists and laypersons. Peering through a telescope renews the sense of wonder we all had as children. In short, it is a tool that sparks the imagination in us all.

Who is responsible for this marvelous creation? Ask this question of most people and they probably will answer, "Galileo." Galileo Galilei did, in fact, usher in the age of telescopic astronomy when he first turned his telescope, illustrated in Figure 2.1, toward the night sky. With it, he became the first person in human history to witness craters on the Moon, the phases of Venus, four of the moons orbiting Jupiter, and many other hitherto unknown heavenly sights. Though he was ridiculed by his contemporaries and persecuted for heresy, Galileo's observations changed humankind's view of the universe as no single individual's ever had before or has since. But he did not make the first telescope.

So who did? The truth is that no one knows for certain just who came up with the idea, or even when. Many knowledgeable historians tell us that it was Jan Lippershey, a spectacle maker from Middelburg, Holland. Records indicate that in 1608 he first held two lenses in line and noticed that they seemed to bring distant scenes closer. Subsequently, Lippershey sold many of his telescopes to his government, which recognized the military importance of such a tool. In fact, many of his instruments were sold in pairs, thus creating the first field glasses.

Other evidence may imply a much earlier origin for the telescope. Archaeologists have unearthed glass in Egypt that dates to about 3500 B.C., while primitive lenses have been found in Turkey and Crete that are thought to be

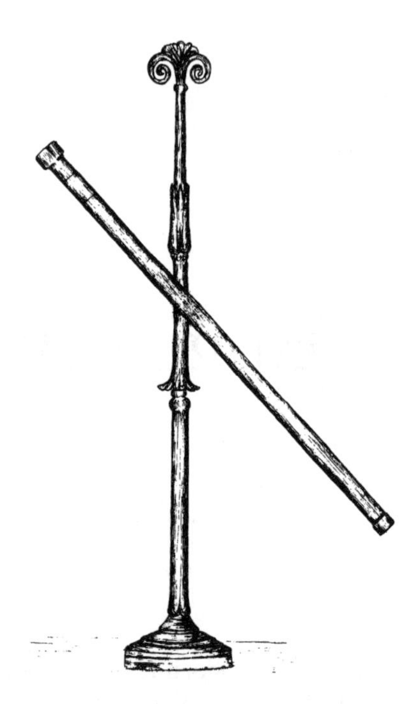

Figure 2.1 *Artist's rendition of Galileo's first telescope. Artwork by David Gallup.*

4,000 years old! In the third century B.C., Euclid wrote about the reflection and refraction of light. Four hundred years later, the Roman writer Seneca referred to the magnifying power of a glass sphere filled with water.

Although it is unknown if any of these independent works led to the creation of a telescope, the English scientist Roger Bacon wrote of an amazing observation made in the thirteenth century: "... Thus from an incredible distance we may read the smallest letters ... the Sun, Moon and stars may be made to descend hither in appearance ... " Might he have been referring to the view through a telescope? We may never know.

Refracting Telescopes

Though its inventor may be lost to history, this early kind of telescope is called a *Galilean* or *simple* refractor. The Galilean refractor consists of two lenses: a convex (curved outward) lens held in front of a concave (curved inward) lens a certain distance away. As you know, the telescope's front lens is called the objective, while the other is referred to as the eyepiece, or *ocular.* The Galilean refractor placed the concave eyepiece *before* the objective's prime focus; this produced an upright, extremely narrow field of view, like today's inexpensive opera glasses.

Not long after Galileo made his first telescope, Johannes Kepler improved on the idea by simply swapping the concave eyepiece for a double convex lens, placing it behind the prime focus. The *Keplerian refractor* proved to be far

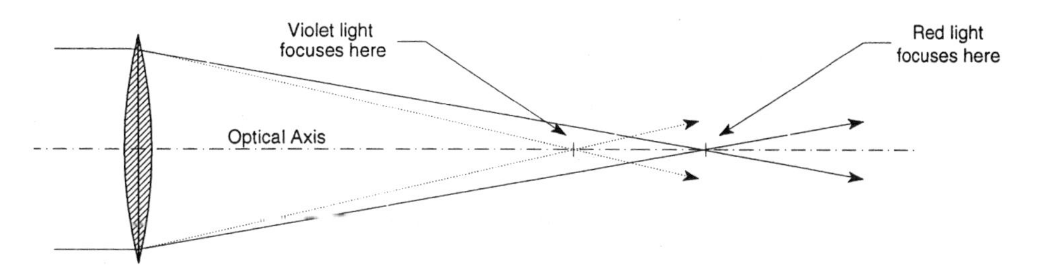

Figure 2.2 *Chromatic aberration, the result of a simple lens focusing different wavelengths of light at different distances.*

superior to Galileo's instrument. The modern refracting telescope continues to be based on Kepler's design. The fact that the view is upside down is of little consequence to astronomers because there is no up and down in space; for terrestrial viewing, extra lenses may be added to flip the image a second time, reinverting the scene.

Unfortunately, both the Galilean and the Keplerian designs have several optical deficiencies. Chief among these is *chromatic aberration* (Figure 2.2). As you may know, when we look at any white-light source, we are not actually looking at a single wavelength of light but rather a collection of wavelengths mixed together. To prove this for yourself, shine sunlight through a prism. The light going in is refracted within the prism, exiting not as a unit but instead broken up, forming a rainbowlike spectrum. Each color of the spectrum has its own unique wavelength.

If you use a lens instead of a prism, each color will focus at a slightly different point. The net result is a zone of focus, rather than a point. Through such a telescope, everything appears blurry and surrounded by halos of color. This effect is called chromatic aberration.

Another problem of simple refractors is *spherical aberration* (Figure 2.3). In this instance, the curvature of the objective lens causes the rays of light entering around its edges to focus at a slightly different place than those striking the center. Once again, the light focuses within a range rather than at a single point, making the telescope incapable of producing a clear, razor-sharp image.

Modifying the inner and outer curves of the lens proved somewhat helpful. Experiments showed that both defects could be reduced (but not totally eliminated) by increasing the focal length—that is, decreasing the curvature—of the objective lens. And so, in an effort to improve image quality, the refractor became longer... and longer... and even longer! The longest refractor on record was constructed by Johannes Hevelius in Denmark during the latter part of the seventeenth century; it measured about one hundred and fifty feet from objective to eyepiece and required a complex sling system suspended high above the ground on a wooden mast to hold it in place! Can you imagine the effort it must have taken to swing around such a monster just to look at the Moon or a bright planet? Surely, there had to be a better way.

In an effort to combat these imperfections, Chester Hall developed a two-element *achromatic lens* in 1733. Hall learned that by using two matching

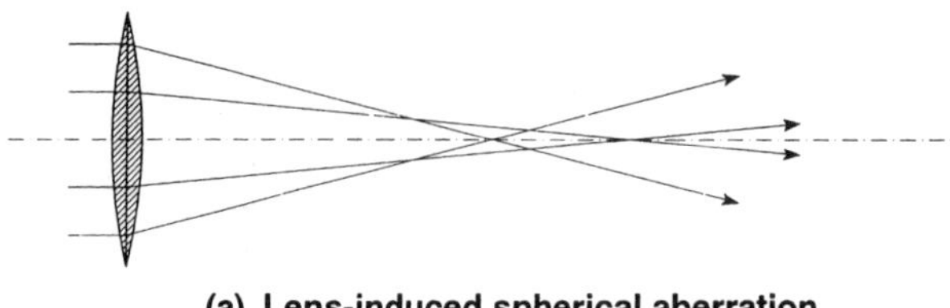

(a) Lens-induced spherical aberration

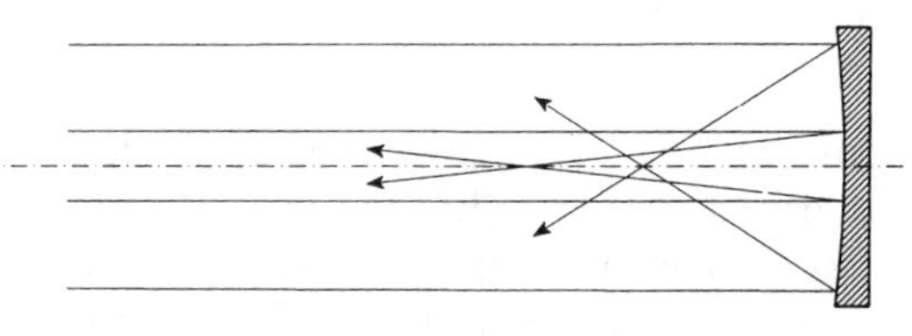

(b) Mirror-induced spherical aberration

Figure 2.3 *Spherical aberration. Both (a) lens-induced and (b) mirror-induced spherical aberration are caused by incorrectly figured optics.*

lenses made of different types of glass, aberrations could be greatly reduced. In an achromatic lens, the outer element is usually made of crown glass, while the inner element is typically flint glass. Crown glass has a lower dispersion effect and therefore bends light rays less than flint glass, which has a higher dispersion. The convergence of light passing through the crown-glass lens is compensated by its divergence through the flint-glass lens, resulting in greatly dampened aberrations. Ironically, though Hall made several telescopes using this arrangement, the idea of an achromatic objective did not catch on for another quarter century.

In 1758, John Dollond reacquainted the scientific community with Hall's idea when he was granted a patent for a two-element aberration-suppressing lens. Though quality glass was hard to come by for both of these pioneers, it appears that Dollond was more successful at producing a high-quality instrument. Perhaps that is why history records John Dollond, rather than Chester Hall, as the father of the modern refractor.

Regardless of who first devised it, this new and improved design has come to be called the *achromatic refractor* (Figure 2.4a), with the compound objective simply labeled an *achromat.* Though the methodology for improving the refractor was now known, the problem of getting high-quality glass (especially flint glass) persisted. In 1780, Pierre Louis Guinard, a Swiss bell maker, began experimenting with various casting techniques in an attempt to improve the glass-making process. It took him close to 20 years, but Guinard's efforts ultimately paid off, for he learned the secret of producing flawless optical disks as big as roughly 6 inches in diameter.

Later, Guinard was to team up with Joseph von Fraunhofer, inventor of the spectroscope. While studying under Guinard's guidance, Fraunhofer experimented by slightly modifying the lens curves suggested by Dollond, which resulted in the highest-quality objective yet created. In Fraunhofer's design, the front surface is strongly convex. The two central surfaces differ slightly from each other, requiring a narrow air space between the elements, while the innermost surface is almost perfectly flat. These innovations bring two wavelengths of light across the lens's full diameter to a common focus, thereby greatly reducing chromatic and spherical aberration.

The world's largest refractor is the 40-inch f/19 telescope at Yerkes Observatory in Williams Bay, Wisconsin. This mighty instrument was constructed by

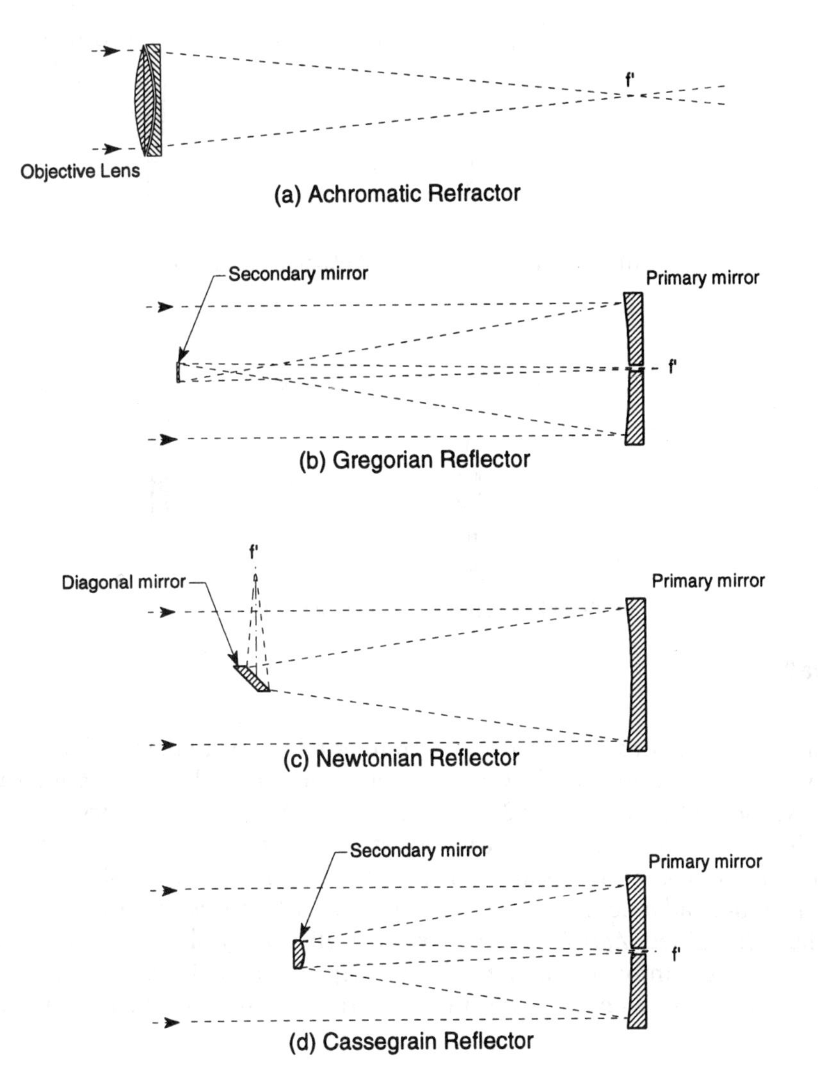

Figure 2.4 *Telescopes come in all different shapes and sizes: (a) achromatic refractor, (b) Gregorian reflector, (c) Newtonian reflector, (d) Cassegrain reflector, (e) Schmidt catadioptric telescope, (f) Maksutov-Cassegrain telescope, and (g) Schmidt-Cassegrain telescope.*

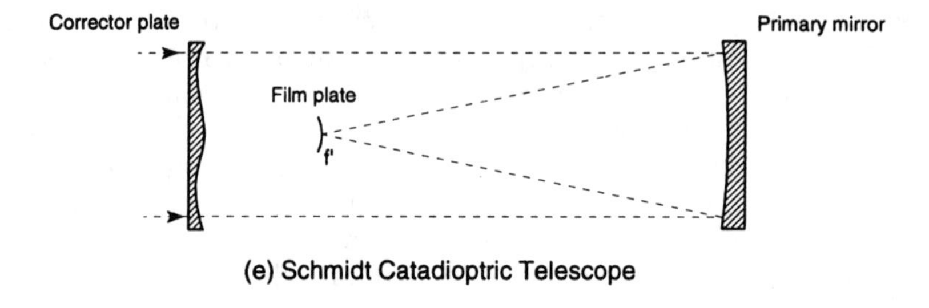

(e) Schmidt Catadioptric Telescope

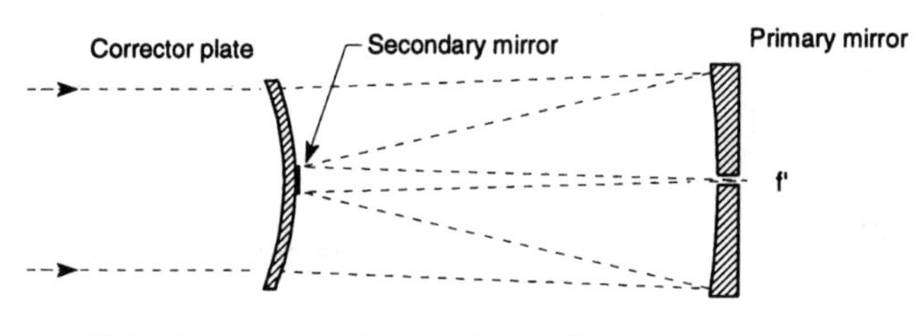

(f) Maksutov-Cassegrain Catadioptric Telescope

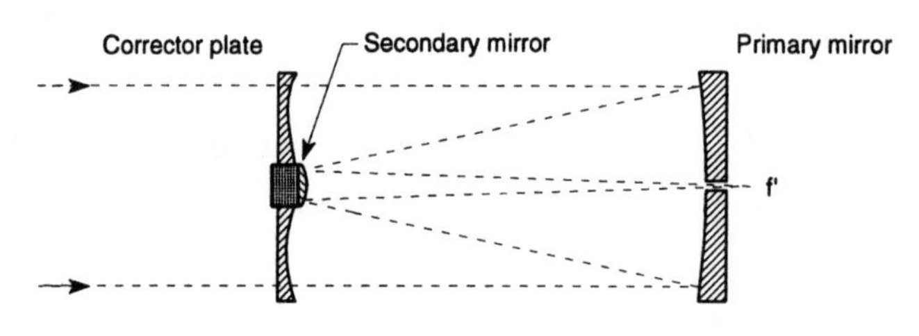

(g) Schmidt-Cassegrain Catadioptric Telescope

Figure 2.4 continued

Alvan Clark and Sons, Inc., America's premier telescope maker of the nineteenth century. Other examples of the Clarks' exceptional skill include the 36-inch at Lick Observatory in California, the 26-inch at the U.S. Naval Observatory in Washington, D.C., and many smaller refractors at universities and colleges worldwide. Even today, Clark refractors are considered to be among the finest available.

The most advanced modern refractors offer features that the Clarks could not have imagined. *Apochromatic refractors* effectively eliminate just about all aberrations common to their Galilean, Keplerian, and achromatic cousins. More about these when we examine consumer considerations in chapter 3.

Reflecting Telescopes

But there is more than one way to skin a cat. The second general type of telescope utilizes a large mirror, rather than a lens, to focus light to a point—not just any mirror, mind you, but a mirror with a precisely figured surface. To understand how a mirror-based telescope works, we must first reflect on how mirrors work (sorry about that). Take a look at a mirror in your home. Chances are it is flat, as shown in Figure 2.5a. Light that is cast onto the mirror's polished surface in parallel rays is reflected back in parallel rays. If the mirror is convex (Figure 2.5b), the light diverges after it strikes the surface. But if the mirror is concave (Figure 2.5c), then the rays converge toward a common point, or focus. (It should be pointed out here that household mirrors are *second-surface* mirrors; that is, their reflective coating is applied onto the back surface. Reflecting telescopes use *front-surface* mirrors, coated on the front.)

The first reflecting telescope was designed by James Gregory in 1663. His system centered around a concave mirror (called the *primary mirror*). The primary mirror reflected light to a smaller concave *secondary mirror*, which, in turn, bounced the light back through a central hole in the primary and out to the eyepiece. The *Gregorian reflector* (Figure 2.4b) had the benefit of yielding

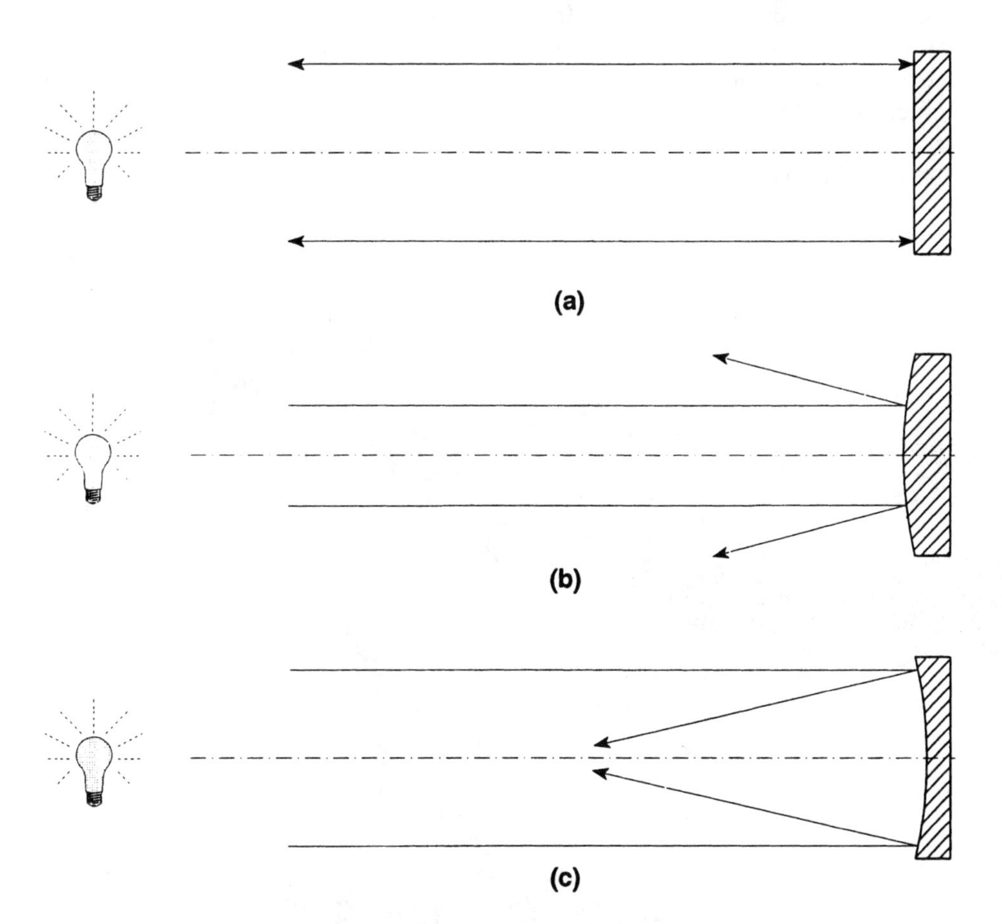

Figure 2.5 *Three mirrors, each with a different front-surface curve, reflect light differently. A flat mirror (a) reflects light straight back to the source, a convex mirror (b) causes light to diffuse, and a concave mirror (c) focuses light.*

an upright image, but its optical curves proved difficult for Gregory and his contemporaries to fabricate.

A second design was later conceived by Sir Isaac Newton in 1672 (Figure 2.6). Like Gregory, Newton realized that a concave mirror would reflect and focus light back along the optical axis to a point called the prime focus. Here an observer could view a magnified image through an eyepiece. Quickly realizing that his head got in the way, Newton inserted a flat mirror at a 45° angle some distance in front of the primary. The secondary, or *diagonal,* mirror acted to bounce the light 90° out through a hole in the side of the telescope's tube. This arrangement has since become known as the *Newtonian reflector* (Figure 2.4c).

The Newtonian became the most popular design among amateur astronomers in the 1930s, when Vermonter Russell Porter wrote a series of articles for *Scientific American* magazine that popularized the idea of making your own telescope. The Newtonian is relatively easy (and, therefore, inexpensive) to make, giving amateurs the most bang for their buck. Although chromatic aberration is completely absent (as it is in all reflecting telescopes), the Newtonian is not without its faults. Coma, which turns pinpoint stars away from the center of view into tiny "comets," with their "tails" aimed outward from the center, is the biggest problem, and is exacerbated as the telescope's focal ratio drops. Optical alignment is also critical, especially in fast optical systems, and must be checked often.

Figure 2.6 *Newton's first reflecting telescope. From* Great Astronomers *by Sir Robert S. Ball, London, 1912.*

The French sculptor Sieur Cassegrain also announced in 1672 a third variation of the reflecting telescope. The Cassegrainian reflector (yes, the telescope is correctly called a *Cassegrainian,* but since most other sources refer to it as a *Cassegrain,* I will from this point on, as well) is outwardly reminiscent of Gregory's original design. The biggest difference between a *Cassegrain reflector* (Figure 2.4d) and a Gregorian reflector is the curve of the secondary mirror's surface. The Gregorian uses a concave secondary mirror positioned outside the main focus, while Cassegrain uses a convex secondary mirror inside the main focus. The biggest plus to the Cassegrain is its compact design, which combines a large aperture inside a short tube. Optical problems include lower image contrast than a Newtonian, as well as strong curvature of field and coma, causing stars along the edges of the field to blur when those in the center are focused.

Both Newton and Cassegrain received acclaim for their independent inventions, but neither telescope saw further development for many years. One of the greatest difficulties to overcome was the lack of information on suitable materials for mirrors. Newton, for instance, made his mirrors out of bell metal whitened with arsenic. Others chose speculum metal, an amalgam consisting of copper, tin, and arsenic.

Another complication faced by makers of early reflecting telescopes was generating accurately figured mirrors. In order for all the light striking its surface to focus to a point, a primary mirror's concave surface must be a parabola accurately shaped to within a few millionths of an inch—a fraction of the wavelength of light. Indeed, the Cassegrain reflector didn't begin to catch on for another two centuries, due to the difficulty in making the mirrors. It has since become the most popular design for large, professional instruments and has spawned several variations, discussed later. Even the simpler Newtonians were rarely used, since good mirrors were just too difficult to come by. Unfortunately, the first reflectors used spherically figured mirrors. In this case, rays striking the mirror's edge come to a different focus than the rays striking its center. The net result: spherical aberration.

The first reflector to use a parabolic mirror was constructed by Englishman John Hadley in 1722. The primary mirror of his Newtonian measured about 6 inches across and had a focal length of 62⅝ inches. But whereas Newton and the others had failed to generate mirrors with accurate parabolic concave curves, Hadley succeeded. Extensive tests were performed on Hadley's reflector after he presented it to the Royal Society. In direct comparison between it and the society's 123-foot-focal-length refractor of the same diameter, the reflector performed equally well and was immeasurably simpler to use.

A second success story for the early reflecting telescope was that of James Short, another English craftsman. Short created several fine Newtonian and Gregorian instruments in his optical shop from the 1730s through the 1760s. He placed many of his telescopes on a special type of support that permitted easier tracking of sky objects (what is today termed an *equatorial mount*—see chapter 3). Today, the popularity of the Gregorian reflector has long since faded away, although it is interesting to note that NASA chose that design for its highly successful Solar Max mission of 1980.

Sir William Herschel, a musician who became interested in astronomy when he was given a telescope in 1722, ground some of the finest mirrors of his day. As his interest in telescopes grew, Herschel continued to refine the reflector by devising his own system. The *Herschelian* design called for the primary mirror to be tilted slightly, thereby casting the reflection toward the front rim of the oversized tube, where the eyepiece would be mounted. The biggest advantage to this arrangement is that with no secondary mirror to block the incoming light, the telescope's aperture is unobstructed by a second mirror; disadvantages included image distortion due to the tilted optics and heat from the observer's head. Herschel's largest telescope was completed in 1789. The metal speculum around which it was based measured 48 inches across and had a focal length of 40 feet. Records indicate that it weighed something in excess of one ton.

Even this great instrument was to be eclipsed in 1845, when Lord Rosse completed the largest speculum ever made. It measured 72 inches in diameter and weighed in at an incredible 8,380 pounds. This telescope (Figure 2.7), mounted in Parsonstown, Ireland, is famous in the annals of astronomical history as the first to reveal spiral structure in what were then thought to be nebulae and are now known to be spiral galaxies.

The poor reflective qualities of speculum metal, coupled with its rapid tarnishing, made it imperative to develop a new mirror-making process. That evolutionary step was taken in the following decade. The first reflector to use a glass mirror instead of a metal speculum was constructed in 1856 by Dr. Karl Steinheil of Germany. The mirror, which measured 4 inches across, was coated with a very thin layer of silver; the procedure for chemically bonding silver to glass had been developed by Justus von Liebig about 1840. Although it apparently produced a very good image, Steinheil's attempt received very little attention from the scientific community. The following year, Jean Foucault (creator of the Foucault pendulum and the Foucault mirror test procedure, among others) independently developed a silvered mirror for his astronomical telescope. He brought his instrument before the French Academy of Sciences, which immediately made his findings known to all. Foucault's methods of working glass and testing the results elevated the reflector to new heights of excellence and availability.

Although silver-on-glass specula proved far superior to the earlier metal versions, this new development was still not without flaws. For one thing, silver tarnished quite rapidly, although not as fast as speculum metal. The twentieth century dawned with experiments aimed to remedy the situation, which ultimately led to the process used today of evaporating a thin film of aluminum onto glass in a vacuum chamber. Even though aluminum is not quite as highly reflective as silver, its longer useful life span more than makes up for that slight difference.

Although reflectors do not suffer from the refractor's chromatic aberration, they are anything but flawless. We have already seen how spherical aberration can destroy image integrity, but other problems must be dealt with as well. These include *coma,* which describes objects away from the center of view appearing like tiny comets, with their tails aimed outward from the cen-

Figure 2.7 *Lord Rosse's 72-inch reflecting telescope. From* Elements of Descriptive Astronomy *by Herbert A. Howe, New York, 1897.*

ter; *astigmatism,* resulting in star images that focus to crosses rather than points; and *light loss,* which is caused by obstruction by the secondary mirror and the fact that no reflective surface returns 100% of the light striking it.

Today, there exist many variations of the reflecting telescope's design. While the venerable Newtonian has remained popular among amateur astronomers, the Gregorian is all but forgotten. In addition to the classical Cassegrain, we find two modified versions: the Dall-Kirkham and the Ritchey-Chretien.

Invented by American optician Horace Dall in 1928, the Dall-Kirkham design didn't begin to catch on until Allan Kirkham, an amateur astronomer from Oregon, began to correspond with Albert Ingalls, then the editor of *Scientific American* magazine. Ingalls subsequently published an article that noted the design as the Dall-Kirkham, and the name stuck. Dall-Kirkham

Cassegrains are designed around what is essentially a simplified classical Cassegrain. While easier to make, it is plagued by strong field curvature and coma.

The Ritchey-Chrétien Cassegrain reflector design is based on hyperbolically curved primary and secondary mirrors, concave for the former and convex for the latter, and both very difficult to make (or, at least, make *well*). The design was developed in the early 1910s by American optician George Ritchey and Henri Chrétien, an optical designer from France. Although some field curvature plagues the Ritchey-Chrétien design, it is totally free of coma, astigmatism, and spherical aberration. Interestingly, Ritchey, who had made optics for some of the instruments used atop Mount Wilson in California, so severely criticized the then-new 100-inch Hooker Telescope for not using the Ritchey-Chrétien design that he was fired and, ultimately, ostracized from American astronomy. George Hale, when conceiving the 200-inch Hale reflector for Mount Palomar, also refused to use the Ritchey-Chrétien design, since it had Ritchey's name on it. Ultimately, however, history has proven that Ritchey was right, since nearly every new large telescope built or designed since the 200-inch has used the Ritchey-Chrétien design.

Finally, for the true student of the reflector, there are several lesser-known instruments, such as the Tri-Schiefspiegler (a three-mirror telescope with tilted optics).

Like the refractor, today's reflectors enjoy the benefit of advanced materials and optical coatings. While they are a far cry from the first telescopes of Newton, Gregory and Cassegrain, we must still pause a moment to consider how different our understanding of the universe might be if it were not for these and other early optical pioneers.

Catadioptric Telescopes

Earlier this century, some comparative newcomers launched a whole new breed of telescope: the *catadioptric*. These telescopes combine attributes of both refractors and reflectors into one instrument. They can produce wide fields with few aberrations. Many declare that this genre is (at least potentially) the perfect telescope; others see it as a collection of compromises.

The first catadioptric was devised in 1930 by German astronomer Bernhard Schmidt. The *Schmidt* telescope (Figure 2.4e) passes starlight through a corrector plate *before* it strikes the spherical primary mirror. The curves of the corrector plate eliminate the spherical aberration that would result if the mirror were used alone. One of the chief advantages of the Schmidt is its fast f-ratio, typically f/1.5 or less. However, owing to the fast optics, the Schmidt's prime focus point is inaccessible to an eyepiece, restricting the instrument to photographic applications only. To photograph through a Schmidt, film is placed in a special curved holder (to accommodate a slightly curved focal plane) at the instrument's prime focus, not far in front of the main mirror.

The second type of catadioptric instrument to be developed was the *Maksutov telescope*. By rights, the Maksutov telescope should probably be called the

Bouwers telescope, after A. Bouwers of Amsterdam, Holland. Bouwers developed the idea for a photovisual catadioptric telescope in February 1941. Eight months later, D. Maksutov, an optical scientist working independently in Moscow, came up with the exact same design. Like the Schmidt, the Maksutov combines features of both refractors and reflectors. The most distinctive trait of the Maksutov is its deep-dish front corrector plate, or *meniscus,* which is placed inside the spherical primary mirror's radius of curvature. Light passes through the corrector plate to the primary and then to a convex secondary mirror.

Most Maksutovs resemble a Cassegrain in design and are therefore referred to as *Maksutov-Cassegrains* (Figure 2.4f). In these, the secondary mirror returns the light toward the primary mirror, passing through a central hole and out to the eyepiece. This layout allows a long focal length to be crammed into the shortest tube possible.

In 1957, John Gregory, an optical engineer working for Perkin-Elmer Corporation in Connecticut, modified the original Maksutov-Cassegrain scheme to improve its overall performance. The main difference in the *Gregory-Maksutov* telescope is that instead of a separate secondary mirror, a small central spot on the interior of the corrector is aluminized to reflect light to the eyepiece.

Though not as common, a Maksutov telescope may also be constructed in a Newtonian configuration. In this scheme, the secondary mirror is tilted at 45°. As in the classical Newtonian reflector, light from the target then passes through a hole in the side of the telescope's tube to the waiting eyepiece. The greatest advantage of the Maksutov-Newtonian over the traditional Newtonian is the availability of a short focal length (and therefore a wide field of view) with greatly reduced coma and astigmatism.

Finally, two hybrids of the Schmidt camera have also been developed: the *Schmidt-Newtonian* and the *Schmidt-Cassegrain* (Figure 2.4g). The Newtonian hybrid remains mostly in the realm of the amateur telescope maker, but since its introduction in the 1960s, the Schmidt-Cassegrain has grown to become the most popular type of telescope sold today. It combines a short-focal-length spherical mirror with an elliptical-figured secondary mirror and a Schmidt-like corrector plate. The net result is a large-aperture telescope that fits into a comparatively small package. Is the Schmidt-Cassegrain the right telescope for you? Only you can answer that question—with a little help from the next chapter, that is.

The telescope has certainly come a long way in its nearly four-hundred-year history, but its history is by no means finished. The age of orbiting observatories, such as the Hubble Space Telescope, has untold possibilities. Back here on the ground, new, giant telescopes, like the Keck reflectors in Hawaii, using segmented mirrors—and even some whose exact curves are controlled and varied by computers to compensate for atmospheric conditions (so-called *adaptive optics*)—are now being aimed toward the universe. New materials, construction techniques, and accessories are coming into use. All this means that the future will see even more diversity in this already diverse field. Stay tuned!

3

So You Want to Buy a Telescope!

So you want to buy a telescope? That's wonderful! A telescope will let you visit places that most people are not even aware exist. With it, you can soar over the stark surface of the Moon, travel to the other worlds in our solar system, and plunge into the dark void of deep space to survey clusters of jewel-like stars, huge interstellar clouds, and remote galaxies. You will witness firsthand exciting celestial objects that were unknown to astronomers only a generation ago. You can become a citizen of the universe without ever leaving your backyard.

Just as a pilot needs the right aircraft to fly from one point to another, so, too, must an amateur astronomer have the right instrument to journey into the cosmos. As we have seen already, many different types of telescopes have been devised in the past four centuries. Some remain popular today, while others are of interest from a historical viewpoint only.

Which telescope is right for you? Had I written this book back in the 1950s or 1960s, there would have been one answer: a 6-inch f/8 Newtonian reflector. Just about every amateur either owned one or knew someone who did. Though many different companies made this type of instrument, the most popular one around was the RV-6 Dynascope by Criterion Manufacturing Company of Hartford, Connecticut, which for years retailed for $194.95. The RV-6 was to telescopes what the Volkswagen Beetle was to cars—a triumph of simplicity and durability at a great price!

Times have changed, the world has grown more complicated, and the hobby of amateur astronomy has become more complex. The venerable RV-6 is no longer manufactured, although some can still be found in classified advertisements. Today, looking through astronomical product literature, we find sophisticated Schmidt-Cassegrains, mammoth Newtonian reflectors, and state-of-the-art refractors. With such a variety from which to choose, it is hard to know where to begin.

Optical Quality

Before examining specific types of telescopes, a few terms used to rate the caliber of telescope lenses and mirrors must be defined and discussed. In the everyday world, when we want to express the accuracy of something, we usually write it in fractions of an inch, centimeter, or millimeter. For instance, when building a house, a carpenter might call for a piece of wood that is, say, 4 feet long plus or minus $^1/_8$ inch. In other words, as long as the piece of wood is within an eighth of an inch of 4 feet, it is close enough to be used.

In the optical world, however, close is not always close enough. Since the curves of a lens or mirror must be made to such tight tolerances, it is not practical to refer to optical quality in everyday units of measurement. Instead, optical quality is usually expressed in fractions of the wavelength of light. Since each color in the spectrum has a different wavelength, opticians use the color that the human eye is most sensitive to: yellow-green. Yellow-green, in the middle of the visible spectrum, has a wavelength of 550 nanometers (that's 0.00055 millimeter, or 0.00002 inch).

For a lens or a mirror to be accurate to, say, $^1/_8$ (0.125 mm) wave (a value frequently quoted by telescope manufacturers), its surface shape cannot deviate from perfection by more than 0.000069 mm, or 0.000003 inch! This means that none of the little irregularities (commonly called *hills* and *valleys*) on the optical surface exceed a height or depth of $^1/_8$ of the wavelength of yellow-green light. As you can see, the smaller the fraction, the better the optics. Given the same aperture and conditions, telescope A with a $^1/_8$-wave prime optic (lens or mirror) should outperform telescope B with a $^1/_4$-wave lens or mirror, while both should be exceeded by telescope C with a $^1/_{20}$-wave prime optic.

Stop right there. Companies are quick to boast about the quality of their primary mirrors and objective lenses, but in reality, we should be concerned with the *final wavefront* reaching the observer's eye, which is double the wave error of the prime optic alone. For instance, a reflecting telescope with a $^1/_8$-wave mirror has a final wavefront of $^1/_4$ wave. This value is known as *Rayleigh's Criterion* and is usually considered the lowest quality level that will produce acceptable images. Clearly, an instrument with a $^1/_8$ to $^1/_{10}$ final wavefront is very good. However, even these figures must be taken loosely since there is no industrywide method of testing.

Due to increasing consumer dissatisfaction with the quality of commercial telescopes, both *Sky & Telescope* and *Astronomy* magazines often test equipment and subsequently publish the results. Talk about a shot heard around the world! Both organizations quickly found out that the claims made by some manufacturers (particularly a few producers of Newtonian reflectors and Schmidt-Cassegrains) were a bit, shall I say, inflated.

In light of this shake-up, many companies have dropped claims of their optics' wavefront, referring to them instead as *diffraction limited,* meaning that the optics are so good that performance is limited only by the wave properties of light itself and not by any flaws in optical accuracy. In general, to be diffraction limited, an instrument's final wavefront must be at least $^1/_4$ wave, the Rayleigh Criterion. Once again, however, this can prove to be a subjective claim.

Telescope Point-Counterpoint

So which telescope would I recommend for you? None of them . . . or all of them! Actually, there is no one answer anymore. It all depends on what you want to use the telescope for, how much money you can afford to spend, and many other considerations. To help sort all this out, you and I are about to go telescope hunting together. We will begin by looking at each type of telescope that is commercially available. The chapter's second section will examine the many different mounting systems used to hold a telescope in place. Finally, all considerations will be weighed to let *you* decide which telescope is right for you.

Binoculars

Binoculars may be thought of as two low-power refracting telescopes that happen to be strapped together. Light from whatever target is in view enters a pair of objective lenses, bounces through two identical sets of prisms, and exits through the eyepieces. Unlike astronomical telescopes, which usually flip the image either upside down or left to right, binoculars are designed to produce an upright image. This, coupled with their wide-field views, adds to their appeal, especially for beginners.

Just as the inventor of the telescope is not clearly known, the identity of whoever came up with the first pair of binoculars is also lost to history. It appears that soon after the refractor was invented, people started to fasten two of the long tubes together to make binoculars. As is true today, their most popular use was for terrestrial, rather than celestial, pursuits. As you can well imagine, trying to support these cumbersome contraptions was difficult, at best. Not only were the long tubes awkward, but keeping them parallel was a tough task. More often than not, early binoculars were plagued by double images.

In 1894, Dr. Ernst Abbe, a German physicist and mathematician working for Carl Zeiss Optics, devised an ingenious "binocular prismatic telescope," as he called it. This signaled two very important advances in binocular design. First, since the light path through the instrument now ricocheted within a prism, the binoculars' physical length could be reduced while maintaining its focal length. A second, side benefit was that the prisms flipped the image right-side up, for a correct view of the world. The modern prismatic binocular was born. Today, binoculars are available in two basic styles depending on the type of prisms used: Porro prism binoculars and roof prism binoculars (Figure 3.1).

All binoculars are labeled with two numbers, such as 7 × 35 (pronounced "7 by 35") or 10 × 50. The first number refers to the pair's magnification, while the second specifies the diameter (in millimeters) of the two front lenses. Typically, values range from 7 power (7×) to 20 power (20×), with objectives measuring between 35 mm (1.5 inches) and 150 mm (6 inches). The next chapter gives you the full story of binoculars, hopefully helping you to choose a pair that is best suited for your needs.

The ultimate in portability, binoculars offer unparalleled views of rich Milky Way star fields thanks to their low power and wide fields of view. As

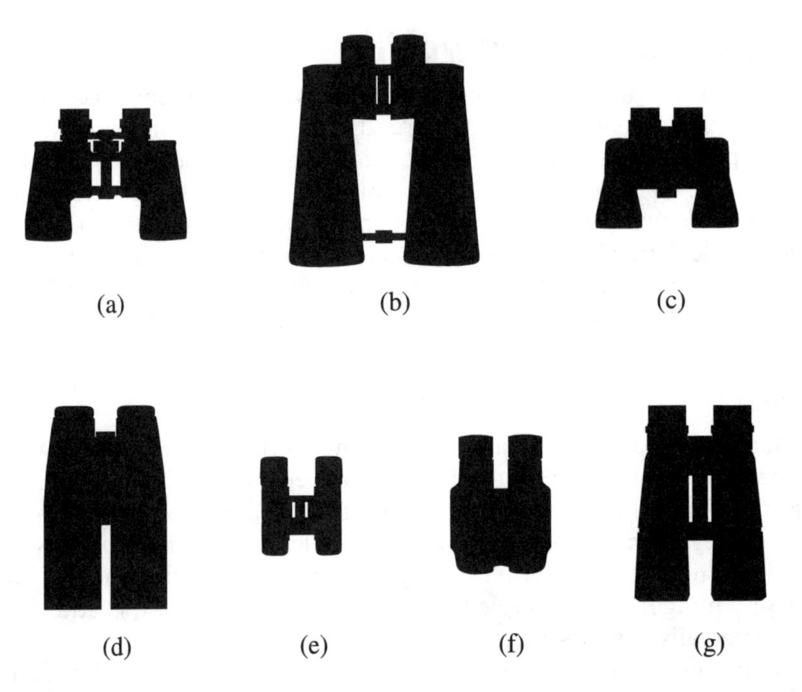

Figure 3.1 *Binoculars come in all different shapes and sizes, though all share one of two basic designs. Silhouettes (a) through (c) show typical Porro prism binoculars, while (d) through (g) illustrate the roof prism design.*

much as this is an advantage to the deep-sky observer, it is a serious drawback to those interested in looking for fine detail on the planets, where higher powers are required. In these cases, the hobbyist has no choice but to purchase a telescope.

Refracting Telescopes

After many years of being all but ignored by the amateur community, the astronomical refractor (Figure 3.2) is making a strong comeback. Hobbyists are rediscovering the exquisite images seen through well-made refractors. Crisp views of the Moon, razor-sharp planetary vistas, and pinpoint stars are all possible through the refracting telescope.

Achromatic refractors. As mentioned in chapter 2, many refractors of yesteryear were plagued with a wide and varied assortment of aberrations and image imperfections. The most difficult of these faults to correct are chromatic aberration and spherical aberration.

If we look in a dictionary, we find this definition of the word *achromatic:*

> ach·ro·mat·ic (ăk'rə-măt/ik) adj. 1. Designating color perceived to have zero saturation and therefore no hue, such as neutral grays, white, or black. 2. Refracting light without spectral color separation.

Figure 3.2 *The 4-inch f/9.8 Celestron C102-HD achromatic refractor. Telescope courtesy of Astrotec.*

Achromatic objective lenses, in which a convex crown lens is paired with a concave flint element, go a long way in suppressing chromatic aberration. These are designed to bring red and blue, the colors at opposite ends of the visible spectrum, to the same focus, with the remaining colors in between converging to nearly the same focus point. It works, too. Indeed, at f/15 or greater, chromatic aberration is effectively eliminated in backyard-size refractors. Even at focal ratios down to f/10, chromatic aberration is frequently not too offensive if the objective elements are *well made.* High-quality achromatic refractors sold today range in size from 2.4 inches (6 cm) up to 6 inches (15 cm). Even though most of the chromatic aberration can be dealt with effectively, a lingering bluish or purplish glow frequently will be seen around brighter stars and planets. This glow is known as *residual chromatic aberration* and is almost always present in achromatic refractors.

How apparent this false color will be depends on the refractor's aperture and focal ratio, and will become more intrusive as aperture increases. In his book *Amateur Astronomer's Handbook* (Dover), author John Sidgwick offers the following formula as guidance.

$$\text{f.l.} \geq 2.88D^2$$

What does this mean? Simply put, residual chromatic aberration will not be a factor as long as an achromatic refractor's focal length (abbreviated f.l.) is greater than 2.88 times its aperture (D, in inches) squared. So a 4-inch refractor will exhibit some false color around brighter objects if its focal length is less than 46 inches (f/11.5), while a 6-inch refractor will show some false color if shorter than 103 inches (f/17.3).

In *Telescope Optics: Evaluation and Design,* authors Harrie Rutten and Martin van Venrooij (Willmann-Bell) suggests an even more conservative value. They suggest that the minimum focal ratio be governed by this formula:

$$f/ratio_{(min)} \geq 0.122D$$

In this case, the aperture D is measured in millimeters. Therefore, a 4-inch (100-mm) achromatic refractor needs to be no less than f/12.2 to eliminate chromatic aberration, while a 6-inch must be at least f/18.3.

Chromatic aberration does more to damage image quality than just add false color. The light that should be contributing to the image that our eye perceives is instead being scattered over a wider area, causing image brightness to lag. Image contrast also suffers due to the scattering of light. Specifically, darker areas are washed out by overlapping fringes of brighter surrounding areas (as an example, think of the belts of Jupiter or the dark areas on the surface of Mars).

Why do I raise this issue? Recently, the telescope market has been flooded with 4- to 6-inch f/8 to f/10 achromatic refractors from China and Taiwan. Their promise of large apertures (for refractors, that is) and sharp images have attracted a wide following. How good are these newcomers? The short answer is that most are actually surprisingly good. Yes, there is some false color evident around the Moon, the planets, and the brighter stars, but for the most part, it is not terribly objectionable. There are a few, however, that are not quite ready for prime time. Each is addressed in chapter 5, but continue here first to get the rest of the refractor's story.

A strong point in favor of the refractor is that its aperture is a *clear* aperture. That is, nothing blocks any part of the light as it travels from the objective to the eyepiece. As you can tell from looking at the diagrams in chapter 2, this is not the case for reflectors and catadioptric instruments. As soon as a secondary mirror interferes with the path of the light, some loss of contrast and image degradation are inevitable.

In addition to sharp images, smaller achromatic refractors (that is, up to perhaps 5 inches aperture) are also famous for their portability and ruggedness. If constructed properly, a refractor should deliver years of service without its optics needing to be realigned (recollimated). The sealed-tube design means that dust and dirt are prevented from infiltrating the optical system, while contaminants can be kept off the objective's exterior simply by using a lens cap.

On the minus side of the achromatic refractor is its small aperture. While this is of less concern to lunar, solar, and planetary observers, the instrument's small light-gathering area means that faint objects such as nebulae and galaxies will appear dimmer than they do in larger-but-cheaper reflectors. In addition, the long tubes of larger achromatic refractors can make them difficult to store and transport to dark, rural skies.

Another problem common to refractors is their inability to provide comfortable viewing angles at all elevations above the horizon. The long tube and short tripod typically provided can work against the observer in some cases. For instance, if the mounting is set at the right height to view near the zenith, the eyepiece will swing high off the ground as soon as the telescope is aimed toward the horizon. This disadvantage can be partially offset by inserting a star diagonal between the telescope's drawtube and eyepiece, but this has disadvantages of its own. Using either a mirror or a prism, a star diagonal bends the light either 45° or 90° to make viewing more comfortable. Most astronomical refractors include 90° star diagonals, as they are better for high-angle viewing, although some instruments that double as terrestrial spotting scopes include a 45° diagonal instead. The latter usually flips everything right-side up, but dims the view slightly in the process.

A particularly popular segment of the achromatic refractor population nowadays are the so-called short tubes. Ranging in aperture from 60 mm to 150 mm, these telescopes all share low focal ratios (low f/ ratios), short focal lengths, and compact tubes. While larger telescopes can require considerable effort to set up, short-tubed refractors are ideal as "grab-and-go" instruments that can be taken outside at a moment's notice to enjoy a midweek clear night. They are designed primarily as wide-field instruments, ideal for scanning the Milky Way or enjoying the view of larger, brighter sky objects. They are *not* intended for high-magnification viewing and are, therefore, not as suitable for viewing the planets as other designs. Also keep in mind that based on the discussion above, these telescopes still obey the laws of optics and, as such, suffer from false color and residual chromatic aberration. Their performance cannot compare with apochromatic refractors that share similar apertures and focal lengths (described in the next section). Of course, they don't share their exorbitant price tags, either. Personally, I feel that short-tubed refractors are ideal *second* telescopes, but if this is your first and only, I would recommend other designs unless you live on the seventh floor of a walk-up apartment building and *need* their portability.

Apochromatic refractors. If *achromatic* means "no color," then *apochromatic* might be defined as "really, really no color, I mean it this time." While an achromat brings two wavelengths of light at opposite ends of the spectrum to a common focus, it still leaves residual chromatic aberration along the optical axis. Though not as distracting as chromatic aberration from a single lens, residual chromatic aberration can still contaminate critical viewing and photography.

Apos, as they are affectionately known to many owners, eliminate residual chromatic aberration completely, allowing manufacturers to increase aperture and decrease focal length. First popularized in the 1980s, apochromatic refractors use either two-, three-, or four-element objective lenses with one or more elements of an unusual glass type—often fluorite, SD (*special-dispersion*), or ED (short for *extra-low-dispersion*) glass. Apochromats minimize dispersion of light by bringing all wavelengths to just about the same focus, reducing residual chromatic aberration dramatically, and thereby permitting shorter, more manageable focal lengths.

Much has been written about the pros and cons of fluorite (monocrystalline calcium fluorite) lenses. Fluorite has been called the most colorful mineral in the world, often displaying intense shades of purple, blue, green, yellow, reddish orange, pink, white, and brown. Unlike ordinary optical glass, which is made primarily of silica, fluorite is characterized by low-refraction and low-dispersion characteristics, perfect for suppressing chromatic aberration to undetectable levels. Unfortunately, it is very difficult to obtain fluorite large enough for a lens, and so telescope objectives (and camera lenses, etc.) are made from fluorite "grown" through an artificial crystallization technology.

In optics, the most popular myth about fluorite lenses is that they do not stand the test of time. Some "experts" claim that fluorite absorbs moisture and/or fractures more easily than other types of glass. This is simply not true. Fluorite objectives work very well and are just as durable as conventional lenses. Like all lenses, they will last a lifetime if given a little care.

While durability is not a problem, there are a couple of hitches to fluorite refractors. One problem with fluorite that is not popularly known is its high thermal expansion. This means that the fluorite element will require more time to adjust to ambient temperature than a nonfluorite lens. Telescope optics change shape slightly as they cool or warm, and this characteristic is more pronounced in fluorite than in other materials.

Other hindrances are shared by all apos. Like most commercially sold achromatic refractors, apochromatic refractors are limited to smaller apertures, usually somewhere in the range of 3 to 7 inches or so. This is not because of unleashed aberrations at larger apertures; it's simply a question of economics, which brings us to their second (and biggest) stumbling block: Apochromats are not cheap! When we compare dollars per inch of aperture, it soon becomes apparent that apochromatic refractors are the most expensive telescopes. Given the same type of mounting and accessories, an apochromatic refractor can retail for more than twice the price of a comparable achromat. That's a big difference, but the difference in image quality can be even bigger. For a first telescope, an achromatic refractor is just fine, but if this is going to be your ultimate dream telescope, then you ought to consider an apochromat.

Reflecting Telescopes

Reflectors offer an alternative to the small apertures and big prices of refractors. Let's compare. Each of the two or more elements in a refractor's objective lens must be accurately figured and made of high-quality, homogeneous glass. By contrast, the single optical surfaces of a reflector's primary and secondary mirrors favor construction of large apertures at comparatively modest prices.

Another big advantage that a reflector enjoys over a refractor is complete freedom from chromatic aberration. (Chromatic aberration is a property of light refraction but not reflection.) This means that only the true colors of whatever a reflecting telescope is aiming at will come shining through. Of course, the eyepieces used to magnify the image for our eyes use lenses, so we are not completely out of the woods.

These two important pluses are frequently enough to sway amateurs in favor of a reflector. They feel that although there are drawbacks to the design, these are outweighed by the many strong points. But just what are the problems of reflecting telescopes? Some are peculiar to certain breeds, while others affect them all.

One shortcoming common to all telescopes of this genre is the simple fact that mirrors do not reflect all the light that strikes them. Just how much light is lost depends on the kind of reflective coating used. For instance, most telescope mirrors are coated with a thin layer of aluminum and overcoated with a clear layer of silicon monoxide for added protection against scratches and pitting. This combination reflects about 89% of visible light. But consider this: Given primary and secondary mirrors with standard aluminum coatings, the combined reflectivity is only 79% of the light striking the primary! That's why special enhanced coatings have become so popular in recent years. Enhanced coatings increase overall system reflectivity to between 90% and 96%. Some say, however, that enhanced coatings scatter light, in turn decreasing image contrast. If so, the decrease is minimal.

Reflectors also lose some light and, especially, image contrast because of obstruction by the secondary mirror. Just how much light is blocked depends on the size of the secondary, which in turn depends on the focal length of the primary mirror. Generally speaking, the shorter the focal length of the primary, the larger its secondary must be to bounce all of the light toward the eyepiece. Tradition has it, however, that the central obstruction is expressed in terms of the percentage of aperture diameter, not area. Therefore, a reflector with a 10% central obstruction by area is referred to as having a 16% obstruction by diameter. Both refer to an obstruction that measures 1.27 inches across. Primary mirrors with very fast focal ratios can have central obstructions in the neighborhood of 20%, even 25%. The only reflectors that do not suffer from this ailment are referred to as *off-axis* reflectors. Classic off-axis designs include the Herschelian and members of the Schiefspiegler family of instruments.

Since the idea of a telescope that uses mirrors to focus light was first conceived in 1663, different schemes have come and gone. Today, two designs continue to stand the test of time: the Newtonian reflector and the Cassegrain reflector. Each shall be examined separately.

Newtonian reflectors. For sheer brute-force light-gathering ability, Newtonian reflectors rate as a best buy. No other type of telescope will give you as large an aperture for the money. Given a similar style mounting, you could buy an 8-inch Newtonian reflector for the same amount of money needed for a 4-inch achromatic refractor.

Newtonians (Figure 3.3) are famous for their panoramic views of star fields, making them especially attractive to deep-sky fans, but they also can be equally adept at moderate- to high-powered glimpses of the Moon and the planets. These highly versatile instruments come in a wide variety of styles. Commercial models range from 3 inches to more than 2 *feet* in diameter, with

Figure 3.3 *The 6-inch Celestron C150-HD Newtonian reflector on a German equatorial mount. Telescope courtesy of Astrotec.*

focal ratios stretching between f/3.5 and about f/10. (Of course, not all apertures are available at all focal ratios. Can you imagine climbing more than 20 feet to the eyepiece of a 24-inch f/10?)

For the sake of discussion, I have divided Newtonians into two groups based on focal ratio. Those instruments with focal ratios less than f/6 have very deeply curved mirrors, and so are referred to here as *deep-dish* Newtonians. Those reflectors with focal ratios of f/6 and greater will be called *shallow-dish* telescopes.

Pardon my bias, but shallow-dish reflectors have always been my favorite type of telescope. They are capable of delivering clear views of the Moon, the Sun, and other members of the solar system, as well as thousands of deep-sky objects. Shallow-dish reflectors with apertures between 3 inches (80 mm) and 8 inches (200 mm) are usually small enough to be moved from home to observing site and quickly set up with little trouble. Once the viewing starts, most amateurs happily find that looking through both the eyepiece and the small finderscope is effortless because the telescope's height closely matches the eye level of the observer.

A 6-inch f/8 Newtonian is still one of the best all-around telescopes for those new to astronomy. It is compact enough so as not to be a burden to transport and assemble, yet it is large enough to provide years of fascination—and all at a reasonable price (maybe not $194.95 anymore, but still not a bad deal). Better yet is an 8-inch f/6 to f/9 Newtonian. The increased aperture permits even finer views of nighttime targets. Keep in mind, however, that as aperture grows, so grows a telescope's size and weight. This means that unless you live in the country and can store your telescope where it is easily accessible, a shallow-dish reflector larger than an 8-inch might be difficult to manage.

Most experienced visual observers agree that shallow-dish reflectors are tough to beat. In fact, an optimized Newtonian reflector can deliver views of the Moon and the planets that eclipse those through a catadioptric telescope and compare favorably with a refractor of similar size, but at a fraction of the refractor's cost. Though the commercial telescope market now offers a wide range of superb refractors, it has yet to embrace the long-focus reflector fully.

While shallow-dish reflectors provide fine views of the planets and the Moon, they grow to monumental lengths as their apertures increase (we old-timers still remember 12.5-inch f/8s!). The eyepiece of an 18-inch f/4.5 reflector is about 78 inches off the ground, while an 18-inch f/8 would tower at nearly 12 feet. That's why most 10-inch and larger Newtonians have comparatively fast focal ratios, falling into the deep-dish category.

Many of these large-aperture Newtonians use thin-section primary mirrors. Traditionally, primary mirrors have a diameter-to-thickness ratio of 6:1. This means that a 12-inch mirror measures a full 2 inches thick. That is one heavy piece of glass to support. Thin-section mirrors cut this ratio to 12:1 or 13:1, slashing the weight by 50%. This sounds good at first, but practice shows that large, thin mirrors tend to sag under their own weight (thicker mirrors are more rigid), thereby distorting the parabolic curve, when held in a conventional three-point mirror cell. To prevent mirror sag, a new support system was devised to support the primary at nine (or more) evenly spaced points across its back surface. These cells are frequently called *mirror flotation systems,* as they do not clamp down around the mirror's rim, thereby preventing possible edge distortions due to pinching.

If big aperture means bright images, why not buy the biggest aperture available? Actually, there are several reasons not to. For one thing, unless they are made very well, Newtonians (especially those with short focal lengths) are susceptible to a number of optical irregularities, including spherical aberration, astigmatism, and coma.

Spherical aberration results when light rays from the edge of an improperly made mirror (or lens) focus to a slightly different point than those from the optic's center. In general, the faster the focal ratio, the greater the need for accuracy. Slower focal ratios are more forgiving. In fact, a parabolic mirror may not be required at all. Some Newtonian reflectors will work perfectly well with spherical mirrors, which are considerably easier (and, therefore, cheaper) to fabricate. In his classic work *How to Make a Telescope* (Willmann-Bell), author Jean Texereau recommends a formula derived by André Couder that calculates the minimum focal lengths for given apertures that will satisfy Rayleigh's Criterion and, therefore, produce satisfactory images. The formula reads:

$$\text{f.l.}^3 = 88.6A^4$$

where A is the telescope aperture and f.l. is the focal length, both expressed in inches. Table 3.1 tabulates the results for some common apertures.

Table 3.1 ***Minimum Focal Length for Spherical Mirrors to Satisfy Rayleigh's Criterion***

Aperture (inches)	Minimum Focal Length (inches)	Focal Ratio (f.l. / A)
3.0	18	6.3
4.5	33	7.3
6.0	49	8.2
8.0	72	9.0
10.0	96	9.6

As you can see, as aperture grows, the minimum acceptable focal length also grows, and quite quickly. Some manufacturers (notably in Russia, Taiwan, and China) supply a few of their telescopes with spherical mirrors. While they seem to work well enough in general, they are all quite close to the minimum values cited in the table. I must also point out that many optical purists feel that Texereau's values are low, that they should all be increased by 30% or so to produce a well-functioning Newtonian telescope based on a spherical mirror. Therefore, approach with wariness any telescope with a focal ratio below the values shown in Table 3.1 that claims to use a spherical mirror.

Astigmatism is due to a mirror or lens that was not symmetrically ground around its center. The result: elongated star images that appear to flip their orientation by 90° when the eyepiece is brought from one side of the focus point to the other. *Coma,* especially apparent in deep-dish telescopes, is evident when stars near the edge of the field of view distort into tiny blobs resembling comets, while stars at the center appear as sharp points. With any or all of these imperfections present, resolution suffers greatly. (Note that coma can be eliminated using a *coma corrector*—see chapter 6.)

Furthermore, if you observe from a light-polluted area, large apertures will likely produce results inferior to instruments with smaller apertures. The old myth is that larger mirrors gather more sky glow, washing out the field of view. That's just not the case, but it is true that they are more sensitive to heat currents and turbulent atmospheric conditions, a frequent byproduct of the overuse of concrete in cities. For that reason, city dwellers might do best with a telescope no larger than 8 to 10 inches in aperture.

Both shallow-dish and deep-dish Newtonians share many other pitfalls as well. One of the more troublesome is that of all the different types of telescopes, Newtonians are among the most susceptible to collimation problems. If either or both of the mirrors are not aligned correctly, image quality will suffer greatly, possibly to the point of making the telescope worthless. Sadly, many commercial reflectors are delivered with misaligned mirrors. The new owner, perhaps not knowing better, immediately condemns his or her telescope's poor performance as a case of bad optics. In reality, however, the optics may be fine, just a little out of alignment. Chapter 9 details how to examine and adjust a telescope's collimation, a procedure that should be repeated frequently. The need for precise alignment grows more critical as the primary's focal ratio shrinks, making it especially important to double-check collimation at the start of every observing session if your telescope is f/6 or less.

There are cases where no matter how well aligned the optics are, image quality is still lacking. Here, the fault undoubtedly lies with one or both of the mirrors themselves. As the saying goes, you get what you pay for, and that is as true with telescopes as with anything else. Clearly, manufacturers of low-cost models must cut their expenses somewhere in order to underbid their competition. These cuts are usually found in the nominal-quality standard equipment supplied with the instrument but sometimes may also affect optical testing procedures and quality control.

Cassegrain reflectors. Though they have never attained the widespread following among amateur astronomers that Newtonians continue to enjoy, Cassegrain reflectors (Figure 3.4) have always been considered highly competent instruments. Cassegrains are characterized by long focal lengths, making them ideally suited for high-power, high-resolution applications such as solar, lunar, and planetary studies. While Newtonians also may be constructed with these focal ratios, observers would have to go to great lengths to reach their eyepieces! Not so with the Cassegrain, where the eyepiece is conveniently located along the optical axis behind the backside of the primary mirror.

The Cassegrain's long focal length is created not by the primary mirror (which typically ranges around f/4) but rather by the convex, hyperbolic secondary mirror. As it reflects the light from the primary back toward the eyepiece, the convex secondary actually magnifies the image, thereby stretching the telescope's effective focal ratio to between f/10 and f/15. The net result is a telescope that is much more compact and easier to manage than a Newtonian of equivalent aperture and focal length.

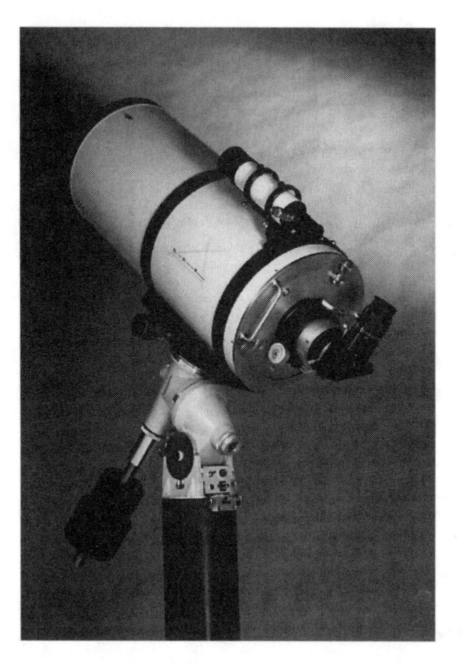

Figure 3.4 *The RC-10, 10-inch Ritchey-Chrétien Cassegrain reflector. Photo courtesy of Optical Guidance Systems.*

Unfortunately, while the convex secondary mirror gives the Cassegrain its great compactness, it also contributes to many of the telescope's biggest disadvantages. First, in order to reflect all the light from the primary back toward the eyepiece, the secondary mirror must be placed quite close to the primary. This forces its diameter to be noticeably larger than the flat diagonal of a Newtonian. With the secondary blocking more light, image brightness, clarity, and contrast all suffer. Second, the convex secondary combined with the short-focus primary mirror make alignment critical to the Cassegrain's proper function and at the same time cause the telescope to be more difficult to collimate than a similar Newtonian. Finally, Cassegrains are prone to coma just like deep-dish Newtonians, making it impossible to achieve sharp focus around the edge of the field of view.

The advantage of the eyepiece's placement along the optical axis also can work against the instrument's performance. The most obvious objection will become painfully apparent the first time an observer aims a Cassegrain near the zenith and tries to look through the eyepiece. That can be a real pain in the neck, although the use of a star diagonal will help alleviate the problem. Another problem that may not be quite as apparent involves a very localized case of light pollution, caused by extraneous light passing around the secondary and flooding the field of view. To combat this, manufacturers invariably install a long baffle tube protruding in front of the primary. The size of the baffle is critical, as it must shield the eyepiece field from all sources of incidental light while allowing all of the light from the target to shine through.

Though Cassegrains remain the most common type of telescope in professional observatories, their popularity among today's amateur astronomers is low. So it should come as no surprise to find that so few companies offer complete Cassegrain systems for the hobbyist.

Catadioptric Telescopes

Most amateur astronomers who desire a compact telescope now favor hybrid designs that combine some of the best attributes of the reflector with some from the refractor, creating a completely different kind of beast: the catadioptric. Catadioptric telescopes (also known as *compound telescopes*) are comparative Johnny-come-latelies on the amateur scene. Yet in only a few decades, they have developed a loyal following of backyard astronomers who staunchly defend them as the ultimate telescopes.

Most lovers of catadioptrics fall into one, two, or possibly all three of the following categories:

1. They are urban or suburban astronomers who prefer to travel to remote observing sites.
2. They enjoy astrophotography (or aspire to at least try it).
3. They just love gadgets.

If any or all of these profiles fit you, then a catadioptric telescope just might be your perfect telescope.

Catadioptric telescopes for visual use may be constructed in either Newtonian or Cassegrain configurations. Three catadioptrics have made lasting impacts on the world of amateur astronomy: the Schmidt-Cassegrain, the Maksutov-Cassegrain, and, most recently, the Maksutov-Newtonian. For our purposes here, the discussion will be confined to these designs.

Schmidt-Cassegrain telescopes. Take a look through practically any astronomy magazine published just about anywhere in the world and you are bound to find at least one advertisement for a Schmidt-Cassegrain telescope (also known as a *Schmidt-Cas* or an *SCT*). As your eyes digest the ads chock-full of mouth-watering celestial photographs that have been taken through these instruments, you suddenly get the irresistible urge to run right out and buy one. Don't worry—you would not be the first to find these telescopes so appealing. In the last few decades, sales of Schmidt-Cassegrains have outpaced both refractors and reflectors to become the most popular serious telescope among amateur astronomers. Though Schmidt-Cassegrain telescopes are available in apertures from 5 inches to 16 inches, the favorite size of all is the 8-inch model.

Is the Schmidt-Cassegrain (Figure 3.5) the perfect telescope? Admittedly, it can be attractive. By far, its greatest asset has to be the compact design. No other telescope can fit as large an aperture and as long a focal length into such a short tube assembly as a Schmidt-Cas (they are usually only about twice as long as the aperture). If storing and transporting the telescope are major concerns for you, then this will be an especially important benefit.

Here is another point in their favor. Nothing can end an observing session quicker than a fatigued observer. For instance, owning a Newtonian reflector, with its eyepiece positioned at the front end of the tube, usually means having to remain standing—sometimes even on a stool or a ladder—just to take a peek. Compare this to a Schmidt-Cassegrain telescope, which permits the observer to enjoy comfortable, seated viewing of just about all points in the sky. Your back and legs will certainly thank you! The eyepiece is only difficult to reach when the telescope is aimed close to the zenith. As with a refractor and a Cassegrain, a right-angle star diagonal placed between the telescope and eyepiece will help a little, but these have their drawbacks, too. Most annoying of all is that a diagonal will flip everything right to left, creating a mirror image that makes the view difficult to compare with star charts.

All commercially made Schmidt-Cassegrain telescopes look pretty much the same *at a quick glance,* but then again, so do many products to the uninitiated. Only after closer scrutiny will the features unique to individual models come shining through. Standard-equipment levels vary greatly, as reflected in the wide price range of Schmidt-Cassegrain telescopes. Some basic models come with an undersized finderscope, one eyepiece, maybe a couple of other bare-bones accessories, and cardboard boxes for storage, while top-of-the-line instruments are supplied with foam-lined footlockers, advanced eyepieces, large finders, and a multitude of electronic gadgets. (As I mentioned before, if you love widgets and whatchamacallits, then the Schmidt-Cassegrain will certainly appeal to you.) Most amateurs can find happiness with a model somewhere between these two extremes.

Another big plus of the Schmidt-Cassegrain is its sealed tube. The front corrector plate acts as a shield to keep dirt, dust, and other foreign contaminants off

Figure 3.5 *The Meade 8-inch LX90 Schmidt-Cassegrain telescope, one of the most sophisticated instruments on the market today.*

the primary and secondary mirrors. This is especially handy if you travel a lot with your telescope and are constantly taking it in and out of its carrying case. A sealed tube also can help extend the useful life of the mirrors' aluminized coatings by sealing well against the elements. (Always make sure the mirrors are dry before storing the telescope to prevent the onset of mold and mildew.)

While the corrector seals the two mirrors against dust contamination, it slows a telescope from reaching thermal equilibrium with the night air and can also act as a dew collector. Depending on local weather conditions, correctors can fog over in a matter of hours or even minutes, or they may remain clear all night. To help fight the onslaught of dew, manufacturers sell *dew caps* or *dew shields.* Dew caps are a must-have accessory for all Cassegrain-based catadioptrics. Consult chapter 7 for more information.

Many of the accessories for Schmidt-Cassegrain telescopes revolve around astrophotography, an activity enjoyed by many amateur astronomers.

Here again, the SCT pulls ahead of the crowd. Because of their comparatively short, lightweight tubes, Schmidt-Cassegrains permit easy tracking of the night sky. Just about all are held on fork-style equatorial mounts complete with motorized clock drives. Once the equatorial mount is properly aligned to the celestial pole (a tedious activity at times—see chapter 10), you can turn on the drive motor, and the telescope will track the stars by compensating for Earth's rotation. With various accessories (many of which are intended to be used only with Schmidt-Cassegrain telescopes), the amateur is now ready to photograph the universe.

What about optical performance? Here is where the Schmidt-Cassegrain telescope begins to teeter. Due to the comparatively large secondary mirrors required to reflect light back toward their eyepieces, SCTs produce images that are fainter and show less contrast than other telescope designs of the same aperture. This can prove especially critical when searching for fine planetary detail or hunting for faint deep-sky objects at the threshold of visibility. Enhanced optical coatings, now standard on all popular models, improve light transmission and reduce scattering. They can make the difference between seeing a marginally visible object and missing it, and they are an absolute must for all Schmidt-Cassegrains. Still, planet watching, which demands sharp, high-contrast images, suffers in Schmidt-Cassegrains. A typical 8-inch SCT has a secondary mirror mounting that measures about 2.75 inches across. That's a whopping 34% central obstruction!

Most 8-inch Schmidt-Cassegrain telescopes operate at f/10, while a few work at f/6.3. What's the difference? On the outside, they both look the same, the only difference being in the secondary mirrors. Are there pluses to using one over the other? Yes and no. If the telescopes are used visually (that is, if you are just going to look through them), then there should be negligible difference between the performance of an f/10 telescope and an f/6.3 telescope when operated at the same magnification. Image brightness is controlled by clear aperture, not by f-ratio.

The faster focal ratio may actually work against the observer. To achieve an f/6.3 instrument, a larger secondary mirror is required (3.5 inches across, compared to between 2.75 and 3 inches across in an f/10 instrument). The larger central obstruction in f/6.3 SCTs cause a decrease in contrast, making them less useful for planetary observation than their f/10 brethren. If you really want to split hairs, there is also a slight difference in image brightness—only about 5%. (Of course, in 8-inch f/6 Newtonian reflectors, the secondary blocks only 1.8 inches of the full aperture, which helps to explain their superior image contrast.) The fast focal ratio also means that to achieve the same magnification, shorter focal-length eyepieces will have to be used. These often have shorter eye relief, making them less comfortable to view through. See chapter 6 for further discussion.

The biggest advantage to using an f/6.3 Schmidt-Cassegrain telescope is enjoyed by astrophotographers. When set up for prime-focus photography (camera body coupled directly to the eyepieceless telescope), exposure time can be cut by 2.5 times to get the same image brightness as an f/10. Of course, image size is going to be reduced at the same time, but for many deep-sky objects, this is usually not a problem. (See also the discussion in chapter 6 about focal-length reducers for Schmidt-Cassegrain telescopes.)

Image sharpness in a Schmidt-Cassegrain is not as precise as that obtained through a refractor or a reflector. Perhaps this is due to the loss of contrast mentioned above or because of optical misalignment, another problem of the Schmidt-Cassegrain. In any telescope, optical misalignment will play havoc with image quality. What should you do if the optics of a Schmidt-Cassegrain are out of alignment? If only the secondary is off, then you may follow the procedure outlined in chapter 9, but if the primary is out, then manufacturers suggest that the telescope be returned to the factory. That is good advice. Remember—although just about anyone can take a telescope apart, not everyone can put it back together!

Finally, aiming a Schmidt-Cassegrain telescope can sometimes prove to be a frustrating experience. This is not the fault of the telescope, but is due instead to the low position of the finderscope (a small auxiliary telescope mounted sidesaddle and used to aim the main instrument). Some SCTs are supplied with right-angle finders, which, while greatly reducing back fatigue, introduce a whole cauldron of problems of their own. See chapter 7 for more about right-angle finders and why you shouldn't use them.

In general, Schmidt-Cassegrain telescopes represent good values for the money. They offer acceptable views of the Sun, the Moon, the planets, and deep-sky objects, and work reasonably well for astrophotography. But for exacting views of celestial objects, SCTs are outperformed by other types of telescopes. For observations of solar system members, it is hard to beat a shallow-dish Newtonian (especially f/10 or higher) or a good refractor, while the myriad faint deep-sky objects are seen best with large-aperture, deep-dish Newtonians. I guess you could say that Schmidt-Cassegrains are the jack-of-all-trades-but-master-of-none telescopes.

Maksutov telescopes. The final stop on our telescope world tour is the Maksutov catadioptric. Many people feel that Maksutovs are the finest telescopes of all. And why not? Maks provide views of the Moon, the Sun and the planets that rival those of the best refractors and long-focus reflectors, and they are easily adaptable for astrophotography (although their high focal ratios mean longer exposures than faster telescopes of similar aperture). Smaller models are also a breeze to travel with.

Maksutov telescopes come in both Newtonian and Cassegrain configurations, the latter being more common. Typically, Maksutov-Cassegrains have slower focal ratios and so are more appropriate for high-powered planet watching, while Maksutov-Newtonians are ideal for deep-sky observing as well as guided astrophotography. But make no mistake—a well-made Mak-Newt will also show stunning planetary views.

Is there a downside to the Maksutov? Unfortunately, yes. Because of the thick corrector plate in front, Maks take longer to acclimate to the cool night air than any other common type of telescope. A telescope will not perform at its best until it has equalized with the outdoor air temperature. To help speed things along, some companies install fans, but the optics are still slow to reach thermal equilibrium. Some Maks also cost more per inch of aperture than nearly any other type of telescope.

To help digest all this, take a look at Table 3.2, which summarizes all the pros and cons mentioned above. Use it to compare the good points and the bad among the more popular types of telescopes sold today.

Table 3.2 ***Telescope Point–Counterpoint: A Summary***

	Point	Counterpoint
1. Binoculars Typically 1.4″ to 4″ apertures	• Most are comparatively inexpensive • Extremely portable • Wide field makes them ideal for scanning	• Low power makes them unsuitable for objects requiring high magnification • Small aperture restricts magnitude limit
2. Achromatic Refractors Typically 2.4″ to 6″, f/5 and above	• Portable in smaller apertures • Sharp images • Moderate price vs. aperture • Good for Moon, Sun, planets, double stars, and bright astrophotography	• Small apertures • Mounts may be shaky (attention, department-store shoppers!) • Potential for chromatic aberration
3. Apochromatic Refractors Typically 2.8″ to 7″ aperture, f/5 and above	• Portable in smaller apertures • Very sharp, contrasty images • Excellent for Moon, Sun, planets, double stars, and astrophotography	• Very high cost vs. aperture
4. Shallow-Dish Newtonian Reflectors Typically 3″ to 8″ aperture, f/6 and above	• Low cost vs. aperture • Easy to collimate • Very good for Moon, Sun, planets (especially f/8 and above), deep-sky objects, and astrophotography	• Bulky/heavy, over 8″ • Collimation must be checked often • Open tube end permits dirt and dust contamination • Excessively long focal lengths may require a ladder to reach the eyepiece
5. Deep-Dish Newtonian Reflectors Typically 4″ and larger, below f/6	• Low cost vs. aperture • Wide fields of view • Large apertures mean maximum magnitude penetration • Easy to collimate • Excellent for both bright and faint deep-sky objects (solar system objects OK, but usually inferior to same size shallow-dish reflectors . . . though not always!)	• Larger apertures can be heavy/bulky, especially those with solid tubes • Even with fast focal ratio, larger apertures may still require a ladder to reach the eyepiece • Very-low-cost versions may indicate compromise in quality • Dobsonian mounts do not track the stars (some have this option, however) • Collimation is critical (must be checked before each use) • Open tube ends permit dirt and dust contamination • Large apertures very sensitive to seeing conditions

(continued)

Table 3.2 ***(continued)***

	Point	Counterpoint
6. Cassegrain Reflector		
Typically 6″ and larger, f/12 and above	• Portability • Convenient eyepiece position • Good for Moon, Sun, planets, and smaller deep-sky objects (e.g., double stars, planetary nebulae)	• Large secondary • Moderate-to-high price vs. aperture • Narrow fields • Offered by few companies
7. Schmidt-Cassegrain (Catadioptric)		
Typically 4″ to 16″ apertures, f/6.3 to f/10	• Moderate cost vs. aperture • Portability • Convenient eyepiece position • Wide range of accessories • Easily adaptable to astrophotography • Good for Moon, Sun, planets, bright deep-sky objects, and astrophotography	• Large secondary mirror reduces contrast • Image quality not as good as refractors or reflectors • Slow f-ratio means longer exposures than faster Newtonians and refractors • Corrector plates prone to dewing over • Potentially difficult to find objects, especially near the zenith (fork mounts only) • Mirror shift
8. Maksutov-Newtonian (Catadioptric)		
Typically 3.1″ to 7″, typically f/6	• Sharp images, approaching refractor quality • Convenient size for quick setup • Excellent for Moon, Sun, planets, and bright deep-sky objects	• High price vs. aperture
9. Maksutov-Cassegrain (Catadioptric)		
Typically 4″ to 16″ apertures, f/12 and higher	• Sharp images, approaching refractor quality • Convenient eyepiece position • Easily adaptable to astrophotography • Excellent for Moon, Sun, and planets; good for bright, small deep-sky objects	• Some are very expensive vs. aperture • Some have finderscopes that are inconveniently positioned • Some models use threaded eyepieces, making an adapter necessary for other brand oculars • Difficult to collimate, if needed • Slow f-ratio means longer exposures and narrow fields of coverage

Support Your Local Telescope

The telescope itself is only half of the story. Can you imagine trying to hold a telescope *by hand* while struggling to look through it? If the instrument's weight did not get you first, surely every little shake would be magnified into a visual earthquake! To use a true astronomical telescope, we have no choice but to support it on some kind of external mounting. For small spotting scopes, this might be a simple tabletop tripod, while the most elaborate telescopes come equipped with equally elaborate support systems.

Selecting the proper mount is *just as important* as picking the right telescope. A good mount must be strong enough to support the telescope's weight while minimizing any vibrations induced by the observer (such as during focusing) and the environment (from wind gusts or even nearby road traffic). Indeed, without a sturdy mount to support the telescope, even the finest instrument will produce only blurry, wobbly images. A mounting also must provide smooth motions when moving the telescope from one object to the next and allow easy access to any part of the sky.

Though Figure 3.6 shows many different types of telescope mounting systems, all fall into one of two broad categories based on their construction: altitude-azimuth and equatorial. We shall examine both.

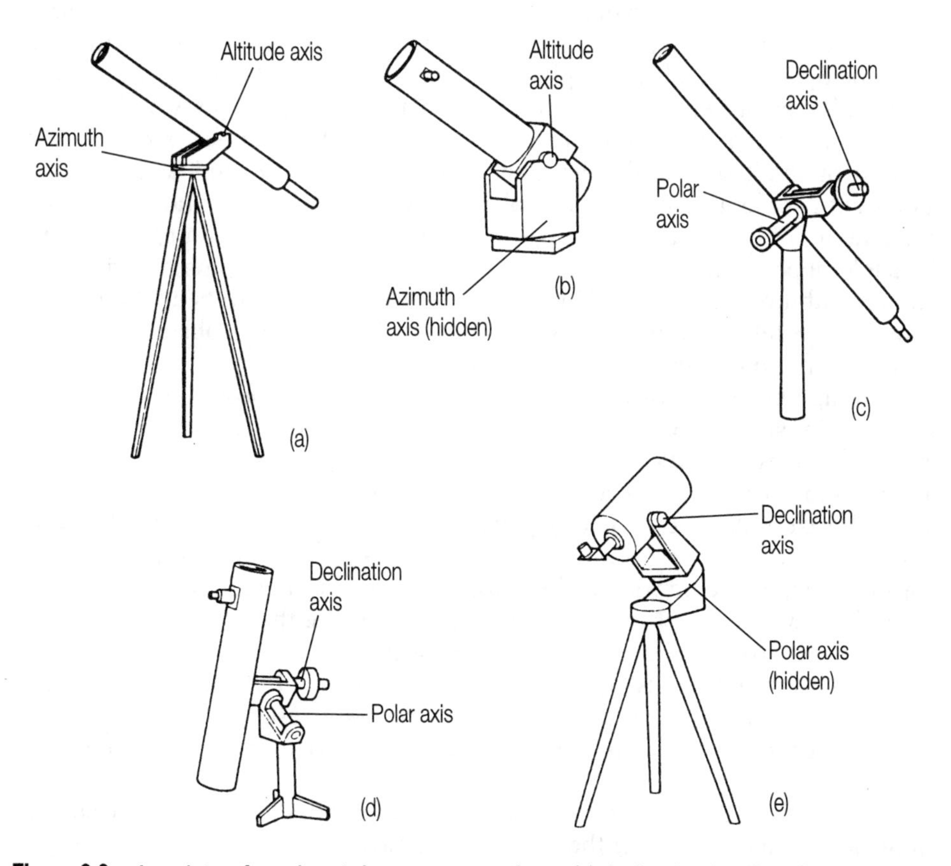

Figure 3.6 *A variety of modern telescope mountings. (a) A simple alt-azimuth mounting for a small refractor, (b) a Dobsonian alt-azimuth mounting for a Newtonian, (c) a German equatorial mounting for a refractor, (d) a German equatorial for a Newtonian, and (e) a fork equatorial mounting for a catadioptric telescope. Illustration from* Norton's Star Atlas and Reference Handbook, *edited by Ian Ridpath, Longman Scientific & Technical.*

Altitude-Azimuth Mounts

Frequently referred to as either *alt-azimuth* or *alt-az*), these are the simplest types of telescope support available. As their name implies, alt-az systems move both in azimuth (horizontally) and in altitude (vertically). All camera tripod heads, for instance, are alt-az systems.

This is the type of mounting most frequently supplied with smaller, less-expensive refractors and Newtonian reflectors. It allows the instrument to be aimed with ease toward any part of the sky. Once pointed in the proper direction, the mount's two axes (that is, the altitude axis and the azimuth axis) can be locked in place. Better alt-azimuth mounts are outfitted with *slow-motion controls,* one for each axis. By twisting one or both of the control knobs, the observer can fine-tune the telescope's aim as well as keep up with Earth's rotation.

In the past 20 years, a variation of the alt-azimuth mount called the Dobsonian has become extremely popular among hobbyists. Dobsonian mounts are named for John Dobson, an amateur telescope maker and astronomy popularizer from the San Francisco area. Back in the 1970s, Dobson began to build large-aperture Newtonian reflectors in order to see the "real universe." With the optical assembly complete, he faced the difficult challenge of designing a mount strong enough to support the instrument's girth yet simple enough to be constructed from common materials using hand tools. What resulted was an offshoot of the alt-az mount. Using plywood, Formica, and Teflon, along with some glue and nails, Dobson devised a telescope mount that was capable of holding steady his huge Newtonians. Plywood is an ideal material for a telescope mount, as it has incredible strength as well as a terrific vibration-damping ability. Formica and Teflon together create smooth bearing surfaces, allowing the telescope to flow across the sky. No wonder Dobsonian mounts have become so popular.

Though both traditional alt-az mounts as well as Dobsonian mounts are wonderfully simple to use, they also possess some drawbacks. Perhaps the most obvious is caused not by the mounts but by Earth itself! If an alt-azimuth mount is used to support a terrestrial spotting scope, then the fact that it moves horizontally and vertically plays in its favor. However, the sky is always moving due to Earth's rotation. Therefore, to study or photograph celestial objects for extended periods without interruption, our telescopes have to move right along with them. If we were located exactly at either the North or South Pole, the stars would appear to trace arcs parallel to the horizon as they move around the sky. In these two cases, tracking the stars would be a simple matter with an alt-az mounting; one would simply tilt the telescope up at the desired target, lock the altitude axis in place, and slowly move the azimuth axis with the sky.

Once we leave the poles, however, the tilt of Earth's axis causes the stars to follow long, curved paths in the sky, causing most to rise diagonally in the east and set diagonally in the west. With an alt-azimuth mount, it now becomes necessary to nudge the telescope both horizontally and vertically in a steplike fashion to keep up with the sky. This is decidedly less convenient than the single motion enjoyed by an equatorial mount, a second way of supporting a telescope.

Equatorial mounts

"If you can't raise the bridge, lower the river," so the saying goes. This is the philosophy of the equatorial mount. Since nothing can be done about the stars' apparent motion across the sky, the telescope's mounting method must accommodate it. An equatorial mount may be thought of as an altitude-azimuth mount tilted at an angle that matches your location's latitude.

Like its simpler sibling, an equatorial mount is made up of two perpendicular axes, but instead of referring to them as altitude and azimuth, we use the terms *right-ascension* (or *polar*) *axis* and *declination axis*. In order for an equatorial mount to track the stars properly, its polar axis must be aligned with the celestial pole, a procedure detailed in chapter 10.

There are many benefits to using an equatorial mount. The greatest is the ability to attach a motor drive onto the right-ascension axis so the telescope follows the sky automatically and (almost) effortlessly. But there are more reasons favoring an equatorial mount. One is that once aligned to the pole, an equatorial will make finding objects in the sky much easier by simplifying hopping from one object to the next using a star chart, as well as by permitting the use of setting circles.

On the minus side of equatorial mounts, however, is that they are almost always larger, heavier, more expensive, and more cumbersome than alt-azimuth mounts. This is why the simple Dobsonian alt-az design is so popular for supporting large Newtonians. An equatorial large enough to support, say, a 12- to 14-inch f/4 reflector would probably tip the scales at close to 200 pounds, while a plywood Dobsonian mount would weigh under 50 pounds.

Just as there are many kinds of telescopes, so, too, are there many kinds of equatorial mounts. Some are quite extravagant, while others are simple to use and understand. We will examine the two most common styles.

German equatorial mounts. For years, this was the most popular type of mount among amateur astronomers. The German equatorial is shaped like a tilted letter T with the polar axis representing the long leg and the declination axis marking the letter's crossbar. The telescope is mounted to one end of the crossbar, while a weight is secured to the opposite end for counterbalance.

The simplicity and sturdiness of German equatorials have made them the perennial favorite for supporting refractors and reflectors, as well as some catadioptrics. They allow free access to just about any part of the sky (as with all equatorial mounts, things get a little tough around the poles), are easily outfitted with a clock drive, and may be held by either a tripod or a pedestal base. To help make polar alignment easier, some German equatorials have small alignment scopes built right into their right-ascension axes—a big hit among astrophotographers.

Of course, as with everything in life, there are some flaws in the German mount as well. One strike against the design is that it cannot sweep continuously from east to west. Instead, when the telescope nears the meridian, the user must move it away from whatever was in view, swing the instrument around to the other side of the mounting, and re-aim it back at the target.

Inconvenient as this is for the visual observer, it is disastrous for the astrophotographer caught in the middle of a long exposure because as the telescope is spun around, the orientation of the field of view is also rotated.

A second burden to the German-style mount is a heavy one to bear: They can weigh a lot, especially for telescopes larger than 8 inches. Most of their weight comes from the axes (typically made of solid steel), as well as the counterweight used to offset the telescope. At the same time, I must quickly point out that weight does *not* necessarily beget sturdiness. For instance, some heavy German equatorials are so poorly designed that they could not even steadily support telescopes half as large as those they are sold with. (More about checking a mount's rigidity later in this chapter.)

If you are looking at a telescope that comes with a German mount, pay especially close attention to the diameter of the right-ascension and declination shafts. On well-designed mounts, each shaft is at least 1/8 of the telescope's aperture. For additional solidity, superior mounts use tapered shafts instead of straight shafts for the polar and declination axes. The latter carry the weight of the telescope more uniformly, thereby giving steadier support. It is easy to tell at a glance if a mount has tapered axes or not by looking at an equatorial's T housing. If the mount has tapered shafts, then the housing will look like two truncated cones joined together at right angles; otherwise, it will look like two long, thin cylinders.

Finally, if you must travel with your telescope to a dark-sky site, moving a large German equatorial mount can be a tiring exercise. First the telescope must be disconnected from the mounting. Next, depending on how heavy everything is, the equatorial mount (or *head*) might have to be separated from its tripod or pedestal. Lastly, all three pieces (along with all eyepieces, charts, and other accessories) must be carefully stored away. The reverse sequence occurs when setting up at the site, and the whole thing happens all over again when it is time to go home.

Fork equatorial mounts. While German mounts are preferred for telescopes with long tubes, fork equatorial mounts are usually supplied with more compact instruments such as Schmidt-Cassegrains and Maksutov-Cassegrains. Fork mounts support their telescopes on bearings set between two short *tines,* or *prongs,* that permit full movement in declination. The tines typically extend from a rotatable circular base, which, in turn, acts as the right-ascension axis when tilted at the proper angle.

Perhaps the biggest plus to the fork mount is its light weight. Unlike its Bavarian cousin, a fork equatorial usually does not require counterweighting to achieve balance. Instead, the telescope is balanced by placing its center of gravity within the prongs, like a seesaw. This is an especially nice feature for Cassegrain-style telescopes, as it permits convenient access to the eyepiece regardless of where the telescope is aimed... that is, except when it is aimed near the celestial pole. In this position, the eyepiece can be notoriously difficult to get to, usually requiring the observer to lean over the mounting without bumping into it.

The fork mounts that come with some SCTs and Maksutov-Cassegrain telescopes are designed for maximum convenience and portability. They are compact enough to remain attached to their telescopes, and fit together into their cases for easy transporting. Once at the observing site, the fork quickly secures to its tripod using thumbscrews. It can't get much better than that, especially when compared with the German alternative.

Fork mounts quickly become impractical, however, for long-tubed telescopes such as Newtonians and refractors. In order for a fork-mounted telescope to be able to point toward any spot in the sky, the mount's two prongs must be long enough to let the ends of the instrument swing through without colliding with any other part of the mounting. To satisfy this requirement, the prongs must grow in length as the telescope becomes longer. At the same time, the fork tines also must grow in girth to maintain rigidity. Otherwise, if the fork arms are undersized, they will transmit every little vibration to the telescope. (In those cases, maybe they ought to be called tuning fork mounts!)

One way around the need for longer fork prongs is to shift the telescope's center of gravity by adding counterweights onto the tube. Either way, however, the total weight will increase. In fact, in the end the fork-mounted telescope might weigh more than if it was held on an equally strong German equatorial.

GoTo mounts. Technically, this is not a separate mounting design, but rather the marriage of a telescope mount (often an alt-azimuth mount) to a built-in computer that will move the telescope to a selected object. How does this magic work? In general, the user simply enters the current time and date into a small, handheld control box and selects his or her location from the computer's city database, then hits the align button. The telescope's computerized aiming system then slews toward two of its memorized alignment stars. It should stop in the general location of the first alignment star, but it is up to the user to fine-tune the instrument using the hand controller until the alignment star is centered in the field of view. Then the align button is pressed to store that position. After repeating the process with the second alignment star, the LCD displays whether the alignment was successful or not. If so, the telescope is initialized and ready to go for the evening; if not, the process must be repeated.

There is no denying that GoTo telescopes have a great appeal. But bear in mind that for a given investment, you can choose a larger, non-GoTo telescope for the same price as a small, computerized instrument. Many who purchase one of the smaller GoTo instruments that seem to be proliferating find it disappointing once they realize the smaller aperture's limitations. Weigh all your options before making a commitment, then decide which is more important to you.

Regardless of the type of mounting, it must support its telescope steadily for it to be of any value. To test a mount's rigidity, I recommend doing the *rap test.* Hit the telescope tube toward the top end with the ball of your palm while looking through the telescope at a target, either terrestrial or celestial. Don't really whack the telescope, but hit it with just enough force to make it shake noticeably. Then count how long it takes for the image to settle down. To me, less than 3 seconds is excellent, while 3 to 5 seconds is good. I consider 5 to 10 seconds only fair, while anything more than 10 seconds is poor.

Your Tel-O-Scope

Astrologers have their horoscopes, but amateur astronomers have "tel-o-scopes." What telescope is in your future? Perhaps the stars can tell us. To be perfect partners, you and your telescope have to be matched both physically and emotionally. Without this spiritual link, dire consequences can result. Here are eight questions to help you focus in on which telescope is best matched to your profile. Answer each question as honestly and realistically as you can; remember, there is no right or wrong answer.

Once completed, add up the scores that are listed in brackets after each response. By comparing your total score with those found in Table 3.3 at the end of your tel-o-scope session, you will get a good idea of which telescopes are best suited for your needs, but use the results only as a guide, not as an absolute. And no fair peeking at your neighbor's answers!

1. Which statement best describes your level of astronomical expertise?
   - a. Casual observer [1]
   - b. Enthusiastic beginner [4]
   - c. Intermediate space cadet [6]
   - d. Advanced amateur [10]
2. Will this be your first telescope or binoculars?
   - a. Yes [4]
   - b. No [8]
3. If not, what other instrument(s) do you already own? (If you own more than one, select only the one that you use most often.)
   - a. Binoculars [1]
   - b. Achromatic refractor [4]
   - c. Apochromatic refractor [11]
   - d. Newtonian reflector (2″ to 4″ aperture) [4]
   - e. Newtonian reflector (6″ to 10″ aperture on equatorial mount) [6]
   - f. Newtonian reflector (>10″ aperture on equatorial mount) [11]
   - g. Newtonian reflector (<12″ aperture on Dobsonian mount) [5]
   - h. Newtonian reflector (12″ to 14″ aperture on Dobsonian mount) [7]
   - i. Newtonian reflector (16″ + aperture on Dobsonian mount) [10]
   - j. Cassegrain reflector (< 10″ aperture) [7]
   - k. Cassegrain reflector (10″ and larger aperture) [11]
   - l. Schmidt-Cassegrain catadioptric [9]
   - m. Inexpensive Maksutov catadioptric (e.g., Meade ETX) [5]
   - n. Expensive Maksutov catadioptric (e.g., Questar) [11]
4. What do you want to use the telescope for primarily? Choose only one.
   - a. Casual scan of the sky [1]
   - b. Informal lunar/solar/planetary observing [4]

   c. Estimating magnitudes of variable stars [6]
   d. Comet hunting [5]
   e. Detailed study of solar system objects [10]
   f. Bright deep-sky objects (star clusters, nebulae, galaxies) [6]
   g. Faint deep-sky objects [8]
   h. Astrophotography of bright objects (Moon, Sun, etc.) [4]
   i. Astrophotography of faint objects (deep-sky, etc.) [9]

5. How much money can you afford to spend on this telescope? (Be conservative; remember that you might want to buy some accessories for it—see chapters 6 and 7.)
   a. $100 or less [1]
   b. $200 to $400 [3]
   c. $400 to $800 [5]
   d. $800 to $1,200 [7]
   e. $1,200 to $1,600 [9]
   f. $1,600 to $2,000 [11]
   g. $2,000 to $3,000 [13]
   h. As much as it takes (are you looking to adopt an older son?) [15]
6. Which of the following scenarios best describes your particular situation?
   a. I live in the city and will use my telescope in the city. [1]
   b. I live in the city but will use my telescope in the suburbs. [6]
   c. I live in the city but will use my telescope in the country. [8]
   d. I live in the suburbs and will use my telescope in the suburbs. [9]
   e. I live in the suburbs but will use my telescope in the country. [10]
   f. I live in the country and will use my telescope in the country. [12]
   g. I live in the country but will use my telescope in the city. [–5] (Just kidding!)
7. Which of the following best describes your observing site?
   a. A beach. [4]
   b. A rural open field or meadow away from any body of water. [8]
   c. A suburban park near a lake or a river. [5]
   d. A suburban site with a few trees and a few lights but away from any water. [6]
   e. An urban yard with a few trees and a lot of lights. [4]
   f. A rural hilltop far from all civilization. [13]
   g. A desert. [11]
   h. A rural yard with a few trees and no lights. [9]
8. Where will you store your telescope?
   a. In a room on the ground floor of my house. [6]
   b. In a room in my ground-floor apartment/co-op/condominium. [5]
   c. Upstairs. [4]
   d. In my (sometimes damp) basement. [4]
   e. In a closet on the ground floor. [5]
   f. I'm not sure. I have very little extra room. [4]
   g. In a garden/tool shed outside. [9]
   h. In my garage (protected from car exhaust and other potential damage). [7]
   i. In an observatory. [11]

Now add up your results and compare them to Table 3.3 on the next page:

Table 3.3 ***The Results Are In...***

If your score is between...	Then a good telescope for you might be...
17 and 30	Binoculars
25 and 45	Achromatic refractor (3.1-inch aperture on a *sturdy* alt-azimuth mount)
35 and 55	"Econo-Dob" Newtonian reflector (4- to 8-inch aperture on a Dobsonian mount) -OR- Achromatic refractor (3- to 4-inch aperture on an alt-azimuth or an equatorial mount) -OR- "Econo-Mak" (e.g., Meade ETX series) for those who absolutely must have a small telescope (e.g., for airline travel)
45 and 65	Newtonian reflector (6- to 8-inch aperture on an equatorial mount) -OR- Achromatic refractor (4- to 5-inch aperture on an alt-azimuth or an equatorial mount) -OR- Maksutov (4- to 6-inch aperture, mount often sold separately)
55 and 75	Newtonian reflector (12- to 14-inch aperture on a Dobsonian mount) -OR- Achromatic refractor (5- to 6-inch aperture on an equatorial mount) -OR- Small apochromatic refractor (2.7- to 3.1-inch aperture) if this will be a second telescope -OR- Schmidt-Cassegrain (8-inch aperture)—for astronomers who own small cars and must travel to dark skies -OR- Maksutov (6-inch or larger aperture, mount often sold separately)
65 and 85	Schmidt-Cassegrain (8- to 11-inch aperture)—for astronomers who own small cars and must travel to find dark skies -OR- Newtonian reflector (10-inch or larger aperture on an equatorial mount)
75 and 90	Cassegrain reflector (8-inch or larger aperture) -OR- Apochromatic refractor (4-inch or larger aperture) -OR- Newtonian reflector (16-inch or larger aperture on a Dobsonian or an equatorial mount) -OR- Maksutov (premium 3.5-inch or larger aperture)

The results of this test are based on buying a new telescope from a retail outlet and should be used only as a guide. The range of choices for each score is purposely broad to give the reader the greatest selection. For instance, if your score was 58, then you can select from either the 45 to 65 or the 55 to 75 score ranges. These indicate that good telescopes for you to consider include a 3- to 4-inch achromatic refractor or a 6- to 8-inch Newtonian reflector on an equatorial mount, a 12- to 14-inch Newtonian on a Dobsonian mount, or an 8-inch SCT. You must then look at your particular situation to see which is best for you based on what you have read up to now. If astrophotography is an interest, then a good choice would be an 8-inch SCT. If your primary interest is observing faint deep-sky objects, then I would suggest the Dobsonian, while someone interested in viewing the planets would do better with the refractor. If money (or, rather, lack of) is your strongest concern, then choose either the 6- or 8-inch Newtonian. Of course, you might also consider selecting from a lower-score category if one of those suggestions best fits your needs.

Some readers may find the final result inconsistent with their responses. For instance, if you answer that you live and observe in the country, you already own a Schmidt-Cassegrain telescope, and you want a telescope to photograph faint deep-sky objects but are willing to spend $100 or less, then your total score would correspond to, perhaps, a Newtonian reflector. That would be the right answer, except that the price range is inconsistent with the answers, since such an instrument would cost anywhere from $600 on up. Sorry, that's just simple economics—but there are always alternatives.

Just what are your alternatives? Who makes the best instruments for the money? Chapter 4 continues with a survey of today's astronomical binoculars, while chapter 5 evaluates today's telescope marketplace and considers some of the best amateur instruments of yesteryear.

4

Two Eyes Are Better Than One

It might strike some readers as odd that I chose to begin the discussion of astronomical equipment in this edition with binoculars rather than telescopes. After all, why use "just binoculars" when there are so many telescopes to be had? If you are limited in budget or are just starting out in the hobby, do *not* even consider buying an inexpensive telescope. Spend your money wisely by purchasing a good pair of binoculars, maybe a tripod and binocular mount, plus a star atlas and few of the books listed in chapter 7. Not only are they great for astronomy, binoculars are perfect for more down-to-Earth activities, such as viewing sporting events, birding, and what I call recreational voyeurism. There will always be time to buy a telescope later.

The fact of the matter is that binoculars remain the best way for someone who is new to astronomy to start exploring the universe that lies beyond the naked eye. Their natural, two-eyed view seems far more instinctive than squinting through a cyclopean telescope (indeed, the word *binocular* is from the Latin for "two eyes"). Increasing the comfort level of an observer not only makes observing more pleasant, but also enhances visual performance. Image contrast, resolution, and the ability to detect faint objects are all improved, typically between 10% and 40%, just by using two eyes instead of one.

Perhaps you are not a novice stargazer, but rather an advanced amateur who has been touring the universe for years, even decades. Great! We share that common bond. Every amateur astronomer should own a pair of good quality binoculars regardless of his or her other telescopic equipment.

Quick Tips

When choosing a pair of binoculars, the first decision to make is magnification. Low-power binoculars are great for wide views of the Milky Way, but higher-power glasses are preferred for specific objects. Keep in mind that as magnification increases, so does the binoculars' weight. This is especially important if you plan on holding the glasses in your hands. Most people can support up to 7 × 50 binoculars with minimal strain and fatigue, but if either the magnification or aperture is increased by much, hand support becomes impractical.

Many manufacturers offer zoom binoculars that can instantly double or even triple their magnification by pressing a thumb lever. Unfortunately, nothing in this life is free, and zoom binoculars pay heavily. To perform this feat, zoom models require far more complex optical systems than fixed-power glasses. More optical elements increase the risk for imperfections and inferior performance. Additionally, image brightness suffers terribly at the high end of their magnification range, causing faint objects to vanish.

Another critical choice is the binoculars' objective lens diameter. When it comes to selecting a telescope, amateur astronomers always assume that bigger is better. While this is generally true for telescopes, it is not always the case for binoculars. To be perfectly matched for stargazing, the diameter of the beam of light leaving the binoculars' eyepieces (the exit pupil) should match the clear diameter of the observer's pupils. Although the diameter of each person's pupil can vary (especially with age), values typically range from about 2.5 mm in the brightest lighting conditions to about 7 mm under the dimmest conditions. If the binoculars' exit pupil is much smaller, then we lose the binoculars' *rich-field* viewing capability, while too large an exit pupil will waste some of the light the binoculars collect. (For a complete discussion on what exactly an exit pupil is, fast forward to the discussion about eyepieces in chapter 6. (Go ahead, I'll wait.)

The exit pupil in millimeters for any pair of binoculars is found by dividing their aperture (in millimeters) by their magnification. If you are young and plan on doing most of your observing from a rural setting, then you would do best with a pair of binoculars that yield a 7-mm exit pupil (such as 7 × 50 or 10 × 70). However, if you are older or are a captive of a light-polluted city or suburb, then you may do better with binoculars yielding a 4-mm or 5-mm exit pupil (7 × 35, 10 × 50, etc.).

Rather than define and discuss exit pupils, some manufacturers specify a binoculars' *twilight factor,* which is a numerical value that is supposed to rate performance in dim light. The twilight factor is calculated by multiplying the aperture by the magnification, then taking the square root of the product. So a pair of 7 × 50 binoculars has an exit pupil of 7.1 mm and a twilight factor of 18.7, while a pair of 10 × 50 binoculars has a 5-mm exit pupil and a twilight factor of 22.4. By definition, the higher the number, supposedly the better the dim light images.

Does this mean that the 10 × 50s are always better for astronomical viewing than 7 × 50s? Maybe, maybe not. I have never put a lot of weight in the twilight factor, since it does not factor in local sky conditions. In my experience, exit pupil is what really tells the tale of how well a pair of binoculars will show the night sky, not twilight factor. Of course, we must also take into account a wide range of other mitigating factors, including lens and prism quality and types of coatings.

Speaking of prisms, modern binoculars use one of two basic prism styles: compact roof prism binoculars and heavier Porro prism binoculars, both shown in Figure 4.1. Which should you consider? All other things being equal, Porro prism binoculars will yield brighter, sharper, more contrasty images than roof prism binoculars. Why? Roof prism binoculars, which were only invented in the 1960s, do not reflect light as efficiently as Porro prisms. By design, roof prisms require one internal surface to be aluminized in order to bounce light through the binoculars. As the light strikes the aluminizing, a bit of it is lost, resulting in a dimmed image. Porro prisms (at least those made from high-quality glass) reflect all of the incoming light without the need for a mirrored surface, allowing for a brighter final image.

The second disadvantage to roof prism glasses is their high cost. Part of the higher expense is the result of the more complex light path that demands much greater optical precision in manufacturing. Unless constructed with great care, the image quality of roof prism binoculars will be unacceptable. Even still, assuming all other things are equal, most roof prism glasses are inferior to Porro prism models for viewing the night sky. Having said that,

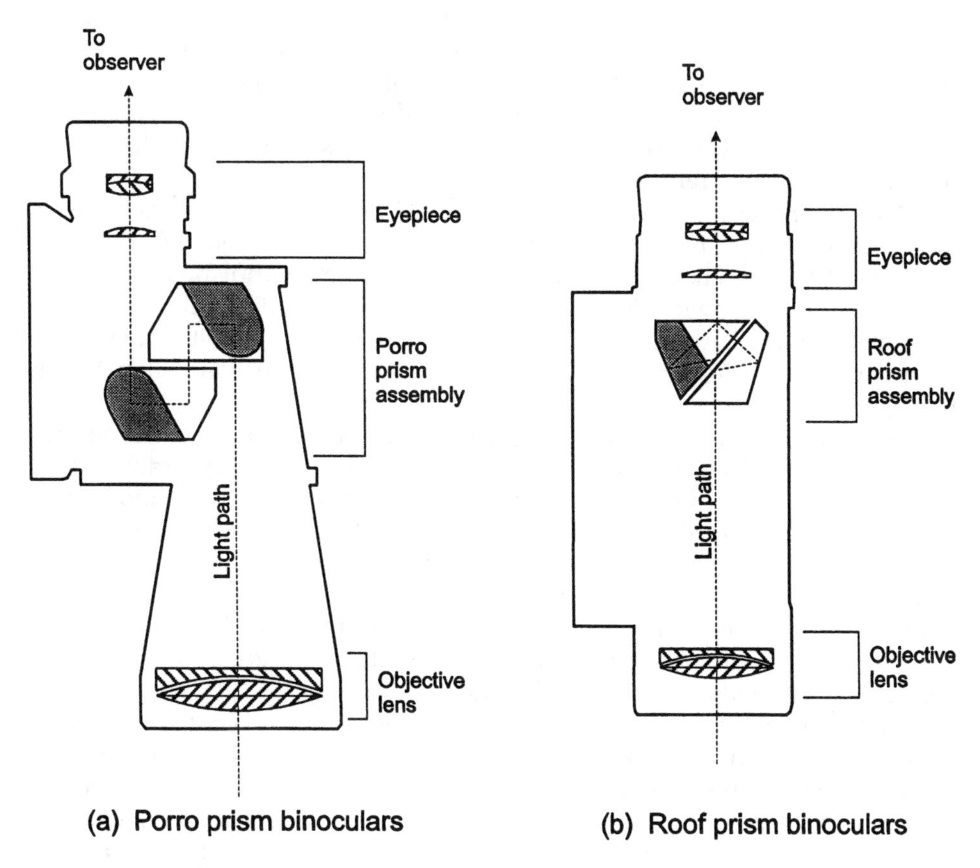

Figure 4.1 *Cross sections through (a) Porro prism binoculars and (b) roof prism binoculars.*

there are still several starworthy roof prism binoculars out there. Some of these are included in the reviews that follow.

If you are considering roof prism binoculars, you will likely come across the phrase *phase coating.* Phase coatings are used to correct for light loss inherent in the roof prism design. As an image is bounced through a roof prism assembly, the horizontal plane is passed slightly out of phase to the vertical plane, causing some dimming and loss of contrast. This so-called phase error can be corrected by applying a special coating to the prism surface. Phase error is of minor consequence, and is often masked by other, more apparent manufacturing flaws.

If you have decided on a pair with Porro prisms, check their type of glass. Better binoculars use prisms made from BaK-4 (barium crown) glass, while less expensive binoculars utilize prisms of BK-7 (borosilicate). BaK-4 prisms transmit brighter, sharper images because their design passes practically all of the light that enters (what optical experts call *total internal reflection*). Due to their optical properties, BK-7 prisms suffer from light falloff and, consequently, somewhat dimmer images. The problem lies in BK-7's *index of refraction,* a measure of how much a beam of light will bend, or refract, when it passes through a medium. Beyond the objective lens, light strikes the wedge-shaped prism face at too steep an angle (referred to as the *angle of incidence*) to be reflected completely through the prism; instead, some is lost. More light is lost as it exits the prism, and again as the light enters and exits the second Porro prism in the set. The higher index of refraction of BaK-4 permits total internal reflection at a steeper angle of incidence, so that all of the image-forming light is fully reflected.

Most manufacturers will state that their binoculars have BaK-4 prisms right on the glasses, but if not, you can always check for yourself. Hold the binoculars at arm's length and look at the circle of light floating behind the eyepieces. This is the previously mentioned exit pupil. The exit pupils will appear perfectly circular if the prisms are made from BaK-4 glass, but diamond-shaped with gray shadows on the edges (because of the light falloff) with BK-7 prisms.

When shopping for a pair of astronomical binoculars, look at several brands and models side by side if possible. Does the manufacturer state that the lenses are coated? Optical coatings improve light transmission and reduce scattering. An uncoated lens reflects about 4% of the light hitting it. By applying a thin layer of magnesium fluoride onto both surfaces of the lens, reflection is reduced to 1.5%. A lens that has been coated with magnesium fluoride will exhibit a purplish tint when held at a narrow angle toward a light. If the coating is too thin, it will look pinkish; if it is too thick, then a greenish tint will be cast. Uncoated lenses have a whitish glint.

Just about all binoculars sold today have optics that are coated with magnesium fluoride, but that does not necessarily imply quality. The words *coated optics* on many of the least expensive binoculars should be assumed to mean that only the exposed faces of the objectives and eye lenses are coated; the internal optics are probably not. For the best views, we need all optical surfaces coated, so *fully coated optics* are preferred. But wait, there's more! It has been found that while a single coating on each optical surface is good, building up that coating in several thin layers is even better at enhancing performance. So-called multicoated lenses reduce reflection to less than 0.5%. These show a greenish reflection when turned toward light. Combining the best with the best, binoculars that have *fully multicoated* optics, which means that every optical surface has been coated with several microscopically thin layers of magnesium fluoride, are cream of the crop for astronomical viewing.

What about ruby-coated lenses? Ruby, or red, coatings are quite the rage in certain circles, although documented evidence of their real benefits remains rather dubious. Manufacturers tell us that they are best at reducing glare in very bright light, such as sunlit snow or sand. Whether this is true, I will leave for others to debate, but ruby-coated binoculars offer no benefit to astronomical viewing and, indeed, may actually work against stargazers.

A well-designed pair of binoculars will have good *eye relief.* Eye relief is the distance that the binoculars must be held away from the observer's eye in order to see the entire field of view at once. If eye relief is too short, the binoculars will have to be held uncomfortably close to the observer, a problem to consider especially if you have to wear glasses. At the same time, if eye relief is excessive, then the binoculars become difficult to center and hold steadily over the observer's eyes. Most observers favor binoculars with eye relief between about 14 and 20 mm. In general, the value tends to shrink as magnification increases.

How about focusing? Most binoculars have a knurled focusing wheel between the barrels. This is called *center focusing.* A few so-called individual-focusing binoculars require that each eyepiece be focused separately. Regardless of the method used, the focusing device should move smoothly. Beware some less expensive binoculars (none listed here, incidentally) that use a *fast-focus* thumb lever instead of a wheel. This latter approach permits rapid, coarse focusing, which may be fine for moving objects but is not accurate enough for the fine focusing required when stargazing.

At the same time, recognizing that most people have one good eye and one bad eye, most manufacturers design their center-focus binoculars so that one eyepiece can be focused differently than the other. This *diopter control,* as it is called, is usually located in the right eyepiece. First, close your right eye and focus the view for your left eye using the center-focusing wheel. Then, using only your right eye, focus the view by twisting the right eyepiece's barrel. With both eyes open, the scene should appear equally sharp.

Binoculars over 10x and larger than 60 mm in aperture are classified as *giant binoculars.* These are especially useful in comet hunting, lunar and planetary examination, and in any application that requires a little more power than traditional glasses. If you have your heart set on purchasing a high-power pair, then some sort of external support (typically a photographic tripod) is a must. But also make certain that the binoculars are tripod adaptable. Usually, that means there is a 0.25–20 threaded socket located in the hinge between the barrels, which makes an L-shaped tripod adapter a must. You will also need a sturdy, tall tripod. Most tripods are geared toward terrestrial viewing, not celestial, and as such do not extend very high. Chapter 7 explains what to look for in a camera tripod and reviews a number of useful binocular mounts. Chapter 8 offers plans on how you can make your own.

Finally, make sure the binoculars fit properly. To check that, twist the binoculars in and out around their central hinge. Notice how the distance between the eyepieces changes as the barrels are pivoted? That distance, called the *interpupillary distance,* must match the distance between the observer's eyes. Now, the average adult has an interpupillary distance between 2.3 and 2.8 inches. For binoculars to work properly, the centers of their eyepieces must be able to spread between those two values. If the eyepieces are just slightly out of adjustment, the observer will see not one but two separate, but distinct, off-center circles. Unless both sets of eyes and eyepieces line up exactly, the most perfect optics, the finest coatings, and the best prisms are worthless.

Equipment Reviews

There are so many binoculars currently available on the market, how can one tell which is better than the rest? Given such a wide variety from which to choose, it should come as no surprise that the answer is not an easy one. Here are capsule reviews of more than 200 different models from more than two dozen companies. Consumers should note that some of the companies listed manufacture their own products, while others simply rebadge goods from other companies, typically from the Far East.

Bausch & Lomb. This company offers dozens of different binocular models in their lineup. Many are sold under the Bausch & Lomb trademark, while many more carry the Bushnell label. Yet for all of this diversity, only a few Bausch & Lomb binoculars pass the astronomical litmus test. Rather, most are designed with other purposes in mind, such as sporting or theatrical events, and other terrestrial pursuits.

Of all the binoculars wearing the Bausch & Lomb name, only some of the glasses in their Custom and Legacy lines are suited to an astronomer's needs. Bausch and Lomb Legacy binoculars range in size from 8 × 22 to 10 × 50, including two zoom variations, but only the Legacy 7 × 35, 8 × 40, and 10 × 50 models offer useful combinations of aperture and magnification. Each has multicoated optics, BaK-4 prisms, and a built-in tripod socket. Weight is typical for their sizes, although many will find their short 9-mm eye relief uncomfortable, especially if you must wear eyeglasses. All have wider-than-average fields of view, with the 7 × 35s sporting a huge 11° field. Yes, that will lead to curvature of field, but my personal philosophy is that I'd rather see a curved edge to the field than no field at all, as would be the case with narrower binoculars.

Bausch & Lomb's Custom 8 × 36 and 10 × 40 binoculars include tripod sockets, multicoated optics, BaK-4 prisms, and have a respectable 19 mm of eye relief. Focusing is rather stiff, as it seems to be for most Bausch & Lomb binoculars, and interpupillary adjustment is limited, although adequate for most people. Overall, not a bad binocular at all, but given their small apertures and resulting exit pupils, there are better binoculars out there for the same money.

Brunton. Although not well known in the astronomical arena, Brunton makes a line of binoculars that might attract the interest of stargazers. While several different models bear the Lite-Tech name, only the two Porro prism models, 7 × 50 and 10 × 50, are mentioned here; the rest are small, roof prism glasses that are not astronomically suitable. Both of the Porro prism Lite-Techs include fully coated optics, BaK-4 prisms, and tripod sockets. Center focusing is quite smooth, although diopter focusing is done by twisting the right objective lens, as opposed to the more typical, turning right-side eyepiece. While the binoculars are nicely designed ergonomically, they are subject to ghost images if a bright object is placed off-center in the field. Viewing the Moon, for instance, can be a haunting experience through Lite-Techs.

Bushnell. A subsidiary of Bausch & Lomb, Bushnell binoculars are a step below most of the parent company's offerings, both in terms of price and quality. Most Bushnell models are aimed toward the low end of the market, although the 8 × 42 and 10 × 42 Bushnell Natureview series rise to the occasion. While they cut corners—not too surprising for their comparatively low cost—they remain good choices for limited budgets. Personally, however, I believe that the Scenix binoculars from Orion represent a better buy in this price category.

Bushnell also makes low-end waterproof binoculars called Marines. The 7 × 50 Marines use individually focused eyepieces to maintain watertightness, are tripod adaptable, and include BaK-4 prisms and fully coated optics. Be aware that these maritime glasses are sold either with or without a built-in range finder; choose the latter, since a range finder is something you neither need nor want for astronomy.

Canon. Famous among photographers for its fine cameras, Canon shook the binocular world a few years back when it introduced its *image-stabilized* (IS) roof prism binoculars. These are no ordinary roof prisms, mind you; rather, they are connected in a flexible, oil-filled housing. Inside each of the barrels is a microprocessor that detects all motions of the observer and sends a controlling signal to each of the prism assemblies. The prisms, which Canon calls Vari-Angle Prisms, then flex to compensate for the movement. Powered by two AA batteries, the stabilization effect is activated by the push of a button on the top of the right-hand barrel. It works like a charm, even in temperatures as low as +14°F (–10°C) and as high as +112°F (+44°C).

Of course, this technology doesn't come cheaply. Even the smallest Canon IS binoculars, only 10 × 30, typically sell for about $500, while their largest, 18 × 50s, go for about three times that price. Is it worth the price? Many owners state emphatically that they are, using such flowery language as "wonderful," "amazing," and simply "wow!" And despite their comparatively small apertures, there is no denying that the views through Canon IS binoculars are superior to those of many larger conventional models.

Those views are incredibly sharp and clear, incidentally. Canon, also renowned for their fine photographic lenses, uses fully multicoated optics to produce what is arguably the sharpest set of binocular optics sold today at any price. Stars remain pinpoints across the entire field through all of the IS glasses. Some feel that the fine optics combined with the stable images is like adding 20 mm to the aperture.

Okay, so do you want my opinion? Don't get me wrong. The technology is terrific, as are the optics. Images are sharp across the entire field of any of the Canon IS glasses. But when I engage the IS feature by pushing a button, the view strikes me as odd, taking on a strange quality that is difficult to describe. The image moves as I move, but seems to do so just slightly out of sync with my motions. In fact, the effect almost makes me feel slightly seasick, though I am prone to that sort of thing. Also, with no built-in tripod socket, the binoculars still have to be lowered every time you want to look back at a star chart. They will also have to be lowered fairly often just to rest, since they are on the heavy side for their respective apertures. Between that and the price, I feel that consumers would be much better off spending the same amount of money on a larger pair of binoculars and a sturdy tripod.

Celestron International. This company currently offers five lines of binoculars suitable for astronomical viewing: WaterSports, Bird Watchers, Ultimas, Giants, and Enduros. Enduro binoculars, the least expensive of the bunch, are easy to hold, although eye relief is rather short for eyeglass wearers. Each is also tripod adaptable, a feature not found in many sub-$100 binoculars. Enduros produce sharp images in the center of the field, but lack the image contrast and brightness of more expensive glasses because they use BK-7 prisms. Still, they make good first binoculars for budding stargazers.

Do your observing from very damp environs? Then Celestron's 7 × 50 and 10 × 50 WaterSport binoculars might be a good choice if you are looking for an inexpensive binocular that is waterproof. WaterSport binoculars, designed for boating enthusiasts, are sealed against moisture intrusion and so will not fog internally after a damp night under the stars. If you have ever accidentally left your binoculars outside all night, only to find them dripping wet with dew the next morning, you'll appreciate these. Celestron tells us that the optics are fully coated, while the Porro prisms are made from desirable BaK-4 glass. Although images are not as sharp as Celestron's Ultimas, they will be acceptable to most, especially considering the low price. The only drawback is that neither model has a built-in tripod socket.

The Bird Watcher (7 × 35, 8 × 40, 10 × 50) and Ultima (7 × 50, 8 × 40, 8 × 56, 9 × 63, 10 × 50) series feature center focus and premium components, and they are tripod adaptable. All have very good mechanical construction, although the Bird Watchers have fairly short eye relief. Optically, images are both bright and sharp, although the fully multicoated optics in the Ultima models are clearly superior. The good image sharpness and eye relief of the Ultimas, however, come at the expense of their fields of view, which are tighter than some similarly sized binoculars. And while the Bird Watchers are good, the Ultimas are arguably the finest sub-$300 binoculars on the market today (Ultima 9 × 63s typically retail for a bit over $300). They come highly recommended.

Last but not least, Celestron's 20 × 80 Deluxe Giant binoculars produce sharp, clear images that compare favorably to those of other glasses of similar aperture and magnification, such as Orion and Swift. Views are relatively free of distortion, although the view softens toward the edge of the 3.5° field. There is a fair amount of false color visible around brighter objects, which is typical of many giant binoculars. Some also complain that the range of eyepiece motion is short, making it difficult to reach focus. I must emphasize that these binoculars are far too large and heavy to hold by hand, necessitating an external support, such as the commercial units listed in chapter 7 or the homemade mounts in chapter 8. A reinforcing bar bridges between the objectives and prisms for added stability, while a permanently attached tripod adapter eliminates the need for a separate bracket. The latter slides back and forth along a rod for easy mounting and balancing. This is a good idea *if* the binoculars will be mounted to a photographic tripod. At the same time, however, it prevents the binoculars from attaching to most commercial binocular mounts.

Eagle Optics. A name well known to birders, Eagle Optics sells several brands of binoculars, including some that bear their own name. Eagle Voyager binoculars come in two astronomically worthy sizes, 8 × 42 and 9.5 × 44. Weighing only 21 ounces, the 8 × 42s also have fully multicoated optics as well as BaK-4 Porro prisms. An eye relief measure of 18 mm makes them suitable for observers who must keep their glasses on. Top this off with a built-in tripod socket and you have a satisfying pair of binoculars for stargazing from suburban and urban settings.

Those who follow the binocular industry may recall that Celestron used to market 9.5 × 44 binoculars that had objective lenses made of extra-low-dispersion (ED) glass for better-than-average color suppression. Well, those binoculars are still here, but now wear the Eagle Voyager badge. Although they have a relatively small exit pupil, only 4.6 mm, these binoculars still deliver good images that are virtually color free (that is, free of *false* color). Fully multicoated optics and BaK-4 prisms round out the package. Like the other three Eagle Optics binoculars listed here, the 9.5 × 44s include a built-in tripod socket.

Edmund Scientific. Now a division of VWR Scientific in Tonawanda, New York, Edmund Scientific offers a house brand pair of imported 7 × 50 binoculars. They feature fully coated optics (a plus), BK-7 glass prisms (a minus), and an average field of view. Images are fine on-center, though some distortion is evident toward the edges. Their aluminum body is sheathed with rubber to help absorb shock and also to help improve the grip. All the standard fare, such as case, strap, and lens caps, is included, although there is no provision for attaching a tripod. While adequate for astronomical purposes, they are priced higher than some superior binoculars from other manufacturers.

A footnote about Edmund Scientific. Those of us who are children of the 1950s, '60s, and '70s fondly remember the Edmund Scientific Corporation (known earlier as Edmund Salvage Company, a dealer in war-surplus optics after World War II). Their catalog was one of the premiere sources for all things astronomical. But times changed and the business shifted, diversifying into two separate entities: Scientifics and Industrial Optics. The Scientifics name and side of the company was sold to VWR in February 2001, while the Industrial Optics Division remains in Barrington, New Jersey, where Edmund started. All of the Edmund products discussed in this book are, in fact, Edmund Scientific products, not Edmund Industrial Optics.

Fujinon. This name has come to be known among photographers for excellent film and camera products over the years, and it is now synonymous with the finest in binoculars, as well. Fujinon binoculars are famous for their sharp, clear images that snap into focus. They suffer from little or no astigmatism, residual chromatic aberration, or other optical faults that plague glasses of lesser design. Binoculars just don't come much better.

While not all Fujinon binoculars are designed for astronomical viewing, their Polaris series (also known as the FMT-SX series; see Figure 4.2) is made with the stargazer in mind. Ranging from 6 × 30 (too small for astronomical viewing) to 16 × 70, all models are waterproofed and sealed, purged of air, and filled with dry nitrogen to minimize the chance of internal lens fogging in high-humidity environments. To maintain their airtight integrity, eyepieces must be individually focused, which is an inconvenience when supporting by hand.

Figure 4.2 *Fujinon 16 × 70 FMT-SX binoculars. Note how the author's daughter, Helen, is supporting the ends of barrels, rather than gripping them around their prisms. Though still a tiring exercise, this is the best way to handhold giant binoculars.*

Another inconvenience is their high price, but many are willing to pay for the outstanding optical quality. The FMT-SX 7 × 50s and 10 × 70s are especially popular among amateur astronomers, and I personally find the 16 × 70s to be outstanding performers from my suburban backyard as well as from darker, rural surroundings. Star images are tack sharp, providing wonderful deep-sky views. I fondly recall a view from the Stellafane convention in Vermont a few years ago of M6 and M7, two open clusters in the tail of Scorpius, as they just glided over some distant pine trees. That view painted a three-dimensional portrait in my mind which is far more lasting and meaningful than any photograph.

If you're working within a tighter budget but still want Fujinon quality, consider their Poseidon MT-SX series. Fujinon MT-SX binoculars are available in 7 × 50, 10 × 50, and 16 × 70 models, retailing for about $100 less than their FMT-SX counterparts. Like their more-expensive brethren, MT-SX binoculars are waterproof and nitrogen filled, again requiring individually focused eyepieces. So what's the difference? For one, eye relief is considerably shorter, only 12 mm in each, or what I might call the uncomfortable range. This may certainly make viewing difficult, especially for eyeglass wearers. FMT-SX binoculars also use flat-field optics to reduce edge-of-field distortion, while MT-SX models do not. But the amount of distortion present in any Fujinon binoculars is so small that few complain.

Fujinon also makes giant-giant binoculars called Large-Size Binoculars. That's an understatement, if ever I heard one! Featuring the same quality construction as the Polaris line, including BaK-4 Porro prisms and fully multicoated optics throughout (referred to as Electron Beam Coating by the manufacturer), Fujinon's Large-Size binoculars are available in either 15 × 80, 25 × 150, or 40 × 150 versions. The 15 × 80s look like slightly larger versions of the 70-mm MTs with straight-through prisms and conventional features. The 25 × 150s come in three models: the least expensive MT-SX, featuring straight-through eyepieces; ED-SX, utilizing ED (extra-low-dispersion) glass in their objective lenses for better color correction; and EM-SX models, including ED objectives and eyepieces mounted in 45° prisms. The 40 × 150 ED-SX binoculars also have ED objectives, but conventional, straight-through eyepieces. Fujinon is justifiably proud of its Large-Size binoculars. After all, Japanese amateur Yuji Hyakutake used 25 × 150 MTs to discover Comet Hyakutake (C/1996 B2), while the 40 × 150s are the world's largest production binoculars (among the world's heaviest too, at more than 40 pounds!).

Finally, Fujinon recently joined the image-stabilized world with their 14 × 40 Techno-Stabi binoculars. I mentioned previously when discussing the Canon IS binoculars that if you are sensitive to motion sickness, you might feel a bit queasy using those. Well, the Fujinons really take to the high seas, with a much larger compensation angle (5° versus 1° for the Canons). Images tend to really float around, making them difficult to center after panning. While this may be ideal for seafarers, it is less desirable for landlubber astronomers.

Optically, although the images are as sharp as the Canons, Techno-Stabi binoculars seem dimmer for reasons that escape me, though it likely has something to do with the coatings. Some slight residual chromatic aberration is also evident around brighter objects, but flare and ghosting are suppressed very well. Perhaps the most annoying feature of the Fujinons is their two-button activation process. You have to push the button that is nearest the eyepieces to turn the power on initially; depress it again to stabilize the image; then press the other button to turn the whole thing off. It's very easy to confuse one button for the other in the dark. The stabilization circuit stays on by itself for one minute after the button is released, after which time, it goes into standby mode. The Canons stay on for five minutes before standing by. I must admit, however, that I like the Fujinon's off-center center-focusing knob, which is adjacent to the right eyepiece.

Kowa. This company manufactures a pair of Porro prism binoculars that come at attractive prices. Kowa 7 × 40 and 10 × 40 ZCF binoculars both have BaK-4 prisms and fully multicoated optics in rubber-armored magnesium bodies. Unfortunately, neither include tripod sockets. Optically, both rate an adequate. Image sharpness in the center of the field is quite good, but that falls off fairly rapidly toward the edges. Contrast is also low, which is surprising given the optical specifications. Eye relief, however, is very good in both models, especially the 7 × 40s (which may actually be a little long at 23 mm).

Leupold and Stephens. Here is another name that is probably more familiar to birders than stargazers. Of the binoculars they offer, their Wind River Porro prism glasses are best suited for our purposes. Two models, 8 × 42 and 10 × 50, include BaK-4 prisms and multicoated optics for sharp, contrasty views that compare favorably with any binocular in the same price range. Unfortunately, they lack a tripod socket, which I feel is an absolute must; but perhaps you don't, in which case these are a fine choice. Focusing is smooth, thanks to an easy-to-grip center wheel, and both models are waterproofed with strategically located O-rings to guard against internal fogging. Eyeglass wearers might find their 16 mm of eye relief a little tight.

Meade Instruments. This company offers several lines of binoculars, many of which are suitable for astronomical viewing. Its Mirage 7 × 35 and 10 × 50 glasses win the title of lowest-cost binoculars in this book, retailing for less than $50. Of course, you shouldn't expect these to deliver the same images as binoculars costing ten times more, but overall, they aren't bad for the price. For the money, you get pretty basic fare, including coated optics and BK-7 prisms. Eye relief is on the short side, at only 14 mm and 13 mm, respectively. Sure, you will see ghosts and flaring around bright objects like the Moon and Venus (dare I call them mirages?), but images are reasonably sharp at the center of the field. Images distort toward the field's edge, however; but again, we're talking about cheap binoculars here. Neither includes a tripod socket, but both are light enough to support by hand without much trouble.

Moving up slightly in the price and quality categories, we have Meade's Infinity and Travel View series. The Infinity line includes 7 × 35, 8 × 40, and 10 × 50 models, while of all that carry the Travel View name, the 7 × 35, 10 × 50, and 12 × 50s are the most astronomically attractive. All include optics that are either fully coated (Travel View) or multicoated (Infinity)—but not both—and BK-7 crown glass prisms. Their fields of view are wider than those in the Meade Mirages, although edge distortion is also more prevalent. Again, eye relief is short, in the 12- to 13-mm range. Like the Mirages, none includes a tripod socket, making it tough to support the 30-ounce Travel View 12 × 50s for any length of time. Still, they represent very good values for budget-conscious astronomers. If you are considering these binoculars, be sure to compare them against Celestron Enduro and Orion Scenix binoculars.

If you must wear glasses while viewing, Meade's 7 × 36, 8 × 42, 10 × 42, and 10 × 50 Safari Pro Porro prism binoculars offer good, but not excessive, eye relief. BaK-4 prisms and multicoated optics combine with smooth mechanicals and wider-than-average fields of view to make these good mid-priced glasses. Images are sharp, though slightly less contrasty than fully multicoated models. Safari Pros also include tripod sockets, judged a plus.

Minolta. Though best known for its cameras, Minolta also manufactures several lines of binoculars. Of those, Activa Standard 7 × 35W, 7 × 50, 8 × 40W, 10 × 50W, and 12 × 50W models are the best. Perhaps the nicest thing about some of the Minolta Activas are their wider-than-average fields of view (signified by the W in their designations). Each includes fully multicoated optics, BaK-4 prisms, a tripod adapter, and center focusing. Image brightness and sharpness are excellent, but while edge sharpness is good, some coma is evident. Eye relief is also good in the 7×, 8×, and 10× models, but a little short in the 12× pair, especially if wearing glasses. Moderately priced, Minolta Activa Standard binoculars represent some of the best buys in binoculars today.

Minolta's Classic II Porro prism binoculars are also worth a quick note. Available in 7 × 35W, 7 × 50, 8 × 40W, 10 × 50W (the Ws again indicating a wider field of view), each includes fully coated optics, BK-7 prisms, center focusing, and a tripod socket. So, even though images are not up to the Activas, given their sub-$100 price, they are some of the best inexpensive binoculars on the market.

Miyauchi. Not exactly a brand name that rolls off the tongue easily, Miyauchi nonetheless manufactures binoculars that are sure to make mouths water. Miyauchi's binoculars display very high-quality construction, with a streamlined, gray finish to their barrels. The individually focused eyepieces and interpupillary adjustment are also very smooth. Some have inferred that the latter may be *too smooth,* causing the binoculars to go out of adjustment, but I have never found this to be a problem with those that I have tested.

The smallest Miyauchis are the 20 × 77 BS-77. With a comparatively small exit pupil of less than 4 mm, the BS-77s come with *semi-apochromatic* objective lenses and beautifully executed one-piece aluminum die-cast bodies. Each binocular barrel is filled with dry nitrogen to minimize internal fogging. BS-77s are available either with straight-through eyepieces or eyepieces that are tilted at a 45° angle for easier viewing. Both versions focus individually, with click stops for precise setting. Star images are sharp and clear, with little distortion detectable toward the edge of the field of view. I was surprised, however, at the amount of false color (i.e., residual chromatic aberration) visible through the Miyauchis during a recent viewing session. To my eyes, it was more prominent than through 70-mm Fujinon and Nikon binoculars.

Miyauchi also makes two 20 × 100 binoculars. The BJ100iB (Figure 4.3), like the BS-77, comes with fully multicoated, semi-apochromatic objective lenses. Its unique look, thanks to a nicely finished aluminum body, also includes interchangeable eyepieces, a nicely integrated, detachable carrying handle, and nitrogen-filled barrels to prevent fogging. Three versions based on eyepiece angle—either straight-through or tilted at 45° or 90°—are available, although here in the United States, only the 45° version is usually seen. A 3× finderscope and a fork mount/pedestal combination are also available separately. I have had a fondness for these binoculars ever since I first looked through a pair at the Winter Star Party in Florida several years ago. They are intelligently designed and come with several features that I wish others would copy.

For a mere $900 more, Miyauchi BJ100iBF binoculars include all of these features plus apochromatic objective lenses. These are designed to eliminate all false color, usually exhibited as halos of purple around the edges of bright objects, which is a common problem with short-focal-length refractors (chapter 5 discusses this problem further, if you wish to look ahead). False-color

Figure 4.3 *Miyauchi BJ100iB 20 × 100 binoculars. Photo courtesy of Selwyn Malin.*

suppression is very good through the BJ100iBF binoculars, although to my eye it is not completely eliminated. Still, the view is very impressive!

Then there are the Miyauchi BR-141 25 × 141 apochromatic binoculars. One of the world's most expensive binoculars, the BR-141 feature two 5.6-inch objective lenses, each with a false-color-suppressing fluorite element. Eyepieces are removable, so that standard 25× magnification can be changed to several magnifications up to 45× with auxiliary eyepieces from Miyauchi. Scanning the skies through a pair of 5.5-inch apochromatic refractors is an amazing trip... if you can afford the first-class ticket! The BR-141s can be purchased with 25× eyepieces, a 3× finderscope, and a mount/tripod combination; or, if preferred, the body may be bought alone.

Nikon. Like several other brands, the name Nikon may be more familiar to photographers than astronomers, but all that is changing fast. Nikon produces several lines of outstanding binoculars, although many are based on the roof prism design, which is more appropriate for terrestrial, rather than celestial, study.

Nikon's Action binoculars, its lowest-price line included here, feature aspherical eyepieces for clear, sharp images over nearly the entire field of view. Some edge distortion is evident, but multicoated optics and BaK-4 prisms make these some of the finest glasses in their price range. Consumers can select from the 7 × 35 Naturalist IV, 7 × 50 Shoreline, 8 × 40 Egret II, 10 × 50 Lookout IV, or 12 × 50 Fieldmaster. All include smooth center focusing, reasonable eye relief and fields of view, and are tripod adaptable. The 7 × 50 and 10 × 50 models are especially noteworthy for their 20 mm of eye relief.

Nikon's 7 × 50 Sports and Marine binoculars, designed for boaters, include sealed barrels and individually focusing eyepieces. Once again, the images are Nikon quality, with only a slight softness to star images around the very edge of the 7.5° field. Some people, however, may grow tired holding these 2.5-pound binoculars by hand, but they are still a little lighter than Nikon's premier 7 × 50 Prostars, described below. The Sports and Marine binoculars are also tripod adaptable with a standard 90° mounting bracket.

Those who own them swear by Nikon's Superior E glasses. Designed more for birders than astronomers, these 10 × 42 and 12 × 50 Porro prism binoculars are just as adept looking skyward at night as they are during the day. The construction quality is second to none, as anyone can tell with a turn of the ultrasmooth focusing knob. The body, constructed of magnesium, is covered with a rubberized material that is a pleasure to grip. And the optics are, well, Nikon. That means images are bright and sharp across the field, and the contrast is excellent. Some users have noted that the eye relief causes some blacking out of the view unless the binoculars are held just so, which can lead to increased fatigue after prolonged use.

To me, the finest Nikon binoculars wear the Criterion badge. For handheld use, Nikon's 7 × 50 IF SP Prostars have almost all the right stuff. Not surprisingly, optical performance is outstanding thanks to fully multicoated objective lenses made of extra-low-dispersion glass, which go a long way in eliminating chromatic and spherical aberrations as well as coma. Like Fujinon

Polaris FMT binoculars, Nikon Prostars come with individually focused eyepieces and are sealed and waterproofed to prevent internal fogging. Their field of view is comparable to the Fujinons, but eye relief is less. Optical performance is arguably a little better than the Fujinons.

The Nikon 10 × 70 IF SP Astroluxes are remarkable giant binoculars. Sharing all the features of their smaller Prostar cousins, Astroluxe glasses have a 5.1° field of view, nearly identical to their Fujinon 10 × 70 rivals. Again, optical performance approaches absolute perfection. Changing eyepieces, Nikon offers the 18 × 70 IF SP Astroluxe giants. All in all, they are identical to the 10 × 70s, but with more magnification and, happily, the same eye relief. The field of view is cut from 5° to 4°, with appropriately dimmer images because of the smaller exit pupil.

It is critical to point out that none of the Nikon Superior E or Criterion beauties come with a standard tripod adapter. Nikon does, however, sell a customized, clamp-on bracket, but it may not work with all of the commercial binocular mounts in chapter 7. Be sure to ask before ordering.

Finally, there is Nikon's 20 × 120 III Bino-Telescope. The name is a bit misleading, as unlike other *binocular telescopes* (such as those by Miyauchi and Vixen), the Nikon does not have interchangeable eyepieces. Still, it goes without saying that the view, spanning 3°, is striking. The Bino-Telescope includes 21 mm of eye relief, fully multicoated optics, BaK-4 prisms, and an integral fork mount. Customized tripods are available separately. These instruments were designed originally for seafaring use, and so they are waterproofed to prevent fogging. Unfortunately, the fork mount was also designed primarily for horizontal viewing. As a result, the binoculars are limited in vertical travel to about 70° above the horizon, at which point they hit the mount.

Oberwerk. Although Oberwerk is not exactly a name with high recognition value, this company nonetheless stands out as offering giant binoculars at small-binocular prices. Their models range in size from 8 × 60 Mini-Giants to monster 25/40 × 100 military binoculars.

Oberwerk 60-mm Mini-Giants are available in a choice of either 8×, 9×, 12×, 15×, or 20× magnification. All are housed in the same rubber-coated body and include BaK-4 prisms as well as optical coatings on all lens surfaces. Images are good considering what you pay for these glasses, just over $100, but they do not compare with the likes of Fujinon or Nikon. Stars in the center of the field are sharp, but become blunted as the eye moves toward the field edge. Eye relief, unfortunately, is quite short in all but the 8× and 9× models, with only 10 mm in the 20× glasses. All models include a standard tripod socket built into the central hinge, an especially important feature because these glasses are no lightweights.

Oberwerk also sells 11 × 56 binoculars, which, like the 60-mm Mini-Giants, include BaK-4 prisms, fully coated optics, and a built-in tripod socket. Performance is also roughly comparable, although images will be slightly (almost imperceptibly) dimmer due to the smaller aperture.

Moving up, Oberwerk's 15 × 70 binoculars have received high praise from many users. Again, these are no Fujinons, but for the money, they are quite respectable. Images are excellent in the center and reasonably sharp to about the outer 25% of the field. Overall, they are quite similar to Orion's Little Giant IIs, but with double the eye relief and a lower cost.

Finally, we come to Oberwerk's pièce de résistance, its 25/40 × 100 Military Observation Binoculars (sold by others as Border Hawk binoculars). Built originally for the Chinese military, these are *big* binoculars, weighing more than 26 pounds. Throw in the optional wooden tripod, and you're toting around more than 42 pounds! For that, you get multicoated optics and BaK-4 prisms, plus the flexibility of two magnifications. A pair of rotating turrets lets users select between two individually focusing Erfle eyepieces that produce either 25× or 40×. Eye relief is tight, however, especially at 40× (a paltry 8 mm). Focus is crisp across the inner half of the view, but stars tend to become distended toward the outer edge. There is little residual chromatic aberration and no detectable flaring or ghosting. Even at 25×, these gargantuans magnify enough to show some remarkable detail in deep-sky objects—detail that is missed in lower-power, though costlier, binoculars. The 40× eyepieces offer enough extra oomph to reveal the rings of Saturn and, just barely, the two equatorial belts of Jupiter—that is, assuming you can find them. Aiming these binoculars is, at best, challenging. While there is no provision for a finderscope, many users have added a unity finder, which is a big help. Still, while these giant giants certainly take some extra effort to set up and use, the views are quite impressive. That, added to a price that is below what some people pay for 7 × 50s, makes these worth strong consideration.

Orion. Few astronomy-related companies are into binoculars in as big a way as Orion Telescopes and Binoculars, which imports several lines of binoculars to satisfy a wide customer base. In the popular 7× to 10× range, Orion has three different grades of Porro prism binoculars, all with BaK-4 prisms, fully coated or fully multicoated optics, and built-in threaded tripod sockets.

Orion's least expensive Scenix models (7 × 50, 8 × 40, 10 × 50, and 12 × 50) feature fully coated optics that are enhanced by multicoated objectives. Their combined features, usually restricted to binoculars costing 30% to 50% more, move Scenix right to the head of the under-$100 crowd. Yes, there is some edge distortion to the view, and eye relief is shorter than some, but optical quality is quite good, especially given the price.

Orion's Intrepid 10 × 60 binoculars offer a little more aperture for a little more money. Featuring a generous 20 mm of eye relief, Intrepids are well suited for observers who must wear eyeglasses when observing. The built-in tripod socket will definitely come in handy, given that these glasses weigh in excess of 3 pounds.

Orion's UltraView series, the next rung up on the quality ladder, feature bright images across the field. I have owned a pair of 10 × 50 UltraViews for several years and am very happy with them. Focusing is smooth, with no binding or rough spots. Eyeglass wearers should especially note the UltraViews' exceptional eye relief. A few owners have complained about the UltraViews' limited amount of eyepiece travel for focusing, but far more stated that these are some of the best under-$200 binoculars on the market. Images are great

Figure 4.4 *Orion's Vista binoculars. Photo courtesy of Orion Telescopes and Binoculars.*

across about 80% of the field, with stars beginning to distort toward the edges. The latest UltraViews also include unique twist-and-lock eyecups in place of the more traditional folding eyecups. I remain undecided as to whether I like them, only because the eyecups are not as soft and pliable as those on my older pair.

Orion's Vista (7 × 50, 8 × 42, 10 × 50) binoculars, shown in Figure 4.4, compare favorably to other moderately priced models. Images are clearer and crisper than through the UltraViews, with little or no edge distortion detectable. Like the UltraViews, eye relief is good and focusing is very smooth. The only real drawback to Vistas is their slightly narrower-than-average fields of view, which is why I ended up with the wider-field UltraViews instead. But then, I like wide-field binoculars. The Vistas, however, have the sharper optics and compare quite favorably to Celestron Ultimas.

Orion also markets 8 × 56, 9 × 63, 12 × 63, and 15 × 63 Mini Giant binoculars. These are ideal for stargazers who need a little extra magnification but still want the option of handheld binoculars for a quick nighttime view out the backdoor before bedtime. The 9 × 63s are an especially good value, with optics that are nearly as sharp as the Celestron 9 × 63 Ultimas and with better eye relief; best of all, they cost only about 60% of what Celestron charges. The 8 × 56 Orion Mini Giants, however, do not quite hold up to the challenge of the Celestron 8 × 56 Ultimas in overall terms of optical and mechanical quality.

Looking for *big* binoculars? Orion sells 11 × 70, 15 × 70, and 20 × 70 Little Giant IIs; 11 × 80, 16 × 80, and 20 × 80 Giants; 15 × 80, 20 × 80, and 30 × 80 MegaViews; as well as the mammoth SuperGiant 25 × 100s. Images in the center of their fields are sharp, but like many giant binoculars, all suffer from edge curvature to some degree. This was especially evident through a pair of the 11 × 70 Little Giant IIs that I tested; star images began to lose the battle about three-quarters of the way out. Still, on-axis sharpness and resolution were quite good. The Little Giant II binoculars are also amazingly lightweight for their size; they are much lighter than Fujinon 70-mm binoculars, for instance. Although a tripod is still needed for long-term use, Little Giant IIs are light enough for quick hand-supported looks.

Not so with the 80-mm Orion Giants. Weighing 77 ounces, a tripod is needed for even a quick glance skyward. Like the Little Giants, the Orion Giants deliver sharp images on-axis, but soften up quickly off-axis. Image brightness is good, however, and focusing is fairly smooth (smoother, in fact, than the Little Giants).

Recently, Orion introduced its MegaView binoculars, manufactured in Japan, in 15 × 80, 20 × 80, and 30 × 80 combinations. Optically and mechanically, MegaViews exceed the Orion Giants. Images are sharp through each, but I am especially impressed with their fields of view and eye relief. The 15 × 80s cover 3.5° of sky (as do the 20 × 80s) and have 20 mm of eye relief (the 20 × 80s have an eye relief of 16 mm). Even the 30 × 80s have 14 mm of eye relief, which is very good for the magnification. All three are bridged by a metal bar with a captive, adjustable tripod post, just like the Celestron Giant Deluxe 20 × 80s. While the built-in post makes a separate tripod L-adapter unnecessary, it also prevents them from easily attaching to several of the parallelogram binocular mounts discussed in chapter 7.

The Orion 25 × 100 SuperGiant binoculars are not for the casual binocularist. This pair of 4-inch glasses comes mounted on a permanently attached platform-type tripod mounting plate, which absolutely requires a sturdy tripod that can be extended high enough for the observer to get *under* the eyepieces when aimed at steep angles. Few tripods qualify, though some are discussed in chapter 7. Unlike the center-focusing Little Giant, Giant, and MegaView binoculars, the eyepieces of the 25 × 100 SuperGiants must be focused individually. Readers in Europe should also note that Helios Stellar Observation Binoculars are virtually identical to Orion Giants and SuperGiants. Helios, however, includes 30 × 80s as well as 14 × 100 and 20 × 100 versions. Optical performance, again, can be expected to match the Orions, with good images on-axis, but fairly soft images off.

Parks Optical. This company is famous for its high-quality reflectors, but it also has an impressive assortment of binoculars from which to choose. Parks's 7 × 50 and 10 × 50 GR series (short for *gray rubber*) are good introductory binoculars, though they are surpassed in quality by Parks's 8 × 42 and 10 × 52 ZWCF glasses. All include BaK-4 prisms and full multicoatings. Some edge curvature is evident in both, but the ZWCF yield the sharper image. Parks also sells 12 × 50 ZWCF binoculars, which, unlike their smaller siblings, only come with BK-7 prisms and fully coated optics.

Parks also imports Giant binoculars. Four models—10 × 70, 11 × 80, 15 × 80, and 20 × 80—all include fully multicoated optics and BaK-4 prisms. Each comes with *winged* rubber eyecups that block light from the sides and includes a tripod adapter and a hard carrying case. Prices are a little higher than Orion's Giant binoculars, and although they look quite similar, the Parks perform better, especially toward the edge of the field.

Finally, Parks offers a 25 × 100 Giant that line for line appears to be the twin to Orion's SuperGiant, except for its higher price. Like the SuperGiant, it features fully multicoated optics, BaK-4 prisms, and a mounting plate attached to the bottom of the barrels. Weight, eye relief, and field of view are also identical. Limited information on performance indicates that the Parks 25 × 100s work quite well, but they absolutely require a very sturdy tripod to be appreciated fully.

Pentax. Best known for its cameras, Pentax offers some exciting Porro prism binoculars that are well suited for astronomical viewing. Their new PCF V Porro prism binoculars replaced the PCF III series a few years ago. Each of the seven models (7 × 50, 8 × 40, 10 × 40, 10 × 50, 12 × 50, 16 × 60, and 20 × 60) has a center-focus thumbwheel as well as a locking mechanism and a click-stop diopter adjustment, a very useful feature. And their rubberized barrels combined with excellent ergonomics make the PCF Vs a pleasure to hold by hand.

All Pentax PCF V binoculars are a pleasure to look through as well. Their multicoated optics produce good-quality views of the night sky. Centralized star images are tack sharp, but tend to soften toward the outer third of the field, as with most binoculars in this price class. All have generous eye relief.

Pro-Optic. The house brand of Adorama Camera in New York City, Pro-Optic glasses offer yet another choice in the ever-expanding arena of moderately priced binoculars. Their smallest binoculars, called Provis, come in three versions: 7 × 50, 8 × 40, and 10 × 50. Each has BaK-4 Porro prisms and fully multicoated optics for good on-axis performance. Off-axis softens, however, due to curvature of field and coma. It is curious that the 8 × 42s cost more than the larger 7 × 50s and 10 × 50s, but given that all Provis binoculars have reasonable eye relief, feature a built-in tripod socket, come with a hard carrying case, and retail in the low $100s, they are a good buy.

The Pro-Optic 70-mm and 80-mm Giant series have each gotten a good amount of attention. Both series include BaK-4 prisms and greenish, fully multicoated optics, as well as built-in tripod sockets. With retail prices well under $200, the two models in the 70-mm class, 11 × 70 and 16 × 70, both get high marks for giving a good bang for the buck. Most agree that the 11 × 70s are the better of the two, with superior eye relief and image quality. Some degradation is apparent around the edge of the field, certainly more than you would find, say, in Fujinons—but at less than half the price, this doesn't come as much of a surprise. My only complaints about the 11 × 70s are that they have a limited field of view of only 4° (versus 4.5° for, say, the Orion Little Giants) and poorly designed eyecups, which impress me as being too long. When fully extended, the eyecups can make viewing the entire field at once difficult without scrunching your head a little. The 16 × 70s are also competent binoculars, but lack eye relief and, in my opinion, image *zip* (sharpness and contrast).

Once again, the biggest attraction of the 11 × 80 and 20 × 80 Pro-Optic 80-mm Giants is the price, which is some $150 less than many comparable brands. The field of view is a little smaller than the others and the construction may not be quite as refined, but overall, both represent excellent values. Just be sure that you have a suitable way of supporting them, since each weighs in excess of 5 pounds. Also consider eye relief, which is only 14 mm for each. By comparison, Orion 11 × 80 Giants and Swift Observers have an eye relief of 16 mm, while Parks Deluxe 11 × 80s have 20 mm.

Finally, Pro-Optic Ultra Giant binoculars include 14 × 100 and 25 × 100 models. Like similar models from Orion Telescopes and Parks Optical, the Pro-Optics look pretty much like overgrown conventional binoculars, with their leatherette-covered barrels and center-focusing eyepieces. To give them the extra support they need, both have built-in platforms with several tapped holes for attaching them to a sturdy photographic tripod. Stars brighter than about 3rd magnitude are surrounded by hints of false color through both of the Ultra Giants, while the unforgivingly bright planet Venus shows tinges of red and green when viewed just slightly either side of focus. The Moon's limb is also falsely colored, yet the image itself remains quite crisp. And once again, Adorama trumps the competition by offering Pro-Optic giants for several hundred dollars less.

Pro-Optics and Oberwerk binoculars, mentioned earlier, are bringing the excitement of big-binocular astronomy to those with small budgets. The best part is that thanks to these little-known brands, you can still join in the fun even if you can't afford the likes of a Nikon, a Miyauchi, or a Fujinon.

Steiner. Based in Germany, Steiner offers several lines of high-end binoculars. If you are looking for combat-ready binoculars, then these are for you! Steiner's 7 × 50, 9 × 40 (also called Big Horns), and 10 × 50 Military/Marine binoculars are built to meet rigorous U.S. military standards. Focusing is done with Steiner's so-called *Sports Auto-Focus System.* They state that once each eyepiece is focused for the observer, objects anywhere from 20 feet to infinity will be sharp and clear. Consumers should note that while this system works for some, it can cause eye strain in others. Once focused, however, images are quite bright thanks to fully multicoated optics and BaK-4 prisms. Built-in tripod sockets complete the 7 × 50 and 10 × 50 packages, a necessary feature as the 7× and 10× binoculars each weigh 37 ounces, near the tops of their respective classes. The 9 × 40s do not include a tripod socket.

Steiner's highly regarded 7 × 50 Admiral Gold also tips the scale at 37 ounces, as well as the bank. The manufacturer tells us that the Admiral's images are much brighter than average thanks to "Steiner's HD (High Definition) optics, which provide incredible brightness at night and good glare reduction in sunny conditions." They further state that light transmission is 96%, "the highest ever recorded." Whether the boasting is true can be debated, but there is no denying these fine 7 × 50s have 22 mm of eye relief and more than 7° fields of view. Images are, in fact, quite bright and very sharp, with good contrast right to the edge of the field. Still, I feel that the view is perhaps not quite as precise as those through the Nikon Superior E or Canon IS lines.

NightHunter 7 × 50s, 8 × 56s, 10 × 50s, and 12 × 56s also meet Mil-Spec requirements for construction. A little lighter than the Military/Marine and Admiral Gold models, 50-mm NightHunters weigh in excess of 30 ounces nonetheless; the 56-mm NightHunters are more than 40 ounces, which some may find too heavy to support by hand. Like other Steiners, images are very good from center to edge; but unlike the others, NightHunters do not include a tripod socket. Again, before purchasing any of these, especially unseen by mail order, heed the earlier caution about the autofocus system. Try before you buy!

Steiner also makes two 80-mm giant binoculars called Senators. Available in 15 × 80 and 20 × 80 variants, each includes the same rubber-coated barrels, fully multicoated optics, and BaK-4 Porro prisms; the only real difference is the focal lengths of the individually focusing eyepieces. Steiner's marketing literature states that the 15 × 80s are supportable by hand, but I would recommend placing a solid tripod underneath instead. Fortunately, both include tripod sockets. Images are high in quality, although taking advantage of that quality might be hindered by the glasses' short, 13-mm eye relief.

Swarovski. This company makes a number of fine binoculars, with several suitable for astronomical viewing. Its Habicht binoculars (7 × 42 and 10 × 40) can only be referred to as classics, since their center-focusing stepped barrels look just like binoculars did back half a century ago. But while they may look old, Habichts incorporate the latest technology, including BaK-4 prisms and fully multicoated optics, to ensure a bright view. Waterproofed, nitrogen-filled barrels guard against fogging. Both are available wrapped in black leather or green rubber armor. Unfortunately, neither has a built-in tripod socket.

Swift Instruments. Established in 1926, Swift offers one of the world's largest and most varied selections of binoculars. Most are designed for earthly pastimes such as birding and yachting, though a few are also appropriate for more ethereal pursuits.

Looking first at *hand-holdable* binoculars, several Swift models have prisms of BK-7 glass, making them somewhat less desirable from an astronomical vantage point. Of those, the Swift Aerolite binoculars are probably the best. Available in five models—7 × 35, 7 × 35W, 7 × 50, 8 × 40, and 10 × 50—all include BK-7 Porro prisms and fully coated optics, but lack tripod adapters. Center focusing is fairly smooth and even, unlike some of the competition's *rapid-focus* units. The 7 × 35Ws have a 9.5° field of view, but at the expense of eye relief, which is a mere 9 mm. The others, with more traditional fields of view, have far better eye relief. Although image quality leaves something to be desired for those who consider themselves aficionados, Aerolites are fine for anyone just starting out.

Other Swift binoculars worthy of consideration include the 7 × 50 Sea Hawk, a nice but heavy pair of binoculars with individually focused eyepieces, fully coated optics, and BaK-4 prisms; the 8 × 40 Plover, with fully coated optics and BaK-4 prisms; the 10 × 50 Sea Wolf, with BK-7 prisms and fully coated optics; and the 10 × 50 Cougar, with fully coated optics and BaK-4 prisms. All include tripod sockets. Of these, I favor the Cougars, but there are other Swift binoculars that are even better suited for astronomers' needs.

Standing above the crowd are Swift's UltraLite glasses (7 × 42, 8 × 42, 8 × 44ED, 9 × 63, and 10 × 42). All offer fully multicoated optics, BaK-4 Porro prisms, built-in tripod sockets, and retractable rubber eyecups. The 8 × 44EDs add the benefit of better color correction, thanks to objective lenses made from extra-low-dispersion glass. The images that each produce are sharp almost out to the edge, with fields of view that are about average for the size and magnification. Eye relief values on the 7 × 42s, 8 × 42s, 8 × 44s, and 9 × 63s are very good, but the 10 × 42s can prove uncomfortably short at only 13 mm.

Three multicoated Swift Porro prism birding binoculars are also well-suited for astronomical viewing. The Swift 8.5 × 44 Audubon and 10 × 50 Kestrel glasses offer fields of view that are among the widest available thanks to their five-lens ocular system. Both are among the finest moderately priced binoculars in their respective size classes, with the only real shortcoming a lack of eye relief, which is only 9 mm, which is among the shortest of any in the 7× to 10× size class. The Kestrels are a bit better at 15 mm.

The Audubon ED 8.5 × 44 binoculars have all of the features found on the non-ED Audubons (including no tripod socket), but with the added benefit of fully multicoated optics. Their biggest plus is that the objective lenses are made from ED glass, like the 8 × 44ED UltraLites above, which gives them exceptional color correction. Their field of view also measures a relatively wide 8.2°, but their eye relief, at 14.5 mm, is short.

In the giant-binocular arena, Swift has the 11 × 80 Observer, the 15 × 60 Vanguard, and the 20 × 80 Satellite models. All three come with multicoated optics, built-in tripod sockets, and smooth center-focusing thumbwheels, all pluses. Images through the Observer and Satellite are clear and crisp, roughly comparable to similarly priced giants from Celestron and Orion. The Vanguards, however, are inferior, since they use BK-7 prisms.

Takahashi. World famous for its exquisite apochromatic refractors, Takahashi also makes an unusual pair of 22 × 60 Astronomer fluorite binoculars. One look and you can tell that these are not just common binoculars. Outwardly, they look like conventional Porro prism binoculars covered in black leatherette that have been surgically connected to a pair of toy telescopes that extend a few inches out front. Each of the two extended binocular barrels are wrapped in white-on-black star maps, further exaggerating the toylike appearance. Very strange!

But as silly as they look, these Takahashis are no toys. Rather, they are among the sharpest pairs of binoculars available. This really shouldn't come as a surprise, since they are, in effect, a pair of 2.4-inch f/5.9 Takahashi apochromatic refractors in binocular form. Few refractor connoisseurs would disagree that Takahashi apos are among the finest. And just like those, Astronomer binoculars show amazingly clear, crisp images. False color, which plagues the vast majority of binoculars, is utterly annihilated through these, save for a faint tint of yellow around the lunar limb (and then, to my eyes anyway, only apparent during the gibbous and full phases).

Now, admittedly, these are small binoculars for their magnification, generating only a 2.7-mm exit pupil. Recommending these appears to fly in the face of earlier advice, and while you can't exceed the limitations of the aperture, the small exit pupil does afford good image contrast that seems to make up for the disadvantage. Chapter 6 discusses the exit pupil and its implications more thoroughly, but in short, this result lends itself nicely to searching for close-up study of objects to about magnitude 10 or so. Jupiter's north and south equatorial belts and four Galilean satellites are all quite apparent, while Saturn's rings can also be distinguished from the planet's gaseous globe. The vast majority of Messier objects should also be within their grasp, as are dozens of NGC objects.

The Takahashi Astronomers include individually focusing eyepieces with 18 mm of eye relief, satisfactory for most who must wear glasses. They also come with a tripod adapter that bolts to a socket built into the center hinge. Unfortunately, although Takahashi doesn't offer a customized finder for the Astronomers, their narrow, 2.1° field really demands one. I would recommend attaching one of the 1× unity finders discussed in chapter 7.

Vixen Optical Industries. This company makes three uniquely designed giant telescopes that really stand out from the crowd as excellent values. All have 45° prisms, which prove to be a mixed blessing when binoculars are tilted skyward. While fine for viewing up to about a 45° elevation, the angle quickly becomes a pain in the neck as the binoculars approach the zenith.

The 80-mm Vixen BT80M-A, in effect a pair of 3.1-inch f/11.25 refractors, comes with a pair of 25-mm orthoscopic eyepieces that generate 36×. But what makes these truly unusual is that the removable eyepieces can be swapped for any 1.25-inch diameter eyepiece, for an almost unlimited range of magnifications. Given a set of high-quality eyepieces, such as those supplied, the image quality through these binoculars is exceptional, with little evidence of residual chromatic aberration or field curvature. Swapping the supplied 25-mm orthoscopics for a pair of some of the *super-mega-ultra* eyepieces detailed in chapter 7 gives you quite the starship!

To support the BT80M-A, Vixen recommends their Custom D alt-azimuth mount on an aluminum tripod. The Custom D mount includes smoothly operating manual slow-motion controls on both axes and a counterweight to help offset the balance disparity when the binoculars are in place. A dovetail plate for attaching them to the Custom D mount is permanently mounted to the barrels, but a 0.25–20 tapped hole is also provided for connecting them to a standard photographic tripod. Although the Custom D mount is heavily constructed, one owner complained that it still shakes under the weight of the binocular telescope. Another inconvenience is that since the tripod does not have a center post to adjust height, the only way to change the height of the binoculars is to extend the tripod's legs.

Vixen also makes three very impressive pairs of HFT-A Giant 125-mm binoculars. One is supplied with a pair of 20× eyepieces, another with 30× eyepieces, while the third includes a pair of 25× to 75× zoom eyepieces. The eyepieces are continuously variable, but have detents at 25×, 37×, 50×, and 75×. Focusing is done individually, as it is with nearly all monster binoculars, by twisting each eyepiece barrel. Each zoom eyepiece's magnification is also set individually by twisting a second ring on each barrel. Although the binoculars do not come with a finderscope, a dovetail base is supplied that will accept any Vixen-manufactured finder, a must-have accessory in my opinion.

Each of these heavyweights is, in effect, a pair of 4.9-inch f/5 achromatic refractors mounted in tandem on a fork mount. The fork arms can be oriented vertically or may be tilted to offset the binoculars from the center of the tripod, making it much easier to view through the binocular eyepieces when the glasses are tilted at angles greater than 45°. At first, the fork mounting appears to be too lightweight to support the binoculars, but it proves adequate for the task. Despite some creaking when the binoculars are moved left or right, the alt-azimuth mount moves very smoothly. Panning back and forth or up and down is a real pleasure, especially along the Milky Way.

Star images are sharp edge to edge across the entire view, regardless of magnification. Coma and curvature of field, two common aberrations of binoculars, especially zoom models, are absent. Imagine seeing in the same field the sparkling stellar jewels of the sword of Orion, with M42 set in the center, or the bespangled beauty of the Double Cluster in Perseus in an overflowing field of stars—truly awe-inspiring! Although coma and curvature of field are nowhere to be found, false color, known as residual chromatic aberration, interfered with the view of bright objects. Celestial sights such as the Moon, Jupiter and its moons, and Saturn's beautiful rings are all sharp (yes, the rings are even discernible at 25×, and can be suspected at 20×), but were marred by unnatural, purplish edges.

Zeiss (Carl). Long known as one of the world's premier sources for fine optics, Zeiss continues the tradition by offering some of the finest, and most expensive, binoculars of all. For instance, the Zeiss 7 × 42, 7 × 50, 8 × 56, and 10 × 40 B/GA T* ClassiC roof prism binoculars include conventional center focus and fully multicoated optics, but no tripod socket. While the eye relief is acceptable but not great in the 8× and 10× ClassiCs, all produce unparalleled images. Forget about coma, field curvature, or other aberrations that plague lesser binoculars. None of those is found here, only crisp stars scattered across a velvety dark field.

Zeiss also offers 15 × 60 B/GA T* ClassiCs, but with Porro prisms instead of roof prisms, as in the smaller models. Few ever complain about the optical quality of Zeiss optics, and you won't hear any complaints from me, either. These semigiants yield nothing less than perfect star images across their 4.3° field, just as we have come to expect from all Zeiss binoculars. Their 15 mm of eye relief is very good for this high a magnification, although observers who must wear thick-lensed eyeglasses might have difficulty taking in the full field even with the rubber eyecups folded down.

Zeiss Victory binoculars replace their old B/GA DesignSelection line of roof prism binoculars. The 8 × 40, 8 × 56, 10 × 40, and 10 × 56 models in this series all come with fully multicoated optics, center-focusing eyepieces, and are sealed with O-rings to prevent internal lens fogging. Each features a wider-than-most field of view, a heavier-than-most weight, and the highest price of any similar-size binoculars mentioned here. Ahh, but the view is spectacular. They are a little rich for my blood, but perhaps not yours.

Finally, Zeiss 20 × 60 BS/GA T* image-stabilized binoculars offer a unique approach to the problem of quivering hands. Unlike those from Canon and

Fujinon, the patented Zeiss stabilization system does not require batteries to power it. In fact, it contains no electronics whatsoever; instead, it's a clever system that relies on mechanics alone. Press a button and you hear a clunk, indicating that the stabilization mechanism is activated. The button releases tension on an internal *cardanic* spring, which is connected to the framework that holds the prism assembly in place. With tension released, the prisms can move in any and all directions, compensating for the observer's motions while maintaining optical alignment. It's a very effective system, although the button must stay depressed for the stabilization to continue working. Thankfully, Zeiss includes a tripod socket for attaching them to a camera tripod, since the binoculars are still heavy—more than 3.5 pounds. Optically, the images are very good, but keep in mind that these are only 60-mm binoculars. As such, even though the view is sharp, it will be on the dim side. The same investment (or less) will also net 80-mm Fujinon, 100-mm Miyauchi, or 125-mm Vixen binoculars, all of which are more astronomically suited.

The binocular universe seems to be much more personal than the view through a telescope. Maybe it's because we are using both eyes rather than one, or perhaps it's simply because we are looking up, rather than down or to the side. In any case, no amateur astronomer should leave home without a pair of trusty binoculars at his or her side. Even if you own a telescope, you may find, as I have, that many of the truly memorable views of the night sky come through those binoculars. Finally, and this is perhaps the most important point in this entire chapter to keep in mind, just because you may not be able to afford an expensive pair of binoculars, it doesn't mean that *your* binoculars can't take you on many wonderful voyages. Begin that journey tonight.

5

Attention, Shoppers!

It is time to lay all the cards on the table. From the discussion in chapter 3, you ought to have a pretty good idea of what type of telescope you want. But we have only just begun! There is an entire universe of brands and models from which to choose. Which is the best one for you, and where can you buy it? These are not simple questions to answer.

Let's first consider where you should *not* go to buy a telescope. This is an easy one! Never buy a telescope from a department store, consumer-club warehouse, toy store, hobby shop, or any other mass-market retail outlet (yes, this includes those 24-hour consumer television channels) that advertises a "400 × 60 telescope." First, what exactly does that mean, anyway? These confusing numbers are specifying the telescope in a manner similar to a pair of binoculars; that is, magnification (400 power, or 400×) and the aperture of the primary optic (in this example, 60 millimeters, or 2.4 inches).

These telescopes can be summed up in one word: garbage. In fact, the late George Lovi, a planetarium lecturer and astronomy author for years, is credited with describing these instruments as "CTTs," short for "Christmas Trash Telescopes." This brings us to the Golden Rule of Telescopes: **Never buy a telescope that is marketed by its maximum magnification.**

Remember, you can make any telescope operate at any magnification just by changing the eyepiece. Telescopes that are sold under this ploy almost always suffer from mediocre optics, flimsy mounts, and poor eyepieces. They should be avoided!

Many of the telescopes mentioned throughout this chapter come outfitted with finderscopes, eyepieces, and other accessories. If you find an item that you are unfamiliar with, chances are good that it is defined in a later chapter. A discussion of various types of eyepieces can be found in chapter 6, while finderscopes and a plethora of other accessories are discussed in detail in chapter 7.

So where *should* you go to buy a telescope? To help shed a little light on this all-consuming question, let us take a look at the current offerings of the more popular and reputable telescope manufacturers from around the world. All of these companies are a big cut above those department-store brands, and you can buy from them with confidence.

I have chosen to organize the reviews by telescope type, with manufacturers listed in alphabetical order. Department-store brands and models have been omitted, since they really fall more in the toy category than scientific instrument. Most toy telescopes come with plastic tubes, plastic lenses, weak mountings, and horrible eyepieces. It is disconcerting to see that some of the reputable companies mentioned in this chapter are beginning to sell such telescopes with their brand names emblazoned across the telescope tubes. Don't be fooled; these are no better than the others being purposely left out of this review. Those small-aperture refractors and reflectors are universally inferior to the better-made telescopes detailed in this chapter. (But what if you already have one of those telescopes? See chapter 8 for ways to make it better.)

In the course of writing this book, I have looked through more than my fair share of telescopes, but it takes more than one person to put together an accurate overview of today's telescope marketplace. To compile this chapter, as well as the chapters to come on eyepieces and accessories, I solicited the help and opinions of amateur astronomers everywhere. Surveys were distributed in print and online. I was quickly flooded with hundreds of interesting and enlightening replies, and I have incorporated many of those comments and opinions into this discussion.

You can play a role in the fourth edition of *Star Ware* by completing the survey form at the back of the book. At the same time, as new telescopes are introduced, visit the Star Ware 3.5 section of my web site at www.philharrington.net for new reviews and further thoughts.

Mysteries of the Orient

One of the biggest changes to hit the amateur telescope market since the second edition of this book was published a scant four years ago is the tidal wave of foreign-made telescopes that has taken the entire world by storm. For years, most low-end department-store telescopes were constructed in China or Taiwan, but it wasn't until recently that companies began to target the amateur-level telescope market. Now, most name-brand companies offer many instruments from the Far East.

Of the dozens of owner reports that I have received, and the many imported telescopes that I have personally tried and tested, I would rate most as good. Very few of these imports deserved an unacceptable grade. On the other side, I would also rate very few as excellent. Generally, people seem to think very highly of the 3.1-inch f/5 achromatic refractors made in China and sold by Celestron (FS-80WA), Orion (ShortTube 80), and Pacific Telescope Company (Skywatcher 804), among others. Other favorite imported refractors include the 4-inch f/9.8 (Celestron C102-HD, for instance). Please refer to each particular review for specific likes and dislikes.

Imported 4.5-, 6-, and 8-inch Newtonian reflectors are also extremely popular. Most of these nearly identical instruments come from either China or Taiwan, although some importers prefer to use their own optics. Again, optical quality is usually quite good. Each typically comes mounted on a German equatorial mount that is a clone of mountings sold by Vixen Optical Industries of Japan, but at a small fraction of the cost. Again, please refer to specific models.

Why do I mention all this here? To educate the consumer! Name companies will often charge more than independent dealers for the *identical* instrument simply because of the recognition that their name carries. My best advice is to shop around, being very careful to compare apples with apples. In many cases throughout this chapter, I try to guide the reader by pointing out that a telescope by Company A is very nearly identical to an offering from Company B to help make those comparisons a bit easier. And while the money that you save may not appear to be that much, it may well be enough to purchase another eyepiece or accessory that you might not otherwise afford.

A final caveat. Reports from owners around the world indicate that there is also some variability in optical quality among some imports. It now appears that a second tier of companies has sprung up that are now copying the clones to varying degrees of success. Therefore, it is absolutely imperative that you know the retailer's return policy before you purchase.

Refracting Telescopes

Although many small refractors continue to flood the telescope market, the discussion here is usually limited to only those with apertures of 3 inches and larger (yes, with a few notable exceptions). As a former owner of a few 2.4-inch (60-mm) refractors, I found that I quickly outgrew those instruments' capabilities and hungered for more. Why? At the risk of possibly offending some readers, let me be brutally honest: most 2.4-inch refractors will produce marginal views of the Moon and maybe Venus, Jupiter's satellites, Saturn, and possibly a few of the brightest deep-sky objects, but nothing else. They are usually supplied on shaky mountings that only add to an owner's frustration. If that is all that your budget will permit, then I strongly urge you to return to the previous chapter and purchase a good pair of binoculars instead.

Interestingly, while the apertures and focal lengths of reflectors and catadioptric telescopes are usually expressed in inches (at least here in the metrically challenged United States) or centimeters, refractors are most often spoken of in terms of millimeters. I can't explain the discrepancy, except that perhaps the larger numerical values make consumers feel they are getting more for their money.

Refractors are further divided into two categories to make comparisons a little easier. The first section discusses achromatic refractors; the second section apochromatic refractors.

Achromatic Refractors

You will recall from chapter 3 that an achromatic refractor uses a two-element objective lens to minimize chromatic aberration. Some false color, called residual chromatic aberration, remains, however, unless the objective's focal length is very long.

Almost universally speaking, all refractors sold by mass-market retailers are of poor quality. The telescopes discussed here are a cut above those, but again, some are a better cut than others.

Apogee, Inc. This manufacturer offers several short-focus achromatic refractors from 3.1 to 4 inches in diameter. Smallest of the group is their G185 3.1-inch f/5. Overall, images are fairly decent at low power, but not on the level of the Orion ShortTube 80, the Celestron FS80-WA, or even the smaller Meade ETX-70AT. Bright stars, planets, and the Moon are spoiled by residual chromatic aberration. A few owners report that their instruments also seem prone to spherical aberration, which is exacerbated by using too high a power with what should be considered a low-power-only instrument. Overall construction of the G185 is unrefined, with far too much play in the focuser and a lens cap that keeps falling off. The G185 is sold with a horribly weak camera tripod, which must be upgraded.

The Apogee 80-mm f/7 refractor costs twice what the f/5 goes for, but also offers better color correction and sharper images. Featuring a black anodized metal tube, a carrying handle, and an integrated 2-inch rack-and-pinion focuser, the instrument does not come with a finder, an eyepiece, a star diagonal, or a mount, but it does have a mounting block for attaching the instrument to a heavy-duty camera tripod. Consumers should note, however, that for not much more money, they can purchase an imported 4-inch refractor, such as the Celestron C102-HD, which comes fully equipped with a German equatorial mount, a finder, an eyepiece, and a star diagonal.

By far, the most talked-about achromatic refractor in the Apogee lineup is the 4-inch f/6.4 Widestar. Like the 3.1-inch f/7 mentioned above, the Widestar's mechanical construction is quite good. Its aluminum tube is nicely anodized black, while the internally geared rack-and-pinion focuser works very smoothly. Note that like the 80-mm f/7 Apogee, the Widestar is sold without a mount. It is well suited, however, for the Vixen Great Polaris, the Celestron CG5, and the Losmandy GM-8 equatorial mounts.

Optically, the Widestar gives great views at low power, making it ideal for scanning the Milky Way. It should come as no surprise that given its fast focal ratio, there is considerable residual chromatic aberration around brighter objects. Most agree, however, that the views are striking nonetheless. Some owners have also complained about a small amount of spherical aberration in their instruments.

Apogee also has a 5-inch f/6.5 instrument that they call the Widestar 127. This instrument comes with a 2-inch focuser, an 8 × 50 finderscope, and a 25-mm Plössl eyepiece. The supplied German equatorial mount is supported on an aluminum tripod. I would be very leery of such a fast achromatic refractor, as its performance is likely to be plagued by abundant residual chromatic aberration as well as spherical aberration.

Borg. No, you have not stumbled upon part of a *Star Trek* script. Borg Telescopes (more correctly, Oasis Borg, a division of TOMY), from Japan, have nothing to do with the alien race from that television and movie series, apart from sharing the rather peculiar-sounding name. Actually the word *Borg* comes from the Japanese words for telescope (*Bo-enkyo*) and equipment (*Do-gu*), which have been shortened into Borg. Borg telescopes are modulized, designed so that several parts are shared by different telescope models and apertures (not unlike the way that the science fictional Borg assimilate others into their "collective"). This lets the user configure a telescope to suit his or her particular needs, rather than having to take what comes right off the shelf.

All Borg telescopes belong to one of three series: Series 80, Series 115, and Series 140. Series 80 Borg refractors are based around an 80-mm diameter tube and come in five variations, including 3-inch f/6.6 and 4-inch f/6.4 achromats (the other three are apochromatic refractors, detailed later in that section). Series 115 Borg achromats, built around a 115-mm diameter tube, include 4-inch f/4 and 4-inch f/6.4 instruments, the latter having the same optics as the Series 80 version. The third series, Series 140, contains a pair of 6-inch apochromats, also discussed later.

Both series share the same philosophy, which is to package the optical components into the smallest, lightest tubes possible without sacrificing performance. That's their biggest selling point: travelability. Each tube assembly is designed to break down into short individual pieces, which is especially convenient for traveling by air (solar eclipse chasers, take note!). The modular design also means that many components are shared across two or all three telescope series, keeping production costs down. But this can also cause some consternation, since you need to screw *everything* together: the lens cell screws onto the tube, which in turn screws onto one or more extension tubes, which then attach to the focuser, and so on. It takes some getting used to.

All Borg refractors use helical focusers, which require that you turn the eyepiece to bring the image into focus (as opposed to turning a knob, as with the more common rack-and-pinion focusers). Although I am not a fan of most helical focusers, those on the Borg refractors turn very smoothly, just as if you were focusing a camera lens. And because the helical mechanism is internal, the eyepiece does not turn as it is focused, which would be terribly inconvenient on a refractor. The focuser also has a rubber grip ring, which adds a nice touch. Note, however, that the focuser friction increases with added weight.

Optically, images through Borg achromatic refractors are quite good, given their fast focal ratios. One owner reports that residual chromatic aberration and false color are "not horrible, but noticeable; actually the amount of false color is fairly small" and is restricted to the brightest sky objects. Images are sharp and contrasty, delivering satisfying views of such showpieces as the rings of Saturn, the moons and belts of Jupiter, and innumerable double stars. Overall, Borg refractors fill a market segment that lies somewhere in between the inexpensive refractors from the Far East and high-end apochromatic refractors that cost $1,000 or more than a Borg.

Celestron International. In the last two decades, Celestron, best known for its Schmidt-Cassegrain telescopes, has branched out into other telescope markets. In the refractor category, Celestron imports instruments from the Far East that range in aperture from 2 to 6 inches.

At the low end of the Celestron refractors reviewed here are the Firstscope 80 AZ and Firstscope 80 EQ, a pair of 3.1-inch f/11.4 refractors. Both have the same optical tube assembly, which includes a fully multicoated objective lens assembly housed in a glossy black steel tube. Optical performance is actually quite good for the aperture and price of these instruments. The alt-az version includes a 45° erect-image diagonal for upright terrestrial viewing, while the equatorial version includes a 90° star diagonal. Focusing is smooth, but even the steadiest hand will cause the view to jump because of the weak mountings that come with the instruments. The alt-azimuth version, actually the sturdier of the two, includes a pair of manual slow-motion controls for steplike tracking of sky objects. The equatorial version also includes manual slow-motion controls and can be outfitted with an optional clock drive, if desired. A pair of setting circles adorn the equatorial mount but, like in most instruments of this genre, are little more than decorative.

For those interested in a superportable terrestrial spotting scope that can do double duty as a low-power, rich-field instrument, consider the Celestron 80 Wide View spotting scope. This 3.1-inch f/5 achromatic refractor measures only 13.5 inches long and is designed to accept standard 1.25-inch eyepieces and accessories. The air-spaced doublet objective lens is fully multicoated for high contrast and good resolution. Keep in mind, however, that at f/5, false color (e.g., residual chromatic aberration) will detract from views of bright objects such as the Moon. Also, the short 400-mm focal length is meant to operate at low power only; this is not intended as a planetary telescope. The 80 Wide View comes with 10- and 20-mm Plössl eyepieces, a 45° star diagonal, a 6 × 30 finder, and a German equatorial mounting. A built-in adapter plate also allows easy mounting to a camera tripod.

Capitalizing on the rave reviews of its NexStar Schmidt-Cassegrain telescopes (reviewed later in this chapter), Celestron has mated the 80 Wide View to a computerized GoTo mount to create the NexStar 80GT. Like the NexStar 5 and 8 SCTs, the NexStar 80GT rides on a one-armed mount attached to a computerized base. Inside, a microprocessor contains data and finding instructions for more than 4,000 sky objects, including the Moon, the planets, and a host of deep-sky objects. While I am not a big fan of GoTo telescopes for newcomers to astronomy (since I feel that most who are interested in the science are probably also interested in learning their way around the night sky), there is no denying that a telescope like this has a certain appeal for some people.

The NexStar 80GT is optically identical to the 80 Wide View, so let's concentrate on the mechanics. The mounting's single support arm and base, which contains the computer drive itself, are attached to an adjustable aluminum tripod. The eight AA batteries needed to power the NexStar's brain are held in a vinyl power pack attached to the telescope's mount. The instrument's hand controller, featuring a keypad and directional arrows, plugs into the mounting's arm with a standard RJ-11 telephone-type connector. Oddly, there is no on-off power switch; as soon as the battery pack and hand controller are plugged in, the latter's display lights up and readies for use. (Just remember to unplug the batteries after use, or they will drain quickly.) See the description of the NexStar 114 reflector later in this chapter for further insight.

The Nexstar 80GT comes with 10-mm and 25-mm Kellner eyepieces, a 90° star diagonal, and a 1× Starpointer unity finder, which projects a red dot against a small window for aiming the telescope. A version of the same telescope, but without the computerized aiming system, is also sold under the name NexStar 80HC, the *HC* standing for hand controller, which includes east-west and north-south motions only. (Celestron also sells the 2.4-inch f/12 NexStar 60GT, which comes on the same mounting, but is below the aperture threshold set for this book. Tasco markets the same instrument as the Star-Guide 60.)

Celestron's 4-inch f/9.8 C102-HD appears to be an almost exact replica of their former GP-C102, a product imported from Vixen Optical in Japan. The Vixen-made C102 was considered to be a fine instrument, both mechanically and optically, for visual observations as well as some rudimentary astrophotography. But it also sold for around $1,800. Can the substantially cheaper, Chinese-made C102-HD be as good?

Of the many owners that I have heard from, most agree that the C102-HD works quite well, especially when considering its cost. One owner thought the optics were not quite up to the quality of the original Vixen's, but were suitable for beginners on tight budgets. Most owners were less critical, typically saying that the telescope produced "sharp images and almost no false color." False color is inevitable around such unforgiving objects as Venus and the limb of the Moon through an achromatic refractor of this sort, but detail in the clouds of Jupiter and the rings of Saturn are quite sharp, while double stars snap into focus. Deep-sky objects also show well, but the small aperture does prove limiting.

The focusing mechanism, so critical to producing sharp views through a telescope, is smooth, though not as refined as some instruments. The supplied 6 × 30 finderscope is also marginal in my opinion. You might consider supplementing the C102-HD's finderscope with a small 1× unity finder, such as those sold by Celestron, Tele Vue, and Orion Telescope Center (the Telrad, long a favorite among amateurs, is too large for this small an instrument).

The CG-4 mounting that is supplied with the C102-HD is a close copy of the original Vixen mounting. Many would probably be hard pressed to tell them apart if they were placed side by side, but in the dark their differences begin to come to light. While the CG-4's equatorial head is reasonably well made, the all-aluminum tripod is not as sturdy as the original wooden-legged tripod. Slow-motion controls in both right ascension and declination are smooth, however, with little binding noticed. Adjustment screws let the owner tune gear tension, should that ever become necessary.

More recently, Celestron began to import from China a compact f/5 version of the C102 called the 102 Wide View. Measuring only 21 inches long and weighing just 5 pounds, the 102 Wide View looks just like a shrunken C102-HD, with the same focuser, finderscope, and other accoutrements. But it's quite a different story optically. One would expect that an achromat with such fast focal ratio would be plagued with residual chromatic aberration, and indeed it is prevalent around brighter objects, but most owners report that they like the images that the telescope produces very much. Perhaps the biggest appeal is the wide field of view. Attached to a sturdy camera tripod or a light-duty equatorial mount, such as Celestron's imported CG-4, the 102 Wide View offers some spectacular views of wide open star clusters such as the Pleiades, the Alpha Persei Cluster, and the Milky Way star clouds (but, of course, it should not be considered comparable to, say, an Astro-Physics Traveler apochromat). Not surprisingly, it doesn't do well on the planets; but then again, it's not designed with planets in mind. Jupiter, for instance, shows some of its belts, but it appears to be surrounded by a purplish corona when viewed through the Wide View and a 7-mm Pentax eyepiece. But that aside, a telescope like the 102 Wide View makes the perfect grab-and-go second telescope for anyone who cannot afford (or who is unwilling to spend) the cost of an apochromat.

The 102 Wide View is available on an alt-azimuth mounting as the Celestron C102-AZ. Slow-motion controls on both axes work well. Apart from the problems inherent to aluminum tripods that were mentioned previously, the mounting is adequate for the telescope's size and weight.

The big brother to the C102-HD is the Celestron CR150-HD, a 6-inch f/8 refractor. Now, I know that the purists in the crowd will immediately balk at that focal ratio. Recall that the discussion in chapter 3 stated that achromatic refractors will suffer from false color due to residual chromatic aberration if their focal ratios are too fast (i.e., too low an f-number). According to the formulas there, a 6-inch achromat should be no less than f/17.3 in order to eliminate this problem. That's more than 8.5 feet long! So how good could an f/8 achromat be? Frankly, I had my doubts, too.

Many owner reports have been received and nearly all have given the CR150-HD high marks for optical quality, noting a decided *lack* of false color on all but the brightest sky objects (e.g., the Moon, Jupiter, etc.). A few also added that whatever false color there was appeared to lessen as magnification increased. One owner from Toronto, Canada, wrote: "The false color virtually disappears on Saturn at 240×, leaving a clean sharp image with a wonderful amount of high contrast detail. Cassini's division could be followed all around the ring, the polar region is an olive color, and hints of detail could be seen in the equatorial belt. The image is quite comparable with those in my 8-inch and 10-inch Newtonian reflectors. On deep-sky objects, the CR150 is close to my 8-inch in performance. The most spectacular view was the Orion Nebula, M42, whose wings spread as wide as I've ever seen them." Note, however, that even when residual chromatic aberration is not visible to the eye, it still affects image sharpness, since light at different wavelengths focuses at slightly different distances.

As strong a performer as the CR150-HD is optically, it is weak mechanically. The biggest problem, as with nearly all tripod-mounted instruments from the Far East, is the tripod legs. In a word, they are terrible, causing undue flexure, twist, and torsion during use. It is truly a shame that a telescope with as much potential as the CR150 is coupled to such a weak mounting. Evcn though other mechanical components, such as the focuser, work smoothly, they cause the mounting to wobble when used.

D&G Optical Company. D&G Optical is a small company that specializes in achromatic refractors ranging in aperture from 5 to 10 inches at either f/12 or f/15. D&G also makes larger refractors on a custom-order basis. All optics are handmade by D&G and are tested to exacting standards. The 5- and 6-inch objectives receive coatings of magnesium fluoride as standard, while coatings on the 8-inch and larger objectives are available at extra cost.

Unlike many refractors, whose objective lenses are mounted in nonadjustable cells, D&G objectives come with a push-pull cell. This allows the owner to align the lenses' optical axis precisely, resulting in better images and a telescope that stays in collimation even with constant transport.

Viewing through a 6-inch f/12 D&G refractor a while back, I couldn't help but notice some evidence of residual chromatic aberration around bright stars. Noticeable, yes, but not so intense as to distract me from the overall high levels of sharpness and contrast of the images. Epsilon Lyrae, the Double-Double quadruple star in Lyra, for instance, was easily resolved, while globular star clusters burst into swarms of individual stars.

All D&G refractors also come with oversized rack-and-pinion focusing mounts complete with adapters for 1.25- and 2-inch eyepieces. The focuser on this particular instrument was certainly not of the same quality as those from Vixen, Astro-Physics, and Tele Vue. Yes, it did rack the eyepiece in and out adequately, but its feel was more reminiscent of older focusers made from cast, rather than machined, components.

Discovery Telescopes. While Discovery concentrates mainly on its made-in-house Newtonian reflectors, it also offers an imported 3.5-inch f/5.6 refractor called the System 90. Like the Orion ShortTube 90, the System 90 includes a fully coated objective lens, a 6 × 30 finderscope, and a reasonably smooth rack-and-pinion focuser. See the description under "Orion Telescope Center" for thoughts on optical performance. The only downside is the 45° star diagonal that comes with the System 90, which is much less convenient to use astronomically than a 90° diagonal.

Helios. Although not exactly a well-known brand here in the United States, Helios has been marketing telescopes in Europe for over three decades. Several astronomical refractors bear the Helios name, although you have already met some of them under different noms de plume. The Helios Startravel 80, for instance, is the equivalent of the Celestron 80 Wide View and Orion Short-Tube 80, both 3.1-inch f/5 refractors. Its big brother, the Startravel 102, is a 4-inch f/5 twin to the Celestron 102 Wide View.

Five imported Helios Evostar achromatic refractors range in size from 3.1 to 6 inches. Like so many cloned refractors nowadays, these are near twins of such popular refracting telescopes as Orion's AstroView 90, SkyView Deluxe 90, and AstroView 120, and the Celestron C102-HD and CR150-HD. The Evostar 80 (a 3.1-inch f/11.25) and 90 (a 3.5-inch f/10) are available on either alt-azimuth or German equatorial mounts and include 10- and 25-mm Kellner eyepieces and a 6 × 30 finderscope. The Evostar 90 can also be purchased on a deluxe EQ-3 equatorial mount (seemingly identical to Celestron's CG-4), which is preferred to the standard EQ-2 (similar to Celestron's CG-3) version. Optical and mechanical performance can be expected to be comparable, as well.

The three larger Evostar refractors, the 102 (a 4-inch f/9.8), 120 (a 4.7-inch f/8.3), and 150 (a 6-inch f/8) come with 10- and 20-mm Plössl eyepieces, the same 6 × 30 finderscope, 2-inch focusers, and 1.25-inch eyepiece adapters. Some dealers offer the 102 and 120 on either a standard (EQ-3) or deluxe (EQ-4, similar to Celestron's CG-5) mount; absolutely get the EQ-4 if you purchase this telescope. The Evostar 150 is only available with the EQ-4 mount.

Konus. This is another European name that only recently broke into the United States marketplace. But while their promotional literature proclaims their heritage as European, it should be noted right up front that their astronomical telescopes have their origins in the Far East.

While many of their refractors fall more into the toy category rather than scientific instruments, a few are worth noting here. The 3.1-inch f/5 Konus Vista 80 is effectively a rebadged Orion ShortTube 80, so I will refer you to its discussion later in this section. Moving up slightly in aperture, the Konuspace 910 is a nice instrument for those looking to enter the hobby but who must restrict themselves to a telescope that can be carried in one hand. The instrument's 3.5-inch f/10.1 achromatic objective lens performs quite well when trained on the Moon and the naked-eye planets, as well as on brighter deep-sky objects. Some false color exists around the Moon's limb, Venus, and Jupiter, but for the most part this is not objectionable. The star test reveals that the telescope's optics are quite good, with no sign of spherical aberration or astigmatism. Mechanically, the Konuspace 910 rides on a typical German equatorial mount made in the Far East. As can be expected, the mounting tends to wobble whenever the telescope is touched, such as during focusing. Fortunately, the two Kellner eyepieces that come with the telescope are nearly parfocal, keeping focusing to a minimum.

The Konuspace 102 is another of the ubiquitous 4-inch f/9.8 achromatic refractors with Far East origins. While the optics in these instruments are surprisingly good, the Konuscope 102 stands apart from the rest because of its poor mounting (the same as on the Konuspace 910), which is terribly underdesigned for an instrument of this size and weight. The result is a telescope that can never reach its full potential. Consumers would do far better by purchasing the same telescope from another company that marries it to a proper mounting.

The Konusuper 120 comes far better equipped, but still has its weaknesses. The 120 is built around a 4.7-inch f/8.3 objective lens mounted on a German equatorial mount atop an adjustable, aluminum tripod. Reviews of the Konusuper 120 seem to indicate some potential quality-control concerns. Its near twin, the Orion AstroView 120, produces consistently sharp images, though with some evidence of residual chromatic aberration around brighter objects. Some Konus owners, however, report additional problems with spherical aberration and strong astigmatism. As with the Orion scope, the Konusuper comes with a 6 × 30 finderscope, a 90° star diagonal, and two eyepieces. The tube colors are different, however. The Orion is painted a glossy white, while all Konus instruments sport their trademark bright orange. At least you won't misplace one at night!

One of the newest additions to Konus's line of telescopes is their Konusky Evolution 150, a 6-inch f/8 achromatic refractor that bears more than a passing resemblance to the Celestron C150-HD. The Konus instrument also sports 10- and 25-mm Plössl eyepieces, a 9 × 50 finderscope, and a CG5-like German equatorial mount. Comparable performance can be expected, with some residual chromatic aberration inevitable around brighter objects.

Meade Instruments. The Meade line of Digital Electronic Series (DS) refractors and reflectors was introduced in 1999 as the company's next generation of small telescopes. Six instruments make up the DS line, including the 3.1-inch f/11.3 DS-80EC and the 3.5-inch f/11.1 DS-90EC refractors, as well as two reflectors, discussed later. (Two smaller refractors, the DS-60EC and the DS-70EC, are not included here, as they fall below my self-imposed aperture threshold of 3 inches.) Unlike other, simpler instruments, the DS telescopes boast advanced electronics as well as the ability to be outfitted with GoTo computer control that will actually steer the telescope to selected objects. *Sky & Telescope* magazine proclaimed this innovation as nirvana for new stargazers. Is it really? I'm a bit more skeptical.

Both the DS-80EC and the DS-90EC come with two 1.25-inch eyepieces (a 25-mm Kellner, which Meade calls a Modified Achromat, and a 9-mm Huygens), a 6 × 30 finder, an extendable aluminum tripod, and an electronically controlled altitude-azimuth mount with motorized altitude and azimuth motions and tracking. Their price tags are quite steep for the apertures, especially given the liberal use of plastics, but much of their cost is likely tied up in the telescopes' electronics.

I tested a DS-90EC with the optional Autostar hand controller, which contains a database of approximately 1,500 sky objects (ordered by object type and ranging from solar system members to traditional star clusters, nebulae, and galaxies, as well as the ridiculous, such as black holes). It's a simple matter to unplug the hand controller that came with the instrument and substitute the Autostar controller.

On the plus side, the telescope's tripod is surprisingly steady, with a nicely designed flip lever for adjusting each leg's length. The telescope's slewing rates, adjustable from very slow to quite fast, work very smoothly. The plastic rings for attaching the finder are nicely designed, with adjustment screws set at 90°

angles to one another (that is, the two screws on the front ring compared to the two on the back ring). This makes finder alignment much more intuitive. And the Autostar was also easy to calibrate, requiring that you tell it your location, the time and date, and a couple of other parameters.

The downside to the DS refractors, however, is a bit longer. First and foremost was the view. Images through all that I have used do not show a consistent focus across the field of view. In the case of the DS-90 that I tested, standing back and looking into the eyepiece finally showed me why: The fault was in the plastic star diagonal. It seems that the star diagonal supplied with my telescope was out of alignment (i.e., crooked), making sharp focus impossible to achieve. I have subsequently checked more than one dozen of the DS star diagonals and have yet to find one that was assembled properly. The star test proved illuminating, as well. Even with the star diagonal out of the system, defocused star images were triangular, implying that the objective must be pinched in the plastic lens cell.

The Autostar aiming system that is available optionally with DS refractors works quite well. Although the selected objects are not always centered in view, they are usually close enough to the eyepiece field to be easily found. The Autostar's self-guided tour mode lets the telescope select the targets for you. The telescope seemed to make good choices more than 90% of the time, demonstrating that the programmers must have a good handle on the telescope's abilities. Often, however, the telescope would not take the shortest route, causing the wires connecting the Autostar to the telescope to become entangled.

Overall, the electronics of the DS refractors, and especially the Autostar, are amazingly good for the price. But as impressive as they were, the bottom line for any telescope should be its optical performance—which, in the case of the Meade DS scopes, is disappointing.

In late 2001, Meade came out with the DS-2070AT, a 2.8-inch f/10 refractor on a one-armed computer-controlled mounting that appears to be the twin of Celestron's NexStar 80GT. The biggest improvement over the original DS models is that the drive motors and their associated cables are all internal to the mounting. This eliminates the concern I mentioned above about the telescope becoming tangled during operation. The only external cable is now the one that runs from the telescope to the Autostar hand controller. No doubt a wireless version will be coming along from either Meade or Celestron at some point. The DS-2070AT includes two 1.25-inch eyepieces, a small finderscope, and, unfortunately, the same plastic star diagonal as the original DS refractors.

Meade has adapted a small fork mount and GoTo electronics to the 2.8-inch f/5 ETX-70AT achromatic refractor. For a small, on-the-go GoTo telescope, the ETX-70 is a far better choice than Meade's DS instruments, in my opinion. The short tube assembly rides between the fork arms of the Meade's popular ETX-90 Maksutov telescope (reviewed later in this chapter), which is a far steadier mounting than that which comes with the DS series. Happily, the ETX-70AT comes standard with a battery-driven, internally wired clock drive (which uses six AA batteries) and Autostar built into its mounting (no twisted wires!). A variety of add-on accessories are also available; some you will need, others you may not. Unless you live in extremely dry environs, the optional screw-on dew cap is a must-have feature. A hard case is also available for those who are really on the go.

Regarding the Autostar, of the more than 1,400 objects in the instrument's database, 11 are quasars and one is Pluto, all 12th magnitude or fainter. None will be visible through a 2.8-inch telescope. In fact, since the ETX-70AT also has such a short focal length—only 350 mm—it is really not a telescope for *any* of the planets. It is far better suited for casual scans of the Milky Way and trips across wide star fields (which do not typically require GoTo technology; instead, just sit back and sweep the sky). The fast f/5 optical system will also produce strong residual chromatic aberration when aimed toward bright objects. But perhaps the biggest detriment to planet watching will be the short focal-length eyepieces that are needed to reach a high enough power to see any planetary detail. A 7-mm eyepiece will produce only 50× by itself, and only 100× with a 2× Barlow lens. Remember the rule: as most eyepieces' focal lengths shrink, so, too, do their comfort factor, known as exit pupil. If planets are what you crave, you'd better look elsewhere; but for an extremely portable instrument with the power to GoTo, the ETX-70AT is a good choice.

Meade also sells two 4-inch f/9 achromats, the 102ACHR/300 and 102ACHR/500. Both use the same imported optical tube assemblies. Overall, the optical quality of this refractor (made in Taiwan) is not bad, although as we should expect, the view does suffer from residual chromatic aberration. A purplish halo can be seen around even moderately bright stars and is strikingly apparent around the likes of the Moon, Jupiter, and Venus. At f/9, the Meade 102ACHR may show slightly more false color than the Celestron C102-HD, which operates at f/9.8, but the difference is negligible.

Turning to deep-sky objects, the Meade 102ACHR refractors do quite well for their aperture. Image contrast is quite good on such showpiece targets as the Orion Nebula and M13, while close-set double stars are well resolved. Again, some false color disturbs the view of brighter doubles, such as Albireo.

The LXD 300 mount is supported by an adjustable aluminum tripod and features manual slow-motion controls on both the right-ascension and declination axes. A polar alignment scope is available separately, as is a DC-powered clock drive. Overall, the mount is at its upper weight limit with the 102ACHR tube assembly.

The LXD 500 version, which adds about $100 to the price, is certainly a better choice. Again, manual slow-motion controls are found on both axes, while a DC-powered motor drive and polar scope are available separately. The tripod, the weak link in the system, is the same as on the LXD 300, but the mounting itself features larger components for more stable support.

Despite the inherent problems with the mountings and residual chromatic aberration, I believe the Meade 102ACHR scopes to be good choices for anyone looking for a competent 4-inch refractor at a surprisingly low price. Now all you need to do is to choose between it and Celestron's C102-HD (and its many clones). Which is better? Overall, they are quite comparable, and so I have no problem recommending them both. I would probably base my choice on availability.

Meade has also recently introduced a larger pair of very interesting achromatic refractors on a new, computerized German equatorial mount. The 5-inch f/9 and 6-inch f/8 LXD55 refractors each include multicoated optics, an 8 × 50 finderscope, a 26-mm Super Plössl eyepiece, an adjustable aluminum tripod, an illuminated polar-alignment scope, and, best of all, the Autostar GoTo aiming system as standard equipment. The Autostar technology is tried and true, and should work very well, but as of this writing, the telescopes themselves are too new to draw any conclusions. Their relatively fast focal ratios tell me to expect some leftover chromatic aberration, but if Meade's quality-assurance program ardently checks the instruments, then these could represent a very good buy. I am a little leery, however, about the mounting's ability to support the weight of the instruments, especially the 6-incher. As more becomes known of these promising instruments, I will be sure to post findings on my web site, http://www.philharrington.net.

Murnaghan Instruments. Based in Florida, Murnaghan Instruments offers four achromatic refractors under the subsidiary name e-Scopes. Again, all have their roots in the Far East.

Murnaghan and e-Scopes sell three versions of a 90-mm f/11 refractor. We are told that all share the same optical tube assembly, which is based around a fully coated achromatic objective lens. The slow focal ratio keeps false color to a minimum, though images do suffer from some spherical aberration. Image quality is not bad at lower magnifications, but, as with so many other small refractors, it degrades above about 120× to 150×. The ValueScope 90 includes a steel tube, reasonably good 1.25-inch focuser, a so-so 6 × 30 finderscope, and may be purchased with either an alt-azimuth mount or a German equatorial mount. Both are poor. The same optical tube assembly is available on Murnaghan's imported advanced German equatorial mount as the Astron 90, which is a far better choice.

Murnaghan, through e-Scopes, also sells a 4-inch f/9.8 refractor called the Astron 102. The optics of the 102 can be expected to show more false color than the smaller Astron 90, but the extra bit of aperture is welcome nonetheless. Optics are of reasonable quality, though perhaps not quite up to the level of, say, those found in Celestron's C102. The equatorial mounting is also not as sturdy as the CG-4 that comes with the C102, but it is still adequate for visual observations.

Orion Telescope Center. This California firm imports several refractors and sells them under the Orion name. Let's begin with their ShortTube 80, a very popular 3.1-inch f/5 achromatic refractor. Images are surprisingly good, with some spectacularly wide fields of the Milky Way star clouds possible, although a fair amount of color fringing is evident around brighter stars because of its fast focal ratio. Potential purchasers should also note that the instrument is not really suitable for high-powered planetary views. The general consensus is that while it's not an ideal first telescope, the ShortTube 80 is almost universally proclaimed by its owners as the perfect grab-and-go second telescope. Its light weight and small size make it a great choice for those who own larger, more cumbersome telescopes that might prove too unwieldy to take out for a quick midweek observing session after a long day at work or school. Given that it comes with a built-in camera-tripod adapter plate, a detachable 6 × 26 *correct image* finderscope (I was sorry to see them discontinue the original 6 × 30 finderscope), and two Kellner eyepieces, the Orion ShortTube refractor is one of the best buys in small rich-field achromatic telescopes today. But the 45° diagonal that comes with the telescope is almost universally condemned by owners. Replacing it with a 90° star diagonal brightens the view greatly. Some have complained about so-so optics and some glare, but most owners are happy with the telescope. Note that Orion offers the ST80 either with or without its EQ-1 German equatorial mount. Normally, a mount of this sort is too wobbly to support a telescope adequately, but given the small, lightweight ST80, it is fine for visual use. An optional clock drive is also available. Instruments that are similar to the ST80 include the Celestron Firstscope 80-WA and the Pacific Telescope Company's Skywatcher 804, reviewed elsewhere in this section.

Looking for a little more aperture in an easy-to-carry achromatic refractor? Consider the Orion ShortTube 90, the 3.5-inch f/5.6 big brother of the ST-80. Most agree that residual chromatic aberration is visible but not intrusive, except, perhaps, when viewing the Moon at low magnification. But it quickly diminishes as magnification is increased. Owners report good views of brighter deep-sky objects, as well as the planets. One observer commented that he was able to spot both a large equatorial festoon on Jupiter's atmosphere as well as Cassini's Division in Saturn's rings. But one should keep in mind that this is not designed as a planetary telescope but rather for wide-field scanning.

One strike against the ST90 is its 6 × 30 finderscope. While its performance is typical for the size, its location directly over the focuser is poor. Many users place the ST90 on a camera tripod or other alt-azimuth mount, which means that the star diagonal must be twisted out of the way to look through the finder's eyepiece. The telescope also includes a 90° star diagonal, which, while better than the ST80's 45° terrestrial prism, still leaves room for improvement. Most owners noticed sharper views and better contrast after it was replaced with a better unit (such as a premium, mirror-star diagonal). But I have no complaint about the 26-mm Sirius Plössl eyepiece that comes standard, an excellent choice for wide-field viewing. Like the ST80, the ST90 is sold with or without a German equatorial mount. The EQ-2 mount that comes with the ST90 is fine for visual use with such a light instrument, but it is overtaxed under the load of some other Orion telescopes, detailed below.

Orion also offers two conventional 90-mm refractors. The f/10 Explorer 90 comes outfitted with 10- and 25-mm 1.25-inch Kellner eyepieces, a 6 × 30 finderscope, and a 45° erect-image prism diagonal. The multicoated optics are adequately sharp, with decent images produced during star testing. Some minor residual chromatic aberration can be seen around brighter objects, but not to the degree of the ST90. Mechanically, the Explorer's focuser is fairly smooth, though some have mentioned minor irregularities as they rack it in and out. The Explorer 90 rides atop Orion's AZ-3 alt-azimuth mounting, a sturdy platform that offers dual slow-motion controls on both axes. As you may have inferred already, I prefer these sorts of mounts over small equatorials, because they are almost always steadier. True, you can't install a clock drive and they don't have setting circles, but the slow-motion controls are usually smooth enough to let observers keep things pretty well centered. And let's face it, the setting circles on most small German equatorial mounts are more decorative than functional.

The same 90-mm f/10 optical tube assembly is combined with the EQ-2 German equatorial mount to create the Orion AstroView 90. While the optics remain reasonably good, the EQ-2 mount adds nothing but shakes and jitters. As is typical of many imports, the mounting itself is underdesigned and the aluminum tripod is weak. As a result, you would do better either to spend less money and get the Explorer 90, the same amount of money on a 6-inch Dobsonian-style Newtonian reflector, or save up some extra money and get a 4-inch refractor.

Orion's AstroView 120, a 4.7-inch f/8.3 achromatic refractor from Synta, has attracted a good amount of attention since its release a few years ago. And rightfully so, since it is an excellent bargain. Optical quality impresses me as sharp and clear. Images remain sharp beyond 300×, with pinpoint stars surrounded by well-defined diffraction rings. Again, there is the characteristic purple fringing on the Moon and brighter objects that one might expect from a fast achromat, but I doubt you will find it objectionable unless you are extremely picky. If so, then you would do best to avoid achromats as a breed and move onto another type of telescope. The 2-inch focuser, which accepts 1.25- and 2-inch eyepieces, is a welcome addition that lets owners upgrade to a larger star diagonal and enter the world of premium eyepieces. Focusing is smooth, with no backlash detected.

The AstroView 120 is fine for looking at the planets, and it is also a good choice if you are looking for a refractor as your only instrument. Some readers, however, may be looking for a second instrument to complement their existing equipment. While many may be drawn to 3.1- or 4-inch f/5 refractors, Orion offers a larger rich-field refractor that is certainly worth consideration. The AstroView 120ST puts a 4.7-inch f/5 optical package into a sleek-looking, black metal tube that includes all of the features of the standard AstroView; only the focal ratio differs. True, false color will be more apparent, but it is surprisingly tolerable. I do offer one caveat, however. If you are looking for a transportable telescope—to carry on board an airplane, for instance—the AstroView 120ST is probably *not* for you. But for a relatively compact, lightweight instrument (the total weight with the mount is 36 pounds) that gives some great low-power views, then this is your next instrument.

Both AstroView 120s come with the EQ-3 German equatorial mount, roughly the equivalent of Celestron's CG-4. At the risk of repeating myself, this mount is undersized in my opinion, especially given the weight of the f/8.3 optical tube assembly.

Pacific Telescope Company. Based in Richmond, British Columbia, the Pacific Telescope Company sells many Skywatcher brand refractors manufactured by China's Synta Optical Technology Corporation.

Least expensive of the Pacific Telescope Company's refractors included here is the 3.1-inch f/5 Skywatcher 804, which bears more than a passing resemblance to Orion's ShortTube 80 and Celestron's FS80-WA. The difference? The most obvious is the tube color! The 804 is a deep blue, while Orion's version is white and Celestron's black. Other differences? The basic packages vary some, with more or fewer eyepieces of different focal lengths. The Skywatcher 804 is sold as an unmounted spotting scope with a tripod mounting block or mated to their AZ3 alt-azimuth mount or EQ1 light-duty German equatorial mount. While the latter can be outfitted with a clock drive, it does not support the instrument as well as the AZ3.

In the 3.5-inch aperture bracket, Pacific Telescopes has the Skywatcher 909, an f/10 instrument that is available either on their AZ3 alt-azimith or EQ2 light-duty German equatorial mount. The alt-az mount strikes me as the more stable, although the latter is sufficiently strong to support the metal-tubed optical assembly for visual observations at magnifications at or below about 150×. Unfortunately, the AZ3 version comes with a 45° diagonal, which as you know by now, is not suitable for astronomical viewing.

Skywatcher refractors also include the 4-inch f/5 Skywatcher 1025, the 4-inch f/9.8 Skywatcher 1021, the 4.7-inch f/5 Skywatcher 1206, the 4.7-inch f/8.3 Skywatcher 1201, and the 6-inch f/8 Skywatcher 150/1208. In the interest of space, and rather than simply reiterate their pros and cons here, you may read about those instruments under Celestron (look for the C102 and CR150) and Orion Telescopes (AstroView 120), where they are discussed in detail. One word of warning. Just be sure you are comparing equivalent telescopes and, especially, mountings. Table 5.1 may help sort things out.

Photon Instruments. Based in Mesa, Arizona, Photon Instruments is another small start-up company that sells telescopes both from other manufacturers, such as Meade and Celestron, as well as their own brand, which are actually imported from the Far East.

The smallest instrument to carry their house brand is a 102-mm f/9.8 refractor called, appropriately, the Photon 102. Like the Celestron C102, among others, the Photon 102 includes a 6 × 30 finderscope and a 1.25-inch rack-and-pinion focuser. But while Celestron includes a single 20-mm Plössl eyepiece with the C102, Photon throws in 10- and 25-mm Plössls. Optically, the Photon

Table 5.1 ***Mounting Equivalencies***

Synta	Celestron	Orion	Pacific
AZ3	Alt-az	AZ-3	AZ3
EQ-2	CG-3	EQ-2	SimPlex
EQ-3-2	CG-4	EQ-3	SkyScan
EQ-5	CG-5	—	StediVue
EQ-6	—	—	EQ-6

102's performance can be expected to be comparable to the C102, providing reasonably sharp views of wide star clusters, brighter nebulae, and the planets.

The Photon 127, a 5-inch f/9 instrument, is unique in size and focal length amid the vast ocean of imported 4.7- and 6-inch achromatic refractors. Made in China, the Photon 127 features a fully multicoated objective lens held in an aluminum tube painted white and includes a nice 2-inch rack-and-pinion focuser, a 2-inch star diagonal, a 1.25-inch adapter, and a 7 × 50 finderscope. Like most of the other imports, the Photon 127 is secured to a German equatorial mount perched atop an aluminum tripod.

Optically, the Photon 127 does a good job on the Moon, the planets, and brighter deep-sky objects. Spying Jupiter through his Photon 127, one respondent thought the view was "outstanding; excellent detail, more than five bands and very sharp, even when the scope was pushed to 350-power." Another owner offered these comments on some springtime objects: "The double star Epsilon Boötis was easily split with the 10-mm eyepiece, while M3 was partially resolved along its edges with the same eyepiece."

Stellarvue. Owned by amateur astronomer Vic Maris, Stellarvue has established itself as the supplier of some of the highest quality short-tube refractors sold today. Like many companies, Stellarvue relies on other sources for their optics. But unlike much of the competition, Maris actually designs, assembles, and tests each instrument by hand himself. This lets him control quality very closely, ensuring that each instrument that goes out works as it should.

The best-selling telescope in Stellarvue's lineup is the 3.1-inch f/6 AT1010, which comes in several variations and, as a result, prices. Looking like a slightly longer version of the Orion ShortTube 80, the Stellarvue AT1010 includes a red-dot unity finder, but eyepieces are sold separately. The instrument comes with a 2-inch rack-and-pinion focuser and comes either mountless, supplied with a tripod mounting adapter block, or on a German equatorial mount. The latter, a clone of Orion Telescope's SkyView Deluxe, is quite sufficient to support any version of this lightweight instrument.

The Stellarvue AT1010 performs amazingly well for its size. Owners rave about the images, with one saying that "the quality is good, portability can't be beat, and the views are fantastic." While some minor false color exists around brighter objects, it is less than through the many generic 80-mm f/5 short-tube achromatic refractors sold today. Spherical aberration is also nearly nonexistent. As a result, although this is not an instrument for the planets, it holds its own surprisingly well when magnification is pushed. Jupiter, for instance, shows several belts clearly, while several double stars, such as the Double-Double in Lyra, are also resolvable. Not bad for what is supposed to be a rich-field telescope. Indeed, the Stellarvue 80-mm f/6 reigns as king of the short tubes. True, it costs more than the many imported, mass-produced f/5 short-tube refractors, but all owners whom I have spoken with agree that the extra money is worth it.

Stellarvue also sells a 3.1-inch f/9.4 refractor known as the AT80/9D, which, like the f/6 above, comes in several versions, differing by accessories.

Again, all are assembled personally by company owner Vic Maris from hand-picked optics. The AT80/9D comes standard with a 2-inch focuser and a red-dot LED unity finder and can be purchased with several combinations of eyepieces, mounts, and focusers. The least expensive comes without any eyepieces or mount, while the deluxe version has 9- and 32-mm Plössl eyepieces, a Barlow lens, and a German equatorial mount.

Geared for more conventional viewing than the f/6, the Stellarvue AT80/9D does a great job on brighter sky objects. There is some minor color fringing around objects such as the Moon and Venus, but it really does not degrade the image noticeably. Indeed, Jupiter displays a sharp, color-free disk crisscrossed by several bands at 80×, while Cassini's Division in Saturn's rings can also be spotted easily. Once again, Stellarvue charges more for its instruments than some other companies, but all who have spent the extra money seem well pleased with their investments.

Stellarvue has also recently introduced a trio of 4-inch refractors. The Stellarvue 102D is a conventional 4-inch f/6.9 achromat, which shows the expected purplish glow of secondary chromatic aberration around brighter objects, but somewhat less so than, say, the Celestron Wide View. The other two are described under the "Apochromatic Refractors" heading later in this chapter. Rather than use a generic rack-and-pinion focuser, the 102D features the JMI NGF-3 Crayford focuser, with the superb NGF-1 available optionally. Each objective lens set is mounted in a separate, removable lens cell for easy cleaning, another thoughtful touch.

Unitron. Since its founding in 1952, Unitron has become widely respected for its excellent achromatic refractors. Chances are that if an amateur astronomer in the 1950s or 1960s wanted a high-quality refractor, he or she would have bought a Unitron. Although it has much more competition today than it did back then, Unitron still produces fine instruments that compare favorably with others in their size classes.

A quick glance through this market segment reveals that most manufacturers produce achromatic refractors between f/5 and f/10. While this lets the telescopes' tubes shrink in length, the shorter focal length makes the instruments more prone to residual chromatic aberration. Unitron, on the other hand, has chosen to stay with the classic approach of longer focal ratios. Though the instruments can be significantly longer, they are less apt to suffer from this aberration and so typically produce superior views of the planets and double stars.

Unitrons range in aperture from 2.4 to 3 inches. Each telescope features fully coated optics, a rack-and-pinion focusing mount, two or more 1.25-inch eyepieces, a small (too small) finderscope, and, in some cases, a solar projection screen. Mountings include both alt-azimuth and German equatorial designs on wooden tripods. The equatorials are especially well crafted and sturdy, and may be outfitted with optional motor drives. Another nice option for Unitron scopes is the Unihex rotary eyepiece turret, which holds up to six eyepieces at any one time. To change power, rotate the turret until the desired eyepiece is in position.

Recently, Unitron introduced ImageTrac, a manual or computer-driven GoTo system designed specifically for its 3-inch equatorial refractors. Once attached to the mounting, ImageTrac will steer the instrument at a rate of 1.6° per second to any of the many preprogrammed objects. The ImageTrac hand controller attaches to any Windows-based computer via an RS-232 port. Using standard protocols, ImageTrac can be guided with several popular astronomy programs, such as the Sky, Sky Map Pro, and others. ImageTrac software for Windows is also included with the system.

Vixen Optical Industries. This is a highly regarding manufacturer of refractors, reflectors, and telescopic accessories. In the category of achromatic refractors, they offer several conventional and a few not-so-traditional instruments.

Let's begin with Vixen's smallest refractors, the Custom 80M, a 3.1-inch f/11.4 achromatic refractor, and the 3.5-inch f/11.1 Custom 90. (Vixen also offers a 60-mm refractor, too small to be mentioned in this book.) The Custom 80M is sold in the United States by the country's only authorized Vixen dealer at present, Orion Telescopes and Binoculars, as the VX80, but otherwise it is the same instrument as the one available throughout the world. As with all Vixen instruments, the optical quality of both the Custom 80s and Custom 90s is excellent for their apertures, although the small apertures limit their use to the Moon, the planets, and brighter sky objects. If you are looking for a high-quality small refractor, both represent very good choices. Consumers should note, however, that for about the same price as the Custom 80M, they can purchase a 4-inch Chinese clone refractor (that is, cloned after Vixen products). Are they as well made as the Vixen? Absolutely not. Optically and mechanically, the Vixens are superior, but that superiority may only be detectable to an experienced eye. Also note that some distributors around the world offer these telescopes with 0.965-inch eyepieces, which are not as preferable as standard 1.25-inch oculars. Be sure you know what you are getting before purchasing.

Looking for the same high quality, but with a little more aperture? The Vixen GP 102M is a 4-inch f/9.8 achromatic refractor that comes mounted on the acclaimed Great Polaris (GP) German equatorial mount and an aluminum tripod. The rack-and-pinion focusing mount takes 1.25-inch eyepieces, but it will also accept 2-inch eyepieces with an optional adapter. As the owner of an older Celestron C102, which is essentially the same instrument that was previously imported into the United States by Celestron, I can attest to the Vixen model's optical excellence with enthusiasm. Color correction is very good for a sub-f/10 achromat, with star images that are both sharp and clear. Both planetary and deep-sky views are exceptionally sharp, with any evidence of spherical aberration all but undetectable. Once again, the question: better than a less expensive Chinese clone refractor? Yes, but again, you will pay several hundred dollars extra for that superiority.

Vixen also offers three refractors that feature unusual *Petzval* objective lenses. A Petzval objective consists of four separate lens elements set in two groups, a long focal-length pair in the front and another farther along, close to the focuser, that serves as a focal reducer and field flattener. It's a complex design that is tough to make well because of the critical nature of the optical alignment. When it is done well, this arrangement reduces residual chromatic aberration and eliminates field curvature.

This same approach is also employed in some of the instruments made by Tele Vue, Pentax, and Takahashi. Of course, just because another manufacturer uses one of these complex lens designs does not necessarily make it apochromatic. The Tele Vue NP-101, for instance, includes a special-dispersion (SD) lens pair in the front and a fluorite doublet in the rear group, while the Takahashi FSQ-106 prefers two fluorite/special-dispersion doublets. Both eliminate false color completely and work exceedingly well. Vixen, however, does not specify the type of glass they use in their application. The result, while good, is not apochromatic. Vixen calls it a *neo-achromat* (NA).

Vixen NA refractors come in three apertures: a 4.7-inch f/6.7, a 5.1-inch f/6.2, and a 5.5-inch f/5.7. Here in the United States, only the 4.7-inch GP NA120SS is available through Vixen's lone distributor, Orion Telescopes, with the nameplate VX120. Call it what you will, Vixen NA refractors feature well-corrected optics that produce very sharp views. False color, however, is certainly visible around on bright objects, with stars such as Rigel and Vega bathed in a purplish glow. But even with this, image sharpness is quite impressive.

The cost of these instruments hovers in the realm of slightly smaller apochromatic refractors, although the Vixen NA refractors do offer the added benefit of extra aperture. Orion Telescope only imports the 4.7-inch NA120SS on the lighter GP (Great Polaris) mount into the United States, although others also offer it in the sturdier GP-DX mount, which is preferred for high-power viewing. The 5.1-inch NA130SS is available on either the GP mount or the preferred GP-DX mount, while the NA140SS is sold on Vixen's largest and most impressive mounting, the Atlux. Consumers can also purchase just the telescope tube assemblies, then mate them later with a mounting purchased separately.

Apochromatic Refractors

If you are looking for the best of the best, then this section is for you. Apochromatic refractors, with their multiple-element objective lenses made of exotic materials, are generally considered to be as close to an optically perfect telescope as one can get. Images are almost always tack sharp and high in contrast, offering spectacular views of both the planets and deep-sky vistas alike. But this perfection comes with a steep price tag. Is one right for you?

Aries Instruments. Located in Kherson, Ukraine, Aries is a recent name in the sphere of high-end refractors that was only recently dominated by the United States and Japan. In only a few short years, its three apochromatic refractors have established themselves as more than capable of competing with the most elite amateur telescopes in the world.

All three—a 6-inch f/7.9, a 7-inch f/7.9, and an 8-inch f/9—are built around two-element objectives. Images are reported as being sharp, contrasty, and free of spurious color. Mechanically, Aries refractors are also very good, with very good overall workmanship. Each instrument includes an oversized focuser with 1.25- and 2-inch eyepiece adapters, 8 × 50 finderscopes, retractable dew caps, and mounting rings. Suitable mounts must be purchased separately.

Are they are good as, say, Astro-Physics' or Takahashi's three-element objectives? Perhaps not quite, but they are very impressive nonetheless, and stand as a testimony to how lens design continues to improve. Not long ago, this level of image correction could only be achieved with three- or four-element objective lenses. But the evolution—indeed, revolution—in lens design continues. Prospective consumers should note that Aries instruments are comparably priced to TMB refractors but are less expensive than both Astro-Physics and Takahashi.

Astro-Physics. This is a name immediately recognizable to the connoisseur of fine refractors on rock-steady mounts. Owners Roland and Marjorie Christen introduced their first high-performance instruments in the early 1980s and effectively revived what was then sagging interest in refractors among amateur astronomers. Now, two decades later, Astro-Physics refractors (Figure 5.1) remain unsurpassed by any other apochromat sold today.

Currently, four refractors make up the Astro-Physics line. The heart of each is the superb three-element apochromatic objective lens. Two crown meniscus elements combine with a third lens made of an advanced fluorite ED (extra-low-dispersion) glass to eliminate all hints of false color (residual chromatic aberration) and other optical aberrations, both visually and photographically.

Figure 5.1 *Astro-Physics Starfire apochromatic refractor.*

Least expensive is the 105EDFS, a 4.1-inch f/6 refractor measuring only 19 inches long, aptly named the Traveler. Like the larger Starfires, the performance of the Traveler is outstanding! I can vividly recall the view I once had of the Veil Nebula complex in Cygnus through a Traveler owned by some friends. With a 35-mm Panoptic eyepiece, the three main pieces of the Veil, spanning nearly 3°, fit into a single eyepiece field! The nebula's resolution and brightness looked just like a photograph, but I was actually seeing this live! Now, years later, that single observation remains one of the highlights of my astronomical career. The Traveler is made for large nebulae (such as the Orion Nebula) and widespread open clusters (the Pleiades and the Double Cluster immediately come to mind). But that doesn't mean that the Traveler is a low-power scope only. With a high-quality, short-focal-length eyepiece in place, the planets also show some amazing detail.

The Traveler is as superb mechanically as it is optically. Especially impressive is its oversized rack-and-pinion focuser made of aluminum and brass. Its inside diameter measures 2.7 inches, designed to let advanced astrophotographers use medium-format cameras with minimal vignetting. Adapters are supplied for both 1.25- and 2-inch eyepieces.

For those looking for a little larger aperture and longer focal length, the 130EDFS 5.1-inch f/6 Starfire should be given serious consideration. Like its smaller sibling, the 130EDFS Starfire is an exceptional instrument, both optically and mechanically. Images are breathtakingly sharp and clear, with no sign of false color or spherical aberration. Mated to a suitable mounting, this telescope is a great compromise in terms of aperture versus portability. The larger 6-inch Astro-Physics scopes are great, don't get me wrong, but they can be a little cumbersome to set up, especially for a quick observing session. But the lighter, more compact 130EDFS is easy to take out and set up, and to later tear down and store, while also giving a little more oomph (a technical term) than the Traveler.

The largest refractor in the Astro-Physics fleet is the 155EDFS 6.1-inch f/7 Starfire, the pinnacle in its aperture class. Because of its fast focal ratio, the 155EDFS measures only 41 inches long with the dew cap fully retracted, making it a truly portable telescope. No trace of false color is evident in either instrument, even on such critical objects as the planet Venus. Stars show textbook Airy disks, planetary contrast is tack sharp, and resolution is crisp and clear. Both use the same oversize rack-and-pinion focuser as the Traveler for effortless focusing without any hint of binding or shift, even in subfreezing temperatures. All machining is done by Astro-Physics in its factory, ensuring top-notch quality control on all components.

Finally, there is the 155EDF 6.1-inch f/7 astrograph, a variation of the 155EDFS. This instrument uses the same three-element objective lens as the 155EDFS but also includes a two-element, 4-inch field flattener lens in the rear of the optical tube assembly to eliminate astigmatism and field curvature in photographs. That is especially important if you are contemplating medium-format photography, where the edges would otherwise show less-than-pinpoint star images.

Notice that I have not mentioned anything about Astro-Physics mountings. That is because none of the telescopes come with mountings; instead, you may select from their line of homegrown German equatorial mounts to suit your needs. Each of the four German equatorial mounts that bear the Astro-Physics name is discussed later in this chapter, along with other telescope mountings from several other manufacturers.

For either visual or photographic use, Astro-Physics apochromatic refractors are the finest of their kind, but be forewarned that because of the high demand and limited production, delivery will likely take more than a year.

Borg. As mentioned earlier in the section covering achromatic refractors, Borg instruments are unique in the astronomical world. Unlike other telescopes, which, once you buy them, cannot readily be upgraded optically, Borg refractors feature modular aluminum construction that is easy to improve as finances permit. Borg telescopes are all built around three series, or model lines. All Series 80 telescopes are built around 80-mm (3.1-inch) tubes, Series 115 Borgs are based around 115-mm (3.5-inch) tube components, and Borg Series 140 refractors use 5.5-inch tubes. All are designed to break down into short individual pieces.

Images through Borg refractors are striking. Little, if any, false color is detectable around even critical sky objects, such as Venus and the limb of the Moon, while star tests show no indication of damaging aberrations. Planets and double stars focus precisely and with good image contrast thanks to the instruments' impressive objective lenses, which use ED glass elements stacked in one or more groups, depending on the configuration you select. All Borg apochromats begin as ED doublets (that is, two lens elements of extra-low-dispersion glass), but can end up having as many as six elements in three lens groups, depending on whether you add on their focal reducer or field flattener, or both. The field flattener is especially important if you are contemplating astrophotography.

Meade Instruments. Meade offers five apochromatic refractors. The least expensive is the 102APO/500, a 4-inch f/9 model placed on the imported LXD 500 German equatorial mount that is well accessorized with an 8 × 50 finderscope, a 2-inch mirror star diagonal with a 1.25-inch adapter, and a 26-mm Super Plössl eyepiece. When put to the test, color suppression is certainly better than a traditional achromat, including Meade's own 102ACHR instruments, although some false color can still be detected around bright objects, such as Venus and the lunar limb. Admittedly, these are very unforgiving objects that present the greatest challenge to an apochromatic refractor. Yet many other manufacturers (all charging far more for their instruments) have conquered the problem. Resolution and image contrast through the Meade are also fairly good, though again not up to the benchmark set by Astro-Physics or Takahashi. Meade tells us that the optical tube assembly is the same as that used with their Model 102ED refractor, described below.

The imported LXD 500 mount includes dual-axis manual controls with knobs that are rather difficult to grasp, especially when wearing gloves. Overall, the LXD 500 is adequate for visual observations, although some internal play adds to the mount's woes. An optional polar-alignment scope is available, as are single- and dual-axis 12-volt DC motorized clock drives.

The other four Meade apochromatic refractors are among the most sophisticated instruments on the market today. The 4-inch f/9 Model 102ED and 5-inch f/9 Model 127ED ride atop the LXD 650 German equatorial mount and Meade's Standard Field Tripod, while the 6-inch f/9 Model 152ED and the 7-inch f/9 Model 178ED include the larger LXD 750 mount and Giant Field Tripod. All four combinations offer a good mix of stability, portability, and convenience. The tripods are especially noteworthy. Unlike the stamped-aluminum design used on the LXD500 mount (and just about all Chinese and Taiwanese mounts sold today), the LXD 650 and LXD 750 tripods are based around three tubular, extendable legs that are braced together solidly with a spreader that locks them in place. These same tripods are used on Meade's catadioptric telescopes, described later.

Both the LXD 650 and LXD 750 come with manual slow-motion controls but can be easily outfitted with dual-axis, DC-powered motors. Unfortunately, the drives require 18 volts to power them rather than a more convenient 12 volts, although Meade sells a separate adapter to run the drive motors off a 12-volt cigarette adapter. The optional Electronic Drive System includes fast-speed (32×) slewing and slow-motion (2×) control that are quite smooth, with little or no backlash. A second, optional drive, the Computer Drive System, adds several incremental drive speeds as well as GoTo capability with a library of some 64,000 objects. Both drives are easily attached to the mount's control panel and include permanently programmable periodic error correction (PPEC), which will be appreciated by astrophotographers. A description of periodic error correction can be found later in this chapter in the discussion of Schmidt-Cassegrain telescopes.

You may have noticed that all Meade apochromatic refractors have higher focal ratios than some others. The reason is simple. Secondary spectrum can be reduced in two ways in a refractor. One is simply to stretch the instrument's focal length until it goes away, which is why achromatic refractors are usually (but certainly not always) slower than apochromats. Another way to deal with the problem is with an advanced, multiple-element objective lens made of exotic materials. Meade apochromats try to strike a happy medium by combining a slower optical system with a two-element objective lens of ED glass. The resulting images yield pinpoint stars and good field contrast, characteristics of better refractors. When viewing brighter targets, however, some minor blue fringing becomes evident, as does some slight astigmatism and spherical aberration. This led one respondent of my survey to describe these as "semi-apochromatic" telescopes; perhaps not as good as some other apochromats but better than similar-aperture achromatic refractors. Some owners also report that they have had to exchange their telescopes more than once before they received one that they thought was optically satisfactory.

The only mechanical problem evident in the overall design is in the focusing mount, which is not as smooth as some of the competition's mounts. The problem was especially noticeable in earlier production scopes, but has since been improved. But again, most of the competition also costs considerably more. Indeed, you may pay more for their optical tube assemblies than for an entire Meade apochromat. Meade apos represent a bargain in the apo arena, but double-check on your dealer's return policy just in case (a good bit of advice for all major purchases, astronomical and otherwise).

Stellarvue. As mentioned earlier in this chapter, Stellarvue recently introduced a trio of 4-inch refractors. The least expensive, the achromatic 102D, has already been discussed. The semi-apochromatic 102EDT uses a triple-element 4-inch f/6.1 objective to cut color noticeably. Still, some color is inevitable around brighter objects. Viewing through an early prototype at the 2001 Riverside Astronomy Expo, I noted yellowish rim along the limb of the crescent Moon, while the tough test object Vega showed a slight purplish tinge.

The Stellarvue 102APO, a full-fledged apochromat, promises color-free images. Although too new to have been fully evaluated before press time, it does bear watching. As with the Stellarvue 102D achromat, the 102EDT and the 102APO feature the JMI NGF-3 Crayford focuser, with the superb NGF-1 focuser available optionally. Stellarvue's thoughtful touch of securing the objective lens sets in separate, removable cells, which aids cleaning, is also carried over. Best of all, the 102EDT and the 102APO are considerably less expensive than most of the competition, making these excellent choices for people looking to break into the apochromatic world on a budget.

Takahashi. The apochromatic refractors that carry the Takahashi brand have well-deserved reputations for their excellent performance. In fact, to many they represent the pinnacle of the refractor world; they just don't come much better. One owner puts it this way: "Many telescopes feel flimsy and thin, but Takahashi telescopes feel like they are honed from solid metal." The only drawback is that this superior quality does not come cheaply!

The FS line of instruments includes the 2.4-inch f/5.9 FS-60, the 3-inch f/8 FS-78, the 4-inch f/8 FS-102, the 5-inch f/8.1 FS-128, and the 6-inch f/8.1 FS-152. All Takahashi FS refractors use doublet objectives—one element is made of fluorite, the other of low-dispersion flint—for exceptional color correction and image clarity. What makes these objectives unique is that unlike other manufacturers, who place the fluorite element on the inside, Takahashi places it as the outermost element. This unique positioning claims to produce even better results than before. While many optical experts feel this may be little more than an advertising scheme, there is no denying that the view is absolutely amazing (as it always has been with Takahashi apochromats). Indeed, the optics are so sharp that owners routinely report using magnifications in excess of 100× per inch of aperture, well beyond the 60× per inch ceiling mentioned in chapter 6 and elsewhere in this book. With absolutely perfect optics, however, rules can be broken. Planetary detail is simply stunning. Stars don't just draw into a focus, they *snap* into focus with a certainty not seen in many other telescopes. Optically, these telescopes are masterpieces; mechanically, they are art.

The Takahashi FCL-90, also known as the Sky 90, is a wonderful 3.5-inch f/5.6 grab-and-go instrument for astronomy on the run. Measuring only 16 inches end to end with its lens shade retracted, the Sky 90 is small enough to tuck into a briefcase yet has enough quality to show some spectacular views of the night sky. The heart of the instrument is a doublet objective with a fluorite element. This design suppresses spherical aberration to extremely low levels, although some false color can be seen around such unforgiving objects as Venus, Vega, and Sirius. Coupled with the 50-mm Takahashi LE eyepiece, the Sky 90 squeezes an incredible (and incredibly sharp) 5° into a single field of view—great for viewing, say, the sword of Orion, the Andromeda Galaxy, or the star clouds in Sagittarius and Scorpius. But remember, this is still just a 3.5-inch telescope, so while it shows bright objects with amazing clarity, its aperture is limiting. The Sky 90 will accept all accessories (finderscopes, camera adapters, etc.) designed to fit the FS series of refractors. The price is comparable to a Tele Vue TV-102, which also offers excellent optical performance (although perhaps not quite the image contrast), but the TV-102 has added aperture.

The 4.2-inch f/5.2 Takahashi FSQ-106 has a modified Petzval optical design that builds its objective lens around four separate lens elements set in two groups. Two of those four elements, one in the forward group, the other belonging to a widely spaced pair toward the back of the tube, are made of fluorite for excellent image contrast and false-color elimination. Astrophotographers will especially appreciate the FSQ-106's huge 4-inch focuser, which will let them use large-format cameras with little or no vignetting, a common problem with other like-sized apochromats. The focuser can also be rotated a full 360° without shifting focus, making it much easier to look through a camera's viewfinder when aimed at odd angles.

While one or two reports of problems with the instrument have been received, specifically pointing toward the lens cell, most agree, as one owner put it, that "the FSQ-106 is spectacular. I have never, I mean never, seen better contrast. The sky was ink black, stars sharper than pinpoints right to the edge. It is amazing what a 4-inch can do on objects like M13." But be aware that the FSQ-106 costs considerably more than the competition, including Takahashi's own FS-102. Of course, those other instruments cannot brag about their wide, flat photographic fields, which makes the FSQ-106 a stand-out performer for astrophotographers.

The largest Takahashi refractors belong to the FCT line. FCT telescopes are designed around triplet objectives that include a fluorite element to yield uncompromising optical performance. Astrophotographers should be especially interested in the photographically fast 6-inch f/7 FCT-150. And if you are in the market for the ultimate large-aperture refractor, then the 8-inch f/10 FCT-200 is for you. Both feature huge 11 × 70 finderscopes. These telescopes are as close to optical perfection as you are liable to find.

Although Takahashi telescopes are sold without mounts, most owners prefer to support their babies on Takahashi mounts, which are listed later in this chapter (see Table 5.2).

Tele Vue, Inc. This company manufactures several outstanding short-focus refractors for the amateur. The views they provide are among the sharpest and clearest produced by any amateur telescope, with excellent aberration correction.

The least expensive of the Tele Vue refractors is the Ranger, a compact 2.8-inch (70-mm) f/6.8 instrument. Though technically below the 3-inch threshold

I set earlier, the Ranger's sharp, wide-field performance warrants its mention. The Ranger features a two-element objective lens, with the front element made of crown ED glass and the rear of high-index flint. All surfaces are coated to reduce internal reflections. The net result is proclaimed by the manufacturer as "semi-apochromatic"; that is, not as completely color free as a true apochromatic refractor, but better than a standard achromat. From optical testing I've done, the Ranger stands up well, even with short-focal-length eyepieces. Images are crisp and stray-light reflections are nonexistent, even though, surprisingly, there are no internal light baffles. The 1.25-inch helical focuser and tube are made from machined aluminum, the tube coming nicely finished with either ivory-white paint or a brass coating. While I am not a proponent of helical focusing mounts, Tele Vue attacks the design intelligently by including a sliding drawtube for coarse adjustments. The Ranger includes a 20-mm Tele Vue Plössl eyepiece, a 90° mirror diagonal, and a mounting plate for attaching the instrument to a camera tripod. No finder is provided, though the Ranger's wide field at low power makes one unnecessary.

The Ranger's more affluent older sibling is the Pronto. While both the Ranger and the Pronto use the same semi-apochromatic objective lens, the appointments offered with the Pronto are more lavish. Chief among these is its excellent 2-inch rack-and-pinion focuser, one of the smoothest available. Another nice feature is the mounting block that attaches the instrument to any standard camera tripod. Unlike the Ranger's mount, which is fixed in place, the Pronto's sliding mounting ring allows the user to balance the instrument after attaching a camera or other accessory. Also included with the Pronto are a 2-inch mirror star diagonal (a 1.25-inch adapter is included) and a 20-mm Tele Vue Plössl.

Looking for a little more aperture? Then, consider the 3-inch f/6.3 TV-76 or 3.3-inch f/7 TV-85, the smallest apochromatic refractors offered by Tele Vue. Both are built around two-element objectives of ED glass. Optical performance is outstanding, with little or no false color detectable around such demanding objects as the Moon, Venus, and the bright star Vega. Views of deep-sky objects are consistently sharp and clear. Like all Tele Vue scopes, both are a joy to use for both their optical and mechanical excellence. You will need to supply a tripod or other support, as one is not included with either telescope.

Tele Vue has adapted the TV-85 into a strange, two-eyed beast called, appropriately, the Bizarro. In effect, the Bizarro is a TV-85 that is missing about 5 inches off the eyepiece end. In its place is one of Tele Vue's binocular eyepiece holders, which enables observers to drink in the great views of the TV-85 with both eyes instead of just one. And while a binocular viewer can be added to a standard TV-85, a field-cutting transfer lens must also be inserted in order to focus images. The Bizarro, on the other hand, has been optimized specifically for two-eyed viewing. And what a great view it is, too! It's easy to spend hours simply scanning through bright star fields with the Bizarro, which almost gives the observer the feeling of flying. But be forewarned that this ticket comes with a high price—some $800 above the standard TV-85.

Tele Vue sells two, nearly twin 4-inch apochromatic refractors, the NP-101 (the NP is short for *Nagler-Petzval*) and the TV-102. Introduced in 2001, Al

Nagler's NP-101 is a 4-inch f/5.4 apochromatic refractor, the same as its predecessors; the TV-101, the Genesis-SDF, and the original Genesis. The objective consists of four lens elements set in two groups (a Petzval design, hence the product name) that include an internal fluorite lens and two special-dispersion glasses. The TV-102's 4-inch f/8.6 objective is a simpler two-element design that features one element made from a material similar to fluorite, while the second uses special-dispersion glass.

Both telescopes produce tack-sharp images of the Moon, the planets, stars, and brighter deep-sky objects that are nearly devoid of the residual chromatic aberration that plagues achromatic refractors. Objects such as globular cluster M13, the Lagoon Nebula M8, and the open clusters in the Double Cluster all take on sparkling personas against their brilliant star-filled backdrops. Binary and multiple stars, such as Epsilon Lyrae, the famous Double-Double quadruple star, are also well resolved. Finally, the tell-all star test displays perfect sets of diffraction rings on either side of focus in both cases, just as quality telescopes should.

How do they compare? While the NP-101's four-element objective lens system suppresses color a bit better than the TV-102, the difference is minor. The shorter focal length of the NP-101 makes it more suitable for wide-field viewing than the TV-102, but some owners have reported that the TV-102 is superior on the planets. Again, that's an awfully close call.

The NP-101's aluminum tube comes only in ivory, while the TV-102 comes in either ivory or polished brass, the latter sold under the name Renaissance. The NP-101 includes very smooth 2-inch focusers and star diagonals, 1.25-inch eyepiece adapters, a tripod mounting ring, and 20-mm Tele Vue Plössl eyepieces, but no finderscopes. The TV-102 features the same focuser but is sold without an eyepiece, a star diagonal, a mounting ring, or a finderscope. All are sold mountless. Tele Vue's sturdy Gibraltar alt-azimuth mounting was specifically designed with these instruments in mind. If you are looking for an equatorial mount, consider one by Astro-Physics, Losmandy, Takahashi, or Vixen.

TMB Optics. Those amateur astronomers who frequent various Internet newsgroups, such as sci.astro.amateur, will likely recognize the name of Thomas Back of Cleveland, Ohio, one of the most knowledgeable optics experts in the group. Back, who is famous for his discriminating taste in optical excellence, has teamed up with Markus Ludes of Reifenberg, Germany, to create and manufacture some of the finest apochromatic refractors for sale today. Back designs the optics and then arranges through Ludes to have the instruments constructed in Russia. The tubes are then mated to premium mounts manufactured by Astro-Physics, Takahashi, and others, based on owner preference. Optical tube assemblies are available in ten standard apertures that range from the modest 3.1-inch f/6 and f/7 Apo-80 to the impressive 10-inch f/9 Apo-254, and even larger, custom-made instruments. I can only imagine the views through those monsters!

While I have never seen the 10-inch TMB refractor, I have had a chance to look through several smaller models and can only say that the images are absolutely textbook perfect. A 4-inch TMB refractor, for instance, withstood the rigors of the star test with a perfection that I have seen few other telescopes do before. The planets, double stars, and other sky objects all showed an amazing level of both sharpness and contrast. It was a truly amazing telescope.

The instruments' mechanics are every bit as impressive as their optics, with silky smooth focusers, retractable dew caps, and other amenities that we have come to expect on the finest instruments. This quality does not come cheaply, mind you, but TMB Optics does offer yet another option for those considering a refractor from Astro-Physics, Tele Vue, or Takahashi.

Vixen Optical Industries. A highly regarded name in astronomical telescopes, both in its home country of Japan as well as throughout the world, Vixen offers two lines of apochromatic telescopes. Unfortunately, only a limited number of products are available for sale in the United States through their sole importer, Orion Telescope Center, but stargazers in other countries have a wide selection of quality instruments from which to choose.

Vixen's ED line of refractors feature two-element, multicoated objective lenses of extra-low-dispersion glass combined with another element of flint glass. Color correction is very good for an ED apochromatic objective, though most experts feel that these, as a breed, fall slightly short of those that use fluorite. On tough objects, such as the Moon, Venus, and the brightest stars, some minor color fringing is evident, but it is certainly tolerable. Owners consistently sing the praises of the telescope's image sharpness, which produces textbook Airy disks and diffraction rings. Readers in the United States will recognize the Vixen GP 102-ED (4-inch f/6.5) and GP 114-EDSS (4.5-inch f/5.3), both sold through Orion, but two other Vixen ED refractors are also available elsewhere in the world: the 80-ED (a 3.1-inch f/9) and GP-DX 130-EDSS (a 5.1-inch f/6.6). If you are planning to try some guided astrophotography with either the 102 or the 114 (and who wouldn't, given those focal ratios), be sure to opt for the heavier duty GP-DX mounting over the standard Great Polaris mount. The GP-DX is the standard mounting for the 130.

Two of the best, though I feel least appreciated, fluorite-based apochromatic refractors on the market today are the Vixen 3.1-inch f/8 GP FL-80S and 4-inch f/9 GP FL-102S (the 4-inch is sold in the United States through Orion Telescope Center under the name VX102-FL). As good as the ED Vixens are, these fluorite instruments are on a higher plane. Owners all report brilliant images and good contrast, two important traits in premium refractors.

The 80-mm Vixen fluorite refractor comes with a quality 1.25-inch focuser, while the 102 and larger fluorite and ED Vixens feature a high-quality 2-inch rack-and-pinion focuser. All work very smoothly with no detectable backlash or wobble. Some owners have cited problems focusing with certain eyepieces, but this problem seems to revolve around some extension rings that come assembled to the telescope. One owner writes, "When I first took out my new 102ED at night, I could not get it to focus at all. The next day I noticed it had 'extra parts,' such as an extender and a couple of odd rings threaded to the telescope housing. I removed these three unnecessary parts, and the scope works like a dream."

William Optics. Here is another new name in the apochromat game, the first contestant from Taiwan. Today, owner William Yang offers two semi-apochromatic refractors, one apochromat, and several larger, customized instruments. The former pair, the 3.1-inch f/6 Megrez 80 and the 4-inch f/6.8 Megrez 102, make up the Megrez line. Both are built around two-element objective lenses made of special-dispersion (SD) glass to minimize residual chromatic aberration. Although color suppression is better than might be expected from an achromat, some color still does come through. As a result, Yang refers to these with the nebulous term *semi-apochromatic*. Both models include extremely smooth 2-inch focusers, 25-mm Plössls, telescoping dew shields, and pairs of mounting rings. Each metal tube has a pearlized white finish that glistens in daylight. Both are very well made, both physically and optically. Some early quality problems with astigmatism and pinched optics have since been addressed.

William also makes several larger, custom instruments, up to 8 inches in aperture. Each includes a smoothly operating 2-inch focuser and other conveniences. They are not stock items, but perhaps, as production gears up, they will be in time. If so, judging by the William instruments I have seen, they would merit strong consideration, should this be your market.

Right about now, you might be thinking, "Okay, who makes the best apochromatic refractor?" The argument has traditionally been between Astro-Physics, Takahashi, and Tele Vue, but now we must also consider contenders from Aries, Borg, TMB, Vixen, and William Optics. Which is the *absolute* best? Sorry, but I feel that it is too close to call. Ask their owners and all will swear that *their* telescope is the best. The important thing to note is that all consistently receive glowing reviews and few, if any, complaints. It's the same as asking which car is better: Mercedes-Benz, BMW, or Lexus? Chances are they are all pretty darn good, if you can afford them!

Reflecting Telescopes

Like refractors, reflecting telescopes can be further subdivided based on their optical design. By far the most popular type of reflector on the market today is the one originally designed by Sir Isaac Newton, and so the first part of this section addresses Newtonian reflectors. The second part of this section looks at Cassegrain reflectors, which are far more limited in number. Finally, I have created a third category for *exotic reflectors,* which include a few unique instruments that are worth singling out.

Newtonian Reflectors

As mentioned in chapter 2, Isaac Newton was not the first person to design a telescope based around a concave mirror for focusing light, but his design remains the most popular by far. For many, a Newtonian represents the ideal first telescope, offering a large aperture at a comparatively modest price.

As we will see below, Newtonians come in many different shapes and sizes on a variety of mountings, but they do seem to fall into three general categories. *Econo-Dobs,* ranging in aperture between 4.5 and 10.0 inches, are generally characterized by cardboard or metal tubes and Dobsonian-style altitude-azimuth mounts made from laminated particleboard. Most come with one or two Kellner or Plössl eyepieces, and some also include small finderscopes. The solid tubes can make hauling the equipment difficult, especially at larger apertures, but also tend to keep the optics in collimation.

This brings up the argument of which makes the better telescope tube, cardboard or metal. Metal is clearly more durable, although it can also dent. Cardboard would likely absorb any soft blow without a problem, but it has its own disadvantages. What happens when untreated cardboard gets wet? It turns to mush and sags. Happily, all of the telescopes listed here that use cardboard tubes have been coated with several coats of paint, sealant, or other material that will slow moisture penetration.

Personally, I prefer cardboard over metal. The reason is simple. Metal tends to absorb heat much more readily than cardboard. As a result, a metal telescope tube, when brought outdoors from a warm building, will radiate more absorbed heat than a cardboard tube, in effect placing the primary mirror at the bottom of a stovepipe. The result are dancing images that will be difficult to focus until everything has reached thermal equilibrium. True, cardboard-tubed telescopes must also reach thermal equilibrium, but they seem to take less time doing so.

Contrast econo-Dobs with *primo-Dobs,* which come in apertures from 8 up to over 30 inches. The primo-Dobs with low-riding mounts made of superior plywood, precision focusers, top-quality mirrors of Pyrex glass, and open tubes represent the current state of the Dobsonian art. The open-truss design greatly reduces the instrument's overall weight, making transport much easier than solid-tubed telescopes. But at the same time, collimation must be checked and reset every time the telescope is reassembled. Most are supplied without finders or eyepieces, which stands to reason since most who buy these have already owned other telescopes and have probably accumulated a fair share of accessories along the way. Expect to pay a premium price for these instruments.

The third category of Newtonians comes on equatorial mounts. Actually, we can further divide this group into two subsections. In the first, we would include all of the imported equatorial Newtonians, typically from China, Russia, and Taiwan. These usually come with one or two eyepieces and a 5 × 24 or 6 × 30 finderscope and are often sold on one of the same mounts used with many of the achromatic refractors discussed earlier. Some of these low-end equatorially mounted Newtonians are known as *catadioptric Newtonians,* or *short-tube Newtonians,* because they squeeze a relatively long focal length into a short tube. Magic? Not really. Instead, a corrector lens (akin to a built-in Barlow lens; see chapter 6) is usually set permanently into the focuser's tube to stretch the instrument's f/4-ish primary mirror to an effective focal length of between f/7 and f/9. This has good and bad points. First, the good, which should be obvious. By using a shorter-focal-length mirror, the telescope's tube can be cut in half, making it easier to carry. Also, by reducing the weight, the burden placed on the mounting is reduced. The disadvantages include the need for a larger secondary mirror, increasing the telescope's central obstruction and decreasing image contrast. Optical distortions, such as chromatic aberration (unseen in traditional reflectors), can also be introduced by these auxiliary lenses. Short-tube reflectors also use very fast optics (i.e., low f-number) and, as such, must be collimated much more precisely than a traditional Newtonian with the same focal ratio. Many newcomers are not aware of this, however, and so their instruments may work poorly.

Comparatively few Newtonians belong to the second subgroup, which features premium equatorial mounts. These instruments are far heavier than the others. Still, for advanced amateurs who crave that style of telescope, offerings from JMI, Excelsior Optics, and Parks Optical fill this market niche.

Bushnell. This subsidiary of Bausch & Lomb is probably most familiar to amateur astronomers as a company that sells the types of telescopes that most of us tell beginners to avoid. Recently, however, Bushnell decided that maybe it was about time that it gained a little respect in the amateur marketplace, and so it introduced three better-quality Newtonians for the more serious amateur. All three retain Bushnell's tendency to name their instruments as they would binoculars (i.e., by magnification and aperture), terminology that confuses consumers and is most often associated with Christmas Trash Telescopes (CTTs).

The Bushnell Voyager 100 × 4.5″ bears a striking resemblance to the Edmund Astroscan, an instrument that has been around for more than two decades and which itself took its basic design from amateur telescope makers who used bowling balls at the bases of their small reflectors. Outwardly, the only real difference between the Bushnell and the Astroscan is the color (blue versus red) and the tube material (metal versus plastic), but how do they compare in performance?

First, some facts. The Voyager 100 × 4.5″ is a 4.5-inch f/4.4 instrument housed within a metal sphere. It comes with two 1.25-inch eyepieces, a 27-mm Plössl and a 5-mm Symmetrical Ramsden, and a die-cast cradle mount for tabletop use (but which may also be mounted to a camera tripod). The instrument's standard rack-and-pinion focuser works smoothly and is superior to the Astroscan's roller mechanism. The tube is sealed with an optical window, helping to keep the primary and diagonal mirrors clean. Unfortunately, it is also precariously close to the front end of the tube, leaving it more susceptible to dew, damage, and finger smudges than the Astroscan's slightly recessed window. No finder is supplied with the Bushnell, but given that the supplied 27-mm eyepiece takes in an impressive 2.7°, that shouldn't be too much of a hindrance.

Unlike the Astroscan, the Bushnell has three adjustment screws that allow owners to tweak the collimation of the diagonal mirror. Neither permits users to adjust the primary mirrors, however.

Ignoring company literature, which is often emblazoned with planetary photographs, the Voyager 100 × 4.5″ is designed primarily as a wide-field instrument. This is *not* a scope for the planets, despite the fact that the 5-mm eyepiece that comes with the telescope produces a magnification of 100×. Images at that magnification are very poor and plagued with spherical aberration, making focusing nearly impossible. Why? It turns out that the primary mirror has a spherical curve, rather than a parabola, which is a must at this fast a focal ratio. Spherical aberration increases rapidly with magnification. Therefore, while low-power views of Milky Way star clouds are clear, higher-power views prove fuzzy and frustrating.

The Bushnell Voyager line also includes two larger Newtonians. The Voyager 48 × 6″ (a 6-inch f/8) and 48 × 8″ (an 8-inch f/6) both embrace the overall econo-Dob philosophy of similar instruments offered by Discovery Telescopes and Orion Telescope, both of which are discussed later in this section. That principle is to make a serviceable instrument that packs the largest aperture into a usable package but leaves off the fancy bells and whistles.

Imported from Taiwan, the Voyagers feature steel tube assemblies, which are more durable than the cardboard tubes used by some of the others but will also retain image-distorting heat longer. As with the others, their Dobsonian mounts are made from laminated particleboard. Each is outfitted with a parabolic plate-glass mirror, a rack-and-pinion focuser, a 6 × 30 finderscope, and a 25-mm Kellner eyepiece.

Unfortunately, the Voyager Dobsonians are a bit too "econo." Optically, their performance is only lukewarm, in part due to their larger-than-necessary, four-vane diagonal mirror mount, which cuts into image contrast and softens the view. Then there are the mechanics. Left-to-right movement in azimuth is very stiff. Indeed, depending on its angle, the entire telescope, mount and all, may tip. The altitude bearings are smaller than just about any other econo-Dob on the market, which leads to stiff vertical motion, as well. I have also found that some of the cradle cutouts on the mount's size boards are a little undersized or off-center, pinching the altitude bearing and impinging motion.

Celestron International. Long known for its Schmidt-Cassegrain instruments, Celestron sells several small reflectors. The Firstscope 114 EQ features a 4.5-inch f/8 spherical mirror mounted in a steel tube. The FS 114 EQ also comes with a 1.25-inch focuser and 10-mm and 20-mm Kellner eyepieces, as well as a Star Pointer red-dot unity finder, which is more useful than the 5 × 24 finderscopes that used to come on this model. Optically, the Firstscope 114 and Star Hopper 4.5 instruments are surprisingly good, but the FS's equatorial mount is rather shaky.

The Celestron Firstscope 114 Short, like the Orion ShortTube 4.5 discussed later, packs a longer focal length instrument into an 18-inch-long tube. Two metal rings hold the FS 114 Short's thin-walled metal tube to the mounting. There are slow-motion controls on both the right-ascension and the declination axes, while a battery-powered clock drive is an available option. The telescope also comes with a terribly inadequate 5 × 24 finderscope and two equally poor 1.25-inch eyepieces (a 20-mm Huygens and 6mm of unknown design). In sum, while the FS 114 Short is not a bad choice for a new stargazer, the same money will also purchase a 4.5-inch Dobsonian-mount reflector, which is both sturdier and easier to set up.

The Celestron NexStar 114GT, also marketed as the Tasco StarGuide 114CT, is another 4.5-inch f/8.8 short-tube reflector. The NexStar 114GT (which is made in China) comes on an adjustable aluminum tripod and includes 10- and 25-mm, 1.25-inch SMA (Kellner) eyepieces and the Celestron Star Pointer red-dot unity finder. Setup out of the box takes no more than 10 minutes. Simply spread the legs of the preassembled tripod and screw the one-arm mount on top.

As I mentioned earlier, I am not a huge fan of the short-tube Newtonian design, but to be fair, I conducted a head-to-head test between the Celestron NexStar 114 and a conventional 4.5-inch reflector (the Orion SkyQuest XT4.5, reviewed later). The Orion handily beat the Celestron in terms of low-power performance. Indeed, the Celestron was plagued with coma when I used it with generic 25-mm SMAs and Plössls. Interestingly, coma was far less apparent when I tried my Tele Vue 26-mm Plössl. (I should also note that coma was not a problem with any of the three eyepieces through the XT4.5.) But when it came to testing with higher magnifications, the Celestron did far better than the Orion. Jupiter and Saturn, for instance, were much sharper through it than the XT4.5, as were star images. Indeed, the star test showed that the Celestron had excellent optics. These findings back up what several owners have told me—that their Celestron short-tube reflectors produce nice views of Jupiter, Saturn, the Moon, and brighter deep-sky objects.

What makes the NexStar so intriguing is the GT suffix, which stands for GoTo. Once set up and initialized, the NexStar should take the user anywhere in the sky that he or she wants to go, simply by entering the destination into the hand controller. But alas, my test sample proved unworkable. After inputting time and date (which must be reentered each time the telescope is initialized), location, and other parameters, the telescope should slew to the first alignment star. Mine did not, instead going into what I call a *death slew*. It either tried to go past the zenith or just turned forever in azimuth. If left unattended, it would have eventually either pulled out the power cord or broken off the RS232 plug from the hand controller.

It seems that there are two problems with some of the NexStar instruments. The first is power drain. A set of eight AA batteries will be fully drained after only a night's worth of use. To save the expense of buying new batteries every clear night, use an AC adapter. The other problem is "GoTo-ability." To address this, Celestron has recently introduced a new, hopefully improved hand controller. Once these hurdles are met, the GoTo function works fairly well, but it may take some time getting to that point.

The largest equatorially mounted Newtonians in Celestron's menagerie are the 6-inch f/7 C150-HD and 8-inch f/5 G8-N, a pair of short-tubed Newtonian reflectors. Like the Firstscope 114 and NexStar 114GT, both depend on an *auxiliary correction lens* in the system to stretch their deeply curved, short-focus mirrors to longer effective focal lengths. Several owners have commented that their instruments suffer from curvature of field, astigmatism, chromatic aberration, and spherical aberration. All of these flaws are undoubtedly accentuated, if not directly caused by, the shortened optical design. That's not so good, putting a cap on the instrument's maximum magnification and, therefore, their versatility. In these two cases, a standard Newtonian would be a better, albeit more cumbersome, choice.

Discovery Telescopes. Here's a new name for an old acquaintance. Discovery began doing business back in 1991 under the name Pirate Instruments. That was a short-lived retail venture, after which the company turned its efforts to the world of wholesale, supplying Dobsonian-style Newtonian reflectors to both Orion Telescopes and Celestron. In late 1998, the tide seemed to turn again, as Bill Larsen and Terry Ostahowski decided to reenter the retail arena, this time as Discovery Telescopes. Discovery's Dobsonian-style telescopes are built right in their Oceanside, California, factory.

Optical quality is a strong point behind all of Discovery's Dobsonian-style reflectors. Their least expensive line, known as DHQ, include 6-inch f/8, 8-inch f/6, and 10-inch f/4.5 or f/5.6 apertures. Right out of the box, these instruments show that Discovery is thinking. The mirrors and their metal mountings are removed from the tubes for shipment, but they are color-keyed so that the owner can reinstall them easily and orient them as they were in the factory. As a result, collimation is either close to or right on the mark the first time.

Each DHQ telescope consists of a cardboard tube that sits on a Dobsonian mount made from laminated kitchen-countertop material. Motions in both altitude and azimuth are very smooth, with no indication of binding or slipping. One problem often encountered with Dobsonians is the issue of vertical balance. Adding a finderscope or simply changing from a lighter to a heavier eyepiece will often cause a poorly balanced telescope to shift in altitude, causing the target object to go whizzing out of the field of view. To help fix this problem, each telescope's balance point can be adjusted by sliding the mount's altitude bearings along a pair of metal tracks on either side of the tubes. Adding a finder? Slide the tube forward. Removing it later? Just slide the tube back. It's a simple but effective solution.

Discovery Premium DHQ reflectors come in four apertures—10-inch f/6, 12.5-inch f/5, 15-inch f/5, and 17.5-inch f/5—and are essentially upscale versions of the DHQ models that add both aperture and amenities to the same basic design. Like the DHQ line, all PDHQ optical components are made of Pyrex glass and housed in a spiral-wound cardboard tube painted black. Each instrument includes two Plössl eyepieces and your choice of a Telrad or 7 × 50 finderscope, as well as a nice 2-inch Crayford-style focuser. Most notably, the PDHQs are the only solid-tube Dobsonians to offer enhanced aluminizing on both mirrors, oversize altitude bearings (still on sliding tracks), and nicely crafted rocker boxes made from Appleply plywood, noted for its fine grain structure and furniture quality appearance.

The only downside to the PDHQ reflectors are their tubes' weight and girth. Even the 10-inch telescope's 64-inch-long tube assembly tilts the scales at 57 pounds, while the 84-inch-long 17.5-inch telescope weighs in at a massive 115 pounds. That is a lot of tube to carry around! If, however, you can store the telescope in a shed or barn, and simply wheel it out on some sort of cart or dolly, then Discovery PDHQ telescopes represent excellent values.

Discovery most recently added six large-aperture, Truss Design Dobsonian instruments (Figure 5.2) to their dossier. Ranging in size from a 12.5-inch f/5 to a 24-inch f/4.5, these instruments are designed to go head-to-head with the likes of StarMaster, Obsession, and Starsplitter. A unique feature of its design is that the eight truss tubes that serve as each instrument's backbone come assembled in pairs. To put everything together, slip each pair's end brackets over threaded studs on the mirror box and upper cage assembly and screw them into place with large thumbscrews. Unlike the standard-bearing Obsession instruments (discussed later in this chapter), which use captive hardware to hold the truss together, Discovery's design uses loose, black thumbscrews, which I judge less convenient.

Optically and mechanically, Discovery Truss Design telescopes are very good values, costing hundreds less than similar products. Optically, owners rave about them, giving their instruments consistently high marks during star testing. Like the PDHQ telescope, all Truss Design reflectors feature optics made of Pyrex glass. All of the primary and secondary mirrors in Discovery Truss Design telescopes feature enhanced aluminizing and overcoating for 96% reflectivity, yielding brighter images. Not surprisingly, images of deep-sky objects as well as the planets are also superb. The views I had of Jupiter and Saturn through a particular 12.5-inch Truss Design instrument that I tested for *Astronomy* magazine were especially impressive. Apochromatic refractors may be considered *planetary telescopes,* but there is no substitute for aperture on nights of steady seeing.

Figure 5.2 *Discovery Telescopes' 12.5-inch f/5 Truss Design Newtonian reflector on a Dobsonian alt-azimuth mount.*

The only optical problem I detected with that 12.5-inch was a glow or flare that seemed to invade one edge of the view, especially when I teamed it with my 12-mm Nagler Type 2 eyepiece. No matter which way I looked through the eyepiece, tilting my head back and forth, the glow persisted. I finally discovered that the problem was due to the fact that the upper cage assembly is too short. In other words, the focuser is so close to the front end of the telescope that light can shine over the edge of the assembly and right into the focuser. Placing a light shield fashioned from a piece of black poster board eliminated the problem, which was more pronounced from my suburban backyard than from a dark, rural setting. Discovery now offers an optional light shield to solve the same problem. You may not need it, but it's nice to know it is available if you do.

Finally, all Discovery truss models come with a few extras that the competition either charges for or misses altogether. One is a nylon light shroud that wraps around the truss tubes to effectively seal out stray light from infiltrating the light path. Simply wrap the shroud around the tubes and press together the long strip of Velcro. Another nice touch is the small nylon bag that is designed to fit over the diagonal mirror when the telescope is not in use to keep it from getting dusty. The Discovery also comes with a nylon carrying bag for the eight truss tubes, another handy accessory. You will probably want to purchase a pair of specially designed wheeled handles to attach to the telescope's rocker box so that the instrument can be moved around like a wheelbarrow.

Edmund Scientific. Now a division of VWR Scientific Products, Edmund's share of the astronomical marketplace had dwindled over the last quarter century, with many of their once-popular products no longer sold, instead replaced by a new, limited assortment of astronomical gear. Sad to see an icon pass.

The mainstay of Edmund's astronomical products for the past twenty-five years has been the 4.25-inch f/4 Astroscan 2001 rich-field Newtonian. The Astroscan is immediately recognizable by its unique design, which resembles a bowling ball with a cylinder growing out of one side. The primary mirror is held inside the 10-inch ball opposite the tube extension that supports the diagonal mirror and the eyepiece holder. The telescope is supported in a three-point tabletop base that may also be attached to a camera tripod.

The primary mirror, advertised as 0.125 wave at the mirror surface, yields good star images when used with the provided 28-mm RKE eyepiece. As magnification increases, however, image quality degrades. This should come as no great shock, since small deep-dish Newtonians are really not suitable for high-power applications. Although the Astroscan does not come with a finder in the classic sense, it does include a peep sight that permits easy aiming of the telescope.

Excelsior Optics. Looking for one of the best planetary telescopes on the market today? If so, Excelsior Optics might just be for you! Readers of astronomical periodicals may recognize the name of Maurizio Di Sciullo, from Coconut

Creek, Florida, for his outstanding lunar and planetary photographs, many of which are considered to be the finest ever taken by an amateur astronomer. Recently, he began to offer two 10-inch Newtonian designs commercially. For planet watchers, there is the E-258, a 10-inch f/8 long-focus reflector, while for those who prefer a more general-use instrument, the E-256 fits the bill. Both show an intelligence in their overall engineering seen in few, if any, other commercially sold telescopes today.

What makes these instruments so special? For openers, their mirrors. Each uses hand-figured optics coated with enhanced aluminizing (each with 92% reflectivity). The primary mirrors are supported in unique open cells that promote good air flow and cooling while evenly supporting the full-thickness Pyrex mirrors. The sizes of the matching secondary mirrors are kept to a minimum in order to lessen their intrusion into the light path. (Recall that any obstruction will diminish image contrast; the larger the obstruction, the greater the problem.) The E-256 uses a 1.63-inch diagonal, blocking only 17% of the aperture by diameter, while the E-258 includes a 1.25-inch diagonal, for a 12.5% obstruction. Compare that to a typical 10-inch f/5.6 reflector, which requires between a 1.9- and 2.3-inch diagonal (19% to 23% obstruction), or a 10-inch Schmidt-Cassegrain telescope with a 3.7-inch secondary mirror (37% obstruction). Each diagonal mirror is supported on an adjustable curved-vane spider mount to eliminate diffraction spikes (those starburst patterns seen around brighter stars in reflectors that use traditional three- or four-vane spiders). Finally, Excelsior's small diagonals are made possible thanks to the use of very-low-profile Crayford focusers from either JMI or Van Slyke Engineering.

Another key to good planetary views is turbulence. While we can do little to improve the stability of Earth's atmosphere above our heads, a properly designed telescope tube can help decrease internal heat currents. Excelsior uses aluminum alloy tubes for light weight and durability, but then lines them with flat-black cork to prevent any heat from the aluminum affecting the view. As I say, an intelligent design.

Helios. A familiar name to European amateurs, Helios sells several equatorially mounted Newtonian reflectors imported from the Far East. Apertures range in size from 4.5 to 8 inches.

The Helios Skyhawk-114, a 4.5-inch f/8.8 instrument, appears to be a twin to Orion Telescope's ShortTube 4.5 reflector, so I'll refer you to that section for general thoughts. One obvious difference between the two is that the Orion version comes with a wooden tripod, which is preferred over the aluminum version sold with the Skyhawk-114, though given the light weight of the instrument, performance differences are probably negligible. The Helios model includes 8- and 25-mm Plössl eyepieces (1.25-inch diameter), a poor 5 × 24 finderscope, and a mediocre 2× Barlow lens.

Speaking of standard Newtonians, Helios's 4.5-inch f/8 Explorer-114 is a nice beginner's telescope that deserves strong consideration for those who are looking for an equatorially mounted instrument that is easy to carry, set up, and use. The Explorer-114 features all-metal construction, includes a competent 1.25-inch focuser, and is mounted in rotating tube rings set atop a well-made

German equatorial mount with a built-in polar-alignment scope. The supplied 6 × 30 finderscope leaves something to be desired, as do the 10- and 25-mm Kellner eyepieces, but at least they give newcomers a jumping-off point. Images are typically quite good for the aperture, providing sharp views of brighter deep-sky objects, the Moon, and the planets. Manual slow-motion control knobs come standard, while an optional clock drive makes it possible for owners to try their luck at guided piggyback astrophotography.

The 6-inch f/5 Helios Explorer-150 Newtonian includes 10- and 25-mm Plössl eyepieces, a 2-inch focuser, a 2× Barlow lens, a 6 × 30 finderscope, and an equatorial mount with built-in polar scope. Unlike the spherical mirror found in the 114, which has a slower focal ratio, the Explorer 150 includes a parabolic primary, an absolute must for this fast a focal ratio. Unfortunately, a larger-than-expected secondary mirror holder interferes with image contrast, never allowing observers to realize the optics' full potential. On the plus side, the mount is certainly sturdy enough for prime-focus photography of the Moon, but is not really capable of long exposures through the telescope, even with the optional clock drive. Your neck will also appreciate that the tube assembly is attached to the mounting with a pair of metal rings, which allow the eyepiece to be rotated to a convenient angle.

Helios also sells the Apollo-150, a short-tube 6-inch f/6.7 Newtonian that also includes 10- and 25-mm Plössl eyepieces, a 2× Barlow lens, a 6 × 30 finderscope, and the EQ-3 mount with built-in polar scope. Like the Celestron C150-HD and others, the Apollo-150 uses a built-in *flattener/corrector* lens set just in front of the diagonal to stretch the focal length of the deeply curved primary mirror to f/6.7.

The 8-inch f/5 Helios Explorer-200 appears to be an inflated version of the Explorer-150. Like the 6-inch, the Explorer-200 includes a parabolic primary mirror, 10- and 25-mm Plössl eyepieces, a 2-inch focuser, a 2× Barlow lens, and the EQ-3 mount with built-in polar scope, plus a bigger, better 9 × 50 finderscope. While image quality is fairly good, the scope suffers from the same problem as the Explorer-150: a too-large central obstruction that reduces image contrast. The mounting is also a bit overtaxed with the weight of the Explorer-200. Helios's EQ-4 mount would be a better choice; unfortunately, it is not offered as an option.

Helios also rebadges Synta-made Dobsonians with the name Skyliner. The Helios version comes with 10- and 20-mm Super eyepieces, apparently slightly modified Modified Achromats, or Kellners, and a 9 × 50 finderscope, which helps balance things better than the standard 6 × 30. Further details are found under the Pacific Telescope discussion later in this section.

Jim's Mobile, Inc. (JMI). This company produces some of the most innovative large-aperture Newtonian reflectors and aftermarket accessories sold today. JMI's three Next Generation Telescopes (NGT, for short) Newtonian reflectors break the Dobsonian mold by combining each instrument with a split-ring equatorial mount (the same type of mount used with the 200-inch Hale reflector at Palomar Observatory). Three models currently make up the NGT line of telescopes: a 6-inch f/5, 12.5-inch f/4.5, and an 18-inch f/4.5. Sibling resemblance is unmistakable, with each instrument sporting an open-truss tube.

Image quality through the NGTs is consistently excellent thanks to their fine optics. Of those I have looked through, all have had great optics, with image resolution limited only by seeing conditions and aperture, not by mirror imperfections. Each mirror is secured in a fully adjustable cell that holds collimation very well even after the telescope has been disassembled and subsequently reassembled. The 12.5-inch and 18-inch NGTs each have a rotatable upper assembly that lets users turn the finderscope / focuser to a more comfortable viewing angle. Bear in mind, however, that this may cause collimation to wander a bit. The 6-inch NGT is constructed so that the entire optical tube assembly rotates in a permanently mounted cage that is set within its equatorial mount.

Let's look closer at each. Although sized like a beginner's instrument, the NGT-6 is aimed more toward those who have been into the hobby for a while. Its short focal length means that the overall assembly is both compact (35 inches) and lightweight (24 pounds complete), making it easy to carry and transport. Standard features include JMI's 1.25-inch RCF (Reverse Crayford Focuser) focusing mount, a 26-mm Celestron Plössl eyepiece, a rotating optical tube assembly, snap-fit truss rods, a folding rocker assembly for compact storage and transportation, and a 9-volt, dual-control clock drive and drive corrector with hand controller.

The NGT-6's optics are very good, some of the finest I have seen in a commercial 6-inch Newtonian. They yield sharp views of deep-sky objects and planets alike and are highly recommended. Indeed, planets are amazingly sharp for such a short-focus instrument. Mechanically, the NGTs split-ring mount tracks the sky just fine, though it is a little bouncy due, in all likelihood, to the light weight of the instrument. Tightening the axes will help some, but some shakiness will still remain. To assemble the NGT-6, six truss poles, made from an amalgamate of nylon and fiberglass, snap onto brass fittings. There are actually two sets of brass fittings, letting users place the eyepiece assembly either closer or farther from the primary. The closer position pushes the prime focus point up into the focuser, making it more suitable for astrophotography. The farther position is more suitable for visual use. The RCF focuser, while fairly smooth, has limited travel (range of focus), however, and as such, may not focus all eyepieces.

In order to gain access to the eyepiece when aimed at different points in the sky, the entire optical tube assembly is designed to rotate within the split-ring mount. Specifically, a thin lip along the outer edge of the metal mirror tub is captured within three small blocks of plastic, which lets the user turn the telescope left and right. Unfortunately, the NGT's tub ring is easily bent, making rotating the telescope by hand uneven and difficult.

Be advised also that the instrument rides very low to the ground, making it necessary to place the NGT-6 on a sturdy table or JMI's own tripod mount. The latter is very well designed, and is exceptionally steady.

The NGT-12.5 (Figure 5.3) is a far more stable, serviceable instrument than the 6-inch telescope. Standard accessories include a 9 × 50 finderscope, a dual-axis clock drive, a more versatile drive corrector that runs on either AC or 12-volts DC, a piggyback camera mount, and JMI's own 2-inch NGF-DX2 Crayford

Figure 5.3 *The NGT-12.5, a sophisticated 12.5-inch f/4.5 Newtonian reflector on a horseshoe-style equatorial mount. Shown with optional cloth shroud covering the open-truss tube assembly.*

focuser and NGC-miniMax digital setting circles. The optics for the NGT-12.5 come from Nova Optics, well regarded among amateur astronomers for producing quality mirrors. The latitude range of the NGT-12.5's mount, between 30° and 58° (north or south), can also be expanded with an optional latitude adjustment kit.

Finally, the NGT-18 offers everything that the NGT-12.5 includes, but adds a motorized focuser, NGC-MAX digital setting circles, built-in lifting handles (*wheeley bars*) for easier lifting and transport, a mirror cooling fan, and a quartz drive corrector with CCD auto-guider interface. Optics come from Galaxy Optics, which, like Nova, is well known for high-quality mirrors.

While JMI offers a wide variety of accessories for its telescopes, two are absolutely required, especially if you observe from light-polluted or damp environs. One is a black shroud that wraps around the telescope's skeletal structures to prevent light and wind from crossing the instrument's optical path. It also slows dewing of the optics, a problem common to all open-truss telescopes, not just the NGTs. The other must for the two larger NGTs is a tube extension that sticks out beyond the telescope's front. The extension serves to slow dewing of the diagonal mirror, which lies *very* close to the end of the nose assembly, as well as to prevent stray light from scattering into the diagonal and washing out the scene.

Konus. Headquartered in Pescantina, Italy, Konus sells a number of Newtonian reflectors that stand out from the crowd thanks to their distinctive orange tubes. Each Konus reflector is a combination of a Chinese-made tube assembly mated with Italian-made optics.

Like the company's achromatic refractors, Konus divides its reflectors into several different series. At the low end are two Konuspace reflectors, which accept 1.25-inch eyepieces and are mounted on small, wobbly German equatorial mounts. Clock drives are available, but they are only accurate enough to keep the target *somewhere* in the field of view. The Konuspace 114, a traditional 4.5-inch f/8.7 Newtonian, is adequate optically but really too heavy for its mount. The larger, even heavier 5.1-inch f/6.9 Konuspace 130 comes on the same overburdened mount and should also be avoided.

The Konusmotor 500 is a fast, 4.5-inch f/4.4 rich-field telescope that makes a nice, low-power, grab-and-go telescope, but it's not really designed for viewing the planets. The two Huygens eyepieces that come with the instrument are poor and must be replaced almost immediately. A clock drive on the right-ascension axis is included as standard equipment, though it is not terribly accurate.

If you are looking for a small, usable Konus Newtonian, consider the 4.5-inch f/7.9 Konusuper 45. The Konusuper 45 comes mounted on a sturdy German equatorial mount and includes two eyepieces and a 6 × 30 finderscope. Manual slow-motion controls are included, while a battery-powered clock drive is also available optionally. Images are very good for the aperture, making this the instrument of choice among the smaller Konus reflectors.

The 6-inch Konusuper 150 is supplied on the same German equatorial mount as the Konusuper 45, and it works quite well mechanically. Unfortunately, like the similar Celestron C150-HD, a built-in focal extender is used to stretch the instrument's focal ratio to f/7.9; in the process, it can hamper optical performance.

Finally, the newest Newtonian reflector added to the Konus lineup is the 8-inch f/5 Konusky Evolution 200. Now this is a nice telescope, with decent optics mounted in a steel tube that rides atop a heavier-duty German equatorial mount (the equivalent of the Celestron CG-5). Konus gets high marks for mating these two. The Konusky Evolution 200 comes with two 1.25-inch eyepieces, a 9 × 50 finderscope, manual slow-motion controls, and, like other Konus instruments, an extendable aluminum tripod. Bear in mind that the fast focal ratio will require frequent collimation checks. With everything aligned properly, the Konusky Evolution 200 should produce some very nice views.

LiteBox Telescopes. These ultralight, truss-tube Dobsonians hail from Hawaii, where owner Barry Peckham custom-builds instruments to customers' specifications. LiteBox instruments are available in three apertures: 12.5-inch, 15-inch, and 18-inch, each ranging in focal length from f/4.5 to f/6. The catch is that you supply the optics, purchased separately from any of a number of different companies, including Galaxy, Nova, Pegasus, Meade, and Discovery. Peckham then takes your optics and builds a telescope around them, complete with a 2-inch Crayford focuser and fully adjustable mirror cells.

As the name implies, LiteBox telescopes are designed to minimize weight while maximizing transportability. Each can break down into several lightweight pieces for easy transport in smaller cars than one might normally expect for the aperture class. Peckham accomplishes this feat by using 12-mm Baltic birch plywood for all wooden components, including the rocker box, ground board, and upper tube assembly, with each corner reinforced with corner bracing. A *virtual counterweight* (i.e., a big spring) helps to counterbalance the telescope in much the same way as Orion's SkyQuest instruments are, thereby letting Peckham shrink the overall size of the ground board. Hard to believe, but these and other techniques make an 18-inch reflector airplane transportable!

How does it work? Is the LiteBox *too* lightweight? Motions in both altitude and azimuth are smooth, thanks to the use of Teflon pads and Ebony Star laminate material. Overall construction is excellent, although they impress me as more unstable than heavier Dobsonians. Nevertheless, all of the LiteBox owners that I have spoken with report that their instruments are wonderful to use. There is no denying that if ultraportability is your desire, then a LiteBox is the only way to fly!

Mag One Instruments. Looking for something different in a Newtonian reflector? Consider a Mag One telescope. Owner Peter Smytka has created four PortaBalls, an 8-inch f/6, a 10-inch f/5, a 12.5-inch f/4.8, and a 14.5-inch f/4.3, some of the most unique telescopes to come along in years.

The PortaBall looks like the big brother of Edmund's Astroscan. Rather than copying conventional design, the PortaBall centers around a hollow sphere that houses the primary mirror. The fiberglass sphere, reinforced with a central flange, is hand sanded and finished with a smooth coat of white paint.

Mirrors for PortaBalls come from Carl Zambuto Optics, acclaimed by owners for their excellence. Since the spherical mounting completely encloses the back of the primary, a conventional mirror cell is impractical. Instead, each primary mirror is glued in its mounting with small blobs of silicone adhesive, a technique that will prove troublesome when the mirror needs to be realuminized. A pair of black anodized metal rings and a short tube segment of a composite material make up the telescope's upper tube assembly, which holds the focuser and the secondary mirror. Six (as opposed to the usual eight) aluminum truss tubes join the upper assembly to the primary-mirror ball. Each PortaBall also comes with a Rigel Systems QuikFinder unity finder as well as a unique *Helical-Crayford focuser* (1.25 inches on the 8-incher, 2 inches on the larger models), which permits both coarse and fine eyepiece adjustment.

Several options are available for each of the three PortaBalls, including electrical packages that add a battery-operated fan to help cool the primary mirror to ambient temperature, as well as antidewing elements on the QuikFinder and the secondary mirror. The optional tube baffle and cloth light shroud are must-have options if you use the PortaBall anywhere near light pollution, while a custom-fit telescope cover is good protection against dampness if you will be leaving the telescope outdoors after you retire for the evening.

Drawbacks to the PortaBall? Owners could find only a few. One is that digital setting circles cannot be adapted to the unusual mounting design. Another is the issue of balance. Although Smytka custom builds each unit to maintain its aim with its owner's heaviest eyepieces, or no eyepiece, in the focuser, very heavy eyepieces, such as some from Tele Vue, Meade, and Pentax, might throw the balance off, especially when aimed toward the horizon. Finally, the price of PortaBalls, which is higher than conventional Dobsonians, may turn some people off. But most who have purchased a PortaBall say they are well worth the extra cost.

Meade Instruments. There is no denying that Meade's introduction of computer-driven, introductory-level telescopes back in 2000 stirred up a lot of interest, both among new stargazers as well as seasoned amateurs. Digital Series (DS) refractors were discussed earlier in the chapter, but now it is time to review the line's two Newtonian reflectors. Both consist of conventional f/8 Newtonian optical tube assemblies, the DS114-EC with a 4.5-inch aperture, while the DS127-EC measures 5.1 inches across. Both also include 5 × 24 finderscopes and a pair of eyepieces (25-mm Kellner and 9-mm Huygens). The optical performance of both reflectors is also quite good, with sharp images and little evidence of spherical aberration, astigmatism, or other flaws. But like the DS refractors, this performance is hindered by their shaky mounts. Owners would also be well advised to replace the latter eyepiece with one of similar focal length but better design (such as a Plössl).

The real appeal to the DS reflectors is the Autostar, the optional hand controller that plugs into the telescope mount's control panel. With it substituted for the standard hand controller that comes with each telescope, the Autostar will steer the instrument to any of more than 1,500 sky objects programmed into its database. Follow the directions on the Autostar's digital display to input the time, the date, and the location, let it slew to two alignment stars, and it's ready to go. Simply select an object, press go, and the telescope automatically moves to that object's calculated location. Observers can select objects themselves or put the telescope in the tour mode and let it decide what to look at. As mentioned in the "Achromatic Refractor" section earlier in this chapter, I found the Autostar's pointing accuracy in a DS90-EC refractor to be quite good.

Meade's newer 4.5-inch f/8.8 DS2114-ATS is the company's first short-tube reflector and appears to be the long-lost twin of their competitor Celestron's NexStar 114GT, reviewed earlier. Like the Celestron, the DS2114-ATS is held on a one-armed altitude-azimuth mount atop an aluminum tripod. You would be hard pressed to tell them apart, even from close up. But the resemblance goes even deeper. Like the Celestron, Meade uses an internal lens to extend the focal length of the telescope to f/8.8, which I cautioned against earlier. These lenses tend to introduce optical problems that are not seen in conventional Newtonians. My advice is to tread cautiously when considering any reflector using this design. The DS2114-ATS also comes with two eyepieces, a small finderscope, and the Autostar hand controller. A big plus to the new DS-2000 design is that the drive motors and associated cables are now internalized, making for a sleeker, neater package that is much easier to set up than the original Digital Series (DS) instruments.

Meade also offers two larger Newtonian reflectors called Starfinders. The 12.5-inch f/4.8 Starfinder comes on a Dobsonian-style alt-azimuth mount, while the larger 16-inch f/4.5 Starfinder can be purchased either on a Dobsonian mount or on a clock-driven German equatorial mount. All versions include consistently excellent Pyrex mirrors, which have always been the Starfinders' strongest suit. All that I have viewed through have delivered excellent images, a sentiment echoed by most owners. Whether viewing the planets or deep-sky objects, targets always appear tack sharp and have very high image contrast.

Neither of these telescopes should be considered a small investment in terms of price and, especially, mass. These are big telescopes. Both have tubes made of spiral-wound cardboard that are much more cumbersome to carry and set up than pricier truss-tube Newtonians. The solid tubes also confound mirror cooling, a problem further accentuated by mirror cells fashioned from fiberboard that block air from reaching the mirror from behind. While good protection against dust, this design will slow the optics from reaching the ambient outdoor temperature, possibly for hours.

The Starfinder Dobsonian mounts are also made of fiberboard, covered with a plastic countertop-style laminate. For movement in azimuth, Teflon pads slide against smooth plastic laminate, while in altitude, a pair of nylon pads ride against each hard-plastic trunnion. Both up-and-down and left-to-right motions are smooth, with little binding. Each comes with a 26-mm Super Plössl eyepiece, a far-too-small 5 × 24 finderscope, and a plastic 1.25-inch focuser. A deluxe version of each is available that adds a 9.7-mm Super Plössl eyepiece and substitutes a plastic 2-inch focuser and a 6 × 30 finderscope. The upgrade is worth the extra cost, in my humble opinion.

The 16-inch equatorial Starfinder comes strapped to a German equatorial mount that rides low to the ground. The straps make it possible to rotate the tube for easier access to the eyepiece, which can otherwise end up in some pretty odd angles. They also hold the tube more securely than if it were bolted directly to the mount, as it was in older Starfinder 16s. But the mount impresses me as undersized for the telescope; indeed, I've seen a few that wobbled badly. The Starfinder's DC-powered worm-gear drive runs on six AA-size batteries and is passable for visual use as well as some simple astrophotography.

Murnaghan Instruments. Although their primary products involve accessories for CCD imagers, Murnaghan Instruments entered the telescope business when they bought rights to the name of the Coulter Optical Company in the mid-1990s from the estate of James Braginton, Coulter's original owner. The original Coulter began to produce primary and secondary mirrors back in the 1970s, but the company really made its mark in 1980 with the introduction of its 13.1-inch Odyssey I, the first commercially sold Dobsonian-style reflector. When Murnaghan took over the company name, it picked up pretty much where the original left off, with heavy red optical tube assemblies atop black bases. Perhaps due in part to quality and delivery problems, as well as ever-increasing competition from Far Eastern sources, Murnaghan suspended

Odyssey production in 2001, deciding instead to concentrate on its own line of imported, low-end reflectors and refractors under a different name, eScopes.

The ValueScope 4.5 (a 4.5-inch f/8) and Astron 6 (a 6-inch f/5) make up the current eScopes line of Newtonians. The 6-inch model features a parabolic primary mirror, while the 4.5-inch is spherical. Both include 6 × 30 finderscopes and come fitted into their respective mounts with rotating rings to help position the focusers at more comfortable angles. Beware the 4.5-inch, which includes a German-style mount that is little better than CTTs. The larger mounting supplied with the 6-inch, however, is strong enough to hold the telescope's weight.

Murnaghan gives some hint that the Coulter name might be resurrected once again, though it offers no timetable or other expectations. Perhaps the name will already have been released by the time you read this. If so, I would approach them with caution, given the problems of the past.

Novosibirsk Instruments. The name literally translates as "New Siberia," but Novosibirsk is a Russian manufacturing consortium that makes TAL astronomical telescopes, as well as some decidedly nonastronomical optical products. TAL Newtonian reflectors range in aperture from the TAL-ALKOR and TAL-M, at 2.5 and 3 inches, respectively, to the 6-inch TAL-2 and TAL-2M. All come very well outfitted.

The 3.1-inch f/6.6 TAL-M tilts the scales at a significant 26 pounds. Much of that weight comes from the telescope's metal tube, equatorial mount, pedestal, and feet. The TAL-M delivers good optical performance considering its size, although one respondent received one with a pinched primary mirror, causing some image distortion at first. The problem, similar to that seen in some Orion SkyQuest XT Dobsonians, was resolved by removing the mirror and cell and then loosening the three clips that hold the mirror in the cell just a bit. Other mechanical components perform very well, with smooth motions and smooth focusing.

One of the most unique traits of this instrument is its unusual finderscope, which is built into the helical focusing mount, with the same eyepiece serving for both it and the telescope. To use the finder, start with the barrel of the finderscope all the way in. This causes the finder's star diagonal to intersect the eyepiece field. To switch the view to the telescope itself, pull the finderscope's barrel out toward the front of the tube. This will move the finder's diagonal out of the way, aligning the eyepiece with the telescope's secondary mirror. An interesting design, but like all right-angle finderscopes, it takes a little getting used to (see chapter 7 for further thoughts).

Larger TAL reflectors include the 4.3-inch f/7.3 TAL-1, the 4.7-inch f/6.7 TAL-120, the 6-inch f/5 TAL-150, and the 6-inch f/8 TAL-2. Each features all-metal construction and comes on a German equatorial mount atop a steel pedestal, reminiscent of Newtonians by Criterion, Optical Craftsmen, and Cave Optical Company, all renowned for their telescopes in the 1960s. Indeed, in this day and age of electronic this and computerized that, the TAL foursome is a refreshing trip back to basics. The equatorial mounts used for the 4.3- and 4.7-inch models are one and the same, while the 6-inchers use a somewhat beefier version. All are well designed, and sturdy enough to support the weight of their instruments.

The four TALs have rotatable tubes held onto their mounts with pairs of hinged metal rings, which lets the observer turn the tube around so that the eyepiece is set at a convenient viewing angle. Standard equipment includes two 1.25-inch eyepieces (10-mm and 25-mm Plössls come with both 6-inch models, while the others include a 25-mm Plössl and a 15-mm Kellner), as well as a 2× Barlow lens, six color filters, a 6 × 30 finderscope, and a heavy wooden case (too heavy to be useful for transport unless you are also a weightlifter!). One of the most unique accessories included with TAL instruments is the solar projection screen that screws onto the end of the declination shaft. Once the tube is rotated so that the Sun's image is projected onto the screen, the screen will track with the telescope as the Sun moves across the sky. A simple idea, but clever in that no other instruments have one.

Optical quality of the first TAL reflectors seemed rather variable. Some were very good, while others were reportedly quite poor. Happily, Novosibirsk seems to have evened things out nicely, with consistently good optics now coming out of New Siberia. The 6-inch f/5 TAL-150 features a parabolic mirror, a must for that fast a focal ratio, while the other, slower TALs have spherical mirrors. A parabolic mirror is certainly preferable, yet owners report good image quality nonetheless. The biggest problem with a spherical mirror—spherical aberration—will be a factor, but usually only at high magnifications. If you stay under, say, 200×, you probably won't notice any image degradation.

Indeed, if you are looking for a solidly made telescope in this size range, and don't mind spending a little more than you might on an econo-Dob, then a TAL reflector is a very good choice indeed.

Obsession Telescopes. These are among the finest large-aperture, alt-azimuth Newtonians sold today. They are famous for their sharp optics, clever design, fine workmanship, and ease of assembly and use. All of these attributes add up to a winning combination.

Five models make up the Obsession telescope line: 15-, 18-, 20-, 25-, and 30-inchers! Mirrors are supplied by Nova Optics, Galaxy Optics, or other suppliers, if desired. The secondary mirrors feature a nonaluminum dielectric coating that the manufacturer states has 99% reflectivity. The primaries come with either standard aluminizing (89% reflectivity) or, for an extra charge, enhanced aluminizing (96% reflectivity).

The overall design of the Obsessions (Figure 5.4) makes them some of the most user-friendly large-aperture light buckets around. Their open-truss tube design allows the scopes to break down for easy transport to and from dark-sky sites, but keep in mind that we're still talking about BIG telescopes that are much more unwieldy than smaller instruments. Once at the site, the 15-, 18-, and 20-inch Obsessions can be set up by one person in about 10 to 15 minutes without any tools; the bigger scopes really require two people but are still quick to assemble. First, insert the eight truss poles into their wooden blocks mounted on the outside of the mirror box, then place the upper cage assembly on top. Pass four tethered pins through mating holes and latch down on the

Figure 5.4 *The Obsession-15, an outstanding 15-inch f/4.5 Newtonian reflector on a state-of-the-art Dobsonian-style alt-azimuth mount.*

quick-release skewers. That's all there is to it. This no-tool design is especially appreciated by those of us who know what it is like to drive to a remote site, only to find out that the telescope cannot be set up without a certain tool inadvertently left at home. Even nicer is the fact that all hardware remains attached to the telescope, so it is impossible to misplace anything. The only way to lose Obsession hardware is to lose the telescope!

Another plus is the Obsession's low-profile DX-3 Crayford-design focusing mount supplied by JMI. The DX-3 is one of the finest focusers on the market today. And since the eyepiece is kept closer to the tube, a smaller secondary mirror can be used to keep the size of the diagonal (and the central obstruction) to a minimum.

Many conveniences are included with Obsessions. For instance, a small 12-volt DC pancake fan is built into the primary mirror mount to help the optics reach thermal equilibrium with the outside air more quickly. Another advantage is that each instrument comes with a pair of removable metal wheelbarrow-style handles attached to a set of rubber wheels (except the 15-incher, though wheelbarrow handles are available optionally). The handles quickly attach to either side of the telescope's rocker box, letting the observer roll the scope around like a wheelbarrow. Of course, these handles do not solve the problem of getting the scope in and out of a car. If you own a van, a hatchback, or a station wagon with a low rear tailgate, the scope may be rolled out using a makeshift ramp; but if it has to be lifted, two people will be required.

It should come as no surprise, given the aperture of these monsters, that their eyepieces ride high off the ground. When aimed toward the zenith, the eyepiece of the 20-inch scope, for instance, is at an altitude of about 95 inches. Unless you are as tall as a basketball player, you will probably have to climb a stepladder to enjoy the view. A tall ladder is a must for looking through the bigger Obsessions; the eyepiece of the 30-inch scope (when aimed at the zenith) is a towering 11 feet off the ground. Like all open tubes, a cloth shroud is recommended when the telescope is used in light-polluted locations; Obsession offers optional custom-fit black shrouds for all of their instruments.

Nothing is small about the Obsession scopes. Besides large apertures, considerable weight, and substantial girth, their cost is quite heavy. Yes, they are expensive, but after all, they are a lot of telescope—and they are still the standard by which all other Dobsonian-style Newtonians are judged. For the observer who enjoys looking at the beauty and intrigue of the universe and has ready access to a dark sky, using Obsession telescopes can really become obsessive.

For those who thrive on computerized telescopes, Obsession offers their optional Nightrider Track-n-Go system. Manufactured by Jim's Mobile, Inc. (JMI), the Nightrider marries a Vixen SkySensor 2000 digital setting circle system to a system of wheels, motors, and rollers that the owner installs onto the telescope's Dobsonian base. Not an inexpensive system, but for those who have longed for this sort of an option, the Nightrider is the answer to your prayers. (Note that StarMaster, Obsession's primary competitor, offers a similar system for their instruments.)

Orion Optics (UK). Not to be confused with Orion Telescope Center of California, Orion Optics of Great Britain sells several different Newtonian reflectors for amateur astronomers. All include metal tubes, 1.25-inch rack-and-pinion focusing mounts, traditional aluminum mirror mounts, and German equatorial mounts. Each telescope tube is secured to its mounting with a pair of well-made metal rings that are intended to allow the user to rotate the tube until the eyepiece is in a comfortable position. The primary and secondary mirrors are all made in-house by Orion, with most respondents commenting on the optics' high quality.

Least expensive of Orion Optics telescopes are those in its Europa line, which are available in 4.5-inch f/8 (Europa 110), 6-inch f/5 or f/8 (Europa 150), 8-inch f/4 or f/6 (Europa 200), or 10-inch f/4.8 (Europa 250) apertures. These instruments are created by the intercontinental marriage of British-made optics (except for the 4.5-inch, which uses imported optics) with Taiwanese equatorial mounts and adjustable aluminum tripods, the latter being similar to the U.S. Orion's SkyView Deluxe mount. Each instrument comes with a 6 × 30 straight-through finderscope and two 1.25-inch eyepieces, while a 10 × 50 finderscope and a 12-volt DC clock drive are available optionally. Two-inch rack-and-pinion focusing mounts are also offered at an additional charge on the 6-, 8-, and 10-inch Europas. Owners consistently report that the Europa telescopes operate very well optically, apart from some spherical aberration with the 4.5-inch. Mechanically, the mounting is really only adequate for the 4.5- and 6-inch instruments; it is too weak and shaky for the extra weight of the larger apertures.

Though more expensive, Orion Optics' GX line of Newtonian reflectors come mounted on Vixen's far sturdier Great Polaris (GP) German equatorial mount atop a steel pedestal. Consumers may choose from three apertures: a 6-inch f/5 or f/8 (GX150), an 8-inch f/6 (GX200), and a 10-inch f/4.8 (GX250). The GP mount offers a good degree of stability for visual observations and perhaps some rudimentary astrophotography with the addition of dual-axis stepper motors. If long-exposure astrophotography is planned, however, it would be best to upgrade to Vixen's heavier duty GP-DX mount at the time of purchase. Either Vixen mount also offers the capability of computer control. All optical and mechanical components, apart from the Japanese mounting, are manufactured in Orion's shop in Crewe. Each GX telescope also comes with a 6 × 30 finderscope and a 1.25-inch focuser, though upgrades to 10 × 50 finderscopes and 2-inch focusers are also available.

Orion Optics' largest instrument is the 12-inch f/5.3 DX300 Newtonian reflector. Like the others, the DX300 includes a metal tube and rotating rings, but it also comes with an upgraded 2-inch rack-and-pinion focusing mount for larger eyepieces and a 10 × 50 right-angle finderscope. The supplied Vixen GP-DX mount includes manual slow-motion controls and is designed to accept an optional dual-axis, stepper-motor drive system. Overall, owners give the mounting very high marks for stability. The DX300, however, is a heavy instrument which is best left in place rather than transported to distant observing sites.

Orion Telescope Center. This company offers several Newtonian reflectors with the Orion name. Most popular of all are the Orion SkyQuest XT4.5, XT6, XT8, and XT10 Dobsonians. All four get rave reviews from most owners.

Each telescope is built around a plate-glass mirror that is coated with standard aluminizing. Many amateurs tend to look down upon plate-glass mirrors, condemning them as taking too long to adjust to changes in temperature (such as experienced when taking a telescope outside from a warm house). While it is true that Pyrex has better thermal properties, the plate-glass mirrors in the XTs seem to adjust very quickly (certainly within 30 to 60 minutes, depending on aperture).

Other mechanical attributes include rack-and-pinion focusing mounts that slide smoothly and evenly thanks to thin internal Teflon pads. The 4.5- and 6-inch models come with 1.25-inch focusers, while the 8- and 10-inchers include 2-inch focusers (1.25-inch adapters are included). Each instrument's diagonal mirror is mounted on an adjustable four-vane spider mount. Adjustment requires a small Phillips screwdriver, which proves more inconvenient than a tool-less design. Several owners have mentioned some frustration at the initial collimation of their XTs, resulting in poor initial images. Improper adjustment of the diagonal was often the culprit.

Speaking of optics, the mirrors themselves are consistently good. Once everything is properly collimated, images have been acceptable through all that I have looked through personally. The XT4.5, with its spherical mirror, works best at magnifications below 100×, while the others, with their parabolic mirrors, can handle higher magnifications more readily.

All four XT telescopes include black steel tubes riding atop black, laminate-covered particleboard. If you are out under truly dark skies, the telescope will literally disappear, save for the white Orion lettering. Concern over metal tubes retaining heat has been raised by some, but again, acclimatization seems no worse than with similar, cardboard-tubed telescopes. Motions in both altitude and azimuth, while not as silky smooth as on premium Dobsonians, are certainly fine for most observers.

What about the "CorrecTension Friction Optimization" feature? Pretty fancy language for a couple of springs, but it does work. The springs serve a dual purpose. Their primary purpose is to exert enough downward force to press the side bearings into the ground board, increasing the friction between the two. They also keep the telescope's tube and ground board together, letting users carry the two together. That's handy with the 4.5- and 6-inch XTs, though some will find the XT8 too heavy to move as a single unit. The 58-pound XT10 is best moved in stages.

As far as accessories go, two eyepieces, a 10- and a 25-mm Plössl, are included with each of the four instruments. While not up to the quality of Tele Vue Plössls, these provide a good starting point for new telescope owners. Each XT also comes with a finderscope. The 4.5-, 6-, and 8-inch SkyQuests include 6× finders, while the 10-inch includes a much better 8 × 50 unit.

So how do they compare against similarly priced instruments? All are competent designs, and most amateurs would be happy with any of them. One of the biggest pluses of the larger Meade Dobsonians is that their optics are made of Pyrex, which is preferred over plate glass for its lower coefficient of thermal expansion. Another often-overlooked point in favor of Pyrex is that it is stronger and less prone to cracking or fracturing than plate glass, a consideration if many people will be handling the instrument. Having said that, most owners comment on the high quality of their XT's optics, material notwithstanding. Let's make it clear; there is no reason at all why a plate-glass mirror cannot be of the highest quality. Orion Skyquest XTs and the Skywatcher instruments are about equal in overall terms, but I would have to say that Discovery Dobsonians have the best overall finish and construction of the lot. I also strongly recommend Stargazer Steve's excellent reflectors.

Orion's ShortTube 4.5, a 4.5-inch reflector, has an effective focal length of f/8.8, but it is folded into tubes about half as long as standard telescopes. Like Celestron's Firstscope 114 Short and NexStar 114GT, among others, the effective focal length is stretched by permanently installing a Barlow/corrector lens in the base of the focusing mount, allowing a compact package to house a longer-focal-length system. Review the discussion at the beginning of this section for the pluses and minuses of this design. This particular instrument is about on par with the Celestron FS114 Short, using a very similar German equatorial mount and featuring comparable eyepieces and finderscope.

Orion offers two 5.1-inch Newtonians. The f/6.9 SpaceProbe 130 includes a 6 × 30 finderscope, as well as 10- and 25-mm, 1.25-inch Kellner eyepieces. Its spherical primary mirror is acceptable at lower magnifications, but begins to weaken at magnifications beyond about 100×. At the same time, the benefit of the increased aperture over a 4.5-inch telescope is outweighed by the instrument's weak German equatorial mount. Rather than this instrument, consider the Orion XT4.5 or XT6 Dobsonians or the Orion SkyView Deluxe 6.

The SpaceProbe 130ST is more than a short-tube version of the regular SpaceProbe. Unlike many short-tubed Newtonians, which use a built-in corrector lens to stretch their focal lengths, the 130ST is a true rich-field telescope (RFT), its deep-dish primary curved to an f/5 focal ratio. As such, this instrument is ideal for wide views of extended objects, such as the Orion Nebula, Andromeda Galaxy, and the Pleiades. The parabolic primary mirror performs well, but *must* be accurately collimated to function at its best. Like many other Orion instruments, the 130ST is supplied with 10- and 25-mm Plössl eyepieces, a 6 × 30 finderscope, and the same light-duty equatorial mount as the longer model. In this case, however, the mounting is suitably strong for most uses, since the telescope weighs so little.

Finally, Orion offers the 6-inch f/5 SkyView Deluxe 6 and 8-inch f/4 SkyView Deluxe 8. Optics are consistently quite good in the SVD 6, with owners reporting fine views of Messier objects and brighter members of the NGC. The optics in the 8-inch, however, are often more prone to performance problems. Both come with parabolic, plate-glass primary mirrors. Their elliptical diagonal mirrors are held in adjustable, three-vane spider mounts.

Each instrument includes 9- and 25-mm Plössl eyepieces, a 6 × 30 finderscope, a Moon filter, and a smooth 1.25-inch, rack-and-pinion focuser. Since equatorial mounts can twist a telescope around at some pretty odd angles, the SVDs' tubes are held to their mounts with pairs of metal rings that are hinged so that the tubes can be rotated. The mount itself is suitably sturdy for the 6-inch, but very wobbly under the weight of the heavier 8.

Pacific Telescope Company. Based in British Columbia, Pacific Telescope Company teams with Synta Optical Technology Corporation to offer a variety of imported reflectors, as well as many refractors, which were previously listed.

Three 4.5-inch Skywatcher Newtonians belong to the Skywatcher fleet. The Skywatcher 1145 is a compact, f/4.4 rich-field telescope that makes a great grab-and-go instrument for those on-the-run observing sessions. Supplied with a small German equatorial mount and aluminum tripod, the 1145 is best suited for low-power viewing of broad star clouds and the like, as opposed to planetary studies. While images are quite good, be aware that instruments of this sort must be collimated accurately. It should also be noted that the early 1145s came with spherical mirrors and, as a result, spherical aberration. More recently, Synta began to use paraboloidal mirrors in the 1145, which is what you want.

Outwardly, the Skywatcher 1141 looks identical to the 1145, but is labeled as a Catadioptric-Newtonian, since it incorporates a correcting lens in the focuser to stretch the focal length to f/9. Assuming everything is in proper collimation, images are reasonably good, though still not as good as a conventional Newtonian. One problem is that the corrector lens introduces chromatic aberration, something never seen in Newtonians. The 1141 rides on the EzUse mount, which is too flimsy for high-powered viewing.

In my opinion, if you are looking for a general purpose 4.5-inch reflector, the f/8 Skywatcher 1149 is a better choice than the 1141. To support the longer, heavier tube, the 1149 is coupled to the slightly beefier SimPlex equatorial mount, which is satisfactory but not terribly noteworthy. Optics are fairly decent, although spherical aberration begins to blur the view above about 100×.

Unfortunately, all three of the 4.5-inch instruments only come with poor 5 × 24 finderscopes. While not a terrible drawback with the 1145 because of its inherently wide field, it is debilitating with the others.

Skywatcher also includes a trio of 5.1-inch Newtonians that mirror the variety of their 4.5-inch models. Of these, the Skywatcher 13065P is the most noteworthy. Featuring an f/5 paraboloidal primary, the 13065P is a nice, compact instrument for quick, wide-field viewing with minimal aberrant intrusion. All mechanical parts are well made, with a smooth 1.25-inch focuser, a nicely designed diagonal mirror mount, a spring-loaded finderscope mount for quick adjustment, and an easy-to-adjust primary mirror cell. The f/7.7 Skywatcher 1301, a short-tube Catadioptric-Newtonian, uses an auxiliary lens permanently mounted in the focuser to increase the instrument's effective focal length to 39 inches (1,000 mm). But as with the 1141 above, spherical aberration and field curvature are both introduced in the process. Finally, the Skywatcher 1309 is a conventional f/7 Newtonian that relies on a spherical mirror to focus images into the eyepiece. As a result, users can expect a realistic magnification limit of about 100× before spherical aberration degrades the view. All three of these instruments come mounted on the SimPlex German equatorial mount and an aluminum tripod. Though adequate for viewing, the combination is not sturdy enough for any serious attempts at astrophotography.

Synta makes and Pacific Instruments markets three 6-inch models, as well. Now, I've always had a soft spot for RFT instruments. They can produce some amazing views of the night sky, especially star clusters and nebulae. From dark skies, the views are just wondrous. That's why the Skywatcher 15075P has a special appeal. With its f/5 parabolic primary mirror, the 15075P has enough light-gathering ability to show all of the Messier objects as well as many NGC members, including such challenging sights as the Rosette Nebula, the North America Nebula, and the California Nebula (assuming properly collimated optics, dark skies, a high-quality eyepiece, and a narrowband or oxygen-III nebula filter; see chapter 7). A 25-mm Plössl eyepiece will produce 30× and a field wide enough to encompass nearly 2° of sky (assuming an eyepiece with a 50° apparent field of view; see chapter 6). That's more than enough to take in the entire Pleiades cluster with room to spare. Like the 13065P above, the 15075P includes nicely designed mechanics.

Built around a parabolic primary mirror, the Skywatcher 2001P is an impressive 8-inch f/5 Newtonian reflector that features a 2-inch focusing mount, a 9 × 50 finderscope, and well-designed primary and secondary mirror cells that promote rapid optical cooling. Supporting it all is the StediVue German equatorial mount (a clone of Celestron's CG-5) and an adjustable aluminum tripod. Once again, the tripod is the weak link in the telescope's entire system. Wooden legs are available from third-party companies, such as

Canada's NatureWatch, and go a long way to steadying the instrument. The Skywatcher 2001P yields some very nice images of deep-sky objects as well as acceptable views of the planets. All in all, this is a competent telescope that could give owners years of viewing pleasure.

Pacific Instruments also sells three Skywatcher Dobsonian-mounted Newtonian reflectors from Synta: a 6-inch f/6.7, an 8-inch f/6, and a 10-inch f/4.8. All Skywatcher Dobsonians feature 2-inch rack-and-pinion focusers that are threaded to accept a camera's T-adapter for photography, 6 × 30 finderscopes, and 10- and 25-mm Plössl eyepieces. As part of my research for this edition of *Star Ware*, I anonymously purchased and tested an 8-inch Skywatcher Dobsonian back in 2001 and found it to be a very good product. Mechanically, the mounting moved a little stiffly in azimuth until I replaced the three plastic pads on the ground board with furniture slides (available at hardware or home supply stores). The telescope also proved to be a little tail heavy. As I aimed toward a sky object, the telescope wanted to swing toward the zenith. Balance is very close, however, as I found that simply hanging my car keys from the stem of the finderscope was enough to tilt the scales in my favor.

Optically, the 8-inch worked well once everything was collimated. Mine arrived with the secondary mirror seriously out of alignment, but thankfully, the spider mount was easily adjusted. Unfortunately, my instrument was supplied without any instructions, but collimation was still easy to adjust by following the instructions in chapter 9. Once cooled to ambient temperature and properly collimated, optical performance is excellent. I surveyed Jupiter's Red Spot, Saturn's Cassini's Division, and a trove of deep-sky objects, all with good results. Star tests revealed some slight spherical aberration, but certainly within acceptable limits for an introductory telescope.

Parallax Instruments. This is one of few companies around to offer large-aperture, longer-focal-length Newtonian reflectors. Five high-quality telescopes highlight their line, ranging in size from the 8-inch f/7.5 PI200 to the 16-inch f/6 PI400. Each instrument boasts a diffraction-limited Pyrex primary mirror coated with enhanced aluminum. Overall manufacturing quality is excellent. In addition, all Parallax telescopes include 2.7-inch, low-profile eyepiece focusing mounts (with adapters for 1.25-inch and 2-inch eyepieces); aluminum tubes; and 8 × 50 finders. Their low-profile focusers combined with higher focal ratios mean that Parallax telescopes can use small diagonal mirrors, creating the smallest central obstruction of any reflector on the market. This is a *big* plus for those primarily interested in viewing the planets. The only extras needed are a couple of eyepieces and an adequate mounting, supplied either by Parallax or another company (refer to the section titled "Mounting Concerns" later in this chapter).

Parks Optical. Both optically and mechanically, Parks Newtonians are among the best in the business, although their cost and weight may be a bit heavy for many hobbyists to bear. Parks Newtonians come in apertures ranging from 4.5 to 16 inches and in no fewer than five different versions, or systems, as Parks describes them. All optical assemblies feature fiberglass tubes held in rotating rings (an important convenience feature) and poised on German equatorial mounts (except for the Companion, which is mounted on a tripod).

The 4.5-inch f/5 Companion is Parks's smallest reflector. As the name implies, this little instrument is the perfect travel partner, whether to a local star party or halfway around the world. The views through the Companion are simply awe inspiring. Though designed with wide-field views of Milky Way star clouds in mind, the high quality, parabolic primary mirror will even give some good views of the brighter planets, as well. Measuring only 22 inches long, the Companion comes with a 25-mm Kellner eyepiece and, a 6 × 30 finderscope; it is mounted on a heavy-duty photographic tripod. But while the Companion is wonderful to view through, the need to lock and unlock the tripod's altitude and azimuth axes can get a little wearing. The price can be a little wearing, as well, with the Companion costing more than many 8-inch Dobsonian instruments.

The high quality of the Companion carries through to all Parks instruments, both large and small. The least-expensive equatorially mounted Parks telescopes, the 6- and 8-inch Precision and Astrolight Newtonians, include all-metal mirror cells and 1.25-inch rack-and-pinion focusers, both throwbacks to an earlier era but as functional today as they were when introduced. Standard accessories with each instrument include a lone 25-mm Kellner eyepiece and a 6 × 30 finderscope. Most will want to upgrade both quite quickly. Parks Precision reflectors come on very basic German equatorial mounts on steel pedestals, while the Astrolight models are supported by somewhat more sophisticated mounts, similar in general quality to those imported by Celestron and others. Clock drives are available separately for each. Both mounts are adequate for visual observations, but neither is sturdy enough for long-exposure astrophotography through the telescope.

If wide-field views of deep-sky objects are your forte, then the Parks Nitelight series is especially noteworthy. Nitelights range from 6 to 16 inches in aperture, all with an amazing f/3.5 focal ratio. This makes them the fastest telescopes sold, greatly reducing exposure times required to record faint-sky objects. The 6- and 8-inch Astrolight Nitelights are available on undriven, tripod-mounted equatorial mounts, while the Superior Nitelight models, including 8-, 10-, 12.5-, and 16-inchers, are supplied with very sturdy (though heavy) clock-driven German equatorial mounts. Astrolights also come with 6 × 30 finderscopes, while the Superior models include 8 × 50 finders. All include one or two eyepieces, depending on the model.

At the opposite end of the cost and weight spectrum are Parks's massive 8-, 10-, and 12.5-inch Superior telescopes, as well as 12.5- and 16-inch Observatory instruments. While outstanding in their quality, these heavy scopes are only recommended for amateurs who do not need to travel to dark-sky sites—or who enjoy weight lifting! For instance, the 8-inch f/6 Superior totals 183 pounds, with the mount alone contributing 125 pounds. If you think that's heavy, try the 16-inch Observatory model, which weighs in at 755 pounds! No one ever accused Parks telescopes of being lightweights in any sense of the word.

See also the Parks H.I.T. Series of convertible Newtonian/Cassegrain telescopes, reviewed later in this chapter.

Sky Valley Scopes. Based in Snohomish, Washington, Sky Valley Scopes makes custom-built, large-aperture Newtonian/Dobsonian reflectors. Sky Valley owner Ken Ward has worked in the auto-body, custom-painting, and boat-building industries for more than 30 years. He has applied techniques and materials from all three divergent fields to create some of the most colorful telescopes sold today. All are made from a combination of hardwood, honeycomb-core fiberglass, fiberglass cloth, and carbon fiber. While famous for its rigidity, the stiffness of fiberglass alone may be more prone to transmitting vibrations than, say, wood, which has excellent damping properties. Ward's use of honeycomb-core fiberglass sandwiched with fiberglass cloth should help dampen vibrations.

Sky Valley telescopes come in apertures from 12 to 18 inches. All are open-truss Dobsonians available in two varieties: less-expensive Ultra Lights and costlier Rotating Tubes. The mirror box of the Ultra Lights is similar to other, plywood-based Dobsonians, except that it uses honeycomb and fiberglass materials. All attachment points are reinforced to prevent cracking or other stress-related problems. The more advanced Rotating Tube models enable the observer to turn the entire tube assembly within the lower mounting box to bring the eyepiece to a more comfortable position—a great idea. Ward describes the assembly: "The lower tube fits in a ring assembly made of fiberglass/epoxy laminate with eight bearings on the inside to support the rotation of the optical assembly. A frame made of aluminum bar stock crosses underneath the center of the lower tube to provide support for the central ball bearing that the tube pivots on and supports the weight of the optical tube assembly."

The upper assemblies that hold the diagonal mirrors and focusing mounts are also made from honeycomb-core fiberglass/epoxy and fiberglass cloth. The eight aluminum truss poles that attach the upper assembly to the mirror box are secured with plastic thumbscrews, a no-tools plus. Unfortunately, the hardware is not held captive to the telescope, making it easy to drop one or more items on the ground (or onto the mirror, if left uncovered) during assembly. As might be expected, the Teflon and Ebony Star bearings offer very smooth motions.

Sky Valley Scopes are not inexpensive, but few custom-built things in life are. The workmanship, however, is outstanding. Indeed, Ward won first place for workmanship in the telescope competition at the 1994 Table Mountain Star Party, which led him to start Sky Valley Scopes. Each telescope takes him no less than 150 hours to complete, not including curing and drying time. If you are looking for a one-of-a-kind instrument, then Sky Valley Scopes might be your choice.

Stargazer Steve. Many of us who are children of the 1960s and 1970s might well have owned a 3-inch f/10 Newtonian reflector manufactured by Edmund Scientific. It seemed to be the quintessential first telescope of the era. Although the Edmund 3-inch f/10 is no longer made, Steve "Stargazer Steve" Dodson, an award-winning amateur telescope maker from Sudbury, Ontario, has resurrected, enlarged, and improved the design to produce the 4.25-inch f/7.1 SGR-4. The SGR-4, seen in Figure 5.5, is an excellent first telescope for young

Figure 5.5 *The 4.25-inch f/7.1 Stargazer Steve SGR-4 Newtonian reflector, one of the best telescopes for the next generation of stargazers. Photo courtesy of Stargazer Steve.*

astronomers. Images of the Moon, the planets, and the brighter sky objects are sharp and clear, thanks to its parabolic primary mirror. Many other 4.5-inch reflectors use primary mirrors with spherical curves, which, as mentioned in chapter 3, can lead to spherical aberration.

The telescope tube, made from vinyl-coated cardboard and painted a flat black on the inside to minimize stray-light reflections, is held on a simple altitude-azimuth mount of birch plywood and maple, finished with clear varnish. The mounting and tripod are both lightweight yet sturdy, and are easily carried by young and old alike. Motions in both altitude and azimuth are smooth. A wooden knob adjusts the tension on the altitude axis, which is important when using eyepieces of different weights. The SGR-4 comes with a 1.25-inch diameter 17-mm Kellner eyepiece and a Rigel QuikFinder unity finder for aiming.

Stargazer Steve also offers 4.25-inch f/10 and 6-inch f/8 Newtonian/Dobsonian kits for those who wish to assemble their own instrument from precut parts. Now, before you hold up your hands and say "Never," you should know that each kit includes all parts and hardware, plus a manual and even a step-by-step VHS videotape. Everything is precut for easy assembly, so all you need to do is put the kit together and finish the wood with paint or stain. If you start the project in the early afternoon, the telescope will see its first light that night.

The 4.25-inch telescope kit includes the optics, a cardboard tube, a rack-and-pinion focuser, a birch plywood mount, and Teflon bearings. A 10-mm Kellner eyepiece is supplied, as are two rings that the user sights through when aiming. The deluxe 6-inch kit comes with a parabolic primary mirror, a 17-mm Plössl eyepiece, and a Rigel QuikFinder unity finder (described in chapter 7). Both telescopes work very smoothly. Star tests reveal consistently excellent optical quality, producing sharp and clear images that are a delight to behold. They easily surpass many of the better known econo-Dobs on the market and come highly recommended.

StarMaster Telescopes. This company is attracting a lot of attention in the world of telescopes. Owned and operated by Rick Singmaster of Arcadia, Kansas, StarMaster Telescopes come in a wide variety of sizes to suit nearly anyone looking for a well-made, Dobsonian-mounted Newtonian reflector.

The least expensive StarMaster telescopes belong to their ELT series, the ELT standing for "Economical, Lightweight, and Transportable." Consumers can select between two instruments and three focal ratios: an 11-inch f/4.3 or f/5.4, and a 12.5-inch f/5. Each features an open-tube assembly made from four rubber-coated aluminum poles. Erecting the instrument in the field is a simple operation. Each telescope breaks down into three basic groups: the mirror box/rocker box, the upper cage, and four poles. Plug the poles into the mirror box, tighten them, and top them off with the upper cage. The only drawback is that the thumbscrews that hold the whole thing together are untethered and so may be dropped inadvertently (though I should note that they do not have to be completely unscrewed during assembly). Fortunately, the mirror box remains covered during assembly, so there is no chance of the thumbscrews falling on the primary.

Both instruments come with mirrors from Carl Zambuto Optics, renowned for its consistent excellence. Each primary is coated with standard aluminizing (89%), while the secondary mirrors have 96% enhanced aluminizing. The view through each ELT that I have seen has been excellent, with sharp, contrasty images regardless of target.

In addition to the excellent optics, each StarMaster ELT features striking oak-veneer plywood construction and comes with a JMI DX-2 Crayford-style focuser, a Rigel QuikFinder unity finder, primary and secondary mirror covers, and a cloth light shroud. Carrying handles are also supplied, although there is no provision for wheelbarrow handles, as there are on many larger Dobsonians, including others from StarMaster. Extra-cost options include upgraded focusers and factory-installed digital setting circles.

For the intermediate-to-advanced amateur, StarMaster also offers five Truss telescopes: a 14.5-inch f/4.3, a 16-inch f/4.3, an 18-inch f/4.3, a 20-inch f/4.3, and a 24-inch f/4.2. Each is based on the now-familiar eight-tube truss design riding on large, semicircular bearings and set upon a low-riding Dobsonian base. The 14.5- through 18-inch mirrors come from Carl Zambuto Optics, while the 24-inch mirror is supplied by Pegasus Optics. The 20-inch can be built around either manufacturer's mirrors. Primaries are aluminized and overcoated using a technique called *ion-assisted deposition* (IAD) while all diagonal mirrors have 96 to 97% enhanced coatings. The manufacturer states that this IAD technique provides higher reflectivity than standard aluminizing. To prove their worth, each telescope comes with certification attesting to its optical quality. In fact, Singmaster personally tests each instrument before it is sent to the customer.

Indeed, each and every StarMaster that I have viewed through has produced amazingly clear images. I can recall several memorable views I had through a 16-inch that were just breathtaking. No matter what I looked at, the image was about as good as I have ever seen it through that aperture. Globular clusters like M13 sparkled, while nebulae shone with an eerie glimmer usually only seen in photographs. And the planets Jupiter and Saturn also revealed a level of detail missed through most apochromatic refractors because of their limited apertures.

Each truss-mounted StarMaster comes with the 2-inch DX-2 Crayford focuser by JMI (an excellent choice, although the DX-1 and Starlight Feathertouch focusers are also available at extra cost), a Telrad one-power aiming device, a cloth light shroud, a dust cover for the secondary mirror, a primary-mirror cover, and a carrying case for the truss tubes. Unlike most open-style Dobsonians, the truss tubes that come with StarMaster Dobs are preassembled in pairs, which makes assembly go quickly. Each truss model features furniture-quality oak-veneer plywood construction finished with a durable coating of clear polyurethane and rubber-coated aluminum truss tubes. Motions in both altitude and azimuth are smooth thanks to the Teflon-on-Formica bearing surfaces.

A unique standard feature of 14.5-inch and larger StarMasters is a detachable mirror cell. By first removing the mirror (the heaviest component of a large-aperture reflector), carrying the telescope about becomes much easier. Once at the chosen observing site, the mirror cell is then reattached to the telescope with little ill-effect on collimation. A great idea! Detachable wheelbarrow-style transport handles are also available, but at extra cost.

Each StarMaster Truss telescope can be purchased with digital setting circles or a full computer-driven GoTo mounting, the first commercial Dobsonian to have offered this option. The latter, called Sky Tracker, is amazing to watch. As with other GoTo telescopes, once initialized, just enter the object of interest, press the button, and the telescope slews automatically to its location with amazing accuracy. Yes, it's an expensive option (about $2,200), and as much as I frown on such things (to me, half the fun is in the hunt), I must admit that the Sky Tracker works very well.

Starsplitter Telescopes. Located in Thousand Oaks, California, Starsplitter Telescopes offers a number of different Newtonian reflectors for amateur astronomers. All are mounted on Dobsonian mounts that, while solidly made, do not exhibit quite the level of excellence in finish as those by Obsession or StarMaster. Teflon pads riding on Ebony Star Formica let the telescopes move smoothly in both altitude and azimuth. In addition, each comes with a low-profile focusing mount.

One look at the Starsplitter Tube 8-inch f/6 and 10-inch f/6 models and you can tell that these are not like most other solid-tube Dobsonians. Both include cardboard tubes painted black inside and out, on Dobsonian mounts made from Appleply plywood, which are also nicely finished and sealed against moisture. As mentioned earlier in this chapter, Appleply plywood is famous among woodworkers for its fine grain structure and furniture-quality appearance. The tubes can be easily rotated in their cradles, allowing observers to adjust the focuser's location to a comfortable angle, which can be a great convenience. Other niceties include an open aluminum primary-mirror cell that is easily adjustable, lifting handles, and a four-vane, fully adjustable diagonal mirror holder. Large, circular side bearings let the telescope move easily in altitude, while the side-to-side motion in azimuth is equally smooth. The standard 1.25-inch rack-and-pinion focuser is adequate for the task, although upgrades to higher-quality focusers are available. Eyepieces and finderscopes are not supplied, so be prepared to purchase them immediately. Starsplitter does offer a Telrad as an extra-cost add-on, if you wish.

Starsplitter Tubes come with your choice of generic optics or mirrors from Carl Zambuto (the latter adding a few hundred dollars to the price). Both work quite well, although I would give a nod to the Zambuto mirrors as being the better. Still, many reports from Tube owners praise Starsplitter's brand of mirrors. Jupiter's bands, Saturn's rings, and Cassini's Division all come through clearly, as do deep-sky objects galore. Unless you already own a finderscope and a set of eyepieces, my recommendation would be to get the scope with the Starsplitter optics and invest the difference in those accessories.

As I have said before, give careful consideration before choosing between a tubed versus truss Dobsonian. While the 8-inch Tube is light and small enough for just about anyone to handle, the 10-inch weighs a total of 70 pounds, divided evenly between tube / cradle and rocker box. That's analogous to Discovery Telescope's Premium DHQ instruments, but heavier than a comparable truss-style instrument. Oh, and speaking of the Discovery Premium DHQ models, which are quite similar in style to the Starsplitter Tubes, they lack the convenience of a rotatable tube, but they do add the ability to adjust their balance in altitude. The 10-inch Discovery Premium DHQ also comes with two Plössl eyepieces, a 2-inch Crayford-style focuser, and your choice of a Telrad or a 7 × 50 finderscope for about the same price as the eyepieceless, finderless 8-inch Starsplitter Tube.

Starsplitter Compacts, including an 8-inch f/6, a 10-inch f/6, and a 12.5-inch f/4.8, all feature mirrors supplied by either Swayze, Nova, or Zambuto Optics, as chosen by the buyer (Zambuto garnering the best reputation). Starsplitter Compacts look like no other commercially available telescope. While most reflectors use either a solid tube or an eight-pole truss design, Starsplitter Compacts are based around a unique two-pole frame that was first conceived and used by Thane Bopp, an amateur telescope maker from Missouri.

The Boppian approach is the ultimate in portability. The open design also has the advantage of letting the optics acclimate to the outside temperature rapidly, although the process is slowed some by the closed bottom of the mirror box. Of course, with the optics so exposed to the air, mirror dewing is a problem on damp nights. The optional nylon light shield will help to slow dewing as well as reduce glare and stray light from ruining image contrast, but one respondent noted that the shroud occasionally sagged into the light path.

Collimation of the primary mirror is achieved by turning three thumb knobs on the bottom of the mirror box. Personally, I found that the lack of a tube made collimation difficult, since I had no frame of reference to judge the concentricity of the optical assembly when looking through a sight tube. This should, however, become easier as time goes on and the user gains experience. A holographic laser collimator wouldn't hurt either (see chapter 7).

The Compact II line of Starsplitter telescopes are available in four apertures (an 8-inch f/6, a 10-inch f/6, a 12.5-inch f/4.8 or f/6, and a 15-inch f/4.5), while Starplitter's Compact IV reflectors come in 8-, 10-, and 12.5-inch models. All feature mirrors from either Zambuto, Swayze, or Nova Optics. The Compact IIs follow the more conventional eight-pole truss design utilized by many others, while the Compact IVs use four parallel tubes to bridge the gap between the mirror box and upper cage assembly. Some may question whether the four-tube arrangement is stable or will twist and flex in use. While a rectangle is certainly not as stiff as a triangular truss, the Compact IVs should be fine unless they are heavily laden with accessories.

In both cases, each truss tube is attached to the mirror box by sliding it into a mating aluminum tube that is bolted to the inside of the mirror box. Although the design works, it's not nearly as elegant as that used by Obsession, StarMaster, or, for that matter, the Starsplitter II line (reviewed later). The biggest problem is the potential for dropping something directly onto the primary mirror, since the mirror box's protective wooden cover must be removed *before* the truss tubes can be attached. Although the hardware is captive, I still get nervous leaning over a fully exposed primary while tightening or loosening thumbscrews. Wooden clamps secure the tubes to the focuser / diagonal upper assembly, a simple, yet functional method.

Along with more conventional truss designs, the Compact II and IV scopes use traditional four-vane spider mounts for their diagonal mirrors, which prove to be more stable than the curved-vane diagonal holder used on the Compacts. Their cylindrical top ends also help prevent the diagonal from dewing over and make attaching the optional light shroud easier. Although a DC-powered pancake fan is built into each Compact II mirror box, the box itself is closed at the bottom save for a small hole for the fan itself. This design impresses me as ill-conceived, since the closed bottom will impede cooling of the primary, even with the fan on. Obsession, StarMaster, Discovery, and, again, the Starsplitter II line use open mirror cells to promote faster thermal adjustment.

Starsplitter also makes two lines of *big* telescopes: the Starsplitter II and Starsplitter II Lite. The Starsplitter II series includes six instruments from a 15-inch f/4.5 to a 30-inch f/4.5, while the four Lites range from a 20-inch f/4 to 28-inch f/4.5. All primary and secondary mirrors come from either Nova, Galaxy, Pegasus, or Swayze Optics. All four make decent optics, but if I had to pick one, I would probably go with Galaxy Optics, followed closely by Nova. Starsplitter II telescopes are designed around open-truss tubes that allow for comparatively easy storage. All come with removable wheelbarrow handles for moving the instrument around prior to setup. Their no-tool design uses captive hardware, making separate tools such as wrenches and screwdrivers unnecessary for setting up and tearing down the instrument. The truss tubes are held to both the mirror box and the secondary-mirror assembly with thumbscrews and mating wooden blocks, a simple, usable design. If the overall width of the telescope is a concern, you should know that the truss tubes ride inside the mirror box, rather than outboard as on Obsession telescopes, adding another 3 or 4 inches.

From my research for the first edition of this book in 1994, I purchased an 18-inch Starsplitter II, reasoning that it was the largest telescope that could fit into my car at the time. So how has it held up? As I reported in the second edition of *Star Ware* (1998), everything continues to work quite well. The telescope takes about 10 minutes to set up and another few minutes to collimate by turning one of three large knobs behind the mirror. The diagonal mirror, which is so critical to the telescope's overall performance, is held in a conventional spider mount, requiring a straight-blade screwdriver to loosen and tighten its pivot screws. This conventional system is not nearly as convenient as the innovative designs used by Discovery, StarMaster, and, especially, Obsession. The Galaxy optics in my 18-inch are very good, although their true quality did not come through until I checked and subsequently repositioned the diagonal mirror, which had been installed too far back in the upper cage assembly. With that problem solved, I am now seeing deep-sky objects and planets alike with amazing clarity, contrast, and color.

So, the bottom line: Starsplitter II telescopes are well designed, well-made instruments at some very attractive prices. Although their cabinetry is not quite as refined as Obsession or StarMaster instruments, they are worth strong consideration if you are looking for a large-aperture telescope. The Starsplitter Compact lines, however, seem to cut a few too many corners, though at the same time it must be recognized that they are also less expensive. Still, if I were in the market for a truss-style Dobsonian reflector, but could not afford the cost of a Starsplitter II, StarMaster, or Obsession, then I would give Discovery Telescopes very strong consideration.

Takahashi. In addition to its outstanding line of refractors, Takahashi offers several reflecting telescopes. The company's line of Newtonian reflectors include the 5.2-inch f/6 MT-130, the 6.3-inch f/6 MT-160, and the 7.9-inch f/6 MT-200. Optical performance of all three is typical Takahashi: exceptional! Images are sharp and clear, with good contrast and little evidence of aberrations. Each comes with a smooth rack-and-pinion focuser designed to accept 1.25-inch eyepieces. The MT-130 comes with a 6 × 30 finderscope, while the others include 7 × 50 finderscopes.

Although MT reflectors are sold without mountings, they do include tube holders. Most purchasers seem to naturally select matching Takahashi German equatorial mounts for their instruments. The MT-130 typically rides on the EM-2 mount, which can also be outfitted with a motorized clock drive on the right-ascension axis and a manual slow-motion control on the declination axis. Takahashi describes the EM-2 mount as designed for visual observation, though some astrophotography is certainly possible. The MT-160 requires the larger EM-200 equatorial mount, which includes dual motor-driven axes and is suitable for long-exposure astrophotography. Finally, the MT-200 is best paired with the impressive NJP German equatorial mount, which has dual motor-driven axes and multiple counterweights for fine balancing. It sits atop a steel pedestal, while the others are usually set upon wooden tripods (although pedestals are available optionally). All can be retrofit with computerized control, if desired.

Vixen Optical Industries. This Japanese company offers three short-focus Newtonian reflectors. The smallest, the 5.1-inch f/5.5 GP-R130S, is aimed toward the serious beginner market. Although it costs more than many Chinese and Taiwanese clones, and even though the aperture is somewhat limiting, the sturdy construction and high quality of the GP-R130S optics makes this small Newtonian one that will let the owner hone his or her observation skills as well as experiment with astrophotography. The standard version is supplied with 20-mm Kellner and 5-mm orthoscopic eyepieces and a 6 × 30 finderscope, while a deluxe version includes a clock drive and two Lanthanum LV eyepieces.

Moving up a notch, the GP-R150S adds another 0.8 inches of aperture to create a fine 5.9-inch f/5 telescope that is available in standard or deluxe variants. Both come with a unique back-and-forth sliding focuser, a feature shared with the GP-R130S. I must confess, however, that this telescope is probably not worth the extra money over the GP-R130S. Such a small increase in aperture affords only a fractional increase in the instrument's limiting magnitude, which is of very small consequence. Instead, if you are looking for an equatorially mounted Newtonian, and it just has to be a Vixen, consider the firm's largest instrument.

The GP-R200SS is a stubby 7.9-inch f/4 instrument. That fast focal ratio is perfect for wide-field viewing, terrific for transporting, but will also enhance coma and make dead-on collimation a must. Vixen offers a coma corrector to help ease that problem, but this adds another $100 or so to an already-expensive instrument. Still, owners rave about the images. The R200SS includes a 2-inch rack-and-pinion focuser that also features a built-in T-adapter for attaching cameras directly to the telescope, a nice touch that eliminates the need for a separate camera mount (note that you will still need a T-mount for your camera; see chapter 7).

Each Vixen Newtonian is attached with a pair of rotatable rings to their excellent Great Polaris (GP) German equatorial mount. (The GP-R200SS can also be purchased on the sturdier GP-DX mount, a good upgrade for photography.) The mounting is supported by an aluminum tripod that is adequate for the task, but it's not as vibration resistant as wood.

Vixen products are not inexpensive, especially in this day and age of like-size instruments from China and Taiwan that at first glance appear to be identical in design. But consumers should keep in mind that Vixen quality exceeds these others by a wide margin. Unfortunately, so do their prices. And consumers in the United States should also note that Vixen's lone authorized distributor, Orion Telescope Center, only imports the GP-R200SS at present.

Cassegrain Reflectors

Novosibirsk. From "New Siberia" comes an unusual instrument, the 7.9-inch f/8.8 TAL 200-K Cassegrain reflector. By now, we have come to expect a Cassegrain to look pretty much like the optical path shown back in Figure 2.4d. The TAL 200-K, however, throws in a little twist. Technically, it's a *Klevzov,* a comparatively little known design that uses a spherical primary mirror but adds a corrector lens to help eliminate spherical aberration and coma, in much the same way as the Vixen VC200L does (discussed later in this section).

The corrector lens is actually a part of the secondary mirror assembly. Light bounces from the primary to the secondary, where it passes through a meniscus before it routes back through the centralized hole in the primary and out to the eyepiece. Two distinct advantages to this system over a sealed catadioptric telescope are that since the tube remains open in the front, the optics will cool down more rapidly and there is no corrector plate to fog over in damp conditions.

Like the TAL Newtonians mentioned previously, the TAL 200-K is a solidly made instrument. The optical tube assembly measures only 17 inches long but weighs 27 pounds, which is about the same as the larger Celestron 9.25-inch Schmidt-Cassegrain telescope. Much of that weight lies in the tube itself, a stoic, steel affair painted glossy white. The German equatorial mount and steel pedestal are no lightweights either, weighing 44 pounds when assembled. But what do you expect from Siberia? Those are a stalwart people! The mount also includes a 12-volt AC worm-gear clock drive on the right-ascension axis. A separate transformer, adapting the system to 115-volts AC (household current), is included.

Optically, the TAL 200-K is reported as producing good star images both in focus and out (during the star test), sharper than through a Schmidt-Cassegrain, but not refractor-like pinpoints. Image brightness is good, while contrast between objects and the background sky is roughly comparable to a Schmidt-Cassegrain. That comes as no great surprise, since the TAL 200-K's central obstruction is about 35%, which is comparable to a Schmidt-Cassegrain's. Contrast will also suffer if used under high light-pollution conditions due to the design of the instrument's internal baffling.

The TAL 200-K includes two Plössl eyepieces, a nice 8 × 50 finderscope, a 2× Barlow lens, and a set of color filters, as well as a birch storage case.

Optical Guidance Systems. For the rich-and-famous, diehard amateur astronomer, Optical Guidance Systems offers some impressive telescopes that are worth a look. Their medium-to-large-aperture Classical and Ritchey-Chrétien Cassegrain reflectors rank among the finest instruments in their respective classes.

The Classical design is available with focal ratios around f/16, while the Ritchey-Chrétien variants are around f/8 to f/9. Perhaps the most important benefit of the Ritchey-Chrétien design is its total freedom from coma. This, along with a lower f-number, is especially important if the instrument will be used for astrophotography.

One of the strongest selling points of Optical Guidance System's telescopes is their superb optics, supplied by Star Instruments of Flagstaff, Arizona. To decrease overall weight, lightweight, conical mirrors are made from near-zero-expansion ceramic material, then aluminized and silicon overcoated. Images are striking, with pinpoint stars across the field through their Ritchey-Chrétien instruments. The view is also quite impressive through the Classical Cassegrains, though not quite up to the same standard.

All OGS instruments, regardless of optical design, include oversize focusers from Astro-Physics, which are famous for their smooth, no-backlash movement. Options abound for all OGS Cassegrains. For instance, consumers can elect either aluminum or carbon fiber tubes for their OGS instruments. Carbon fiber has the distinct advantage of not retaining heat like aluminum does, which will keep the focus steady even as the temperature changes as the evening wears on. Mountings, mounting rings, finderscopes and guide scopes, and other customized accessories are also available for all OGS Cassegrain instruments.

Parallax Instruments. Having relocated from North Carolina to Vermont since the last edition of *Star Ware,* Parallax Instruments continues to offer several excellent Cassegrain reflectors. These include four Classical Cassegrains, ranging from the 10-inch f/15 PI250C to the 20-inch f/16 PI500C, and four Ritchey-Chrétien instruments, from the 10-inch f/9 PI250R to the 20-inch f/9 PI500R. As mentioned earlier, the Ritchey-Chrétien design enjoys the distinct advantage of being free of coma across the entire field, making this one of the most ideal reflecting telescope concepts ever devised. This, along with the relatively low focal ratio of f/9, also makes it better suited for astrophotography than the more popular (and less-expensive) Schmidt-Cassegrain catadioptric.

Parallax Cassegrains all come with oversize focusers that accept both 1.25- and 2-inch eyepieces, 7 × 50 finderscopes (except the 20-inch models, which include 11 × 80 finderscopes), mounting rings, handles, counterweights, and, except for the 10-inchers, built-in cooling fans. All primary mirrors are made of Sitall glass, a Russian ceramic glass that is similar in properties to Zerodur, an excellent material for telescope mirrors. Both have excellent thermal properties, keeping cool-down times to a minimum.

Parks Optical. Fans of the Cassegrain telescope hail the H.I.T. series of 6- to 16-inch telescopes by Parks. These might be called Jekyll-and-Hyde telescopes, as they can switch back and forth between a long-focus Cassegrain and an f/3.5 Newtonian, thereby offering the best of both worlds. To permit this versatility, H.I.T. telescopes come with two interchangeable secondary mirrors. Simply remove one, insert the other, check collimation, and the telescope's alter ego is ready to go. H.I.T. scopes are available on a variety of German equatorial mounts with or without clock drives. The 6-inch comes with a too-small 6 × 30 finderscope, while the others feature 8 × 50 finders. Like their Newtonians, Parks sells its H.I.T. convertible instruments in several series, based on the mounting that comes with the instrument.

RC Optical Systems. Owned and operated by Optical Systems of Flagstaff, Arizona, RC Optical makes Ritchey-Chrétien-style Cassegrain reflectors that are just striking. Their optical and mechanical quality is top drawer. Images are free of coma, a common problem with Classical Cassegrains. Like its chief competition, Optical Guidance Systems, RC Optical uses mirrors from Star Instruments, acknowledged as the leading supplier of Cassegrain optics. RC Optical places these fine optics in well-baffled tube assemblies made from carbon fiber for its excellent thermal behavior as well as lighter weight (as contrasted to aluminum and fiberglass). RC Optical completes each telescope with an oversize focuser from Astro-Physics, electronic adjustment and focus of the secondary mirror, and built-in fans to promote active cooling—again, just like OGS.

Complete RC Optics telescope tube assemblies include 10-inch f/9, 12.5-inch f/9, 16-inch f/8.4, and 20-inch f/8.1 models. None are for the faint of heart, however, since prices start at more than $10,000 for the 10-inch. For this, you get a nicely appointed optical tube assembly, but a mounting is extra.

Takahashi. This industry leader manufactures several Mewlon Cassegrain reflectors based on the Dall-Kirkham version of the Cassegrain optical system. You may recall from the brief discussion in chapter 2 that Dall-Kirkham Cassegrains use mirrors with simpler curves than either Classical or Ritchey-Chrétien Cassegrains, yet still produce excellent results. Takahashi Mewlon instruments currently come in four apertures: the 7.1-inch f/12 M-180, the 8.3-inch f/11.5 M-210, the 9.8-inch f/12 M-250, and the 11.8-inch f/11.9 M-300. The M-180 and M-210 sit astride the EM-2 or, optionally, the Takahashi EM-10 German equatorial mount. The M-250 comes on the EM-200 mount standard, while the M-300 is equipped with the EM-500 mounting.

Takahashi's 10-inch f/5 BRC-250 has the fastest focal ratio of any production Ritchey-Chrétien instrument. Think about it: a telescope that is designed to produce a flat field, free of coma, spherical aberration, and astigmatism across an amazing 5° field! Advanced astrophotographers drool over the very thought. The BRC-250 (short for Baker Ritchey-Chrétien) is ideal for large-format photography (i.e., larger film than conventional 35 mm; the larger the format, the finer the resolution of the recorded image), up to 6-inch × 9-inch sheet film, as well as the largest commercially available CCD imagers. Perhaps most incredible of all are the pinpoint star images. If you look at a photo taken with most telescopes, you'll notice how star images appear bloated. The BRC-250 corrects for this defect, obtaining a more natural, aesthetically pleasing appearance across its entire field of view.

The BRC-250 comes outfitted with a 7 × 50 finderscope and an oversize helical focuser, as well as a carbon-fiber tube to ensure that changes in temperature at night do not impact image focus, a problem with some other tube materials.

Takahashi also offers the 8.3-inch CN-212 convertible Newtonian/Cassegrain reflector. This two-way telescope is possible because both the Classical Cassegrain and Newtonian optical designs are based around parabolic primary mirrors. By swapping secondary mirrors, the CN-212 can change between an f/12.4 Classical Cassegrain and an f/3.9 Newtonian. A keyed secondary mirror mount helps to keep the secondary mirror in line, while the mating four-vane spider helps to maintain optical collimation after conversion. This is a fun instrument to use, since it gives you, in effect, the best of both worlds. Coma is evident, as we might expect, but overall, I found the images to be quite good in either configuration. The CN-212 is available on either the EM-2 or the EM-200 mount, the latter being recommended for long-exposure, through-the-telescope photography.

Vixen Optical Industries. One look and you can see that Vixen's 8-inch f/9 GP VC200L Cassegrain is different. Vixen calls its optical system VISAC, which stands for Vixen Sixth-Order Aspherical Catadioptric. Unlike conventional Cassegrains, which only use primary and secondary mirrors to focus light into their eyepieces, the VC200L relies on a compound-curve, parabolic primary mirror, a convex secondary mirror, and a three-element corrector/flattener lens built into the focuser. The secondary is mounted in the front of the open tube in an adjustable four-vane spider mount, similar to those used in traditional Cassegrain reflectors.

Owners comment that the VC200L yields flat, distortion-free images from edge to edge thanks to the corrector lens, while spherical aberration is minimized thanks to the aspheric primary mirror. The result is pinpoint stars from edge to edge, an important consideration for through-the-telescope photography. Resolution is quite sharp, with double stars cleanly split without the distraction of scattered light from spherical aberration. Note, however, that the telescope's optics must be allowed to cool fully before optimal performance, which takes a fairly long time because of the sealed mirror cell (a problem found in all Cassegrain telescopes, not just the Vixen). The only common problem that owners have noted is that visually, the VC200L suffers from low-image contrast due to its large central obstruction—39% by diameter. As a result, this is not the best telescope for planetary observations.

Other notable features on the VC200L include a rack-and-pinion focusing mount, which eliminates image shift common to SCTs, and a slide bar that allows the user to balance the telescope assembly on its mounting. The venerable Great Polaris German equatorial mount perched on an aluminum tripod is used to support the 15-pound VC200L. While adequate, the aluminum tripod is apt to transmit more vibration than wooden legs, which are becoming a rare commodity these days. Also included are a 6 × 30 finderscope, 9-mm orthoscopic and 25-mm Kellner eyepieces, and a 1.25-inch star diagonal. A motorized version, christened the VC200L-SM, includes a dual-axis clock drive and substitutes 5- and 25-mm Vixen Lanthanum LV eyepieces. An optional 2-inch adapter ring is available for both the standard and motorized telescopes, which enables you to use 2-inch star diagonals and eyepieces. Another add-on accessory is the SkySensor 2000 motor-driven pointing system, which is designed to fit any Great Polaris or older Super Polaris mount, not just the one used with the VC200L. The SkySensor will automatically set the telescope in pursuit of any of the 6,500 objects in its built-in database.

Exotic Reflectors

Fitting into neither the traditional Newtonian nor Cassegrain family of reflectors, these unique instruments offer some interesting possibilities for amateurs looking for something a little different to bring along to their club's next star party.

DGM Optics. Many people are under the misconception that mirror-based telescopes can never deliver the same tack-sharp views of the planets seen through high-end apochromatic refractors. Part of this fallacy is due to some less-than-tack-sharp optics, part due to poorly collimated instruments, and part because of large secondary mirrors that lower image contrast. While the first two are caused by either inferior optics or possibly operator error, the third is an intrinsic fault of many designs that need to have a secondary mirror right along the optical axis of the primary. But what if a reflector could be made without *any* central obstruction? Images should be comparable to apochromats; indeed, they could possibly be better in that chromatic aberration would be absolutely nonexistent.

That's exactly the idea behind DGM's line of OA (short for off-axis) reflectors. Rather than have a traditional secondary mirror divert light to the eyepiece, owner Dan McShane has created an instrument that tilts the primary mirror just enough to aim its reflection off axis, up toward a small secondary mirror that is directly under the 2-inch helical focuser, where it bounces into the eyepiece. The trick is in the primary. Rather than using a traditional paraboloid, McShane must cut each of his primaries from larger, parabolic parent mirrors, just as you cut individual cookies from a large flat sheet of dough. Therefore, as long as the parent can be cut without inducing stress fractures, he can produce anywhere from two to four smaller instruments from one larger, well-made mirror. The secondary mirror is permanently mounted and collimated to ensure that it never has to be adjusted by the user, even if bounced around during transport. All in all, a pretty clever execution.

Six models make up the OA arsenal, ranging in size from the OA-4.0 (4-inch f/10.4) to the OA-9.0 (9-inch f/9.6). All mirrors include enhanced (96%) aluminzing and overcoating for optimal brightness. The optics are then mounted in high-quality tube assemblies in a choice of either fiberglass, carbon fiber, reinforced cardboard, or aluminum. All are supported by nicely crafted Dobsonian mounts made from Atlantic birch plywood coated with a clear polyurethane. Bearings are made of UHMW (ultrahigh molecular weight) polyethylene, a smooth polymer, riding on Teflon pads. The resulting motions are very smooth and even in both altitude and azimuth. Each tube is held in a hinged cradle that can be opened, allowing the tube to be balanced and rotated.

In practice, the views really are quite amazing. You'd be hard pressed to tell the difference between the sight of, say, Jupiter or Saturn through an OA-4.0 and a premium apochromat! Images are clear and crisp, with none of the telltale milkiness or mushiness that often plague Newtonians. Given the necessarily steady atmospheric conditions, magnifications can be pushed to better than 100× per inch of aperture before image quality begins to break down, something that even the finest apos have difficulty doing. But as I have long said, a high-quality, long-focus reflector is tough to beat. Eliminate the central obstruction, and you have a killer telescope. Best of all, you can get an OA telescope, mount and all, for about 30% or less of what you'd pay for a similar apochromatic refractor. All you need to do is add a finderscope and eyepiece.

I can only find two drawbacks to DGM off-axis reflectors. First, I am still not a big fan of helical focusers, and so would prefer a Crayford focuser or even a rack-and-pinion model instead. The other drawback—and it's more noticeable in the small models than the larger—is the issue of stability. OA instruments have the three Ls: long, lean, and light. As a result, they tend to sit lightly on the ground. If set up on soft ground, such as grass, they are bound to tilt and lean in ways that the user may not want. Wind will also cause them to shimmy undesirably. Unfortunately, that is a problem with many small Newtonians, although the OA's longer tubes do tend to suffer more.

Takahashi. The company sells a line of specialized hyperbolic photo/visual astrographic reflectors that use hyperbolic concave primary mirrors, instead of the usual parabolic mirrors in standard Newtonians, and four-element field corrector/flattener lens sets to yield sharp, flat photos of wide slices of the sky. The 6.3-inch f/3.3 Epsilon 160, the 8.3-inch f/3.0 Epsilon 210, and the 9.8-inch f/3.4 Epsilon 250 are optimally designed for astrophotography, but all may also be used visually with standard 1.25- and 2-inch eyepieces.

Each of the Epsilon astrographs includes an amazing helical focuser that moves 1 mm per revolution. While a problem for visual use, this degree of accuracy makes it possible to adjust the focuser position in increments of 3 microns (approximately 1/10,000th of an inch) for extremely precise focusing. Graduated marks around the focuser's base make it easy to turn the focuser back to a previously noted position. Once the focuser position is set it can be locked securely. In addition to the fine degree of adjustment, the Epsilon focuser can be rotated a full 360°, a great assist for framing the image without changing the focus setting. Vents in the mirror cell can also be opened to promote more rapid cooling of the primary mirror, then closed to seal out dust. A simple but great idea.

Catadioptric Telescopes

In this edition, I decided to divide this section into two separate parts: Schmidt-Cassegrain and Maksutov instruments. This way, readers can directly compare apples with apples and oranges with oranges.

Schmidt-Cassegrain Telescopes

Celestron International. Celestron is renowned as the first company to introduce the popular 8-inch Schmidt-Cassegrain telescope back in 1970. While only one basic model—the Celestron 8—was sold back then, there are now many variations of the original design from which to choose. They range from bare-bones telescopes to extravagant, computerized instruments. Celestron tells us, however, that all of its 8-inch Schmidt-Cassegrain telescopes share similar optics with the same optical quality, regardless of model and price. All feature Starbright enhanced coatings for improved light transmission and image contrast.

Images through Celestrons are usually good, sometimes excellent. Every now and then, a poor optical tube assembly sneaks through their quality control program, but I must say that I am impressed with the strides they (and Meade) have made since the mid-1980s. Maintaining quality in a mass-produced instrument as sophisticated as a Schmidt-Cassegrain while also trying to maintain an attractive price is a difficult chore.

Due to the comparatively large central obstruction from their secondary mirrors, Schmidt-Cassegrains as a breed generally lack the high degree of image contrast seen in refractors and many Newtonians. It is also important to note up front that both Celestron and Meade Schmidt-Cassegrain telescopes suffer from something called *mirror shift.* To focus the image, both manufacturers have chosen to move the primary mirror back and forth rather than the eyepiece, which is more common with other types of telescopes. Unfortunately, as the mirror slides in its track it tends to shift, causing images to jump. The current telescopes produced by both companies have much less mirror shift than earlier models, but it is still evident to some degree.

Celestron offers several variations of the C8, with equipment levels varying dramatically from one model to the next. At the low end of the price scale are the Celestar 8 and the G-8. Both come equipped with 90° star diagonals, so-so 25-mm SMA (Kellner) eyepieces, and small 6 × 30 finderscopes. Owners quickly discover they need to buy new finderscopes (an 8 × 50 would be nice) and a couple of better-quality eyepieces to make the packages complete.

The G-8, the next generation of the popular Great Polaris-C8 (GP-C8), is mounted on a CG-5 German equatorial mount, a Chinese clone of the Vixen Great Polaris mount that it replaces in Celestron's lineup. That's not necessarily a put-down, although I do think the original Vixen mount was sturdier. Still, the CG-5 mount is certainly up to the task of supporting the Celestron 8's optical tube assembly, at least for visual use. (Some owners have complained about the CG-5's sturdiness, but they more than likely owned the first generation CG-5, which had inferior bearings to the current version.) The mounting is also sufficiently sturdy for short exposures of the Moon and planets, as well as some wide-field, piggyback guided exposures, but if you have your heart set on long exposures through the telescope, you should upgrade to one of the other C8 variants.

The G-8's tube attaches to the mounting on a sliding dovetail plate, which allows users to balance the assembly precisely and easily, very handy for when you upgrade to that 8 × 50 finderscope or a heavy 2-inch eyepiece. Like the Vixen GP mount, the CG-5 can accept a polar-alignment scope for fast, accurate alignment, but unless you are going to try your luck at guided photography, spend the money on a new eyepiece instead. You might, however, want to consider the optional DC-powered motor drive, which runs on four D-cell batteries.

The Celestar 8 is supplied on a fork equatorial mount, a DC spur-gear clock drive, and a *wedgepod.* The wedgepod incorporates in one unit both a fixed-height tripod and the wedge needed for tilting the telescope at the proper angle to let it track the sky. Owners seem divided about the wedgepod concept, with many noting that the mounting was not as stable as they had hoped. Others are fond of the wedgepod. One owner said, "I'm not very gadget oriented, but this telescope is a peach to set up; even I can screw in a couple of bolts." The self-contained clock drive is powered by a single 9-volt alkaline battery, a great feature for observers who spend most of their time under the stars far from civilization. The drive is fine for the visual observer who just wants to keep a target somewhere in the field for an extended period, as well as for shots of the Moon and the planets, and possibly some piggyback guided exposures. Some have even used it successfully for long exposures through the telescope, but most astrophotographers prefer a worm-gear drive.

Introduced in January 2000, the Celestron NexStar 8 features a unique, one-armed fork mount (actually, more like a single, short, broad chopstick). That apparent off-balance appearance makes many people say, "It can't work." It *looks* like the telescope should bounce back and forth like a diver about to spring off a diving board. But looks can be deceiving, and the N8 is proof of that. Although not as sturdy as a biped fork mount, the N8's mount proves sturdy enough for visual observations as well as some basic astrophotography. But while the chopstick mount is surprisingly sturdy, the generic, imported aluminum tripod that sits under it is not. This proves to be the weak link in the NexStar, just as it does for so many other heavy instruments. Fortunately, the N8 can be coupled to Celestron's far sturdier Field Tripod, though this does add several hundred dollars to the investment.

Unlike the experiences I have had with some of the less expensive NexStar refractors and reflectors, the GoTo function seems to work quite accurately with the 8-inch. Owners report that precision is good enough to place each selected target somewhere in the field of a low-power eyepiece, even after a cross-the-sky slew. In all, the N8's onboard computer database has more than 18,000 objects in its memory. All are accessible via a molded hand controller, which mounts neatly on the mount's arm when not in use. The N8 can also be controlled via a computer's serial port by a number of different software programs, including *The Sky, SkyMap Pro, Starry Night Pro, Digital Sky,* and others.

Although most owners are satisfied with the N8, the telescope is not without a few idiosyncracies. One problem is that the mounting's arm is not long enough for the telescope to pass underneath if an f/6.3 reducer/corrector, a 2-inch star diagonal, or a Barlow lens is inserted into the telescope's eyepiece holder. This could potentially lead to a collision if the telescope tries to swing near the zenith as it slews from object to object. Although a clutch on the altitude axis prevents any damage, once the clutch slips, the scope must be realigned with the sky before tracking and GoTo pointing can be resumed. Another complaint is that the NexStar goes through batteries—eight AAs—very quickly, a problem shared by other NexStar telescopes, as well. Fortunately, an AC adapter comes standard, as does a Star Pointer unity finder and a 40-mm Plössl eyepiece.

Then there are the two-armed NexStar GPS models, which couple NexStar technology with the power of the Global Positioning System. The 8-inch f/10 NexStar 8 GPS and 11-inch f/10 NexStar 11 GPS Schmidt-Cassegrains (Figure 5.6) take the ability of the NexStar's GoTo system and couple it to a two-armed fork mount that is outfitted with GPS technology. Both instruments boast carbon-fiber tubes, which are not only lighter than Celestron's conventional metal tubes, but also acclimate to ambient temperature more efficiently. Short of the tube material, the N8GPS's optical tube assembly is the same as the other Celestron 8s mentioned above, while the N11GPS shares optics with the CM-1100, reviewed later in this section. The real showstopper in both cases here is the NexStar mounting. Its database includes full GoTo-ability to any of

Figure 5.6 *The satellite-initiated Celestron NexStar 11 GPS, one of today's most sophisticated Schmidt-Cassegrain telescopes. Shown with optional equatorial wedge.*

40,000 objects (50,000 objects with the 11-inch), all of which can be accessed through the hand controller that clips neatly into one of the fork arms.

Admittedly, other telescopes have all this as well, but what sets the NexStar GPS scopes apart are their built-in electronic compass and GPS technology, making setup a breeze. I saw two early 11-inch models in action at the 2001 Riverside Astronomy Expo in California. Both performed very well. Motions were smooth and quiet, and the aiming accuracy was very good, although the mounting was not quite as sturdy as I had expected. The 11-inch is also available on Celestron's CI-700 mount, discussed later in this chapter, which impresses me as more stable, an important consideration for astrophotographers. And speaking of which, old-fashioned polar alignment and the optional equatorial wedge must also be added for long exposures through the telescope in order to avoid field rotation.

All NexStar Schmidt-Cassegrains feature worm-gear clock drive systems that include an integrated periodic-error-correction (PEC) circuit for greater tracking accuracy. Theoretically, a worm-gear clock drive should track the stars perfectly if it is constructed and polar aligned accurately, but this is not the case in practice. No matter how well machined a clock drive's gear system is or how well aligned an equatorial mount is to the celestial pole, the drive mechanism is bound to experience slight tracking errors that are inherent in its very nature. These errors occur with precise regularity, usually keeping time with the rotation of the drive's worm. The PEC eliminates the need for the telescope user to correct continually for these periodic wobbles. After the observer initializes the PEC's memory circuit by switching to the record mode and guiding the telescope normally with the hand controller (typically a 5- to 10-minute process), the circuit plays back the corrections to compensate automatically for any worm-gear periodicity. The NexStar PEC feature is permanent, in that once the periodic error is compensated for, the instructions will be remembered for subsequent use. The CM1100 and CM1400 instruments detailed later in this section include nonpermanent PEC circuits, requiring that users repeat the steps every time the clock drive is switched on. Permanent or otherwise, PEC will not make up for sloppy polar alignment.

Amateurs who need portability over aperture might consider Celestron's C5, 5-inch f/10 Schmidt-Cassegrain instrument. The C5 is available as an unmounted Spotting Scope, the equatorially mounted G-5, and the computerized NexStar 5. All feature identical optical components that feature fine-annealed Pyrex mirrors with Starbright coatings (standard on all Celestron SCTs). The optics bear magnification well, being equally adept at low-power views of, say, the Lagoon Nebula as well as high-power views of Jupiter, Saturn, and the Moon. Indeed, the only real difference between the models is their mountings. The spotting scope is sold unmounted, but comes with a built-in tripod socket for attaching it to a camera tripod.

The G-5 marries the C5 tube to an imported German equatorial mount that Celestron has christened the CG-3. In reality, the imported CG-3 is a slightly enlarged version of the mountings that come with many department-store telescopes, which left me a little concerned for its stability. But I was pleasantly surprised to find that because of the tube's light weight, the setup is actually quite sturdy. Some short-exposure, prime-focus photography of the Moon may even be possible, although the longer exposures needed for close-up projection photography of the planets would be difficult.

The NexStar 5 shares the same single-arm fork mount and drive system as the NexStar 8. But while the N8 can hit the mounting's base during operation, the shorter N5 tube clears it by a wide margin (except if a camera is attached). The lighter tube is also amazingly steady on the one-armed mount, dampening in a matter of only a second or two after a rap with the ball of my hand. In addition, the NexStar's optical tube assembly is mounted on a dovetail plate, which makes it easy to remove for transport, a nice touch for jet-setting astronomers. Overall, this is an excellent instrument for those who want to set up quickly but also crave the ability to let the telescope do the driving.

For amateurs who want a little *more* aperture than an 8-inch, Celestron offers the G-9.25, which gathers 34% more light. The construction quality of the 9.25-inch f/10 optical tube assembly is high, as we have come to expect from Celestron, with the metal tube painted with the trademark smooth, glossy black finish. Optically, the G-9.25 is a winner, though the instrument seems much more sensitive to thermal equilibrium than an 8-inch. Once, when using a G-9.25 on a cold winter's night, I noticed that the images were dancing around terribly, even though the telescope had been outdoors for close to an hour. But then, almost like flicking a light switch, the images steadied down rapidly and the telescope's true ability shone through. An 8-inch, by comparison, has always struck me as sort of easing into thermal equilibrium, but with this one—BANG, it's there. Quite curious.

The G-9.25 is mounted on the CG-5 German equatorial mount manufactured by Synta in China for Celestron. The optical tube assembly is attached to the equatorial mount's head by means of a 17-inch-long dovetail bar that spans the length of the tube. This freedom lets the user set telescope balance precisely—a big plus when adding cameras or other accessories. That's the good news. The bad news is that the weight of the G-9.25 totally overwhelms the CG-5 equatorial mount. My rap test caused the mount to continue vibrating for an average of 11 seconds after impact, earning the instrument's stability a poor rating.

The G-9.25 strikes me as something of an enigma. On one hand, the optical tube assembly performs very well. Images are sharp and clear, while the star test reveals consistently well-corrected optics. But at the same time, the equatorial mount is far too weak to do the fine optics justice. If this is the scope for you, be ready to replace the mount with something like the Vixen GP-DX or Losmandy GM-8.

The 11-inch f/10 Celestron CM-1100 comes mounted on a sturdy, U.S.-made German equatorial mounting that Celestron calls the CI-700. Optically, the 11-inch Celestron has always been quite good for its aperture and design. Images are sharp and clear, with good detail on such favorites as M42 (all six stars—from A to F—in the Trapezium easily resolved), M51 (nice spiral structure), and M13 (resolved to the core). Because of its long focal length—nearly 2,800 mm—image scale is quite large. This is not a wide-field instrument by any means. The 26-mm Plössl eyepiece that comes standard with the instrument produces a high 107×, making a longer focal length eyepiece a must buy. But then, the long focal length is great for detailed views of the planets and smaller deep-sky objects, such as planetary nebulae.

The telescope and mounting slide together using a dovetail bar, similar to the one on the G-9.25 but much more refined, for great adaptability when adding accessories. Setup takes between 10 and 15 minutes, but keep in mind that this is a heavy instrument that takes an effort to put together in the dark. Not for the faint of heart!

Along with the Plössl eyepiece, the CM-1100 includes a 9 × 50 finderscope, a star diagonal, a polar alignment scope, and a dual-axis clock drive. The latter, which can run on either 110 volts AC or 12 volts DC, tracks the sky at four different rates—sidereal, lunar, solar, and King—and includes a PEC circuit. Celestron's Advanced Astro Master digital setting circles can also be added, if desired.

The CM-1100 was introduced in 1998, replacing Celestron's old and revered CG-11. How do the two compare? Since it's still the same telescope optically as the old, let's look at the mountings. Outwardly, they look almost identical, and while the CM-1100 is a fine mounting for both visual and photographic use, it is just a tiny bit short of the CG-11 in my opinion. For instance, the beefy legs of the G-11 mounting were attached using massive flanges, while the CI-700 has thinner legs that use a *spreader* to keep them rigid. Others complain about the clock drive, saying that the gears sometimes loosen themselves. Unfortunately, Celestron canceled the CG-11 in 1998, although the G-11 mount is still available separately from its manufacturer, Losmandy (also known as Hollywood General Machining). In fact, the CG-11 is also still available as a special-order item from Company Seven, a telescope retailer in Maryland.

By adding extra counterweights to help balance the additional weight, Celestron also uses the CI-700 mounting to support its CM-1400, a 14-inch f/11 Schmidt-Cassegrain telescope. The CM-1400 is truly an observatory-class instrument that can also be made transportable with two or more people. But even then, unless you are an accomplished weight lifter, this is a very heavy instrument, the optical tube assembly alone weighing 49 pounds. Stability of the CM-1400 is surprisingly good, while the images that I saw through one at the 2001 Riverside Astronomy Expo were quite impressive. M51, the famous Whirlpool Galaxy, showed striking spiral structure. I have received a few isolated reports, however, of poor quality optics in C14s, requiring them to be returned to the factory for reworking.

Celestron outfits all of its worm-drive SCTs (the NexStar series, the CM-1100, and the CM-1400) with a Fastar secondary mirror assembly, designed specifically to couple with the ST-237 CCD imaging system manufactured by Santa Barbara Instrument Group (SBIG) and reviewed in chapter 7. While all Fastar-enabled instruments can also be used for regular, visual observations as well as conventional astrophotography, this unique design holds the CCD imager in front of the telescope in place of the secondary mirror. By replacing the specially outfitted secondary-mirror holder with the CCD imager, the Fastar records images at an effective focal ratio of f/1.95 in the 8-inchers or f/2.1 with the others. This, coupled with the tremendous light-gathering ability of a CCD imager, creates an optical system that can record faint objects in a matter of seconds, rather than minutes, or even hours, as with conventional film. The only drawback to this unconventional setup is the large central obstruction caused by the CCD imager (approximately 40% by diameter on the 8-inch), which lowers image contrast even more than usual. Image processing should be able to compensate for much of this ill-effect, but it is still a consideration.

Meade Instruments. In 1972, Meade Instruments opened its doors as a mail-order supplier of small, imported refractors. Meade's first homegrown instruments were 6- and 8-inch Newtonian reflectors, but the company made its mark with the introduction of the Model 2080 8-inch Schmidt-Cassegrain telescope in 1980. Since then, Meade has grown to become the world's largest manufacturer of telescopes for the serious amateur.

Meade's mainstay product is its line of 8-inch Schmidt-Cassegrain telescopes. Today, they offer several variations: the 203SC/300, the 203SC/500, the LX10, the LX10 Deluxe, the LX90, and the LX200GPS, in order of increasing price and sophistication. They all include the same f/10 optical tube assemblies, although the LX200 is also available as an f/6.3 instrument. All include fully "super multicoated" optics. Meade's light transmission and image contrast are comparable to Celestron's, with most owners rating the optical quality as good to excellent. Actually, some of the newest Meade SCTs that I have looked through have really impressed me as being quite sharp (as have several Celestrons). Star tests reveal good consistency from one to the next, with only slight deviations in images from one side of focus to the other. In-focus images, which are really what count, are clear and surprisingly contrasty, given the central obstruction. No, they still won't rival apochromatic refractors when viewing the planets, or even well-made Newtonians, but overall, owners seem quite satisfied. Meade has come a long way since their earliest model 2080s.

Least expensive of the current instruments are the 203SC/300 and 203SC/500 variants, which come mounted on imported LXD300 and LXD500 German equatorial mounts, respectively. Both mounts include manual slow-motion controls and can also accept optional clock-drive motors. And while the optics may be acceptable, both mountings leave something to be desired. Neither impresses me as suitable for the size and weight of the instrument, but the LXD300 is overburdened well beyond its abilities. Indeed, the stronger LXD500 is also pressed for strength under the 8-inch's tube assembly. The telescope is okay for visual use, but it is not nearly steady enough for photographic work. Although it costs a few hundred dollars more, the Meade LX10 is a much better choice. Save your money until you can afford one.

The LX10 is designed for those who need the portability of a fork-mounted Schmidt-Cassegrain but may not have the money (or desire) for an electronics-laden instrument. Although the fork arms are not as substantial as those of the LX90 and LX200, and therefore do not dampen vibrations as well, the LX10 is fine for visual observations and some rudimentary astrophotography. Built into the base of the LX10 mount is a DC-powered, worm-gear, right-ascension clock drive that is powered by four AA-size batteries, while a declination motor drive is available optionally. The LX10 also comes with a

6 × 30 straight-through finderscope and a 25-mm MA (modified achromat, similar to a Kellner) eyepiece. The LX10 Deluxe costs about $150 more, but adds a larger, better 8 × 50 finderscope and an electric declination motor. Although the manufacturer also includes an equatorial wedge with the LX10 and LX10 Deluxe, the mating tripod is sold separately. Many purchase the fixed-height LX10 Field Tripod, but the adjustable-height Field Tripod used with the LX90 and LX200 is much sturdier and well worth the extra cost.

The 8-inch LX90 came along at the end of 2000 to replace the LX50 model line. Overall owner satisfaction with this computer-driven instrument is quite favorable. The f/10 LX90 includes several nice features, including Meade's sturdy Field Tripod, an 8 × 50 finderscope, a 1.25-inch star diagonal, and a 26-mm Super Plössl eyepiece, as well as a solid fork mount. The metal arms are cloaked with plastic covers that conceal Meade's Autostar computerized GoTo system. The LX90's Autostar is powered by eight C-sized batteries (also housed inside the plastic arm covers) or adapters, either from an automobile cigarette lighter plug or a standard 115-volt AC home outlet. The LX90's object database incorporates more than 13,000 deep-sky objects, including all of the Messier, NGC, and IC lists, all major members of our Solar System, as well as 26 asteroids, 15 comets, 50 Earth-orbiting satellites, and 200 blank spaces for users to add other objects that are not listed. The Autostar software is also upgradable directly from the Internet. The LX90's GoTo motors can slew the instrument at up to 6.5° per second, and it moves more quietly than the LX200. In actual use, the LX90's mechanics prove to have very good accuracy, consistently placing objects in the field of low-power eyepieces even after a full-sky slew. Getting there, however, requires a bit more button pushing than with the Celestron NexStar 8, the LX90's chief competition.

Some owners express some minor complaints, such as the focusing being stiff and a little uneven; but overall, most seem very happy with their purchase. Keep in mind, since this is, in actuality, an alt-azimuth telescope, long-exposure astrophotography is not possible without an equatorial wedge, an add-on accessory available from Meade.

As nicely outfitted as the LX90 is, Meade's LX200GPS series of computer-controlled Schmidt-Cassegrain telescopes take the technology to the next level. Both the 8- and larger 10-inch LX200GPSs are available with either f/6.3 or f/10 focal ratios, while the 12-inch version is only sold in an f/10 configuration. All are mounted on computer-controlled fork mounts that automatically align themselves using data from Global Positioning Satellites, like the NexStar GPS scopes. After the telescope is set up, select a target from its built-in listing of more than 125,000 objects, press the GoTo button, and the LX200GPS will automatically slew to it at rates up to 8° per second. The LX200GPS is powered by 12 volts DC, either from eight internally held C-size batteries, a car battery (cables are sold separately), or from a wall outlet with an optional 115-volt AC adapter.

The LX200GPS is normally set up in alt-azimuth configuration, which proves much steadier than when the fork mount is tilted by an equatorial wedge. The only shortcoming is that for long-exposure photography, alt-azimuth-induced field rotation can only be eliminated by using an equatorial wedge or the newly introduced Meade #1220 Field De-rotator. While Meade does not supply a wedge with any of the LX200GPSs, the base is designed to fit on Meade's standard-design wedge or their beefier Superwedge.

Meade touts their system as more accurate than Celestron's by featuring 16-channel GPS coverage, as opposed to the competition's 12 channels. Friends educated in GPS technology tell me that the difference would be negligible. Both include many nice features. Most amateurs would be happy with either instrument, but to me, the Celestrons seem more ergonomically friendly. For instance, their hand controllers clip nicely into one of their mounting's arms. Meade's hand controllers have hooks for hanging, but they can get twisted up more readily. Each Celestron also has a much needed handle on the other mounting arm for easy carrying. Although Meade has two carrying handles, both are located along the edge of the fork arms. That positioning forces the user to carry the telescope with both hands extended in front of his or her body. The Celestrons, however, can be carried in one hand like a suitcase (granted, a heavy suitcase).

The LX200GPS's Smart Drive features a periodic-error-correction (PEC) circuit that lets the user compensate for minor periodic worm gear inaccuracies (see the prior description under the "Celestron" heading earlier in this section). The Smart Drive's PEC remembers the steps needed to compensate for the inaccuracies forever once they are input manually by the user. The Celestron NexStar SCTs also have permanent PEC.

As to choosing between focal ratios, I would almost invariably advise against the f/6.3. The reason is simply a matter of image contrast. Because of the large central obstructions from their secondary mirrors, all SCTs suffer from lower image contrast than refractors and most Newtonian reflectors. Therefore, it makes sense to make the secondary mirror as small as possible. To achieve the faster focal ratio, however, the Meade f/6.3 scopes must use a larger secondary mirror than the equivalent f/10s, thereby decreasing contrast. The 8-inch f/6.3 has a 3.45-inch central obstruction (18.6% by area, 43.1% by diameter), compared to 3 inches for the f/10 (14.1% by area, 37.5% by diameter); the 10-inch f/6.3's central obstruction measures 4 inches (16.0% by area, 40% by diameter), while the f/10's measures 3.7 inches (13.7% by area, 37% by diameter). If focal ratio is of concern to you (attention: photographers), consider purchasing an f/10 Schmidt-Cassegrain telescope and a Meade or Celestron Reducer/Corrector attachment that cuts the focal ratio to f/6.3 (reviewed in chapter 6).

Meade also produces 12-inch LX200GPS and 16-inch f/10 non-GPS LX200 Schmidt-Cassegrain telescopes on overgrown mountings. The latter must be initialized manually, but will then automatically find and track celestial objects, like the others. Impressive in both aperture and weight, these observatory-class instruments are also transportable with difficulty. The 12-inch LX200GPS sits atop the Giant Field Tripod, while the 16-inch is available on either the Supergiant Field Tripod, the Permanent Alt-azimuth Pier, or the Permanent Equatorial Pier, the latter two designed for installation in an observatory. Operation is impressive, and while the optics are usually quite good, their performance still does not compare with a well-engineered Newtonian reflector of similar aperture. Of course, most large-aperture Newtonians are supported by Dobsonian mounts, making long-exposure astrophotography much less practical than with an LX200.

So which brand is better: Celestron or Meade? That question has been pondered by amateurs now for more than two decades. Although both companies have been criticized openly in the past for poor quality, they seem to have cleared up most of the deficiencies. As it is now, both produce about equal-quality optics, although most observers agree that Schmidt-Cassegrains universally produce inferior images to high-quality refractors and Newtonians.

First, let's dismiss the Meade 203SC/300, as that instrument is dramatically undermounted. If that is all your budget can afford, consider an 8-inch Dobsonian-mounted Newtonian or a used 8-inch Schmidt-Cassegrain telescope. In the battle between the Meade 203SC/500 and Celestron's G8, it's a toss-up, since the Meade LXD500 mount is comparable to Celestron's CG-5. Both, however, are at the upper limits of their respective mounting's weight range, so purchase either with that understanding. Again, neither is sturdy enough for any photography apart from some prime-focus lunar shots.

The Meade 8-inch LX10 and Celestar 8 are both competent, basic telescopes, although at a price in excess of $1,000, I am hesitant to label them beginners' telescopes. Personally, assuming the optics are comparable, I like the LX10 better for its improved mounting and more accurate clock drive. Meade's LX10 Deluxe is a nice, midrange instrument that really represents the minimum investment needed if you are interested in getting into through-the-telescope astrophotography of deep-sky objects. You will still need to add several accessories before snapping that first photo, as discussed in chapter 7.

In the race between the NexStar 8 and the LX90, I like the LX90. Its database is more versatile, its mounting steadier; I only wish it had a place to clip in the hand controller. The NexStar is noticeably lighter, however, so if weight is a concern, it might be your better bet. I am still concerned about its one-armed mounting, however, especially for astrophotography. The two-armed Celestron NexStar GPS and Meade LX200GPS models get very high marks simply because of their technology. The idea of tapping into the vast GPS network for initializing a telescope is, well, just plain neat. The NexStar GPS works fine, but cannot draw any comparisons to the Meade system, because I have not seen the latter in operation as of this writing.

For larger mountings, the Meade 10-inch SCT wins handily over the Celestron G-9.25, which seems to be a telescope without a true purpose. What a shame to condemn such a fine optical assembly—arguably the finest Schmidt-Cassegrain telescope in production—to such an undersized mount. The CM-1100 is also a competent performer, and a better choice for serious astrophotography, as is the CM-1400 and the Meade 12-inch LX200GPS. Like the smaller LX200 models, the 12-inch requires the optional wedge for long-exposure photography, while the CM-series German equatorial mounts include that feature as part of their design. The mammoth Meade 16-inch LX200 makes a very good observatory-based instrument, but it is not an easy travel companion.

The end of 2001 also saw Meade reintroduce a line of Schmidt-Newtonian telescopes (Meade had sold Schmidt-Newtonians more than a decade ago, but subsequently canceled those models). This time, Meade offers 6-inch f/5, 8-inch f/4, and 10-inch f/4 optical tubes atop their newly introduced LXD55 computerized German equatorial mount. The mounting, which is similar in appearance to Celestron's CG-5, is secured to an adjustable aluminum tripod. It is probably suitable for the lighter 6-inch, but likely wobbles under the extra weight of the heavier models.

As of this writing, no one has actually looked through one of these telescopes, so I can't comment on their optical performance. But why even consider a Schmidt-Newtonian over a conventional Newtonian reflector? If properly made, a Schmidt-Newtonian should produce sharper star images across a wider portion of the field of view than the equivalent short-focus Newtonian by lessening the detrimental impact of coma. (Coma correctors, discussed in the next chapter, accomplish a similar effect with conventional Newtonians.) Keep in mind, however, that a well-made Maksutov-Newtonian will deliver even sharper optical performance than a well-made Schmidt-Newtonian, although it will cost more money.

The LXD55 mount is powered by eight D-cell batteries held in a separate power pack, by an external 12-volt battery (an auto cigarette lighter plug is available), or by plugging it into a 115-volt AC outlet (again, an optional cord is required). In addition to the computerized mounting, the Meade LXD55 Schmidt-Newtonians include 2-inch, all-metal focusers and 6 × 30 finderscopes.

Maksutov Telescopes

Astro-Physics. Known for its standard-bearing apochromatic refractors, Astro-Physics offers a 10-inch f/14.6 Maksutov-Cassegrain instrument. Using an optical design created by Aries Instruments in the Ukraine, the A-P Mak stands out as one of the finest instruments of its type available today. A look at the list of ingredients tells us why. The primary mirror is made from fused silica (quartz) coated with 96% enhanced and protected aluminizing, and it is held in an open mirror cell made from machined aluminum. The instrument's deeply curved meniscus corrector plate is fully multicoated. The A-P Mak is actually a Gregory-Maksutov, a design that eliminates a separate secondary mirror. Instead, a central spot on the inside of the corrector plate is ground and aluminized for that purpose, with two immediate advantages. First, since there is no secondary mirror mount, central obstruction is minimized; in the case of this instrument, only 23% by diameter. The second benefit is there is never a need to recollimate the secondary. (Other instruments using the Gregory-Maksutov design include the Celestron NexStar 4, the Meade ETX, and the Questar.)

Under working conditions, the A-P Mak is very impressive, with pinpoint star images across the entire field of view. Although the instrument is focused by moving the primary mirror, image shift is minimized with the use of a preloaded ball-bearing assembly that moves the mirror at its center, rather than off-axis, as with most others using this method. The oversized focuser will accept either 1.25- or 2-inch eyepieces. In side-by-side viewing, while I was impressed with the Maksutov's image quality and sharpness, it struck me as not quite up to that of a nearby 6-inch A-P apochromatic refractor. Image brightness, however, was noticeably higher in the Mak.

Celestron International. Going from one extreme to another, we move on from one of the most expensive Maksutovs sold today to one of the least. The Celestron NexStar 4, introduced in late 2000, teams a 4-inch f/13 Maksutov tube assembly with the one-armed NexStar computerized mounting. Tasco offers its own virtually identical unit, the StarGuide 4.

Optically and mechanically, the NexStar 4 has its pros and cons. Most units seem to produce reasonably sharp images, although some owners report eccentric patterns when star testing, which would indicate collimation problems. In focus, Jupiter shows two or more belts, while Saturn's rings display Cassini's Division, although not as clearly as through, say, a 4-inch achromatic refractor. Double stars, such as Castor in Gemini, are cleanly split. Internal baffling, so important to a Cassegrain-style instrument like the NexStar 4, appears to be well designed, as there is no evidence of flaring or image washout.

The NexStar 4 comes with a 25-mm SMA (Kellner) eyepiece, which is fine to start, and a Celestron Star Pointer red-dot unity finder, which is far more useful than the tiny 8× finderscope that comes with Meade's ETX, described later in this chapter. Focusing, accomplished by moving the primary back and forth in the tube, is smooth and surprisingly free of image shift. Like the ETX, a *flip mirror* is built into the telescope directly below the permanent 90° star diagonal. This serves to steer light either toward the diagonal or straight out the back of the instrument, where a camera body can be attached for photography.

Mechanically, the NexStar 4 GoTo mounting suffers the same ills as the other small NexStar models. (Review the comments under the NexStar 114 Newtonian for further thoughts.) I am also disappointed that the telescope tube assembly cannot be easily removed from the mounting, as you can with the ETX. It would make a great spotting scope. Finally, everyone seems to complain about the NexStar's push-on plastic dust cap, which falls off easily, especially in the cold. One or two strips of adhesive-backed felt padding along the edge of the cap solves that problem quickly.

Intes. Telescopes have gone through a number of changes since this Russian manufacturer's instruments were first introduced in 1989. At first they suffered from inconsistent optical quality, but since then they have certainly improved. The Intes line of Maksutov-Cassegrain telescopes include the 6-inch f/12 MK66 and MK67, the 6-inch f/6 MK69, the 7-inch f/10 MK72, and the 9-inch f/14 MK91. All deliver what they promise—high-contrast views of planets and deep-sky objects alike, along with sharp star images across the full field of view. Optical tests, including the tell-all star test, show classic images for an obstructed telescope. As with any telescope, the optics must be fully acclimatized to the ambient air temperature to deliver consistently good results. Intes instruments, and Maksutovs in general, take longer than most designs to cool down, in part due to the thick meniscus corrector plate. Expect cool-down time to take between 60 and 90 minutes, possibly longer. But once the optics are on, they really excel. These are great planetary scopes, but they also make fine deep-sky instruments as well (the MK69 is designed primarily for imaging, though it can also be used visually). Double stars are especially striking through a well-collimated Intes Maksutov, with views rivaling refractors costing twice as much. Best of all, there is no residual chromatic aberration to plague the view.

Rather than focus the image by moving the primary mirror along a track inside the telescope tube, like most other catadioptric telescopes do, the MK67, the MK69, the MK72, and the MK91 all use 2-inch Crayford-style focusers (adapters for 1.25-inch eyepieces are not included, and so must be purchased separately). The obvious benefit is the elimination of mirror shift, a common problem in catadioptric telescopes that focus by moving the primary. Focuser travel is quite limited, however, and may require an extension tube to focus some eyepieces. The MK66, on the other hand, focuses internally by moving the primary back and forth. Though it works well, its images are not quite up to the MK67. Two big pluses to the MK66 are that it is threaded to accept all sorts of accessories designed for Celestron and Meade catadioptrics and it does not have the backfocus problems inherent with the MK67.

Many owners complain about the finderscopes supplied with Intes instruments, commenting that their placement is poor, requiring that the telescope's star diagonal be moved out of the way each time the finderscope is used. Intes finderscopes also have very fine crosshairs, which can make them difficult to see under very dark conditions. Either replace the finderscope with a better model or supplement it with a one-power unity finder, like a Telrad or Rigel QuikFinder.

Intes also makes two Maksutov-Newtonians, the 6-inch f/6 MN61 and the 7-inch f/6 MN71, the former also sold by Orion Telescopes as the Argonaut 6″ GP-DX. Optical quality is top drawer, with both instruments showing beautiful, close-up images of Jupiter and Saturn, as well as wonderfully broad views of deep-sky objects. Contrast is especially noteworthy, thanks to the small central obstruction (19% for the MN61, 20% for the MN71). Diffraction spikes, common to Newtonian reflectors, are also eliminated, since the secondary mirror is mounted directly on the meniscus corrector plate. Despite the fast focal ratio, coma is also nonexistent, producing perfect star points across the entire field of view, especially important to astrophotographers. Focusing is again done with a 2-inch Crayford-style focuser on the MN61, while the MN71 can be purchased with either that or a 2-inch helical focuser. In any case, a 1.25-inch adapter will be needed for either telescope. A dew shield also helps slow the onset of corrector-plate fogging.

Several of the Intes Maksutovs are also available in so-called deluxe versions, which include enhanced (96%) coatings for even brighter images and Astral Sitall, which we are told has better thermal properties than the Pyrex-like LK-5 glass used on the standard models. Expect these extras to tack on $150 or more.

Intes Micro. Although similar in name and products, Intes Micro has no relation to Intes, apart from the fact that it was started in 1993 by a former Intes employee. I thought that only happened here in the United States! Since both companies are located in Moscow, Intes and Intes Micro are actually cross-town rivals. To confuse things even more, many of the Intes Micro model numbers are easily confused with those of Intes. Russian marketing, it would seem, is not terribly inventive. But they do make good telescopes.

Let's begin to clear the air by examining Intes Micro's Maksutov-Newtonian lineup, which includes four apertures, the 5-inch f/6 MN56, the 6-inch f/6 MN66, the 7-inch f/6 MN76, and the 8-inch f/6 MN86 (see the confusing similarity to the Intes Mak-Newts above?). Russian marketing aside, all are world-class performers, just like their Intes half-siblings. Small central obstructions and no distracting diffraction spikes create views that are comparable in nearly every way to similarly sized apochromatic refractors. Star clusters sparkle, nebulae softly glow, and planets show some amazing detail through these instruments. Given good seeing conditions, optical quality is sharp enough to let you go crazy, upping the magnification limit to 80× or more per inch of aperture. But what won't look the same as an apochromat is your credit card bill. The MN56, for instance, retails below $1,000, less than half of a 4-inch apochromat. You can buy quite a few eyepieces and other accessories with that savings!

Owners do point out a few quirks, however. Like the Intes telescopes, the Intes Micro finderscopes are quite poor, with terrible eye relief and inferior focusing. Get a new finderscope from the list in chapter 7. The Crayford focusers are also a bit suspect, with the biggest problem being a lack of back-focus, as in some Intes models. An extension tube will quickly solve the problem, however. Finally, again as with Intes instruments, cool-down time is slow.

Intes Micro also offers several custom-made Maksutov-Newtonians, from the 10-inch MN106 to the 16-inch MN165, all in graphite tubes. Nice instruments to be sure, though their prices are astronomical. The 16-inch f/5 MN165, for instance, sells for more than $20,000. Still, when you compare that price to an apochromatic refractor, maybe it's not that bad after all.

On the Maksutov-Cassegrain front, Intes Micro lists several Alter models that range in size from the 6-inch f/10 Alter 603 to the 16-inch f/10 Alter 1608. Once again, owners sing their praises, especially of the image sharpness—which, as one person put it, "whopped" his Schmidt-Cassegrain. As with the SCT, however, do not expect high levels of contrast, as central obstruction from the secondary mirror is in the realm of 30% (measured by diameter). That is bound to soften things up a bit. Another similarity to Schmidt-Cassegrains is the method of focusing, which moves the primary mirror back and forth along a track inside the tube. Although image shift is negligible through Intes Micro scopes, owners do comment on play (or *lag*) when reversing direction. Again, comments about excellent images after the telescope has stabilized are common.

Finally, pricing may be a stumbling block to these instruments' popularity. Eight-inch SCTs from either Meade or Celestron cost less than the 6-inch Alter 603, and while the Alter tends to produce sharper images, the 8-inch SCT will show more in terms of magnitude penetration and resolution (assuming comparable optics on both sides, of course).

LOMO. Short for the Leningradskoe Optiko Mechanichesckoe Objedinenie (Lenigrad Optical and Mechanical Enterprise), LOMO produces a wide variety of optical equipment, ranging from endoscopes and surgical microscopes to astronomical telescopes. Like Intes and Intes Micro, LOMO produces both Maksutov-Newtonians and Maksutov-Cassegrains.

LOMO's best-known astronomical products, at least here in the United States, are probably its Astele Mak-Newts, ranging in size from the 4-inch f/5.5 Astele 102MN to the 8-inch f/4.6 Astele 203MN. Unique to these instruments are finderscopes that are built right into the secondary mirror housings. In each case, a 30-mm objective lens and a small mirror are mounted right inside the secondary mirror mount! Rotate the housing in one direction, and light from the primary mirror is entering the eyepiece. But turn the housing 180°, and light entering the finder's objective is diverted into the eyepiece instead! A neat design conceptually, but one that does not prove to be very practical. For one thing, just trying to aim the telescope accurately is very difficult, although to be fair, adding a Telrad or other unity finder would help immensely.

Images delivered through an Astele 102MN that I tested for *Astronomy* magazine were extremely sharp, indicating excellent optics. The large central obstruction caused by the built-in finderscope, however, impacted image contrast. The 4-inch Astele 102MN suffers the most, with a 1.75-inch central obstruction—a whopping 44%—while the larger instruments are less affected. The finder's grip also protrudes so little from the corrector plate that trying to turn it at night while not touching the plate (especially with gloves) is nearly impossible.

Five LOMO Astele Maksutov-Cassegrain telescopes range in aperture from the 2.8-inch f/13.2 Astele 70 to the 8-inch f/10 Astele 203. All focus their images by moving their primary mirrors back and forth, like most Schmidt-Cassegrains. Execution is quite good, with few complaints heard about image shift, a common problem with this focusing method. The Astele 70 exhibits the most, while the Astele 95 and larger models have a negligible amount.

I also tested the LOMO Astele 70, 95, 133.5, and 203 Maksutov-Cassegrains for *Astronomy* magazine. The Astele 95 and 133.5 were very good optically, with crisp images and excellent star test results. The 203 showed a small amount of spherical aberration, although not enough to adversely affect performance, except on the critical objects. The small Astele 70 suffered from quite a bit of spherical aberration, however.

All LOMO Maksutov-Cassegrains are sold either without mountings, on German equatorial mounts that are akin to those supplied with many imported refractors and reflectors, or on fork equatorial mounts. Exceptions are the small Astele 70, which can be purchased with a tabletop tripod, and the Astele 95, which only comes on the German equatorial. The German-style mounts are sturdy enough to support the smaller apertures, but they are outweighed by the larger models, especially the Astele 180 and Astele 203. The fork-mounted EL instruments include a 9-volt DC powered clock drive on the right-ascension axis with hand controller, manual slow-motion controls on both axes, and an equatorial wedge, but no tripod. Astele fork mounts will also fit on Meade equatorial wedges.

Meade Instruments. The little Meade ETX-90, one of the most popular telescopes to be introduced in the 1990s, is a 3.5-inch f/13.8 Maksutov-Cassegrain designed for maximum portability while also delivering outstanding images. It certainly succeeds on both counts, and at a terrific price. Images are absolutely textbook, with very good contrast and clear diffraction rings. Focusing is precise with no mirror shift detected, giving some wonderful views of brighter sky objects, such as the Moon and the planets. To quote one owner, "the Meade ETX delivers optical performance well beyond its price class." All that I have seen also exemplify fine optics, a great triumph in low-cost, mass-production optical fabrication techniques. The optics are housed in a deep-purple metal tube that is smooth and nicely finished.

The ETX is available in three versions: the equatorially mounted RA version, the upgradable EC, and the unmounted Spotting Scope. All feature the same optical tube assembly, complete with an integral 90° star diagonal and built-in flip mirror that swings out of the way when the ETX is coupled to a camera. The Spotting Scope also includes a 26-mm Meade Super Plössl eyepiece, a tiny 8 × 21 finderscope, a very nice screw-on dust cap, as well as a 45° erecting prism diagonal for upright terrestrial views. Unfortunately, the finder will likely prove unusable for most, simply because it is mounted so close to the tube. I find it difficult, if not impossible, to look through as the telescope raises in altitude, causing my nose to scrunch up against the eyepiece. It might be best to replace the finder with one of the smaller one-power aiming devices described in chapter 7, or make your own using the plans in chapter 8. A metal plate on the bottom of the tube can be used to attach the ETX Spotting Scope to any standard photographic tripod.

The ETX-90RA includes the dust cap, the eyepiece, and the finder that are supplied with the Spotter. It comes mounted on a miniaturized, clock-driven, fork equatorial mount made mostly from molded plastic. The DC-powered clock drive runs for more than 50 hours on three AA-size batteries. Some respondents commented on the amount of backlash in the drive gears, but once the problem works itself out, the drive tracks well, keeping objects in view for half an hour or more. The model I tested had little problem with the clock drive, but did hit its mount's base when I tried to aim below about –30° declination. The problem is that the optical tube assembly is longer than the fork arms, and so, when the telescope is polar aligned, it can't look below a certain point in the southern sky. What that point is exactly will depend on the angle at which the ETX is tilted for polar alignment, but it worsens as you head south.

Lastly, the ETX-90EC looks the same as the RA model, but it also offers the owner a chance to attach the highly praised Autostar hand controller for full, computerized operation. As mentioned previously, the Autostar was the first computer control sold for small instruments. The ETX-90 version includes more than 14,000 objects in its database, which is certainly more than a 3.5-inch telescope could possibly see even under the best conditions. But we can always dream, can't we? While I am not enthusiastic about the finder, the ETX-90EC is a well-made instrument at an amazingly low price. Complete computer tracking and slewing coupled with outstanding optics in such a compact package would have been unimaginable just a few years ago. Yet here it is.

Despite my enthusiasm for the telescope, there are some drawbacks. Optically, it's great, but mechanically, the ETX-90 leaves something to be desired. I already mentioned the poor finderscope, but another difficulty is focusing. Focusing is smooth, but the small, aluminum knob is very difficult to grasp when looking through the eyepiece. Things are just too close together. The plastic fork mounting is also poor, in my opinion. So, here we have it: the ETX-90 is a very good telescope, containing outstanding optics but in a so-so package. Because of its small size and, especially, poor finderscope, I would not recommend it as a first telescope for a beginner. But for someone who may already own a 6- or 8-inch Dobsonian and wants a nice planetary grab-and-go instrument, the ETX is perfect.

Given the high praise from all corners that the ETX-90 has garnered, Meade introduced the 4.1-inch f/14 ETX-105EC and the 5-inch f/15 ETX-125EC. Could the company do it again, striking gold with two more well-made Maksutovs? The short answer is yes! Like the 90, the larger ETXs earn high scores for consistently fine optics. As one owner put it, his ETX-125 continuously resolved tightly spaced double stars with ease, stars that his larger Schmidt-Cassegrain didn't even show as elongated. Saturn's Cassini's Division is a common sight through all. Star testing demonstrates that like the ETX-90, both the 105 and 125 come with optics that are very close to ideal. Internal baffling very effectively blocks any flaring when viewing bright objects. Only some very small amounts of coma can be detected with long-focal-length eyepieces.

The ETX-105 only comes as a computerized EC model, while the 125 is available as either a spotting scope (a rare find, I am told by dealers) or as the computerized ETX-125EC. Both ECs feature Meade's highly acclaimed Autostar system, which includes the same database as the ETX-90. Both share many of the 90's other pros and cons, as well. One improvement, since the overall instrument is larger, is that focusing is a little easier, although still tight. And the 8 × 25 right-angle finderscope is also easier to look through, but not much easier to aim accurately. Indeed, some owners have said that they find it more difficult. Of course, once the Autostar is initialized, the finderscope becomes academic. Like their smaller sibling, the ETX-105 and the ETX-125 are highly recommended for those who are looking for that second, highly portable instrument that is easier to set up than their other telescope. Neither ETX includes a tripod, but Meade recommends their #883 Deluxe Field Tripod. Personally, I prefer a sturdy photographic tripod, such as those by Manfrotto (see chapter 7).

The views through Meade's 7-inch f/15 LX200GPS Maksutov-Cassegrain impress me as sharper and clearer than through their Schmidt-Cassegrain counterparts. Image contrast is much higher and the focus snaps in, with no appreciable mirror shift evident in the several that I tested. Each Mak includes a built-in fan to help acclimate the optics to the ambient outdoor temperature, speeding up the time between setup and optimal optical performance.

The latest version of the Meade 7-inch Mak mates it with the company's GPS-equipped LX200 mount that automates the onboard computer's initialization. More thoughts about this technology are offered in the "Schmidt-Cassegrain Telescopes" section earlier in this chapter.

Perhaps the Mak's biggest drawback is its tube length. The Maksutov uses the same fork mount as the Meade 8-inch SCTs, but its tube is 3 inches longer (19 inches versus 16 inches), making it impossible to stow between the fork tines during transport and storage. Photographers should also note that the Mak's slow focal ratio of f/15 yields narrower fields and requires longer photographic exposures than 8-inch Schmidt-Cassegrains. But if I were in the market for an 8-inch SCT, I would give the Meade 7-inch Maksutov *very* strong consideration.

Orion Optics (UK). In addition to its Newtonian reflectors, reviewed earlier, Orion Optics of Crewe, England, sells a 5.5-inch f/14.3 Maksutov dubbed the OMC-140. The OMC-140 follows the Gregory-Maksutov philosophy of creating the secondary mirror by aluminizing a centralized spot in the meniscus corrector plate, thereby eliminating the need for a separate mirror mount and minimizing the resulting central obstruction (although, in this case, it is still 33% by diameter).

Owners report telescope performance as quite good. The tell-all star test delivers concentric diffraction patterns on either side of focus, as it should. Some minor image shift during focusing is evident, however, as the primary moves back and forth along an internal track. Perhaps the biggest drawback is that the OMC-140's Pyrex primary mirror is very slow to cool and stabilize after being brought out from a warm home. As a result, even if the optics are exceptional, images will not focus until everything settles down. A fan assembly to promote active cooling is said to be in development, but as of this writing, it is not yet available.

The OMC-140 is sold as an 8-pound optical tube assembly only or on several different mountings. It is best combined with the Vixen Great Polaris mount, but it is light enough for Orion Optics' house-branded equatorial mount (made in Taiwan) for visual use. All versions come with a 6 × 30 finderscope, which is not terribly well made, a generic 1.25-inch star diagonal, and a 25-mm Plössl eyepiece.

Orion Telescope Center. Orion Telescope has thrown its hat into the catadioptric ring with a 6-inch Maksutov-Newtonian that is actually made in Russia by Intes. Orion rebadges its MN-61 as their 6-inch Argonaut GP-DX Maksutov-Newtonian, a winning instrument! As with the MN61, images are just marvelous, very sharp and clear, through this instrument. For instance, the Double-Double quadruple star in Lyra, which often appears mushy with lesser optics, focused precisely into two cleanly split pairs of points. Both the primary and secondary mirrors are coated with enhanced (93%) aluminizing for even brighter, crisper images. Pinpoint stars right across the field, terrific image contrast, and no sign of spherical or chromatic aberration all add up to a terrific telescope. Everything about it just feels solid. Orion makes a good thing even better by mounting the Argonaut's rotatable tube assembly on Vixen's GP-DX German equatorial mount, the finest mounting in its price class. What more could you want? Really, only one thing—a 1.25- to 2-inch eyepiece adapter (which can be purchased separately, of course)—but all in all, Orion has really come up with a winning combination here.

In the summer of 2001, Orion introduced the 3.5-inch f/13.9 StarMax 90, 4-inch f/12.7 StarMax 102, and 5-inch f/12.1 StarMax 127 Maksutov-Cassegrain instruments, all imported from China. Their attractive appearance coupled with remarkably low prices are bound to attract wide attention. Each instrument features a seamless aluminum tube painted with a high-gloss burgundy enamel and includes a padded, soft carrying case, a 1.25-inch star diagonal, a finderscope, and a 25-mm Sirius Plössl eyepiece. Each focuses the image by moving the primary mirror back and forth inside the optical tube assembly, as is the case with most other catadioptric telescopes. Image shift is a possibility, although this is usually only evident at higher magnifications. Like many imported refractors, each StarMax's focuser is threaded to accept a standard camera T-mount for astrophotography, which is very convenient.

I purchased a StarMax 127 right after it was introduced and found its optics to be very good in terms of overall quality. Image brightness is excellent, as is sharpness. A minimal amount of spherical aberration is evident at high power and during the star test. This does little to detract from image quality, however. Saturn, for instance, was sharp and clear when viewed through a 7-mm Pentax XL eyepiece. Contrast is also quite good, especially considering the 30% central obstruction created by the secondary mirror.

Questar Corporation. Questar continues to build exceptional quality Maksutov-Cassegrain telescopes, just as it has ever since 1950. A few years ago, it looked as though Questars were going to become things of the past as the company declared Chapter 11 bankruptcy. But with the business end of things now corrected, Questar begins the twenty-first century on an even keel.

Although their business focus now includes such diverse instruments as long-distance microscopes and surveillance systems, Questar still manufactures its famous 3.5-inch f/15 Standard and Duplex models, as well as the larger Astro Barrel 7-inch f/13.4 and Classic 7-inch f/14.3. All represent the pinnacle in terms of quality, but they're going to cost you! For instance, the Standard 3.5-inch Maksutov from Questar retails for about twice the price of a Meade or Celestron 8-inch Schmidt-Cassegrain!

Questar telescopes come outfitted with many little luxuries that add to the user's pleasure. One of the best is actually very simple: a built-in telescoping dew cap that effectively combats fogging of the front corrector plate. Another welcome plus is the screw-on solar filter that allows safe viewing of our nearest star. No other telescope comes supplied with one as standard equipment. The tabletop fork equatorial mounting is both smooth and rigid, while the built-in clock drive accurately tracks the sky. And the quality of the all-metal assembly is without parallel.

The Questar Standard 3.5 is permanently mounted on the fork mount, while the more expensive Duplex version allows the optical tube assembly to be removed and attached to a standard photographic tripod. Both models include 16- and 24-mm Brandon eyepieces, famous for their image quality. Two mountless spotting-scope versions of the Questar 3.5 are also available. Traveling astronomers should consider the Powerguide II option, which switches the clock drive from 120-volt (220-volt optional) AC to 9-volt DC. The manufacturer states that a common 9-volt alkaline battery will operate the drive for up to 50 hours.

Naturally, the small size of the Questar 3.5 dramatically limits what can be seen through it; but of what is visible, the images are exquisite. This appeals to the many amateurs who prefer image quality over aperture quantity. There is certainly something to be said for that philosophy. Think of it as the difference between a fine painting and a snapshot photograph. For most people, the snapshot adequately shows the scene, although it may miss some of the finer nuances. The painting, on the other hand, reaches deeper to touch the soul of those who appreciate such things. After all, a Questar is as much a work of art as it is a scientific instrument.

Apart from the high cost versus aperture ratio, the only real weakness of Questars in my opinion is actually looked upon as a plus by many people. Instead of equipping the telescope with a separate finder, Questar built one right into each of the instruments (except the 7-inch Astro Barrel). By flipping a lever, the observer can switch back and forth between the finder (4× or 6×, depending on whether or not the built-in Barlow lens is engaged) and the main telescope without ever leaving the instrument's eyepiece. In practice, it takes a lot of getting used to before the telescope can be accurately aimed toward a target, but many people believe this to be a great convenience, so I guess it is largely a matter of personal preference. And as the Meade ETX proves, a separate finder is difficult to incorporate successfully on such a small instrument.

What's that, money is no object? Well, if the stock market has been very, very good to you, then consider a Questar 7. Artwork amplified! The Questar 7 Classic, an enlarged version of the 3.5, includes the same style of built-in focuser and finder, while the Astro Barrel 7 features a Tele Vue Starbeam unity finder, the choice of one Plössl eyepiece, and a traditional focuser that accepts standard 1.25- and 2-inch eyepieces, a big plus in the name of versatility. The others can only accept Questar's own custom-made Brandon eyepieces that screw into their eyepiece holders.

Telescope Engineering Company. This company offers eight high-end Maksutov instruments, ranging in size from 6- to 10-inch apertures, that are the dream of many an amateur. All use tubes made from thin aluminum, properly baffled to lessen stray light infiltration. As with other Maks, cooling of the optics, so critical to image quality, takes longer than with some other designs, but TEC telescopes all include built-in fans to speed the process. The collimation of both the primary and secondary mirrors is factory set, and while there is a way to adjust the secondary, the manufacturer suggests leaving it be unless absolutely necessary.

Six of TEC's eight models are configured as Cassegrains. The smallest of the lot is the TEC-6, a 6-inch f/12 Maksutov-Cassegrain telescope that is compact enough to be easily carried and set up, yet has enough aperture to give objects like M13 in Hercules and M51 in Canes Venatici some personality beyond just glowing blurs. Focusing on this and all of the TEC Mak-Cas instruments is done internally by moving the primary mirror. The instrument accepts 2-inch eyepieces, as well as standard 1.25-inch oculars with the optional adapter. Essentials, such as a finderscope and a mounting, are also sold separately.

A trio of TEC 8-inch Maksutov-Cassegrains (f/11, f/15.5, and f/20) also entice discriminating observers. All three instruments show crisp, clean, pinpoint star images across the entire field of view. Reports of viewing the Moon and the planets at magnifications exceeding 600× serve as testimony to their fine quality. All TEC optics are made from Astrositall, a Russian glass-ceramic material that is similar to Zerodur, known for its stable thermal characteristics. Quartz mirrors are available as an extra-cost option.

Looking for more aperture still? TEC also holds a pair of tens... 10-inch f/12 and f/20 Maks, that is. The f/20 TEC MC250/20 is of special interest to those who live for lunar and planetary views, as the telescope's central obstruction of just under 22% (by diameter) offers excellent image contrast.

For Mak-Newtonian fans, the TEC 7-inch f/6 and 8-inch f/3.5 instruments are well suited for wide-field observations and CCD imaging. The 8-inch offers some especially exciting prospects for astrophotographers who want it all—spacious skies, pinpoint stars, and short exposures. Both share the same construction techniques used so successfully in the Mak-Cassegrain models, but with helical focusers. In all cases, mountings must be purchased separately.

Please bear in mind that TEC instruments are not for everyone. One glance at Appendix A will show that. They are quite expensive, considerably more than many other similarly sized instruments. For the price of one of TEC's 8-inch Maksutov tube assemblies, for instance, you can buy a 4- to 5-inch apochromatic refractor tube assembly, a huge Newtonian, or a fully computerized 10- to 11-inch Schmidt-Cassegrain. Which is best for you depends solely on you and your preferences.

Mounting Concern

When the Beach Boys sang "Good Vibrations," they certainly were not singing about telescope mounts! One of the most common complaints of telescope owners remains dissatisfaction with their mountings. In an effort to remedy this situation, many retrofit their instruments with substantially larger, sturdier support systems. Table 5.2 lists most of the higher quality mounts that are available today. Some are suitable for small, portable telescopes, while others are designed to heft the largest instruments described in this chapter.

Some further explanation is needed. First, the mounts are listed in alphabetical order, then according to how much telescope they can carry steadily. Note that the carrying capacity figures are based on the manufacturers' estimates, not actual trials. For the smaller mounts, these are usually good indications of a mount's usefulness for *visual* observations. If you are interested in doing astrophotography, I would recommend that you reduce these figures by 25% to 30%. The column headed "Type" should be self-explanatory. Here, GEM

Table 5.2 ***Telescope Mounts***

Manufacturer/ Model	Type	Weight	Carrying Capacity	Type of Support	Slow-Motion Controls	Clock Drive	Setting Circles	Polar Scope	DSC or GoTo?	Price Range
Astro-Physics 400GTO	GEM	21+	20	Opt	✓	✓	✓	Opt	GoTo	D
Astro-Physics 600GTO	GEM	27+	40	Opt	✓	✓	✓	Opt	GoTo	D
Astro-Physics 900GTO	GEM	50+	70	Opt	✓	✓	✓	Opt	GoTo	E
Astro-Physics 1200GTO	GEM	91+	140	Opt	✓	✓	✓	Opt	GoTo	E
Celestron CG-3	GEM	19	10	Wood tripod	✓	Opt	✓	Opt	N/A	A
Celestron CG-5	GEM	33	20	Stamped alum tripod	✓	Opt	✓	Opt	DSC Opt	A
Celestron CI-700	GEM	90	60	Tubular alum tripod	✓	✓	✓	✓	DSC Opt	C
Losmandy GM-8	GEM	43	30	Square alum tripod	✓	✓	✓	Opt	Opt	C
Losmandy G-11	GEM	92	60	Tubular alum tripod	✓	✓	✓	Opt	Opt	C
Losmandy HGM-200	GEM	280	125	Tubular alum tripod	✓	✓	✓	Opt	Opt	E
Meade LXD500	GEM	35	20	Stamped alum tripod	✓	Opt	✓	Opt	DSC opt	A
Meade LXD650	GEM	50	30	Tubular steel tripod	✓	Opt	✓	Opt	Opt	C
Meade LXD750	GEM	100	50	Tubular steel tripod	✓	Opt	✓	Opt	Opt	D

Mountain Instruments										
MI-250	GEM	70+	65	Opt	✓	✓	✓	n/a	DSC opt	D
Optical Guidance Systems										
HP-75	GEM	157	110	Opt	✓	✓	✓	n/a	DSC opt	E
Optical Guidance Systems										
OGS-75	GEM or Fork Eq	170+	130	Opt	✓	✓	✓	n/a	DSC opt	E
Optical Guidance Systems										
OGS-100	GEM or Fork Eq	290+	200	Opt	✓	✓	✓	n/a	DSC opt	E
Optical Guidance Systems										
OGS-140	GEM or Fork Eq	830+	300	Permanent pier	✓	✓	✓	n/a	DSC opt	E
Optical Guidance Systems										
OGS-190	GEM or Fork Eq	1,920+	450	Permanent pier	✓	✓	✓	n/a	DSC opt	E
Orion Telescope										
AZ-3	Alt-az	10	6	Stamped alum tripod	✓	n/a	n/a	n/a	n/a	A
Orion Telescope										
EQ-1	GEM	12	6	Stamped alum tripod	✓	Opt	✓	n/a	n/a	A
Orion Telescope										
EQ-2	GEM	18	9	Wood tripod	✓	Opt	✓	n/a	n/a	A
Orion Telescope										
SkyView Deluxe	GEM	24	13	Stamped alum tripod	✓	Opt	✓	✓	n/a	A
Parallax										
125	GEM	39+	50	Wood tripod or pier	✓	✓	✓	Opt	Opt	D
Parallax										
HD150	GEM	115+	75	Pier	✓	✓	✓	Opt	Opt	D
Parallax										
HD200	GEM	130+	120	Pier	✓	✓	✓	Opt	Opt	D
Parallax										
HD300C	GEM	430+	300	Permanent pier	✓	✓	✓	✓	DSC opt	E
Quadrant Engineering										
EQ-2	GEM	90	200	Pier	✓	✓	✓	Opt	✓	E

(*continued*)

Table 5.2 ***(continued)***

Manufacturer/ Model	Type	Weight	Carrying Capacity	Type of Support	Slow-Motion Controls	Clock Drive	Setting Circles	Polar Scope	DSC or GoTo?	Price Range
Quadrant Engineering EQ-3	Fork	40	65	Opt	✓	✓	✓	n/a	✓	D
Software Bisque GT-1100S	GEM	140+	85	Opt	✓	✓	✓	✓	✓	E
Synta EQ-6	GEM	54	40	Tubular steel tripod	✓	✓	✓	✓	Opt	B
Takahashi Teegul	Alt-az	n/s	15	Tripod	n/s	n/a	n/a	n/a	n/a	B
Takahashi P2Z	GEM	n/s	15	Wood tripod	✓	✓	✓	✓	Opt	C
Takahashi EM-2	GEM	23	15	Opt	✓	✓	✓	✓	Opt	C
Takahashi EM-10	GEM	24	20	Opt	✓	✓	✓	✓	Opt	D
Takahashi EM-200	GEM	56	35	Opt	✓	✓	✓	✓	Opt	D
Takahashi NJP-160	GEM	51	65	Opt	✓	✓	✓	✓	Opt	E
Takahashi EM-500	GEM	100	85	Opt	✓	✓	✓	✓	Opt	E
Takahashi EM-3500	GEM	n/s	350	Pier	✓	✓	✓	✓	Opt	E
Tele Vue Panoramic	Alt-az	10	8	Wood tripod	n/a	n/a	n/a	n/a	DSC opt	B
Tele Vue Gibraltar	Alt-az	17	13	Wood tripod	n/a	n/a	n/a	n/a	DSC opt	A

Vixen Custom	Alt-az	10	8	Stamped alum tripod	✓	n/a	n/a	n/a	DSC opt	A
Vixen Custom D	Alt-az	18	13	Stamped alum tripod	✓	n/a	n/a	n/a	DSC opt	A
Vixen Great Polaris	GEM	26 / 46	15	Stamped alum tripod OR pier	✓	Opt	✓	✓	Opt	B
Vixen Great Polaris-DX	GEM	36 / 51	22	Stamped alum tripod OR pier	✓	Opt	✓	✓	Opt	C
Vixen Atlux	GEM	88 / 100	48	Tubular alum tripod OR pier	✓	✓	✓	✓	✓	E
William GT-1	GEM	65	40	Wood tripod	✓	✓	✓	Opt	✓	D
William GT-1HD	GEM	70	50	Pier	✓	✓	✓	Opt	✓	D

n/a: not available.
n/s: not supplied by manufacturer.

indicates German equatorial mount. Be sure to also pay close attention to the type of support. Tripods and piers are, in general, indicative of portability, while permanent piers are for observatory-bound instruments. In many cases, the supports are not included in the base price of the mount. In that case, they are listed as "opt," for optional. Other accessories, such as polar-alignment scopes, digital setting circles (DSC), GoTo computer control, and clock drives are also noted as "opt" if they are available at extra cost. Finally, I have grouped the mounts within several price ranges: A = $500 and less; B = $501 to $1,000; C = $1,001 to $2,000; D = $2,001 to $5,000; and E = $5,000 and up.

From this table, I think you can see that there are a lot of telescope mounts from which to choose. Before making a choice, consider the following advice. First—and you should be well aware of this by now from reading the earlier telescope reviews—stamped aluminum tripod legs are weak. They tend to twist and vibrate more than other designs, transmitting those motions to the mounting and, in turn, the telescope. Tubular metal legs, such as those supplied on Meade's and Celestron's fork-mounted Schmidt-Cassegrain telescopes, are far more stable. This only stands to reason, because a rigid circle is the strongest cross-sectional shape there is (for those who care, a triangle is in second place, while quadrilaterals, such as squares, are a distant third).

If your mount has a stamped aluminum tripod that gives you the shakes, you can modify it to help improve stability. One simple fix seems obvious, but it is often ignored: make sure everything is tight. Screws often loosen just a bit as telescopes are transported and swung to and fro. Connections that are even slightly loose can certainly cause problems. Places to watch: the altitude adjustment and the mount-leg connections. Another trick is to fill the tripod legs with an inert material. Some have suggested sand, but foam insulation from an aerosol can is a better choice. The foam sprays out of the can as a liquid but quickly expands and solidifies, filling the voids in the tripod legs. Best of all, weight gain is negligible. Placing a set of the vibration pads sold by Celestron (see chapter 7) under each leg also helps, as does suspending a weight, such as a plastic gallon container of water, from directly under the mount.

Other fixes are offered by Jeff DeTray, an amateur astronomer from New Hampshire. He points out that in the case of the Celestron CG-5, at least, a very thick lubricating grease installed at the factory tends to gum up the motions. He recommends taking the equatorial mount apart and cleaning off the grease with a solvent (in a well-ventilated area, please!). Before reassembly, feel each interior metal surface for rough and uneven spots. Use a sharpening stone or a fine, cloth-backed sandpaper called Crocus cloth to deburr any rough spots that you find. Then clean the parts again and apply a thin layer of grease (wheel-bearing grease is fine, but keep it light). Assemble everything in reverse order, checking tightness as you go. Lastly, adjust the worm gearing to eliminate any play. Now, admittedly, all this is too much to detail adequately here. Instead, I'll invite you to visit the Star Ware section of my home page (www.philharrington.net), where you will find a link to DeTray's web site, which features step-by-step instructions and photographs.

Even with these improvements, wood is still much better at dampening vibrations than metal. Unfortunately, few companies seem to supply wooden tripods anymore. That being the case, consider either making your own wooden tripod by following the plans in chapter 8 or by purchasing a set of aftermarket legs. NatureWatch of Charlottetown, Prince Edward Island, sells nicely crafted, extendable tripod legs made from oak that are designed specifically as replacements. Purchasers say that the difference is amazing and well worth the price.

Looking at the mounts listed in Table 5.2, several stand out as excellent values for their price and purpose. In the sub-$300 category, the Orion SkyView Deluxe is a good buy for up to 4-inch reflectors and 6-inch reflectors, but with my now-standard caveat about the aluminum legs. Raising the price bar to $500, the Celestron CG-5 and the Meade LXD500 German equatorial mounts are also quite good, again given the caution about the legs. Be sure not to overload them, to prevent twisting and oscillations. The Tele Vue Gibraltar alt-azimuth mount is very nicely crafted, though it lacks the convenience of slow-motion controls. But what's this? The Gibraltar actually has a wooden tripod, available with legs made from either solid ash or walnut!

Upping the ante to $1,000, Vixen's Great Polaris mount is the best buy. The GP-DX mount is also an outstanding product for heavier instruments. Doubling the investment, the Losmandy GM-8 and G-11 are both very well-executed mounts, with very intelligently designed features. (Note that once you figure in the cost of dual-axis drive motors, the GP-DX will end up costing about the same as the GM-8, which comes with drive motors as standard.) And finally, you can't go wrong with a mounting from Astro-Physics, Takahashi, or the Mountain Instruments MI-250, although they are all expensive. Challenging their leads in the portable mount arena are three mounts from Synta and William Optics. I personally used instruments mounted on the William GT-1 and GT-1HD (heavy duty) mounts at the 2001 Riverside Astronomy Expo in California and was extremely impressed with both. Vibration dampening was truly excellent. Early reports say that the Synta EQ-6 is also very exciting, though I have not personally used it. All bear strong consideration.

For permanent observatory installations, mounts from Optical Guidance Systems, Parallax Instruments, and Software Bisque are all magnificent. If it's the right size for your telescope, Software Bisque's Paramount GT-1100S is my personal favorite, although that is certainly not meant to disparage the others in this stratospheric price range, all of which are fine products in their own right.

The Scorecard

With so many telescopes and so many companies from which to choose, how can the consumer possibly keep track of everything? Admittedly, it can be difficult, but hopefully Appendix A will help a little. It lists all of the telescopes mentioned above, sorted by *price range*. Placement within each range is based on "street" prices, not necessarily the manufacturer's suggested retail prices. Frequently, MSRPs are artificially inflated, perhaps in an effort to make consumers believe that they are getting deals.

There are many things to look for when telescope shopping. If you are thinking about buying binoculars or a refractor, make certain that all the optics are fully coated with a thin layer of magnesium fluoride (abbreviated MgFl) to help reduce lens flare and increase contrast. As mentioned before, multiple coatings are the best. For reflectors and catadioptrics, check to see if their mirrors have enhanced aluminum coatings to increase reflectivity. Find out if the telescope comes with more than one eyepiece. Is a finderscope supplied? If so, how big is it? Though a 6 × 30 finderscope might be fine to start, most observers prefer at least an 8 × 50 finderscope; anything smaller than a 30-mm finderscope is worthless. If the telescope does not come with a finder, then one must be purchased separately before the instrument can be used to its fullest potential.

Next, take a long, hard look at the mounting. Does it appear substantial enough to support the telescope securely, or does it look too small for the task? Do what I call the rap test. Hit the side of the telescope tube lightly with the ball of your hand while peering through the eyepiece at a target. If the vibrations disappear in less than 3 seconds, the mounting has excellent damping properties; 3 to 5 seconds is good; 5 to 10 seconds is only fair; while greater than 10 seconds is poor. Remember everything that you and I have gone over in this chapter up to now and, above all, be discriminating.

Without a doubt, the best way of getting to meet many kinds of telescopes personally is to join a local astronomical society. Chances are good that at least one member already owns the telescope that you are considering and will happily share personal experiences, both good and bad. Plan on attending a club observing session, or star party as it may be called. Here, members bring along their telescopes and set them up side by side to share with one another the excitement of sky watching. To find the club nearest you, contact a local museum or planetarium to find out if there is one in your area. If you have access to the Internet, visit my personal web site (www.philharrington.net) for links to the Astronomical League and other directories of astronomy clubs around the world.

Alternatively, if there is no astronomy club anywhere near you, and you have access to the Internet, consider joining one or more online astronomy discussion groups. I'll extend a personal invitation to visit and join my "Talking Telescopes" online discussion group, which boasts thousands of members from around the world. There we talk about our experiences with telescopes of all shapes and sizes. It's a great learning tool for everyone, including me! Again, information can be found at www.philharrington.net.

If you do get to a club star party, be sure to look through every instrument. Bypass none, even if you are not interested in that particular telescope. When you find one that you are considering, speak to its owner. If the telescope is good, he or she will brag just like a proud parent. If it is poor, he or she will be equally anxious to steer you away from making the same mistake. Listen to the wisdom of the owner and compare his or her comments with the advice given in this chapter.

Next, ask permission to take the telescope for a test drive so that you may judge for yourself its hits and misses. Begin by examining the mechanical integrity of the mounting. Tap (gingerly, please) the mounting. Does it vibrate? Do the vibrations dampen out quickly or do they continue to reverberate? Try the same test by rapping the mounting and tripod or pedestal. How rapidly does the telescope settle down?

Working your way up, check the mechanical components of the telescope itself. Does the eyepiece focusing knob(s) move smoothly across the entire length of travel? If you are looking at a telescope with a rack-and-pinion or Crayford-style focusing mount, does the eyepiece tube stop when the knob is turned all the way, or does it separate and fall out? Is the side-mounted finderscope easily accessible?

When you are satisfied that the telescope performs well mechanically, examine its optical quality. By this time, no doubt the owner has already shown you a few showpiece objects through the telescope, but now it is time to take a more critical look. One of the most telling ways to evaluate a telescope's optical quality is to perform the star test outlined in chapter 9. It will quickly reveal if the optics are good, bad, or indifferent.

How should you buy a telescope? Some manufacturers only sell factory-direct to the consumer, while others have networks of retailers and distributors. When it comes time to purchase a telescope, shop around for the best deal but do not base your choice on price alone; be sure to compare delivery times and shipping charges as well! Some of the more popular telescopes, such as those from Celestron, Tele Vue, and Meade, are available from dealer stock for immediate delivery. At the opposite end of the telescope spectrum are other companies whose delivery times can stretch out to weeks, months, or even more than a year! Consult Appendix C for a list of distributors, or contact the manufacturer for your nearest dealer.

Once you decide on a telescope model, it is best, if possible, to purchase the telescope in person. Not only will you save money in crating and shipping charges, but you will also be able to inspect the telescope beforehand to make sure all is in order and as described. Though most manufacturers and distributors strive for customer satisfaction, there is always the possibility of trouble when merchandise is mail-ordered.

"Don't Worry . . . the Check Is in the Mail"

Okay, I admit it. I hold a grudge. A problem that I had with a well-known source of astronomical equipment led to this section of the book. Briefly, an eyepiece that I received from them had a small chip in one of the inner lens elements. I called them, and was told to return it for replacement or refund. I opted for replacement, so I packaged it up and sent it off. Two weeks went by, but nothing happened. I called again and was assured that the faulty eyepiece had been received and that a new eyepiece would be sent to me by week's end. Fourteen more days went by, and still nothing. I called them again, at which time the owner assured me the eyepiece had been sent out a few days earlier and that it should reach me any day. Two more weeks elapsed, and with no eyepiece in hand, I wrote to the owner demanding an explanation. Within a week, I received a refund check and a surprisingly nasty letter stating that the package containing my returned eyepiece arrived damaged because of my negligent packing (earlier, they had told me it arrived in fine shape). Further, the letter continued, they did *not* want me as a customer again because of my "attitude"! The company in question? Sorry, I can't mention it by name, but I can say that they are *NOT* listed anywhere in this book. It is hoped that they will be made conspicuous by their absence!

Happily, this unfortunate experience is the exception, not the rule. The vast majority of astronomical companies are owned and operated by competent, friendly people. They want happy customers (remember, a happy customer is a repeat customer) and guard their good reputations jealously. Most are willing to bend over backward to see that a problem is resolved to the customer's satisfaction. But what can the consumer do if he or she is dissatisfied with a manufacturer or distributor?

Begin on the right foot. Before returning a defective piece of merchandise, always speak to the manufacturer first about the problem. Request instructions for the most expeditious way to return the item for replacement or refund. Conform to the directions precisely, but to protect yourself always follow up the conversation with a letter in which you repeat the nature of the problem as well as the desired outcome. Send the letter by certified mail, return receipt requested, and keep a copy for your records.

Allow the company a reasonable length of time to respond to your complaint, typically 2 to 4 weeks. If after that time a satisfactory resolution has not been reached, write to the company again and inquire as to the delay. State that you expect a response within a given period of time, say, 10 business days. Once again, send the letter by certified mail, return receipt requested, and keep a copy for your records. If there is still no response, call the company and find out the owner's name. Write to him or her directly, recounting all that has happened since the item was ordered.

By now, the predicament should have been resolved, but if it has not, then it's time to take action. The major astronomical periodicals do not have on-staff consumer advocates, yet they do take an active interest in consumer satisfaction with all who advertise in their magazines. Write to them with your complaint, being certain to send a copy to the president/owner of the offending company. In addition, send a copy of the letter to the Astronomical League (see Appendix E for the address). The league is also interested in customer satisfaction and may offer assistance. If you suspect mail fraud, also contact your local postal inspector or complete Form 8165, *Mail Fraud Complaint Questionnaire,* which is available at all U.S. post offices. Return the completed form to the postmaster or mail it to the following address: Chief Postal Inspector, U.S. Postal Service, 475 L'Enfant Plaza SW, Room 3021, Washington, DC 20260-2100. You should also file a complaint with the Federal Trade Commission, 600 Pennsylvania Avenue NW, Washington, DC 20580. If you prefer, complaints may be filed with the FTC on the Internet at www.consumer.gov/ sentinel. I would also like to hear about your problem for future reference. I received several complaints about a particular dealer after the first edition came out, for instance, and dropped its listing in the subsequent editions.

Most consumer advocates recommend charging all mail orders to a major credit card; do not use a check or money order, if possible. Using a credit card gives you certain powers that are not available any other way. On the back of every credit card's monthly statement, you will find instructions that clearly describe the steps to be taken in the event of a consumer problem. Usually, the card requires that the consumer describe in specific detail the exact nature of the problem and provide copies of all receipts and documentation. The charge will then be put in contest until the problem is resolved. If a charge is contested, the consumer is not responsible for any interest that may accrue as a result. When a final determination is made, either a credit will be issued to the charge account or the balance plus interest will be due.

Contesting a charge should be viewed as a measure of last resort. Only put a charge in contest when a bona fide problem exists and the vendor refuses to cooperate. For instance, just because you decided that you don't like an item anymore is not reason enough to contest a charge, but poor quality or workmanship is.

Honest Phil's Used Telescopes

What if you took the pop quiz in the last chapter and found out that the best telescope for your needs was, say, an 8-inch Schmidt-Cassegrain, but you cannot afford to spend $1,500 to get the instrument you want? What can you do? If you needed to buy a car, but could not afford to buy a new one, the odds are that you would check the used car classified advertisements, right? If it works for cars, then why not for telescopes? All other things being equal, an old telescope will work just as well as a new telescope as long as it was treated kindly. *Astronomy* magazine has a small listing of readers' classified advertisements in each issue, while you web surfers can check the used-equipment advertisement services listed in Appendix E. Once again, links are available from my web site.

Look through your local papers' classified advertisements to see if anything strikes your fancy. If possible, restrict your search to an area that is within a day's drive so that you can check out the telescope firsthand instead of relying on a stranger's word. One person's treasure is another person's junk (and vice versa)!

What should you look for in a used telescope or binoculars? Essentially the same things as in a new one. You want to check the instrument both optically and mechanically. Inspect the instrument for any damage or mishandling. Are the optics clean? Did the owner store the telescope properly?

Table 5.3 might help you find a prince among the frogs by listing an inventory of the 10 best telescopes of yesteryear, as voted by survey respondents,

Table 5.3 ***Some of the Best Telescopes of Yesteryear***

1. Astro-Physics 6-inch f / 12 "Super Planetary" apochromatic refractor
2. Cave 6-inch (or larger) f / 8 Newtonian reflector
3. Celestron SP-C102F or GP-C102F, a 4-inch f / 9 fluorite apochromatic refractor
4. Celestron Super C8+, an 8-inch f / 10 SCT
5. Criterion "Dynascope" RV-6 (6-inch f / 8) or RV-8 (8-inch f / 7) Newtonians
6. Meade Research Grade Newtonian reflectors (6-inch or larger)
7. Optical Craftsmen "Connoisseur" 6-inch (or larger) f / 8 Newtonian reflector
8. Quantum 4, a 4-inch Maksutov-Cassegrain
9. StarMaster Oak Classic, a 7-inch f / 5.6 Newtonian reflector
10. Unitron refractors, 3-inch or larger, pre-1980

listed in alphabetical order. Bear in mind that prices may vary greatly for the same instrument depending on its condition.

Congratulations, It's a Telescope!

Whether your telescope is new or used, resist the urge to uncrate your baby immediately once you get it home. Instead, read its instruction manual from cover to cover. When you are done, read it again. Absorb all the information it has to offer. Do everything slowly and deliberately. Remember, the universe has been around for billions of years; it will still be there when you get the telescope together! If you have any questions, call the dealership from whom you bought the instrument. Though the employees there may not know the answer, they should at least have the manufacturer's phone number (if not, check Appendix C). Some companies even have technical assistance lines set up for just such an emergency.

With everything together and the skies finally clear (why is it always cloudy whenever you get a new telescope?!), take your prize outside for "first light." Pick out something special to look at first (I always choose Saturn, if it's up) and enjoy the view! By following all of the steps here as well as the other suggestions found throughout the chapters yet to come, you will be well on your way toward a fantastic voyage that will last a lifetime.

6

The "Eyes" Have It

Have you ever tried to look through a telescope without an eyepiece? It doesn't work very well, does it? Sure, you can stand back from the empty focusing mount and see an image at the telescope's focal plane. But without an eyepiece in place, the telescope's usefulness as an astronomical tool is greatly limited, to say the least.

Until recently, eyepieces (Figure 6.1) were thought of as almost second-class citizens whose importance was considered minor compared to a telescope's prime optic. With few exceptions, many eyepieces of yore suffered from tunnel vision as well as an assortment of aberrations. The 1980s, however, saw a revolution in eyepiece design. In the place of their lackluster cousins stood advanced optical designs that brought resolution and image quality to new heights. With the possible exception of selecting the telescope itself, picking the proper eyepiece(s) is probably the most difficult choice facing today's amateur astronomers.

Although eyepieces (or oculars, if you prefer) are available in all different shapes and sizes, let's begin the discussion here with a few generalizations. Figure 6.2 shows a generic eyepiece with its components labeled. Regardless of the internal optical design, the lens element(s) closest to the observer's eye is always referred to as the *eye lens,* while the lens element(s) farthest from the observer's eye (that is, the one facing inward toward the telescope) is called the *field lens.* A field stop is usually mounted just beyond the field lens at the focus of the eyepiece, giving a sharp edge to the field of view as well as preventing peripheral star images of poor quality from being seen.

While the eyepiece must be sized according to the diameter of the eyepiece optics, the barrel (the part that slips into the telescope's focusing mount) is always one of three diameters. Most amateur telescopes use 1.25-inch diameter eyepieces, a standard that has been around for years. At the same time,

Figure 6.1 *A selection of eyepieces. Left to right are the Pentax 7-mm XL, Sky Instruments Antares 32-mm Erfle, Vixen 17-mm Lanthanum LVW, Gary Russell 19-mm Konig, and Tele Vue 22-mm Panoptic. A generic 10-mm Plössl eyepiece is in the foreground.*

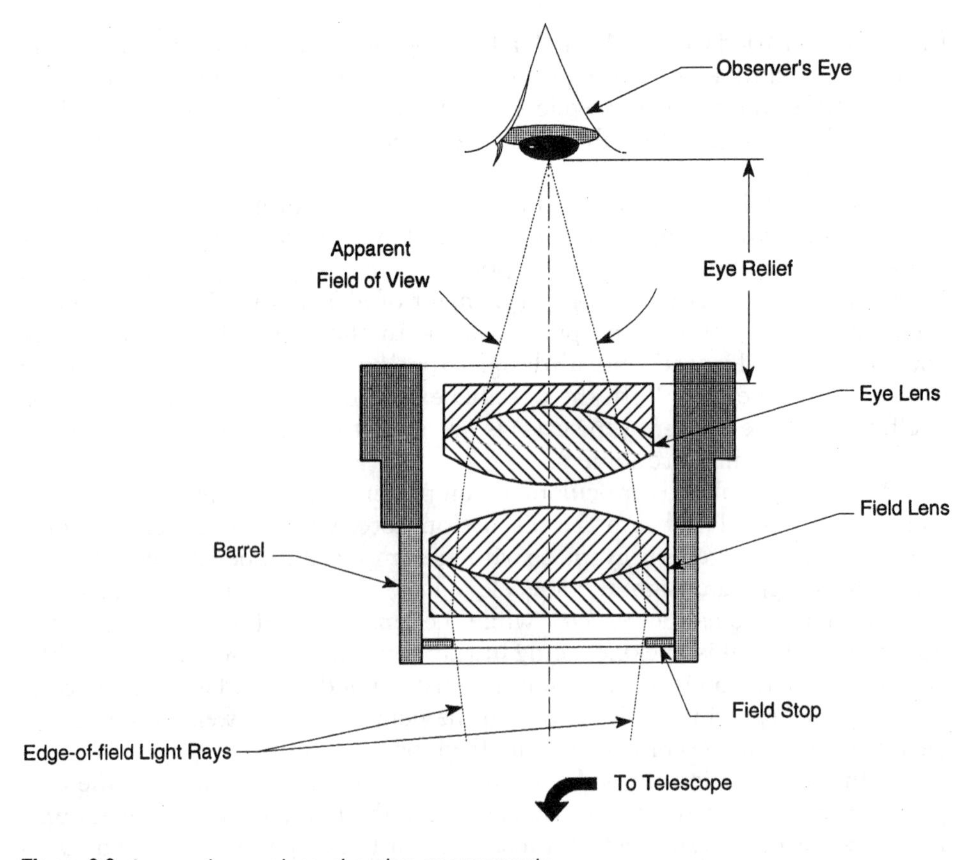

Figure 6.2 *A generic eyepiece showing components.*

many less-expensive, department-store telescopes are outfitted with 0.965-inch oculars. Finally, the astronomical community has seen a recent boon in a whole new breed of giant eyepieces with 2-inch barrels.

Before looking at specific eyepiece designs, we must first become fluent in the terms that describe an eyepiece's characteristics and performance. There are surprisingly few. Perhaps most important of all is *magnification* (or maybe I should say *lack of magnification*). As previously outlined in chapter 1, magnification is equal to the focal length of the telescope divided by the focal length of the eyepiece. Therefore, the longer the telescope's focal length, the greater the magnification from a given eyepiece.

Wouldn't it be nice if there was one specific magnification value that would work well in every telescope for every object in the sky? Sadly, this is simply not the case. For certain targets, such as widely scattered star clusters or nebulae, lower powers are called for. To get a good look at the planets or smaller deep-sky objects (e.g., planetary nebulae and smaller galaxies), higher powers are required. If the magnification is too high for a given telescope, then image integrity will be sacrificed.

Just how much magnification is too much and how much is just right? A good rule of thumb is to use only as much magnification as needed to see what you are interested in looking at. If you own a smaller telescope (that is, 8 inches or smaller in aperture), the oft-repeated rule of not exceeding 60× per inch of aperture is suggested. This means that an 8-inch telescope can operate at a maximum of 480×, but remember this value is *not* cast in stone. It all depends on your local atmospheric conditions and the instrument. Given excellent optics, you may be able to go as high as 90× or even 100× per inch on some nights, while on others, 30× per inch may cause the view to crumble.

On the other hand, larger telescopes (e.g., instruments greater than 8 inches in aperture), especially those with fast focal ratios, can seldom meet or exceed the 60×-per-inch rule. Instead, they can handle a maximum of only 30× or 40× per inch.

The choice of the right magnification is one that must be based largely on past experience. If you are lacking that experience, do not get discouraged; it will come with time. For now, use Table 6.1 as a guide for selecting the maximum usable magnification for your telescope.

Notice how the table tops out at 300× for all telescopes beyond eight inches. Experience shows (and I'm sure that some who are reading this now will disagree vehemently) that little is gained by using more than 300× to view an object, regardless of aperture. Only on those rare nights when the atmosphere is at its steadiest and the stars do not appear to twinkle can this value be bettered.

Another important consideration when selecting an eyepiece is the size of its *exit pupil,* which is the diameter of the beam of light leaving the eyepiece and traveling to the observer's eye, where it enters the pupil. You can see the exit pupil of a telescope or binocular by aiming the instrument at a bright surface, such as a wall or the daytime sky. Back away and look for the little disk of light that appears to float just inside the eye lens.

Table 6.1 ***Telescope Aperture versus Maximum Magnification***

Telescope Aperture		Magnification	
(in.)	(cm)	Theoretical (60×/inch)	Practical
2.4	6	144	100
3.1	8	186	125
4	10	255	170
6	15	360	240
8	20	480	300
10	25	600	300
12.5	32	750	300
14	36	840	300
16	41	960	300
18	46	1080	300
20	5	1200	300
25	64	1500	300
30	76	1800	300

How can you find out the size of the exit pupils produced by your eyepieces in your telescope? Easily, using either of the two formulas shown below:

$$\text{Exit pupil} = D/M$$

where
D = the diameter of the telescope's objective lens or primary mirror in millimeters
M = magnification

or

$$\text{Exit pupil} = F_e/f$$

where
F_e = the focal length of the eyepiece in millimeters
f = the telescope's focal ratio (its f-number)

Knowing the diameter of the exit pupil is a must, for if it is too large or too small, the resulting image may prove unsatisfactory. Why? The pupil of the human eye dilates to about 7 mm when acclimated to dark conditions (though this varies from one person to the next and shrinks as you age). If an eyepiece's exit pupil exceeds 7 mm, then the observer's eye will be incapable of taking in all the light that the ocular has to offer. Many optical authorities would be quick to point out that an excessive exit pupil wastes light and resolution. This is not necessarily the case for the owners of refractors. Say you own a 4-inch f/5 refractor and wish to use a 50-mm eyepiece with it. The resulting exit pupil from this combination is 10 mm. Though the exit pupil is technically too large, this telescope-eyepiece combination would no doubt provide a wonderful low-power, wide-field view of rich Milky Way star fields when used under dark skies.

The key phrase in that last sentence is *when used under dark skies.* If the same pairing was used under mediocre suburban or urban sky conditions, then the contrast between your target and the surrounding sky would suffer greatly. This is due to the fact that the eyepiece is not only transmitting starlight, it is also transmitting skyglow (light pollution)—*too much* sky glow.

What about using a 50-mm eyepiece with a 4-inch f/5 *reflector*? Sorry, not a good idea. With conventional reflectors as well as catadioptric instruments, obstruction from the secondary mirror will create a noticeable black blob in the center of view when eyepieces yielding exit pupils greater than about 8 mm are used. With these telescopes, it is best to stick with eyepieces of shorter focal lengths.

On the other hand, if the exit pupil is too small, then the image will be so highly magnified that the target may be nearly impossible to see and focus. Just as there is no single all-around best magnification for looking at everything, neither is there one exit pupil that is best for all objects under all sky conditions. It depends on what you are trying to look at. Table 6.2 summarizes (rather subjectively) my personal preferences.

Although magnification gives some feel for how large a swath of sky will fit within an eyepiece's view, it can only be precisely figured by adding another ingredient: the ocular's *apparent field of view.* Nowadays, most manufacturers will proudly tout their eyepieces' huge apparent fields of view since they know that big numbers attract attention. Unfortunately, few take the time to explain what these impressive figures actually mean to the observer.

The apparent field of view refers to the eyepiece field's edge-to-edge angular diameter as seen by the observer's eye. Perhaps that statement will make more sense after this example. Take a look at Figure 6.3a. Peering through a long, thin tube (such as an empty paper towel roll), the observer sees a very narrow view of the world—an effect commonly known as *tunnel vision.* This perceived angle of coverage is also known as the apparent field of view. To increase the apparent field of view in this example, simply cut off part of the cardboard tube. Slicing it in half (Figure 6.3b), for instance, will approximately double the apparent field, resulting in a more panoramic view of things.

In the world of eyepieces, the apparent field of view typically ranges from a cramped 25° to a cavernous 80° or so. Generally, it is best to select eyepieces with at least a 40° apparent field because of the exaggerated tunnel-vision

Table 6.2 **Suggested Exit Pupils for Selected Sky Targets**

Target	Exit Pupil (mm)
Wide star fields under the best dark-sky conditions (e.g., large star clusters, diffuse nebulae, and galaxies)	6 to 7
Smaller deep-sky objects; complete lunar disk	4 to 6
Small, faint deep-sky objects (especially planetary nebulae and smaller galaxies); double stars, lunar detail, and planets on nights of poor seeing	2 to 4
Double stars, lunar detail, and planets on exceptional nights	0.5 to 2

Figure 6.3 *Simulated view through eyepieces with (a) a narrow apparent field of view and (b) a wide apparent field of view. Photograph of M33 by George Viscome (14.5-inch f/6 Newtonian, Tri-X film in a cold camera, 30-minute exposure).*

effect through anything less. An apparent field in excess of 60° gives the illusion of staring out the porthole of an imaginary spaceship. The effect can be really quite impressive!

Naturally, eyepieces with the largest apparent fields of view do not come cheaply! Some, especially the long-focal-length models, are quite massive both in terms of weight and cost. Typically, they must be made from large-diameter lens elements and may be available in 2-inch barrels only, restricting their use to medium- and large-aperture telescopes only. Some are so heavy that you may actually have to rebalance the telescope whenever they are used! More about this when specific eyepiece designs are discussed later in this chapter.

By knowing both the eyepiece's apparent field (typically specified by the manufacturer) and magnification, we can calculate just how much sky can squeeze into the ocular at any one time. This is known as the *true or real field of view,* and can be approximated[1] from the following formula:

$$\text{Real field} = F/M$$

where
F = the apparent field of view
M = magnification

[1]Technically speaking, to calculate the real field of view accurately, you also need to take into account the diameter of the field stop. In practice, the formula here is accurate enough for most applications, but if the real field of view needs to be known to the nearest fraction of an arc-minute, it is best to time how long a star located near the celestial equator (Mintaka in Orion's belt is a favorite) takes to cross the eyepiece field. Due to Earth's rotation, the star will appear to move at a rate of 15.05 degrees per hour, or 15.05 arc-minutes per minute of time.

To illustrate this, let's look at an 8-inch f/10 telescope and a 25-mm eyepiece. This combination produces 80× and a 2.5-mm exit pupil. Suppose this particular eyepiece is advertised as having a 45° apparent field. Dividing 45 by 80 shows that this eyepiece produces a real field of approximately 0.56°, a little larger than the Full Moon.

Another term that is frequently encountered in eyepiece literature but rarely defined is *parfocal.* This simply means that the telescope will not require refocusing when one eyepiece is switched for another in the same set. Without parfocal eyepieces, the observer may lose a faint object during the second focusing should the telescope be moved accidentally during the change. Please keep in mind that even when an eyepiece is claimed to be parfocal, that does not mean that it is *universally* parfocal. Two eyepieces of the same optical design, say a 26-mm from Brand X and a 12-mm from Brand Y, that claim to be parfocal are most likely not parfocal with each other. At the same time, two eyepieces of different optical designs may be manufactured by the same company and declared parfocal, but they are not likely to be parfocal with each other.

Finally, a well-designed eyepiece will have good *eye relief.* Eye relief is the distance from the eye lens to the observer's eye when the entire field of view is seen at once. Less expensive eyepieces may offer eye relief of only one-quarter times the ocular's focal length. This is much too close to view comfortably. Some modern designs maintain an eye relief of 20 mm regardless of the eyepiece's focal length, making observing more enjoyable, especially for those who must wear eyeglasses. Of course, there can also be too much of a good thing. Excessive eye relief can make it difficult for the observer to hold his or her head steadily while hovering above the eyepiece.

Image Acrobatics

Since eyepieces can suffer from the same aberrations and optical faults as telescopes, it might be wise to list and define a few of their more common problems. Some have already been defined in earlier chapters, but now we will concentrate on their impact on eyepiece performance. (Of course, if any of the following conditions exist all of the time regardless of eyepiece used, then the problem likely lies with the telescope, not the eyepiece.)

If star images near the field's center are in focus when those near the edge are not, then the eyepiece suffers from *curvature of field.* The overall effect is an annoying unevenness across the entire field of view. Another flaw found in lesser eyepieces is *distortion.* Distortion is most readily detectable when viewing either terrestrial sights or large, bright celestial objects such as the Moon or the Sun. This condition is usually characterized by a warping of the scene in a way that is similar to the effect seen through a fish-eye camera lens.

Chromatic aberration, previously defined in earlier chapters, is nearly extinct in today's eyepieces thanks to the use of one or more achromatic lenses. Still, some less-sophisticated eyepieces, often sold as standard equipment, suffer from this ailment. If an ocular transmits chromatic aberration, the problem will be immediately detectable as a series of colorful halos surrounding all of the brighter objects found toward the edge of the field of view; the center of view is usually color-free.

Spherical aberration has also been all but eliminated in most (but not necessarily all) eyepieces of modern design. If an eyepiece is free of spherical aberration, then a star should look the same on either side of its precise focus point. When spherical aberration is present, however, the star will change its appearance from when it is just inside of focus compared to when it is just outside of focus. This predicament is the result of uneven distribution of light rays at the eyepiece's focal point. Today, if spherical aberration is present, chances are it is being introduced by the telescope's prime optic (main mirror or objective lens) and not the eyepiece. At low powers (large exit pupils), it can also be introduced by the observer's own eye.

Just as with objective lenses and corrector plates, most eyepiece lenses are now coated with an extremely thin layer of magnesium fluoride. Coatings reduce *flare* and improve *light transmission,* two desirable characteristics for telescope oculars. A bare lens surface can reflect as much as 4% of the light striking it. By comparison, a single-coated lens reduces reflection to about 1.5% per surface. As already mentioned in chapter 3's discussion of binoculars, a lens coated with the proper thickness of magnesium fluoride exhibits a purplish hue when held at a narrow angle toward a light. Top-of-the-line eyepieces receive multiple coatings, reducing reflection to less than 0.5%. These show a greenish reflection when turned toward a light.

Eyepiece Evaluation

Galileo, Kepler, and Newton had it pretty easy when it came to selecting eyepieces. Look at their choices! Galileo used a single concave lens placed before the objective's focus. It produced an upright image, but the field of view was incredibly small and severely hampered by aberrations. Kepler improved on the idea by selecting a convex lens as his eyepiece. It gave a wider, albeit inverted view but still suffered from aberrations galore. Progress in eyepiece design was slow in the early years.

Here we will see just how far the art of eyepieces has progressed while evaluating which designs are best suited for the telescope you chose earlier.

Huygens. The first compound eyepiece was concocted by Christiaan Huygens in the late 1660s (just as with new telescope designs, an eyepiece usually bears the name of its inventor). As can be seen in Figure 6.4a, Huygens eyepieces contain a pair of plano-convex elements. Typically, the field lens has a focal length three times that of the eye lens.

In the past, Huygens eyepieces were supplied as standard equipment with telescopes of f/15 and greater focal ratios. At longer focal lengths, these oculars can perform marginally well, although their apparent fields of view are narrow. In telescopes with lower focal ratios, however, image quality suffers from spherical and chromatic aberrations, image curvature, and an overall

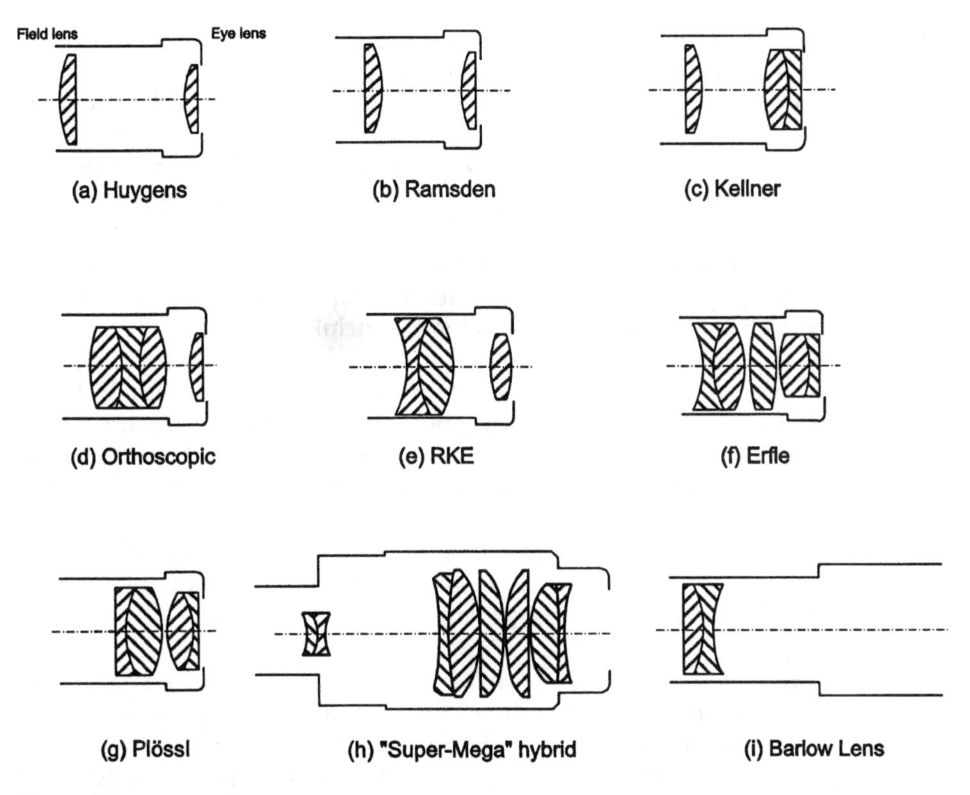

Figure 6.4 *The inner workings of several eyepiece designs from both yesterday and today. Eyepieces shown are (a) Huygens, (b) Ramsden, (c) Kellner, (d) Orthoscopic, (e) Edmund's RKE, (f) Erfle, (g) Plössl, and (h) a hybrid. A typical Barlow lens is shown in diagram (i).*

lack of sharpness. Do you own a Huygens eyepiece? If so, it will probably have an **H** on the barrel; for instance, an **H25mm** indicates a Huygens eyepiece with a 25-mm focal length. You would do well to replace it.

Ramsden. Devised in 1782, this design was the brainchild of Jesse Ramsden, the son-in-law of John Dollond (you may recall him from chapter 2 as the father of the achromatic refractor—small world, isn't it?). As with the Huygens, a Ramsden eyepiece (Figure 6.4b) consists of two plano-convex lenses. Unlike the earlier ocular, however, the Ramsden elements have identical focal lengths and are flipped so that both convex surfaces face each other.

In most cases, the lenses are separated by about two-thirds to three-quarters of their common focal length. This is, at best, a compromise. Setting the elements closer together improves eye relief but dramatically increases aberrations. Going the other way will decrease the design's inherent faults, but eye relief quickly drops toward zero. Therefore, like the Huygens, it is probably best to remember the Ramsden for its historical significance and pass it by in favor of other designs.

Kellner. It took over six decades of experimentation before an improvement to the Ramsden eyepiece was developed. Then, in 1849, Carl Kellner introduced the first achromatic eyepiece (Figure 6.4c). Based on the Ramsden, Kellner eyepieces replace the single-element eye lens with a cemented achromat. This greatly reduces most of the aberrations common to Ramsden and Huygens eyepieces. Sometimes called modified achromats (MA) or super-modified achromats (SMA), Kellners feature fairly good color correction and edge sharpness, little curvature, and apparent fields of view ranging between 40° and 50°. In low-power applications, Kellners offer good eye relief, but this tends to diminish as the eyepiece's focal length shrinks.

Perhaps the greatest drawback of the Kellner design is its propensity for internal reflections. I have often heard of Kellners being referred to as "haunted eyepieces" because of their ghost images, which are especially noticeable on bright objects. But thanks to antireflection optical coatings, this effect is almost eliminated when Kellners are used with 8-inch and smaller telescopes.

Given all of their pluses and minuses, Kellners represent a good buy for budget-conscious owners of small- to medium-aperture telescopes who are in the market for a low- to medium-power eyepiece.

Orthoscopic (Abbe). Introduced in 1880 by Ernst Abbe, the orthoscopic eyepiece has become a perennial favorite of amateur astronomers. As shown in Figure 6.4d, it consists of a cemented triplet field lens matched to a single plano-convex eye lens. What results is close to a perfect eyepiece, harassed by neither chromatic nor spherical aberration. There is also little evidence of ghosting or curvature of field. Orthos offer flat views with apparent fields between 40° and 50° and moderate eye relief that is typically a millimeter or two shorter than the focal length. Color transmission and contrast are superb, especially when combined with today's optical coatings. With either a long focal length for low power, a short focal length for high power, or anywhere in between, orthoscopic eyepieces remain one of the best eyepieces for nearly all amateur telescopes. To my eye, they yield equal or higher image contrast than many of the new-generation super-deluxe eyepieces, especially when aimed at the planets.

Despite their stellar performance, however, few companies offer orthoscopic eyepieces. Most, instead, concentrate on Plössls and other hybrid designs. Edmund's selection is limited to five focal lengths between 25 mm and 4 mm. Be aware that eye relief will become uncomfortably short with short focal lengths.

RKE. From Edmund Scientific comes this twist on the Kellner eyepiece. Instead of using an achromatic eye lens and a single-element field lens, the RKE (short for Rank-modified Kellner Eyepiece after its inventor, Dr. David Rank; Figure 6.4e) does just the opposite. The computer-optimized achromatic field lens and single-element eye lens combine to outshine the Kellner in just about every respect. Actually, performance of the three-element RKE is most comparable to that of the four-element orthoscopic, with a moderate apparent field of view (45° for the RKE) as well as good color correction and image clarity. Both work well at all focal lengths in all telescopes.

Erfle. The Erfle, the granddaddy of all wide-field eyepieces, was originally developed in 1917 for military applications. With apparent fields of view ranging between 60° and 75°, it was quickly embraced by the astronomical community as well. Internally, Erfles consist of either five or six elements; one variety uses two achromats with a double convex lens in between, while a second has three achromats, as shown in Figure 6.4f.

Erfles give observers an outstanding panoramic view of the deep sky. The spacious view takes its toll on image sharpness, however, which suffers from astigmatism toward the field's edge, especially in telescopes with fast focal ratios. For this reason, Erfles are inappropriate for lunar and planetary observations or any occasion that calls for higher magnification. In low-power, wide-angle applications, though, they are very impressive through f/8 and slower systems.

Plössl. One of the most highly regarded eyepieces around today, the Plössl (Figure 6.4g) features twin close-set pairs of doublets for the eye lens and the field lens. The final product is an excellent ocular that is comparable to the orthoscopic in terms of color correction and definition but with a somewhat larger apparent field of view. Ghost images and most aberrations are sufficiently suppressed to create remarkable image quality.

Though it was developed in 1860 by Georg Simon Plössl, an optician living in Vienna, Austria, the Plössl eyepiece took more than a century to catch on among amateur astronomers. Back in the 1970s, when I first became seriously engrossed in this hobby, I heard about a mysterious eyepiece called a Plössl. Try as I might, I could not find much published information on it, other than that I could not afford one! Perhaps the enigma of the Plössl was heightened by the fact that back then, the only company to offer them was Clavé in France, and their distribution was limited. Then, in 1980, the Plössl hit the big time. That year, Tele Vue Optics introduced a line of Plössls that was to start a new eyepiece revolution among amateur astronomers. They were an instant success.

Today, many companies now offer Plössls. Like many telescopes, most of these are imported from Taiwanese and Chinese sources, such as Guan Sheng, Everwin, and Synta. All are quite good, but some are perceptibly better than others. I prefer not to generalize, but in this case you pretty much get what you pay for. Less-expensive Plössls cut corners that the premium brands do not. The more expensive Plössls typically maintain closer optical and mechanical tolerances and include multicoated optics, as well as blackened lens edges and threads. Little things like these can make the difference between seeing a marginally visible object and not. Just because an eyepiece is labeled Plössl, do not automatically assume that it is far better than, say, a Kellner or an equivalent.

Tele Vue Plössls, generally conceded as the finest, are available in eight focal lengths from 8 mm to 55 mm. All but the 55-mm have 1.25-inch barrels;

the 55-mm requires a 2-inch focuser. Each has a 50° apparent field of view, except the 40-mm model, which has a 43° field. The 55-mm makes an especially nice low-power eyepiece if you own a Schmidt-Cassegrain or Maksutov telescope. An integrated rubber eyeguard helps prevent unwanted light infiltration.

For those working on tighter budgets, less-expensive Plössls are still worth consideration. All typically offer focal lengths between about 4 mm and 40 mm, with apparent fields of view in the 40° to 50° range. Price-wise, the Hands-On Optics, Sky Instruments, and Scopetronix Plössls are among the least expensive of the lot, being less than half the price of Tele Vues. Celestron, Meade, and Orion Sirius Plössls are more expensive than the cheapest, but they often have slightly wider fields of view. Owners of less-expensive department-store telescopes should also note that all but the longest focal-length Sirius Plössls are available with 0.965-inch barrels, as well.

Meade's Series 4000 Super Plössls are also very good performers, although technically they are not conventional Plössls; instead, they are a four-element hybrid design. Super Plössls come in focal lengths from 6.4 mm to 56 mm; all but the 40-mm offer 52° fields (the 40-mm has a 44° field of view). All are designed to fit 1.25-inch focusers, except the 56-mm, which requires a 2-inch focusing mount. Super Plössls have excellent performance characteristics, including very good image sharpness across the field. Their foldable rubber eyeguards also make them attractive for amateurs in light-polluted environs.

Super-deluxe-extra-omni-ultra-maxi-mega-colossal . . . Whatever happened to the words *standard* and *regular?* Are they still in the dictionary? Apparently not, judging by today's advertising. Every product, from refrigerators to pet food, is extra special in some way. Top-of-the-line oculars are no different. Each is proclaimed by its manufacturer as something extraordinary. And do you know what? They really are quite good!

These super-duper hybrid eyepieces (Figure 6.4h) were initially intended to meet the demanding needs of amateurs using Schmidt-Cassegrain telescopes and huge Newtonian reflectors, but they also work equally well in refractors and longer focal length reflectors. Most of these company-proprietary designs use multicoated lenses made from expensive glasses to minimize aberrations. Other pluses include huge apparent fields of view, which let users not only see sky objects, but *experience* their grandeur firsthand. Most also offer longer eye relief than other, more generic designs, a big help for those forced to wear glasses. With these eyepieces, the universe has never looked so good!

Not satisfied with the success of their Plössls, Tele Vue led the way in introducing oculars with extremely wide apparent fields of view. Of the original Tele Vue Nagler eyepieces, only the 4.8-mm and 7-mm focal lengths remain. Both are designed to fit 1.25-inch focusers and use a complex seven-element design to produce an incredible 82° apparent field while correcting for astigmatism, chromatic aberration, spherical aberration, coma, and just about every other optical fault.

This is not to say that the original Naglers are not without their faults. One is a loss of contrast when viewing bright objects such as the planets, a common problem with many multiple-element eyepieces. Many of the original Naglers also suffer from the "kidney-bean effect." As you shift your eye around to take in the eyepiece's full field of view, a dark, kidney-bean shaped area becomes noticeable to some. This effect was most evident in longer focal lengths, but not in the remaining 4.8- and 7-mm Naglers.

In an effort to eliminate this latter problem, most of the original Nagler eyepieces were replaced, first by the eight-element Nagler Type 2 design, then later by the eight-element Type 4, the six-element Type 5, and the seven-element Type 6 Naglers. Three eyepieces make up the Nagler Type 4 line, all with the same dramatic 82° apparent field of view as the originals but with better eye relief. The 12-mm Nagler Type 4 is designed to fit both 1.25- and 2-inch focusing mounts thanks to its skirted barrel, while the 17- and 22-mm eyepieces fit 2-inch focusers only. These longer focal lengths are ideal for deep-sky observing, especially for viewing low-surface-brightness objects like diffuse nebulae and face-on spiral galaxies. The images delivered by the Type 4s is even better than through the Type 2s, which suffered from minor reflections when viewing bright stars and planets. Those ghosts have been completely exorcized in the Type 4s.

The summer of 2001 saw the introduction of the new 16-mm Nagler Type 5 as well as Nagler Type 6 eyepieces, in 5-, 7-, and 9-mm focal lengths. All are designed to fit 1.25-inch focusers. Using an updated design and coating process, Type 5 and Type 6 Naglers take advantage of the newer optical glasses that have come along in recent years. These improvements stretch eye relief over their predecessors to 10 mm in the Type 5 and 12 mm in each Type 6. That's still a little short by some standards, but images are quite striking nonetheless! If you need more eye relief, consider eyepieces by Vixen or Tele Vue's own Radian eyepieces. The Nagler Type 5 and all Type 6s feature foldout rubber eyecups, which many prefer to the Instadjust eyecups supplied with Tele Vue's Nagler Type 4 and Radian eyepieces.

Then there is the "Termi-Nagler," as it has been called: the 31-mm Nagler Type 5, the biggest, heaviest eyepiece yet to come from Tele Vue. The Termi-Nagler is tops in many areas, including the huge 82° apparent field of view, the surprisingly flat field even in fast f-ratio telescopes, and, lastly, its cost, at around $600. Ahh, but for those who can afford, the views are just amazing. Looking at the Orion Nebula, for instance, is an experience that words cannot possibly capture. Contrast is excellent, despite the high number of lens elements, and there is no appreciable lens flare even when viewing bright objects.

A couple of words of caution. First, to reach focus, the 31-mm Nagler 5 requires more in-travel than many other eyepieces. If your telescope's focuser cannot be racked in sufficiently, focus will not be achieved. Second, this is a heavy eyepiece, weighing more than 2 pounds. That could be enough to throw off a telescope's balance and cause a Dobsonian's front to drop to the ground. Just be careful when you first use it, especially if you are aiming the telescope fairly close to the horizon.

Imagine paying $600 for a single eyepiece! That's more than many amateurs pay for an entire telescope. Is it worth the price? I have some bad news.

The consensus from all who have viewed with the 31-mm Nagler is, sadly, yes, very much so. Sorry.

Indeed, price-wise, all Nagler eyepieces are among the most expensive eyepieces made, but the view through it is wondrous. You may actually have to move your eye and / or your head around to take in the whole thing. Company literature, independent reviews, and consumers all liken the feeling to peering through a porthole of a spacecraft. (Of course, how many of us actually *know* what a view through a spacecraft porthole is really like?)

Tele Vue Panoptics are also heavyweights in the eyepiece arena, their six elements combining to yield unparalleled low-power views. Panoptics come in five focal lengths, each sporting a 68° apparent field of view. The 15-mm and 19-mm Panoptics feature 1.25-inch barrels, the 22-mm may be used in either 1.25-inch or 2-inch focusers, and the 27-mm and 35-mm fit 2-inch focusers only. Although their fields of view are not as large as the Naglers', eye relief is generally better, an important consideration for those who must wear eyeglasses. Consumers should note, however, that the eye relief of the 15-mm Panoptic is on the short side and might prove troublesome for some observers.

Recognizing that planet fanatics favored other designs over Naglers and Panoptics, Tele Vue came out with their line of 1.25-inch Radian eyepieces. Are these the fabled "perfect" eyepiece? They come awfully close. Each has a comfortable 20 mm of eye relief, allowing even those who must wear glasses to see all of their 60° apparent field of view at once. And what a view it is. Even though the Naglers (and Meade Ultra Wides) have wider fields of view, they are not nearly as sharp. Stars snap into focus through Radians, making them especially nice for studying double stars and open star clusters. Stars remain pinpoints all the way to the edge of the field, as well. As one owner stated in his assessment of star images through his 10-mm Radian, "I don't mean that they are pinpoint 'closer' to the edge . . . I mean that they are pinpoint right up to the point that they leave the field of view." False color and ghost images are also well suppressed, while image brightness and contrast are both excellent. But that's only half of the story. As good as Radians are when trained on stars, they are just as adept when focused on the planets, equaling the performance of nearly any other eyepiece on the market, including some exotic models sold in very limited quantities. Like Nagler Type 4s, all Radians include a click-stop Instadjust eyecup that is pulled in and out to just the right distance for the observer's eye. Wear glasses? Push it all the way in. No glasses? Pull it one or several stops until you see the entire field. Some people love this feature while others hate it. I will remain neutral, except to say that, either way, it doesn't seriously affect what has proven to be some outstanding eyepieces.

Many other companies offer wide-field eyepieces. Meade Super Wide Angle oculars embody six elements that provide an impressive apparent field of 67°, about comparable to Panoptics. The 13.8-mm, 18-mm, and 24.5-mm Super Wides are designed to fit 1.25-inch focusers, while the 32-mm and 40-mm require 2-inch focusers. The two shorter focal lengths yield tack-sharp images across the field. The 18-mm is especially nice because of its comfortable eye relief and small size. The 24.5-mm also performs well, but images near the field's edge are not quite as crisp as with the Tele Vue Panoptic 22-mm or 27-mm, or the Pentax XL 21-mm when used with telescopes that have fast focal-ratios. The two longer focal lengths show further signs of edge distortions, with out-of-focus "seagull" stars evident. Consumers should bear in mind, however, that Meade Super Wides are $70 to $80 less expensive than Panoptics and Pentax XLs.

Meade Ultra Wide Angle oculars come in 4.7-mm, 6.7-mm, 8.8-mm, and 14-mm focal lengths. The two shorter focal lengths are housed in standard 1.25-inch barrels, while the two longer focal-length Ultra Wides feature 1.25-inch / 2-inch step-down barrels. All use a combination of eight elements in five groups to produce a stated apparent field of 84°, approximately equal to the Naglers. Owners expound the eyepieces' excellent performance. Some comments from the mailbag: "The 6.7-mm Ultra Wide has very short eye relief but is a very sharp, high-power eyepiece; works well even on the Astroscan; lighter than the 7-mm Nagler" and "The 14-mm Ultra Wide Angle is the eyepiece that I use most often and is probably the most common in our astronomy club." The Meade Ultra Wide Angle eyepieces are indeed right up there with the best available.

The Japanese camera company Pentax makes several lines of oculars, although many have limited availability, at least here in North America. In addition to 0.965-inch diameter Kellner and orthoscopic eyepieces, they also offer premium-grade Pentax XL eyepieces. Pentax XLs are available in focal lengths of 5.2-, 7-, 10.5-, 14-, 21-, 28-, and 40-mm. All have a 65° apparent field of view, except the 28-mm, which has a 55° apparent field. Only the 40-mm has a 2-inch barrel; the others only fit 1.25-inch focusers. (Don't be fooled by what appears to be a step-down barrel in photos of these eyepieces; the larger diameter is less than 2 inches.) All come with uniquely designed eyecups that are unlike any other available. While most eyecups extend around the eyepiece to shield the observer from stray light, Pentax eyecups look, as one person put it, like an upside-down funnel. To use them, the observer sticks his or her eye (which is slightly convex) *into* the eyecup. It takes a little getting used to, but I have heard of no complaints from those who use them.

For high-power applications, owners agree that the 5.2-, 7-, 10.5-, and 14-mm Pentax XL eyepieces are about as good as they get. Each features excellent eye relief and sharp, contrasty views that compare favorably to Tele Vue Radians (or is it the other way around, since the XLs predate the Radians?). Many respondents to my equipment survey have commented, however, that the three longer focal length Pentax XLs fall a little short of Panoptic and Nagler performance. The 14-mm Radian, for instance, is just as sharp as the 14-mm Pentax XL, but the Pentax's unusually wide housing makes it a little more cumbersome.

Celestron Ultima eyepieces are available in eight focal lengths ranging from 5 mm to 42 mm. Depending on focal length, each incorporates between four and seven optical elements to create fields between a slim 36° in the 42-mm model to a more acceptable 51° in shorter focal length versions. Images are quite sharp with excellent correction against aberrations. Eye relief is also quite good in all focal lengths except the two shortest, 5-mm and the 7.5-mm, which measure only 4 mm and 5 mm, respectively.

More recently, Celestron introduced a line of mid- and long-focal-length mega-eyepieces that they call Axioms. Made in Japan, Celestron Axioms are built around seven separate lens elements, all of which are multicoated to improve image contrast. Each ocular is nicely appointed with a rubber grip ring and a rubber, foldable eyecup. Owners report that images are very good, comparing well to the more expensive Tele Vue Panoptics, though perhaps with slightly inferior contrast. Eyeglass wearers should take heed, however, since eye relief is surprisingly short in all but the 40- and 50-mm Axioms. The shortest Axiom, the 15-mm, suffers with a poor 7-mm eye relief, while the 19-mm, available with either a 1.25- or 2-inch barrel, only has 10 mm of eye relief. Even those of us who do not wear eyeglasses may find these uncomfortable.

If long eye relief is important to you, then Vixen Lanthanum LV eyepieces may be your salvation. With most designs, as focal length shrinks, so does eye relief. As a result, short-focal-length eyepieces require the observer to get uncomfortably close to the eye lens. Not so with these oculars. All Lanthanum LV eyepieces, ranging in focal length from 30 mm to 2.5 mm (the shortest focal length of any eyepiece available, incidentally) feature a long 20-mm eye relief. As a result, the observer can stand back from the eyepiece and still take the whole scene in, making life at the eyepiece much more comfortable. Based on the Plössl, the Lanthanum LV adds a fifth lens element between the eye-lens and field-lens groups as well as an extra element(s) before the field lens. Apparent fields of view are in the 45° to 50° range. Image quality and contrast rank among the highest of any eyepiece, although some feel that the images through Lanthanums are dimmer than through some other designs. Lanthanum eyepieces are imported into the United States by both Orion Telescopes and Celestron, although they do not necessarily carry exactly the same focal lengths.

Looking for a wide-field eyepiece, but prefer to save a little money for other accessories (or possibly just for food!)? Consider the Vixen Lanthanum Superwides. Ranging in focal length from 3.5 to 42 mm, these eight-element oculars all boast 20 mm of eye relief and 65° apparent fields of view. The most pleasant surprise is that they show sharp, contrasty views across nearly the entire field, even through fast-focal-ratio instruments. The 17-mm Lanthanum Superwide, for instance, has now become one of my most-used eyepieces with my 18-inch f/4.5 reflector; its edge correction is quite comparable to Tele Vue Panoptics. All are designed to fit either 1.25- or 2-inch focusers, except the 42-mm, which fits 2-inch focusers only.

Takahashi LE (for "Long Eye Relief") eyepieces are another proprietary design. Available in focal lengths from 5 to 30 mm (1.25-inch barrels) as well as 50 mm (2-inch barrel), Takahashi LE eyepieces are also highly regarded by amateur astronomers. One respondent to the *Star Ware* survey described her 5-mm LE eyepiece as "the only lens that I use for planetary and lunar observing; very sharp, with no ghosting on bright objects like some of my other eyepieces; excellent eye relief." Several others commented on the similar performance characteristics of Takahashi LE eyepieces and Vixen's Lanthanum series. Which is better boils down to a largely subjective judgment; all will serve well.

Brandon eyepieces, offered exclusively by VERNONscope, yield sharp, crisp views with good image contrast across the entire field. Ghosting, prevalent in many lesser eyepieces, is effectively eliminated in four-element Brandons thanks to precise optical design and magnesium-fluoride coatings. Overall performance is comparable to Plössl and orthoscopic eyepieces for observing deep-sky objects but superior for observing the planets. Brandons come in six focal lengths from 8 mm to 48 mm; all have standard 1.25-inch barrels (except the 48-mm, which has a 2-inch barrel) and feature foldable rubber eyeguards. But take heed: for those who plan on using screw-in filters, Brandon eyepieces do not have standard threads to accept other manufacturers' filters. Instead, they can only use VERNONscope filters.

Another company to jump on the bandwagon is Orion Telescope Center. In addition to selling the aforementioned Lanthanum eyepieces, they also feature Orion Ultrascopic oculars. Ultrascopics are available in nine different focal lengths, from 3.8 to 35 mm. All include five lens elements, except the 3.8- and 5-mm versions, which add a two-element negative achromat (in effect, a built-in Barlow lens), like many other hybrid eyepiece designs. Views are ghost-free with pinpoint stars seen across nearly the entire field. A common comment among owners is summed up by one who uses the eyepieces with a large-aperture reflector: "Ultrascopics are excellent; very good images to the edge of the field, even in an f/5.2 Newtonian." All Ultrascopics come with rubber eyeguards and produce a reasonable 52° apparent field of view (the 35-mm has a 49° apparent field of view).

For those on a budget, yet still lusting after low-power 2-inch eyepieces, consider the three long focal length Optiluxe eyepieces from Orion. All have fully multicoated optics, which reduce optical aberrations. The longest focal length of the three, a monster 50-mm, is comprised of five elements in what Orion describes as a "modified Plössl" design. The others, 40-mm and 32-mm, include four elements each. Light transmission and suppression of flare seem very good for such inexpensive eyepieces and coma is minimal, even in f/4.5 telescopes. Although they have narrower apparent fields than premium 2-inchers, they still produce respectable views, especially considering their comparatively low prices. The 32-mm, with its 62° field, is an especially nice eyepiece in f/6 and greater telescopes. If I had to point out a weakness, it is their excessive eye relief, which forces you to hold your head just so to see the field.

Sky Instruments of Vancouver, British Columbia, markets several different types of imported eyepieces, including Plössls and Erfles, under the brand names Antares and Omcon. None attract more attention than their line of Antares Speers-Waler super eyepieces. Glenn Speers created this unique design of Wide-Angle, Long Eye Relief (Waler) eyepiece, packaging it in eight different focal lengths. The 5- to 8-mm zoom is addressed later, so the discussion here will be restricted to the seven fixed focal length eyepieces.

First, Speers-Waler eyepieces look nothing like any other eyepiece on the market. All have short, chromed 1.25-inch diameter barrels (except for the longest of the lot, the 30-mm, which has a short 2-inch barrel), yet most stand very tall. The 10-mm, for instance, is almost 6 inches long. Most of this is the black lens housing, attractively accented with orange lettering to make it easy to read at night.

Performance-wise, the short focal length Speers-Walers are very good. Of these, the 10-mm seems to be everyone's favorite. It compares very favorably to the Tele Vue 10-mm Radian, although it does not have the same long eye relief or produce images that are quite as sharp. There is also some minor false color in the Speers-Waler. On the plus side, however, the Speers-Waler has a wider apparent field of view and is also about $60 cheaper, making it an excellent buy.

The 18-mm and 30-mm Speers-Waler eyepieces are not nearly as sharp at the shorter focal lengths. The former has a distracting amount of field curvature, which is accentuated in faster focal ratio instruments. The 30-mm also has some field curvature, though not to the degree of the 18. In this case, however, eye relief is a surprisingly close 16 mm, which makes it difficult to take in the entire 67° apparent field of view at once. The 24-mm seems to be a better performer than either the 18- or 30-mm, but does not equal the performance of, say, Tele Vue Panoptics.

It should be noted that all Speers-Waler eyepieces require more *in-travel* (turning the focuser into the telescope) than most eyepieces. If the eyepieces that you now own must be racked in to hit focus, then Speers-Walers will probably not focus without modifying the telescope. And while all certainly qualify as wide angle (WA), they do not live up to the long-eye-relief (LER) part of their name. Values range between 13 and 16 mm, even at the longer focal lengths, which is okay, don't get me wrong, but somewhat short by today's standards. Keep that in mind if you must wear glasses.

Apogee sells several imported eyepieces, including Plössls, orthoscopics, and Erfles, as well as a few supermega eyepieces. One, their 30-mm Wide Scan Type II, is a short-barreled, wide-field eyepiece that has received some fairly good press from its users. Most report the eyepiece holds a sharp focus for about two-thirds of the field, after which the image softens quickly. As one respondent says, "It's not for large open clusters." Eye relief, although not specified by the manufacturer, is reported as being surprisingly short for the long focal length. The eyepiece does not include a rubber eyecup, which can make head positioning more difficult.

Apogee Super Easyview eyepieces, with their hourglass-shaped housings, also look a bit eccentric. Three focal lengths are included—3.6, 5.5, and 9.5 mm—all of the same five-element design that is similar to an orthoscopic but with a negative meniscus transfer lens. Reviews say that the 9.5-mm does very well on the planets and double stars. Quoting one owner, "The contrast in the clouds of M42 was spectacular, but what really amazed me were the two extra stars in the [multiple star] Trapezium! I had never seen the fifth and sixth stars so plainly with an 8-inch scope from light polluted suburbs. Using a 7-mm Nagler in the same telescope, I could barely see the fifth star, while the sixth was all but invisible." Another writes that all of "the Easyviews have superb eye relief, are remarkably free of internal reflections, and have good color rendition. Sharpness falls a bit towards the edge." If you are using them for planet watching, the lack of sharpness at the edge is irrelevant, however.

Finally, what about the Collins I3-piece, what I call an "electro-eyepiece"? Unique among eyepieces, the I3-piece, also called the I-Cubed, mates a four-element optical system to a state-of-the-art image intensifier. You have probably seen an image intensifier, also called a nightscope, on television news or in the movies, usually shown with soldiers peering through them. What they are actually looking at is a greenish, phosphorescent display of the dark landscape magically transformed into a well-lit, albeit green panorama. That same technology has now been packaged into the I3-piece. The heart of the I3-piece is its Generation-Three and Generation-Four image intensifier tubes, claimed by the manufacturer to amplify light 30,000 to 50,000 times, coupled to a four-element set of optics from Tele Vue. Net effect? Typically, observers can see stars up to two magnitudes fainter than through a similar, conventional eyepiece. That's like going from an 8-inch telescope to a 20-inch telescope instantly! Collins also sells adapters for CCD video and photography through the I3-piece.

Now, before we all run out and buy one, be aware of the following drawbacks. First, remember that you are not looking through a telescope but rather at a screen. Some liken the experience to watching something on television as opposed to in person. To make matters a little less natural, imagine that you are watching that television show on an old, monochromatic green computer screen. And that screen has some snow, or tiny, momentary specks of light, scattered across it. Finally, did I mention that you are looking at the green television screen through a paper towel roll, a good analogy since the I3-piece only has a 35-degree apparent field of view. It does have very good eye relief, however, and owners quickly point out that the initial strong green color seems to lessen as the evening wears on.

I don't mean to sound like I am coming down hard on the I3-piece, even though I guess I am. The technology is great; there's nothing else like it on the market. The fact that you can have a 25-mm eyepiece and state-of-the-art electronics in a package that measures just over 4 inches long and weighs under a pound is just amazing. (A set of 15-mm optics is also available.) True, this technology does not come cheaply, with the complete eyepiece selling for around $2,000. But then, a 20-inch telescope would cost *much* more than that!

As previously mentioned, all of these super-deluxe-extra-omni-ultra-maxi-mega-colossal eyepieces offer excellent image quality, color correction, and freedom from edge distortions. Their designers are to be congratulated for creating superb eyepieces for the backyard astronomer. On the negative side, however, I must point out that not only are their performances stellar, their prices are, too. Some of these may actually cost more than your telescope! Are they worth the money? In all honesty, I have to say yes, especially if you own a fast (e.g., low-focal-ratio) telescope. But while these top-of-the-line eyepieces are great if you can afford them, they are not absolutely necessary. Many hours of great enjoyment can be yours with less costly eyepieces.

No-frills eyepieces. At the opposite end of the price spectrum are three small companies or individuals who offer several very inexpensive eyepieces, many of which come with high claims. Does their performance meet those expectations?

Cheapest of the cheap are the eyepieces made by Paul Rini and sold through Surplus Shed. Rini's eyepieces include Plössls, Erfles, and RKE designs, ranging from a 16-mm modified Plössl to a 52-mm RKE. The most amazing part is the price. Plössl and Erfle eyepieces normally cost between $50 and $100, maybe more, but no Rini eyepiece costs more than $40! Indeed, many cost less than $20. What's going on here?

For one, Rini uses surplus lenses in his eyepieces, which is not necessarily a bad thing, but it does beg the question of quality control. Premium eyepieces adhere to strict specifications for optical coatings and the like, while surplus optics are much more hit and miss. In general, they work fairly well in slow f-ratio telescopes but poorly in fast telescopes (subjectively, I would say the breaking point is f/8 to f/9). For instance, I purchased a 35-mm Rini Erfle, a 2-inch eyepiece in a dull-finish, glued-together aluminum and plastic housing. I was immediately struck by how large the eye lens is and how precariously close it is to the outside world—no eyeguard or eyecup here. Also, not surprisingly, the eyepiece is not threaded for filters. Through a friend's 8-inch f/10 SCT, we were both pleased with how well the eyepiece performed, although it certainly paled when compared to some top-notch eyepieces in his collection. True, there was a good amount of edge distortion, but overall, images were reasonably sharp. Next came the test with my 18-inch f/4.5 Newtonian. Although both telescopes have similar focal lengths (80 inches for the C8, 81 inches for the 18-inch), eyepiece performance was as different as day and night. In a word: horrible! The Rini proved difficult to focus at all, with perhaps only the inner quarter of the field approximately sharp. The outer three-quarters of the field showed an amazing combination of coma and curvature of field.

Harry Siebert Optics also offers a selection of inexpensive eyepieces, and while they cost more than Rinis, they are also more highly regarded. Their bread and butter are several 1.25-inch and 2-inch Wide Angle and Super Wide Angle eyepieces, which each contain between three and six individual elements. The longer focal lengths (that is, those in the range from 29 to 51 mm) have very good eye relief, though the lack of eyecups may cause image *blackout* unless the observer holds his or her eye just right. One respondent writes, "Star images are very sharp, suppression of false color is very good, and contrast is excellent; I believe that image quality is up there with the best, and in fact puts some more expensive brands to shame." Again, expect the best performance in slower-focal-ratio telescopes.

As focal length shrinks, so does eye relief, to only 7 mm in the 7- to 21-mm Wide Angle eyepieces. That can make for some uncomfortable viewing, especially if you must wear glasses.

Like Rini eyepieces, all Siebert eyepieces have a homemade look and feel, their barrels being a combination of plastic and aluminum. All of their 2-inch eyepieces are threaded for filters, while the 1.25-inch eyepieces are not. The latter will purportedly hold filters in place by friction when slid into the barrel.

Finally, Gary Russell Optics sells seven different 2-inch eyepieces, from a 65-mm Super Plössl to an 11-mm Konig. All have black plastic bodies (Delrin, I believe) that are much lighter in weight than might be expected, a great plus for those who have problems balancing their Dobsonians. Each also comes with a foldable eyecup and are threaded for standard 2-inch filters.

Of all Russell eyepieces, the 19-mm Konig gets the highest marks. I purchased one for the purpose of testing after having been told by one owner that it performs as well as a Tele Vue Panoptic. After using it extensively in my 18-inch reflector and others, I have found that while it is a decent eyepiece, it is no Panoptic. Images are good across perhaps 75% of the field but tend to soften toward the edges. I should be quick to add that the eyepiece would have better edge performance in an f/10 SCT, just like the Rinis. Contrast was quite good and false color was not objectionable.

In summary, while some of these no-frills eyepieces are very attractively priced, most are compromises in quality and consistency. Owners of telescopes with optical systems faster than f/6 (and in some cases, faster than about f/9) should tread cautiously. Another problem of note with surplus eyepieces is that as the surplus supply changes, so do the eyepiece offerings. So, while I have included all of the current inventories in Appendix B, they may well have changed by the time you read this.

Zoom eyepieces. Combining a wide variety of focal lengths into one package, zoom eyepieces typically range in focal length from about 7 to 25 mm. Sounds too good to be true? Unfortunately, it usually is! Most zoom eyepieces are compromises at best. For one thing, aberrations are frequently intensified in zoom eyepieces, perhaps due to poor optical design or because the lenses are constantly sliding up and down in the barrel. Another problem is that their apparent fields of view are not constant over the entire range. The widest apparent fields occur at high power but shrink rapidly as magnification drops. Finally, many are not parfocal across their entire range, forcing you to refocus whenever the eyepiece is zoomed.

In the years since *Star Ware*'s second edition appeared, several companies have begun to offer high-end, 1.25-inch zoom eyepieces. Names that we have come to associate with quality eyepieces, such as Tele Vue, Vixen, and Meade, all feature zoom eyepieces that try to break the adage "zoom is doom." All are essentially the same eyepiece, which is made by Vixen and rebadged accordingly. All have a zoom range from 8 to 24 mm. Are they worth the price? The short answer is no in my opinion, but I do have a qualification. Depending on whom you talk to, these eyepieces are the greatest thing since Lippershey invented the telescope, or they are poor performers that are easily outclassed by conventional eyepieces. Those I have looked through seem to vary in quality, leading me to question the quality inspections that they are put through. In the worst case, images are dim and soft and contrast is blah. In the best case, images are sharp, but they are still dimmer than through a fixed-focal-length eyepiece. All suffer from cramped apparent fields of view and are not precisely parfocal across their entire range. Given the concerns over quality variations, my best recommendation is that you approach the purchase with caution and double-check your dealer's return policy.

A unique, short-range zoom is manufactured and sold by Sky Instruments of Canada. The nine-element Speers-Waler zoom only shifts focal length from 5 to

8 mm, enabling it to maintain an incredible apparent field of view (81° to 89°!). Eye relief is a little short, varying between 11 and 12 mm, but overall, image quality is quite good. Contrast isn't up to that of a simpler, fixed-focal-length eyepiece, but it is still an eyepiece worth strong consideration. Be sure to store the eyepiece in a sealed container after use, since the zoom slide can let dust get into the internals. Like all Speers-Waler eyepieces, detailed earlier in this chapter, the zoom features an unusually long housing and a short, 1.25-inch barrel.

Most recently, Tele Vue introduced the 3-to-6-mm Nagler Zoom. Although it may be adjusted to any point within its range, click stops tell the user when the lens is at exact focal lengths. Unlike all others, which vary in apparent field of view as the eyepiece is zoomed in and out, the Nagler Zoom maintains a 50° field regardless of focal-length setting. It also maintains an eye relief of 10 mm across its range. But the most unusual thing about the Nagler Zoom is its appearance. Most zooms keep the sliding mechanism inside their housings, but the Nagler Zoom has the moving eye-lens assembly on the outside, creating a unique effect that can best be likened to a flattened mushroom. I had the opportunity to test an early production model on some astronomical sites and came away feeling that this was the best zoom I had ever looked through. But even the best, in my opinion, is a compromise. By their very nature, zooms will dim the view as they are turned to their shortest focal lengths, and the Nagler is no exception.

Barlow lens. The Barlow lens was invented in 1834 by Peter Barlow, a mathematics professor at Britain's Royal Military Academy. He reasoned that by placing a negative lens between a telescope's objective or mirror and the eyepiece, just before the prime focus, the instrument's focal length could be increased (see Figure 6.4i). The Barlow lens, therefore, is not an eyepiece at all but rather a focal-length amplifier.

Depending on the Barlow's location relative to the prime focus, the amplification factor can range up to about 3×. For example, remember the 8-inch f/10 telescope and 25-mm ocular that I used to illustrate the concept of real field of view? If we insert the eyepiece into a 2× Barlow lens and then place both into the telescope, the combination's magnification will climb from 80× to 160×. A 3× would boost it to 240×, and so on.

Why use a Barlow lens? The first reason should be obvious. By purchasing just one more item, the observer effectively doubles the number of eyepieces at his or her disposal. But the benefits of the Barlow go even deeper than this. Since a Barlow stretches a telescope's focal length, an eyepiece/Barlow team will yield consistently sharper images than an equivalent single eyepiece (provided that the Barlow is of high quality, of course). This becomes especially noticeable near the edge of the field. Another important advantage is increased eye relief, something that short-focal-length eyepieces always have in short supply. Several of the super-mega eyepieces get their long eye relief and short focal lengths by incorporating a Barlow lens right in the eyepiece's barrel.

When shopping for a Barlow lens, make sure that it has fully coated optics; fully multicoated are even better. Avoid so-called variable-power Barlows, as they tend to perform less satisfactorily than fixed-power versions. Several high-quality Barlows are available in the 1.25-inch range, including those by Tele Vue, the Meade apochromatic Barlow, and Celestron's Ultima. Note that the Celestron Ultima is made to very tight specifications because a number of eyepieces that are slightly more than 1.25 inches in diameter can have problems fitting into it. I have yet to experience any trouble personally, but I will say that there is an audible pop when some of my eyepieces are removed. If you own 2-inch diameter eyepieces, the giant Barlows by Astro-Physics and Tele Vue are your best choices, the Tele Vue being a little less expensive.

Another option is offered by Tele Vue's Powermate series. Not conventional Barlows in the strictest sense, Powermates are more appropriately classified as *image amplifiers*. While traditional Barlows use two-element negative lens sets, the Powermate has both a negative achromatic doublet plus a positive pupil-correcting doublet to correct vignetting (darkening of the field edges) and other optical problems that inevitably crop up when a Barlow is pushed beyond its reasonable limits. The net difference is that a Barlow stretches a telescope's existing focal length while a Powermate effectively creates a new telescope. This may seem to be a fine point, but for the perfectionists in the crowd, especially those who own short-focus apochromatic refractors but who crave highly magnified views of the planets, the difference is noteworthy. Tests show that the Powermates all create nicely magnified views of their selected targets without introducing false color, edge distortions, or other aberrations. Assuming you go into it with a high-quality telescope and eyepiece, the inclusion of a Powermate will only add to the system's magnification, and nothing else.

Powermates come in three magnification factors: 2.5×, 4×, and the original 5×. The 4× fits 2-inch focusers while the others are for 1.25-inch focusers. All are very well made in typical Tele Vue fashion, with chromed barrels, black anodized bodies, and captive setscrews for securing eyepieces.

Table 6.3 lists the most popular Barlow lenses available today.

Focal reducers. While the f/10 focal ratios of most Schmidt-Cassegrain telescopes have great appeal for observers who enjoy medium- to high-power sky views, their long focal lengths make it difficult to fit wide star fields into a single scene. To view large objects, such as the Andromeda Galaxy or the Orion Nebula, many amateurs use focal reducers to shrink a telescope's overall focal length by 37% (from f/10 to f/6.3), thereby increasing its field of view.

Modern focal reducers, listed in Table 6.4, do more than just compress the focal length; they also give pinpoint star images across the entire field of view. Although beneficial to all observers, this is especially attractive to astrophotographers, as the reduction in focal length also cuts down exposures by a factor of 2.5. Not surprisingly, both Celestron and Meade offer focal reducers for their telescopes, each designed with its own brand in mind. Consumers should note, however, that both will also fit each other's SCTs and work equally well, so the choice is yours.

Coma correctors. One look through chapter 5 and it should be clear that Newtonians with low focal ratios have become immensely popular—and with

Table 6.3 ***Barlow Lenses***

Company/ Model	Magnifying Factor	Barrel Diameter	Features
Adorama Pro-Optic Short	2×	1.25″	2.25″ long
Apogee 1.75× Barlow	1.75×	2″	6″ long; 46.5-mm clear aperture
Apogee 2× Barlow	2×	1.25″	3″ long
Apogee 3× Barlow	3×	1.25″	4.38″ long
Celestron #93502 Barlow	2×	0.965″	Less than 3″ long; fully multicoated optics
Celestron #93507 Barlow	2×	1.25″	Less than 3″ long; multicoated optics
Celestron #93506 Ultima	2×	1.25″	Three-element optical design; 2.75″ long; fully multicoated optics
Discovery Telescopes	2×	1.25″	
Earth and Sky Spectiva	2×	1.25″	
Hands-on Optics Short	2×	1.25″	2.88″ long; fully coated optics
Intes 8–01–01	2.4×	1.25″	
Meade #134 Barlow	2×	0.965″	
Meade #122 Telenegative	2×	1.25″	Multicoated optics
Meade #126 Short-Focus	2×	1.25″	2.5″ long; multicoated optics
Meade #127 Variable Barlow	2–3×	1.25″	Engraved magnification scale; multicoated optics
Meade #140 Series 4000	2×	1.25″	Three-element optical design; multicoated optics
Orion #8711 Shorty	2×	1.25″	Multicoated optics; 3″ long
Orion #5121 Shorty-Plus	2×	1.25″	Three-element optical design; multicoated optics; 2.75″ long
Orion #8746 Deluxe	2×	1.25″	Multicoated optics
Orion #8724 Deluxe	2×	2″	Multicoated optics
Orion #8748 VariPower	2–3×	1.25″	Fully multicoated optics
Orion #8725 Ultrascopic	2×	1.25″	Three-element optical design; fully multicoated optics
Orion #8704 Tri-Mag	3×	1.25″	Multicoated optics; 5″ long
Parks Optical 701–30000	2×	0.965″	
Parks Optical 701–30010	2×	1.25″	
Parks Optical 701–30050	2×	1.25″	Short
Parks Optical 701–30015	2×	2″	
Sky Instruments UB2S	2×	1.25″	Short; multicoated optics
Sky Instruments UB3S	3×	1.25″	Short; multicoated optics
Sky Instruments B2S	2×	1.25″	Short; fully coated optics
Tele Vue 2×	2×	1.25″	Fully multicoated optics
Tele Vue Big Barlow	2×	2″	Fully multicoated optics
Tele Vue Powermate	2.5×	1.25″	Four-element optical design; fully multicoated optics
Tele Vue 3×	3×	1.25″	Fully multicoated optics
Tele Vue Powermate	4×	2″	Four-element optical design; fully multicoated optics
Tele Vue Powermate	5×	1.25″	Four-element optical design; fully multicoated optics
VERNONscope Dakin	2.4×	1.25″	Multicoated optics

Table 6.4 **Focal Reducers/Correctors**

Company/Model	% Reduction	Comments
Celestron International		
Reducer/Corrector	37	Four-element design; fully multicoated optics
Lumicon		
Cassegrain Rich Field Viewer	50	Does not correct for field curvature
Meade Instruments		
Series 4000	37	Four-element design; multicoated optics
Focal Reducer/Field Flattener		

them, unfortunately, so has coma. You'll recall that coma causes stars near the edge of an eyepiece's field to appear like tiny comet-shaped blobs instead of sharp points. Coma is an inborn trait of all Newtonians, although it becomes objectionable only when a telescope's focal ratio is f/5 or less.

To help counter coma's deleterious effect, many amateurs use coma correctors. At first glance, coma correctors look just like Barlow lenses; in fact, that is just how they are used. Simply slip an eyepiece into the corrector's barrel and then place the pair into the telescope's eyepiece holder. By refining the light exiting the telescope before it reaches the eyepiece, coma is effectively eliminated. As a side bonus, they also correct for off-axis astigmatism, another common problem with fast Newtonians. But bear in mind that coma correctors are not magic. They do nothing for bad optics. If your telescope has *on-axis* astigmatism, it is most likely caused by a poor-quality primary or secondary mirror.

Two companies, Tele Vue and Lumicon, manufacture coma correctors (see Table 6.5). Although they are basically the same, Lumicon uses two single lenses in their design while Tele Vue's Paracorr (short for "parabolic corrector") is based around a pair of achromats for better correction. The newest Paracorr has what Tele Vue calls a "tunable top." The manufacturer found that an important aspect to optimizing coma correction is the distance between the eyepiece and the Paracorr lens. The Tele Vue Tunable Top Paracorr lets users adjust this spacing by loosening a thumbscrew and turning the top of the Paracorr, which raises and lowers the eyepiece. Some work best with the Paracorr at its maximum height setting, some at minimum height, and some in between. Experiment with your eyepieces for the best results.

Table 6.5 **Coma Correctors**

Company/Model	% Focal-Length Increase	Comments
Lumicon Coma Corrector	0	Uses two single-element lenses; fits 2-inch focusing mounts only
Tele Vue Paracorr	15	Uses two achromatic lenses; fits 2-inch focusing mounts only

Okay, are they worth it? The answer is yes, but with a qualification. As mentioned above, they do nothing to help fix bad optics. At the same time, they do not do much to fix *good* optics, either. In my experience, a coma corrector has a far greater impact when coupled with Plössl, orthoscopic, and other mere mortal eyepieces. In my humble opinion (and that's all this is, mind you), coma correctors make little difference if you are using a top-quality eyepiece, such as a Tele Vue Nagler or a Meade Ultra Wide. These eyepieces are designed to eliminate internal astigmatism, which has the effect of improving off-axis performance. As a result, coma correction strikes me as offering little additional effect. Some people will read this and say that I clearly don't know what I'm talking about. If you take exception to *anything* in this book, visit my Talking Telescopes online discussion group and take me to task. Who knows, maybe you can change my mind.

Reticle eyepieces. For some applications, such as through-the-telescope guided photography, it can be useful (even necessary) to have an internal grid, or reticle, superimposed over an eyepiece's field of view. Reticle patterns, typically etched on thin, optically flat windows, come in a wide variety of designs depending on the intended purpose. The simplest have two perpendicular lines that cross in the center of view, while the most sophisticated display complex grids.

Most reticle eyepieces (Table 6.6) come with strange-looking appendages sticking out one side of their barrels. These are illuminating devices used to light the reticle patterns for the observer. Ideally, an illuminator's light level should be adjustable so the reticle is just bright enough to be seen, but not so bright as to overpower the object in view. Illuminating devices use either small incandescent flashlight bulbs or light-emitting diodes (LEDs), which are both red in color. Several alternately flash the reticle on and off, rather than glowing steadily. The advantage here is that by not keeping the reticle on continuously, fainter stars may be spotted and followed with greater ease. Rate and brightness are both adjustable by the user.

One of the most versatile is Celestron's Micro Guide, shown in Figure 6.5. Built around a 12.5-mm orthoscopic eyepiece, the Micro Guide features a laser-etched reticle that includes a bull's-eye–style target for guiding during astrophotography and micrometer scales for measuring object size and distances, such as the separations and position angles of double stars. Meade's Astrometric eyepiece is quite similar, but it is based around a 12-mm modified-achromat (Kellner) eyepiece. Meade's distinctive CCD Framing Ocular, made from a 12-mm Plössl, has multiple rectangular frames showing popular CCD fields. Note that the scales shown in these three eyepieces are calibrated only for telescopes with 80-inch focal lengths.

Putting the Puzzle Together

Now comes the moment of truth: Which eyepieces are best for you? It is always preferable to have a set of oculars that offers a variety of magnifica-

Table 6.6 ***Reticle Eyepieces***

Company/Model	Focal Length	Optical Design	Type of Reticle	Illuminated?
Celestron International				
#94170	12 mm	Kellner	Double crosshairs	Sold separately (choose from either standard or pulsating, both cordless)
#94169	10 mm	Plössl	Double crosshairs	Sold separately (choose from either standard or pulsating, both cordless)
#94171 Micro Guide	12 mm	Orthoscopic	Multiple scales	Yes, built-in, adjustable LED
Lumicon				
LEPXHR	25 mm	Kellner	Double crosshairs	Yes, built-in, adjustable LED; pulsating or constant brightness settings
Meade Instruments				
12-mm Reticle	12 mm	Mod. Achromat	Double crosshairs	Yes, must be plugged into telescope power panel
12-mm Cordless	12 mm	Mod. Achromat	Double crosshairs	Yes, built-in, adjustable LED
Astrometric	12 mm	Mod. Achromat	Multiple scales	Yes, built-in, adjustable LED
CCD Framing	25 mm	Plössl	Multiple frames	Yes, built-in, adjustable LED
Advanced Reticle	9 mm	Plössl	Double crosshairs and two concentric circles	Yes, must be plugged into telescope power panel
Advanced Cordless	9 mm	Plössl	Double crosshairs and two concentric circles	Yes, built-in, adjustable LED
Orion Telescope Center				
#8482	12 mm	Kellner	Double crosshairs	Yes, must be plugged into telescope power panel
#8481	12 mm	Kellner	Double crosshairs	Yes, built-in, adjustable LED
#8484	12.5 mm	Plössl	Double crosshairs	Yes, built-in, adjustable LED
#8498	10 mm	Ultrascopic	Double crosshairs	Yes, built-in, adjustable LED
#8496	5 mm	Ultrascopic	Double crosshairs	Yes, built-in, adjustable LED
Sky Instruments				
Illuminated Reticle	12 mm	Kellner	Double crosshairs	Yes, built-in, adjustable LED
Van Slyke Engineering				
IRE	25 mm	Mod. Kellner	Center circle with graduated line	Sold separately (choose from either standard or pulsating, both cordless)

tions, since no one value is good for everything in the universe. Low power is best for large deep-sky objects such as the Pleiades star cluster or the Orion Nebula. Medium power works well for lunar sightseeing as well as for viewing smaller deep-sky targets such as most galaxies. Finally, high power is needed to spot subtle planetary detail or to split close-set double stars.

Figure 6.5 *Celestron's Micro Guide illuminated-reticle eyepiece. Photo courtesy of Celestron International.*

Table 6.7 ***Four Telescope/Eyepiece Alternatives***

Dream Outfit	Middle of the Road	Good and Cheap
4″ f / 9.8 refractor		
LP: 35-mm Panoptic	LP: 32-mm Plössl	LP: 25-mm Kellner
MP: 12-mm Nagler	MP: 12-mm orthoscopic	MP: 9-mm Kellner
HP: 5-mm Radian	HP: 8-mm Plössl	HP: —
Barlow lens	Barlow lens	Barlow lens
8″ f / 6 Newtonian		
LP: 21-mm Pentax XL	LP: 25-mm Plössl	LP: 25-mm Kellner
MP: 14-mm Ultra Wide	MP: 10-mm Plössl	MP: 9-mm Kellner
HP: 8-mm Radian	HP: —	HP: —
Barlow lens	Barlow lens	Barlow lens
18″ f / 4.5 Newtonian		
LP: 31-mm Nagler 5	LP: 25-mm Ultrascopic	LP: 25-mm orthoscopic
MP: 12-mm Nagler 4	MP: 9-mm orthoscopic	MP: 10-mm Plössl
HP: 7-mm Pentax XL	HP: 6-mm orthoscopic	HP: —
2-inch Barlow lens	Barlow lens	Barlow lens
Coma Corrector		
8″ f / 10 Schmidt-Cassegrain		
LP: 50-mm Plössl	LP: 40-mm Optiluxe	LP: 25-mm Kellner
MP: 17-mm Nagler	MP: 15-mm Lanthanum	MP: —
HP: 6-mm Radian	HP: 9-mm orthoscopic	HP: 9-mm Kellner
Barlow lens	Barlow lens	Barlow lens
Reducer/compressor		

Who makes what? In Appendix B, I try to sort out the eyepiece marketplace by giving a blow-by-blow account of what eyepieces are sold by which companies. This table also looks at all of the important criteria that should be weighed when deciding what to purchase. In an ideal world, eyepiece quality, eye relief, field of view, and optical coatings would be our chief considerations, but in the real world, most of us must also factor in cost. That is why eyepieces have been collected according to their costs (early 2002 prices).

As an aid to guide your selection, Table 6.7 also offers three different possibilities for four of today's most popular telescope sizes. Each lists eyepieces according to the magnification they would produce (LP = low power, MP = medium power, HP = high power). The first is a dream outfit, where money is no object, the second offers a middle-of-the-road compromise between quality and cost; and the third represents the minimum expenditure required for a good range of eyepieces. Prices are not listed because they can change quickly and dramatically from dealer to dealer; be sure to shop around. If your budget is tight, rely on your medium-power eyepiece and Barlow lens for high-power views.

Keep in mind that these represent only three possibilities for each telescope. You are encouraged to flip back and forth among the eyepiece descriptions, Appendix B, and Table 6.6 to substitute your own preferences. And remember, the only way to learn which oculars are really best for your particular situation is to try them out first. Once again, I strongly recommend that you seek out and join a local astronomy club and go to an observing session. Bring your telescope along and borrow as many different types of eyepieces as possible. Take each of them for a test drive. Then, and only then, will you know exactly what is right for you.

7

The Right Stuff

Congratulations on making it through what might be thought of as telescope and eyepiece obstacle courses. Although the range of choices was extensive, you should now have a fairly good idea about which telescopes, binoculars, and eyepieces best meet your personal needs.

A telescope alone, however, cannot simply be set up and used. First, it must be outfitted with other things, such as a finderscope, maybe some filters, a few reference books, a star atlas, a flashlight, some bug spray . . . well, you get the idea. So get out your wish list and credit card once again. It is time to go shopping in the wide world of "astro-phernalia!"

Let me just take a moment to calm your fears at the thought of spending more money on this hobby. Sure, it is easy to draw up a list of must-have items as you look through astronomical catalogs and magazines. Everything could easily tally up to more than the cost of your telescope in the first place, but is this truly necessary? Happily, the answer is no. Before you buy another item, we must first explore the accessories that are absolutely mandatory, those that can wait for another day, and the ones that can be done without entirely.

A quick disclaimer before going on. There are so many accessories available to entice the consumer that it would be impossible for a book such as the one you hold before you to list and evaluate every item made by every company. As a result, this chapter must limit its coverage to more readily available items. If, as you are reading this chapter, you feel that I have unjustifiably omitted something that you believe is the greatest invention since the telescope, then by all means share your enthusiasm with me. Write your own review and send it to me in care of the address listed in the acknowledgments. I will try to include mention of that item in a future edition.

Finders

After eyepieces (and arguably even before), the most important accessory in an amateur astronomer's bag of tricks is a finder. There is nothing more frustrating to an amateur astronomer than not being able to uncover an object through his or her telescope. That's why most instruments require an auxiliary device, called a *finder,* for pointing. Finders come in two different flavors nowadays: *finderscopes,* which magnify, and *unity finders,* which do not.

Finderscopes (Figure 7.1) are small, low-power, wide-field spotting scopes mounted piggyback on telescopes. Their sole purpose in life is to help the observer aim the main telescope toward its target. While most telescopes come with finderscopes, all are not created equally! What sets a good finderscope apart from a poor finderscope depends on what it is going to be used for.

First, some words of advice for readers who will be using finderscopes only to supplement the use of setting circles or a GoTo computer for zeroing in on sky targets. If this applies to you (and I truly hope it does not, especially after reading my editorial in chapter 10), then the finderscope will probably be used only for locating bright Solar System objects and alignment stars, possibly Polaris to align the equatorial mount to the celestial pole, and perhaps a few terrestrial objects. In this case, just about any will do.

If, however, star hopping is your preferred method for locating sky objects (again, consult chapter 10), then finderscope selection is critical to your success as an observer. The three most important criteria by which to judge a finderscope are magnification, aperture, and field of view. Finderscopes are specified in the same manner as binoculars. A 10 × 50 finderscope, for example, has a 50-mm aperture and yields 10 power. Most experienced observers agree that the smallest useful size for a finderscope is 8 × 50, although like binoculars, they should be matched to the sky conditions. Under suburban skies, an 8 × 50 finderscope (with a 6-mm exit pupil) will penetrate to about 8th magnitude, which is roughly comparable with many popular star atlases. Rural astronomers with darker skies may prefer giant 10 × 70 and 11 × 80 finderscopes. They enjoy the benefit of revealing stars a magnitude or two fainter, but suffer from smaller fields of view.

Figure 7.1 *An 8 × 50 finderscope.*

Some telescopes come with right-angle finderscopes. In these, a mirror- or prism-based star diagonal is built into the finderscope to turn the eyepiece at a 90° angle. Sure, a right-angle prism can make looking through the finderscope more comfortable and convenient, but is it really a good idea? To my way of thinking: NO! There are two big drawbacks to right-angle finderscopes. First, although the view through a right-angle finderscope is upright, the star diagonal flips everything left to right. This mirror-image effect matches the view through a telescope using a star diagonal (Schmidt-Cassegrain, refractor, etc.) but makes it very difficult to compare the field of view with a star atlas. By comparison, straight-through finderscopes flip the view upside down, but do not swap left and right. Personally, I find it easier to turn a star chart upside down to match an inverted view than to turn it inside out in order to match a mirrored image! Special prisms, called *Amici prisms,* can be used in place of an ordinary star diagonal to cancel out the mirroring effect, but these are supplied with relatively few finderscopes, probably because they also cause images to dim.

A straight-through finderscope also permits the use of both eyes when initially aiming the telescope. By overlapping the naked-eye view with that through the finderscope, an observer may point the telescope quickly and accurately toward the intended part of the sky. If you keep both eyes open when using a right-angle finderscope, one eye will see the sky, the other will see the ground!

Table 7.1 lists some of today's better finderscopes.

Deep down inside, most observers agree that right-angle finderscopes are poor substitutes for straight-through designs. At least I assume they must, given the incredible popularity of one-power aiming devices, or unity finders, in recent years. By far the most common of these new-generation sighting contraptions is the Telrad (Figure 7.2), invented by the late amateur telescope maker Steve Kufeld. The Telrad is described as a "reflex sight." Using a pair of AA batteries and a red light-emitting diode (LED), a bull's-eye target of three rings (calibrated at 0.5°, 2°, and 4°) is projected onto a clear piece of glass set at a 45° angle. The observer then sights through the glass, which acts as a beamsplitter, to see the reflected target rings as well as stars shining through from beyond. The brightness of the rings is controlled by a side-mounted rheostat which also acts as an on/off switch.

The Telrad is not necessarily meant to be used in lieu of a finderscope but only to supplement its use. The biggest advantage of the Telrad is that it allows easy aiming of a telescope without any need to flip or twist star charts. Its only disadvantage is that the window tends to dew over quickly in damp environs. But since it is not an optical device, the window may be wiped clear with a finger, a sleeve, or a paper towel. A more permanent solution to the dewing problem is to attach a shield over the window. Simply make a roof over the glass window by bending a large file card that has been painted black over the Telrad from side to side and secure it in place with masking tape.

Rigel Systems offers their QuikFinder, which looks and works like a vertical Telrad. The QuikFinder projects an adjustable-brightness 0.5°-diameter red circle onto an angled window. Unlike any of the other unity finders, the QuikFinder's red circle can be set to pulse off and on. This allows the observer to

Table 7.1 ***Finderscopes***

Company/ Model	Magnification	Aperture	Field of View (degrees)	RA or ST[1]	Features
Apogee					
50RA	8	50	6	RA	Interchangeable 1.25" eyepiece; mounting bracket sold separately
50ST	8	50	6	ST	Interchangeable 1.25" eyepiece; mounting bracket sold separately
70RA	12	70	4	RA	Interchangeable 1.25" eyepiece; mounting bracket sold separately
Celestron International					
93779	6	30	7	ST	Mounting bracket sold separately
93777	6	30	7	ST	Long eye relief; mounting bracket included
93785–8P	7	50	5	ST	Illuminated Polaris finder reticle; mounting bracket included
93783–8	9	50	5.8	ST	Mounting bracket included
Coulter					
8 × 50	8	50	7	ST	Mounting bracket included
8 × 50RA	8	50	7	RA	Mounting bracket included
8 × 50ERA	8	50	7	RA	Mounting bracket included; 90° Amici prism; right-side-up field of view
Discovery Telescopes					
5 × 24	5	24	n/s	ST	Mounting bracket included
6 × 30	6	30	n/s	ST	Mounting bracket included
7 × 50	7	50	n/s	ST	Mounting bracket included
Earth and Sky Products					
Spectiva 6 × 30	6	30	n/s	ST	Mounting bracket included
Lumicon					
Super Finder[2]		50	n/a	ST, 45°, or RA	Mounting bracket and eyepiece sold separately
Super Finder[2]		50	n/a	ST or RA	Mounting bracket and eyepiece sold separately
Meade Instruments					
#637	6	30	7	ST	Mounting bracket included
#825	8	25	7.5	RA	90° finder for ETX
#544 (blue)	8	50	5	ST	Mounting bracket sold separately
#545 (white)	8	50	5	ST	Mounting bracket sold separately

Table 7.1 ***(continued)***

Company/ Model	Magnification	Aperture	Field of View (degrees)	RA or ST[1]	Features
Orion Telescope Center					
#13023	6	30	5.2	ST	Mounting bracket included
#13000	6	26	6.3	ST	Mounting bracket included; right-side-up field of view
#13060	8	50	4.7	ST	Mounting bracket sold separately
Roger W. Tuthill					
50mm RA	7 or 8	50	n/s	RA	Dual crosshairs, 1.25" diameter; focusable eyepieces with rubber eyecups; Amici prism available separately
80mm RA	11 or 15	80	n/s	RA	Dual crosshairs, 1.25" diameter focusable eyepieces with rubber eyecups; Amici prism available separately
ScopeTronix					
FS50	8	50	n/s	ST	Mounting bracket included
Sky Instruments					
Omcon F30	6	30	n/s	ST	Mounting bracket included
Antares F850	8	50	7	ST	Mounting bracket sold separately; translucent crosshairs
Antares FR850	8	50	7	RA	Mounting bracket sold separately; translucent crosshairs
Antares FE850	8	50	7	RA	Mounting bracket sold separately; 90° Amici prism; right-side-up field of view; translucent crosshairs
Omcon F1050	10	50	n/s	ST	Mounting bracket sold separately
Takahashi					
5 × 25	5	25	n/s	ST	Mounting bracket sold separately
7 × 50	7	50	n/s	ST	Mounting bracket and illuminated reticle sold separately
11 × 70	11	70	n/s	ST	Mounting bracket and illuminated reticle sold separately
Vixen Optical Industries					
Achromatic	7	50	6.8	ST	Mounting bracket included

1. RA: right-angle; ST: straight-through.

2. Eyepiece sold separately.

n/s: not supplied by manufacturer; n/a: not applicable.

Figure 7.2 *A one-power Telrad aiming device by Steve Kufeld, one of the most popular accessories among today's amateur astronomers.*

keep the circle's brightness relatively high while aiming at dim stars, a helpful feature. The QuikFinder weighs just 3.4 ounces (as compared to Telrad's 11 ounces) and stands about 4.5 inches tall. Its small footprint, just 1.38 inches, is the most compact of any unity finder currently made, although it also seems to imply a certainly frailty, that it might be easily snapped off. To prevent this from happening, the QuikFinder can and should be removed from its base and stored separately whenever the telescope is moved.

Several of the unity finders in Table 7.2, such as the Celestron Star-Pointer and Tele Vue Qwik-Point, are direct adaptations of an aiming device that Daisy Manufacturing Company markets for their BB guns and pistols. Both are made from black plastic and project single red dots onto partially reflective curved windows. The observer-side surface of the windows is coated with a clear reflective material, while the outer surface is uncoated. Both surfaces reflect images of the red dot, which merge into one when the observer's eye is on-axis. Power to the units is provided by thin, button batteries mounted on the bottoms of the units. Most are adjustable in brightness.

In principle, Orion Telescope's EZ Finder is the same as the other red-dot models detailed above. Its long tube, however, is a little more difficult to sight through when out under the stars. Like the others, its brightness is fully adjustable, in this case, by turning a small side-mounted potentiometer.

Leanest and sexiest of all is Tele Vue's Starbeam. Like the Celestron Star-Pointer, Orion EZ Finder, and Tele Vue Qwik-Point, the Starbeam projects a red LED dot onto a partially reflective window that the observer looks through when aiming the telescope. But while the others are made of plastic, the lean body of the Starbeam is crafted from machined aluminum. Not surprisingly, the Starbeam costs about four times as much as the others. Beauty may be only skin deep, but cost cuts right to the bone, so the Starbeam has never achieved the popularity of some of its ugly duckling cousins.

Finally, a word or two on finderscope mounts. Good finderscopes are secured to telescopes by a pair of rings with six adjustment screws. These

Table 7.2 ***Unity Finders***

Company/Model	Type (target or dot)	Comments
Apogee		
Mars-Eye	Dot	Includes dimmer and mounting plate
Celestron International		
Star-Pointer	Dot	Includes dimmer and mounting plate
Orion Telescope Center		
EZ Finder	Dot	Includes dimmer and quick-release dovetail mount
Rigel Systems		
QuikFinder	Target	Pulsing or steady 0.5° and 2° red circles; includes two dovetail mounting plates
Scopetronix		
LightSight	Dot	Dual brightness setting
Stellarvue		
Red Dot Finder	Dot	Includes dimmer and mounting plate
Tele Vue		
Qwik-Point	Dot	Includes dimmer and mounting plate
Starbeam	Dot	All-metal construction; Includes dimmer and mounting plate
Telrad	Target	0.5°, 2°, and 4° red circles; includes dimmer and dovetail mounting plate

thumbscrews allow the finderscope to be aligned precisely with the main instrument. On some, one of the thumbscrews has been replaced by a spring-loaded piston, which makes alignment quite easy. Let the buyer beware, however, that many smaller finderscopes (primarily 5 × 24 and 6 × 30 models) use single mounting rings with only three adjustment screws. These are notoriously difficult to adjust and even more difficult to keep in alignment. If there is a choice between a single ring or a pair of rings for mounting a finderscope, always select the pair.

Filters

For years, photographers have known the importance of using filters to change and enhance the quality and tone of photographs in ways that would be impossible under natural lighting. Today, more and more amateur astronomers are also discovering that viewing the universe can also be greatly enhanced by using filters with their telescopes and binoculars. Some heighten subtle, normally invisible planetary detail, others suppress the ever-growing effect of light pollution, while still others permit the safe study of our star, the Sun. Which filter or filters, if any, are right for you depends largely on what you are looking at and from where you are doing the looking.

Light-Pollution Reduction (LPR) Filters

As all amateur astronomers are painfully aware, the problem of light pollution is rapidly swallowing up our skies. In regions across the country and around the world, the onslaught of civilization has reduced the flood of starlight to a mere trickle of what it once was. Are we powerless against this beast? Not entirely.

While chapter 10 discusses the dilemma of light pollution in more specific detail, this section will look at some filters that may be used to help counteract part of the problem. First, let's briefly define light pollution. Light pollution is unwanted illumination of the night sky caused largely by poorly designed or poorly aimed artificial lighting fixtures. Rather than illuminating only their intended targets, many fixtures scatter their light in all directions, including up.

There are two kinds of light pollution: local and general. Local light pollution shines directly into the observer's eyes and may be caused by anything from a nearby streetlight to an inconsiderate neighbor's porch light. There isn't a filter made that will counteract this sort of interference. Light-pollution reduction (LPR) filters are much more effective against general light pollution, or *sky glow.* Sky glow, the most destructive type of light pollution, is the collective glare from untold hundreds or even thousands of distant lighting fixtures. It can turn a clear blue daytime sky into a yellowish, hazy night sky of limited usefulness.

Although modern technology caused the problem in the first place, it also offers partial redemption. As indicated in Figure 7.3, many sources of light pollution shine in the yellow region of the visible spectrum, between 550 nm and 630 nm (nm is short for *nanometer,* a very small unit of measure; one nanometer is equal to 10^{-9} meters, or 10 angstrom [Å] units). At the same time, many nonstellar sky objects (planetary nebulae, emission nebulae, and comets) emit most of their light in the blue-green region of the spectrum. Emission nebulae, for example, glow primarily in the hydrogen-beta (486 nm) and oxygen-III (496 nm and 501 nm) regions of the spectrum. In theory, if the yellow wavelengths could somehow be suppressed while the blue-green wavelengths were allowed to pass, then the effect of light pollution would be greatly reduced.

What exactly do light-pollution reduction filters do? A popular misconception is that LPR filters (Figure 7.4) make faint objects look brighter. Not true! LPR filters consist of thin pieces of optically flat glass that are coated in multiple, microscopically thin layers of special optical material. These filters are designed to block specific wavelengths of bad light while letting good light pass. The observer need only attach the filter to his or her telescope (usually either by screwing the filter into the field end of an eyepiece or, in the case of many Schmidt-Cassegrains, inserting it between the telescope and star diagonal). The net result is increased contrast between the object under observation and the filter-darkened background. (Many deep-sky objects also shine in the deep-red hydrogen-alpha [656 nm] portion of the spectrum, a region that is all but invisible to the eye under dim light conditions. Still, most light-pollution reduction filters transmit these wavelengths as well for photographic purposes.)

Light-pollution reduction filters (also called *nebula filters*) come in three varieties: *broadband, narrowband,* and *line filters.* The biggest difference is in

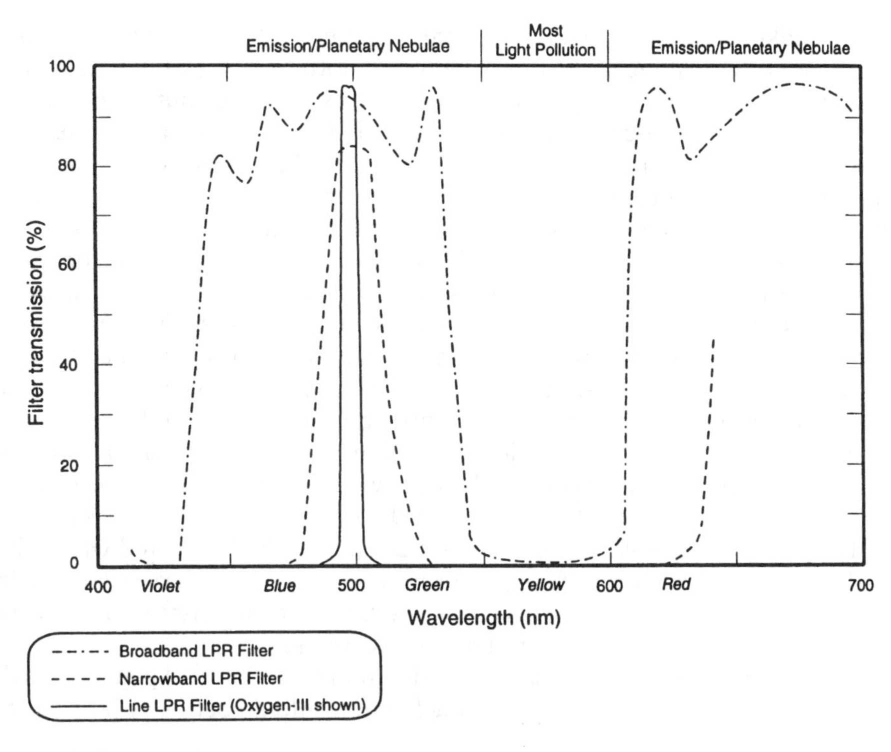

Figure 7.3 *Transmission characteristics of light-pollution reduction (LPR) filters. Chart based on data supplied by Lumicon, Inc.*

their application. As shown in Figure 7.3, broadband filters pass a wide swath of the visible spectrum, from about 430 nm to around 550 nm. Narrowband filters transmit wavelengths of light between about 480 and 520 nm. Line filters have an extremely narrow window, allowing only one or two specific wavelengths of light to pass. No one type of filter is best for everything. Some enhance the object under observation, some have little or no effect, and others

Figure 7.4 *An assortment of LPR filters. Photo courtesy of Lumicon, Inc.*

actually make the target fade or disappear completely! Which LPR filter is best under what circumstances remains one of the hottest topics of debate among amateur astronomers today. Table 7.3 may help sort things out. Each filter is rated on a four-star scale based on performance. One star indicates either negligible or negative results, two stars connote minimal positive effect, three stars indicate a noticeable improvement in appearance, while four stars denote a vast improvement in object visibility and/or detail.

Should you even buy a filter? If you are interested at all in deep-sky observing, then the answer is an unqualified yes! *Which* filter you should buy, however, is not an easy question. The answer depends on where you do your observing from. If you are the captive of a city, then you ought to consider a broadband filter. Though the improvement in deep-sky objects is marginal, broadband filters do help increase the contrast with the surrounding sky, making the view more aesthetically pleasing. All are very close to one another in performance, and you will probably be happy with any of those detailed here. But if I had to rank their performance, the Orion SkyGlow filter seems to work the best, followed closely by the Lumicon Deep Sky, the Thousand Oaks LP-1 broadband filter, the Meade 908B, and the Celestron LPR filters in that order (the Meade and Celestron are virtually identical). From suburban or rural sites, however, broadband filters are nearly worthless.

Observers interested in spotting fine detail in emission and planetary nebulae, regardless of where you are viewing from, would do better by purchasing

Table 7.3 **Light-Pollution Reduction (LPR) Filters: A Comparison**

	Open Clusters	Globular Clusters	Emission Nebulae	Reflection Nebulae	Planetary Nebulae	Galaxies	Comets
Broadband							
Celestron LPR	*	*	**	**	***	*	**
Lumicon Deep Sky	**	**	**	**	***	**	**
Meade 908B	*	**	**	**	***	*	**
Orion SkyGlow	**	**	**	**	***	**	**
Thousand Oaks LP-1	**	**	**	**	***	**	**
Narrowband							
DayStar 300	*	**	***	**	****	*	*
Lumicon UHC	*	*	****	**	****	*	*
Meade 908N	*	*	***	**	****	*	*
Orion UltraBlock	*	*	****	**	****	*	*
Thousand Oaks LP-2	*	*	***	**	****	*	*
Line							
Lumicon H-Beta	*	*	***	*	*	*	*
Lumicon O-III	*	*	***	**	****	*	*
Lumicon Swan Band	*	**	**	**	***	*	****
Meade 908X (O-III)	*	*	***	**	****	*	*
Thousand Oaks LP-3 (O-III)	*	*	****	**	****	*	*
Thousand Oaks LP-4 (H-Beta)	*	*	***	*	*	*	*

either a narrowband or line filter. Each is a vast improvement over an unfiltered telescope as well as with any of the broadband filters. As an added bonus, each narrowband filter also improves the views of Mars and Jupiter, making them more versatile than perhaps even their manufacturers realize! Which is best? Honestly, all work well, but if I had to rate them in order of preference, I would give a slight nod to the Orion UltraBlock, Lumicon UHC, and Thousand Oaks LP-2, followed closely by the DayStar 300 and the Meade 908N. Personally, I would recommend that you let price and availability be your guide; you will be happy with any of them. One exception to this is the Meade 908N, which seemed to outperform the others by a noticeable margin when used with my 4-inch f/9.8 Vixen achromatic refractor.

Finally, what about the line filters? Owners of medium- and large-aperture instruments should also give serious consideration to an O-III filter. Although I personally prefer the narrowband filters, O-III filters have a loyal following that is well deserved. Frequently, they will reveal subtle features in nebulae that remain invisible through the others. Indeed, on many bright emission nebulae, O-III filters outperformed all but the Orion UltraBlock and Lumicon UHC filters. My only complaint is that the O-IIIs tend to darken the background sky too much. The Thousand Oaks LP-3, the Meade 908X, and the Lumicon O-III all work well, but I'll cast my vote for the LP-3.

Lumicon's H-Beta filter and Thousand Oaks LP-4 hydrogen-beta filters work well on a few objects, notably the Horsehead and California Nebulae (two notoriously difficult deep-sky objects), but on most objects they actually have a negative impact! Finally, comets, because of their unique spectra, are best observed using Lumicon's Swan Band filter, designed with comets in mind.

The bottom line is this: LPR filters are wonderful assets for the amateur astronomical community. With them, observers can spot nonstellar objects that were considered impossible to find just a few years ago. But remember, no filter can reveal the beauty of the universe as well as a dark, light-free sky can.

Lunar and Planetary Filters

Most seasoned observers agree that fine details on both the Moon as well as the five planets visible to the naked eye can be significantly enhanced by viewing through color filters. This improvement occurs for two reasons. First, they reduce irradiation, the distortion of the boundary between a lighter area and a darker region on the Moon's or a planet's surface. This effect usually is caused by either turbulence in Earth's atmosphere or by the human eye being overwhelmed by the dazzling image.

Filters also help to increase subtle contrasts between two adjacent regions of a planet or the Moon by transmitting (brightening) one color while absorbing (darkening) some or all other colors. For instance, an observer's eye alone may not be able to distinguish between a white region and a bordering beige region on Jupiter. By using filters of different colors, the contrast between the zones may be increased until their individuality becomes apparent.

Which filters are best for which objects in the sky? Table 7.4 compares different heavenly sights with the results that may be expected when they are

Table 7.4 ***Color Filters: A Comparison***

Object	Filter Type	Benefit
Moon	Moon filter (neutral density)	Reduces brightness of the Moon evenly across the spectrum, making observations easier without introducing false color.
	15 (deep yellow)	Enhances contrast of lunar surface.
	58 (green)	Like deep yellow, green will enhance contrast and detail in some lunar features.
	80A (light blue)	Reduces glare.
	Polarizer	Like the neutral density filter, reduces brightness without introducing false colors.
Mercury	21 (orange)	Helps observer to see the planet's phases.
	23A (red)	Increases contrast of planet against blue sky, aiding in daytime or bright twilight observation.
	25 (deep red)	Same as #23A, but deeper color.
	80A (light blue)	Improves view of Mercury against bright orange twilight sky.
	Polarizer	Darkens sky background to increase contrast of planet. Helpful for determining phase of Mercury.
Venus	25 (deep red)	Darkens background to reduce glare. Some say they also help reveal subtle cloud markings.
	80A (light blue)	Improves view of Venus against bright-orange twilight sky.
	Polarizer	Reduces glare without adding artificial color (especially important for viewing planet through larger telescopes).
Mars	21 (orange)	Penetrates atmosphere to reveal reddish areas and highlight surface features such as plains. (Best choice for small apertures.)
	23A (light red)	Same as #21, but deeper color. (Best choice for medium and large apertures.)
	25 (deep red)	Same effect as #23A, but deeper color. (Best choice of the three for very large apertures.)
	38A (deep blue)	Brings out dust storms on surface of Mars.
	58 (green)	Accentuates melt lines around polar caps.
	80A (light blue)	Accentuates polar caps and high clouds, especially near the planet's limb.
Jupiter	11 (yellow-green)	Reveals fine details in cloud bands.
	21 (orange)	Accentuates cloud bands.
	56 (light green)	Accentuates reddish features such as the Red Spot.
	58 (green)	Same as #56, but deeper color.
	80A (light blue)	Highlights details in orange and purple belts as well as white ovals.
	82A (very light blue)	Similar effect as #80A, though not as pronounced.
Saturn	15 (deep yellow)	Helps to reveal cloud bands.
	21 (orange)	Similar effect as #11, but deeper color.

Table 7.4 ***(continued)***

Object	Filter Type	Benefit
Comets	80A (light blue)	Increases contrast of some comets' tails.
Other	15 (deep yellow)	Helps block ultraviolet light when doing black-and-white astrophotography.
	25 (red)	Reduces impact of light pollution on long-exposure black-and-white photographs taken from light-polluted areas.
	58 (green)	Same as #25. Works well for emission nebulae.
	82A (very light blue)	Suppresses chromatic aberration in refractors.
	Minus violet	Reduces impact of light pollution without dramatically distorting color. Cheaper than broadband LPR filters. Also good for reducing secondary chromatic aberration in achromatic refractors.

viewed through a variety of color filters. The filters are listed according to their *Wratten number* as well as their color. The Wratten series of color filters was created by Eastman Kodak and contains over 100 different shades and hues. Today, it is the industry's standard way to refer to a filter's precise color.

For those just starting out, choose basic colors, such as deep yellow (#15), orange (#21), red (#23), green (#58), and blue (#80A). Though you are free to use photographic filters sold in camera stores, most amateurs prefer color filters designed to screw into the field end of eyepieces. Be sure to check that your eyepieces and the filters both have the same thread size, or they won't screw together. Fortunately, the vast majority do. Also, if you are going to buy a variable polarizing filter, be aware that some designs, such as those by Meade and Celestron, require that the eyepiece slide into a holder that sticks out of the focuser, like a Barlow lens. As a result, some eyepieces will not focus properly if the focuser has limited travel.

Many telescope manufacturers and suppliers offer a wide assortment of standard-threaded color filters. Table 7.5 gives a listing of some of the current offerings.

Solar Filters

Monitoring the ever-changing surface of the Sun is an aspect of the hobby that is enjoyed by many. Before an amateur dares look at the Sun, however, he or she must be aware of the extreme danger of gazing at our star. **Viewing the Sun without proper precautions, even for the briefest moment, may result in permanent vision impairment or even blindness.** This damage is caused primarily by the Sun's ultraviolet rays, the same rays that cause sunburn. While it may take many minutes before the effect of sunburn is felt on the skin, the Sun's intense radiation will burn the eye's retina in a fraction of a second.

Table 7.5 ***Color Filters***

Company/Model	Color Shades/ % Transmission	Diameter	Comments
Celestron International			
Color filters	12 colors/shades	1.25"	Standard thread; 26-mm clear aperture; also available in sets
Moon filter	18% transmission	0.965", 1.25"	Standard thread; 21-mm clear aperture (1.25"); 10-mm clear aperture (0.965")
#93608 Polarizer	Varies between 5% and 30% transmission	1.25"	Eyepiece fits into holder; may result in loss of correct focus in some telescopes with limited focuser travel
Earth and Sky Products			
Moon filter	Not specified	1.25"	Standard thread
Meade Instruments			
Series 4000	12 colors/shades	1.25"	Standard thread; 26-mm clear aperture; dyed-in-the-mass color planetary filters
Series 4000 Moon filter	13% transmission	1.25"	Standard thread; 26-mm clear aperture; dyed-in-the-mass color
#905 Variable Polarizer	Varies between 5% and 25% transmission	1.25"	Eyepiece fits into holder; may result in loss of correct focus in some telescopes with limited focuser travel
Orion Telescope Center			
Color filters	13 colors/shades	0.965", 1.25", 2"	Standard thread; dyed-in-the-mass color; also available in discounted sets
Moon filter	13% transmission	0.965", 1.25"	Standard thread
Variable polarizer	Varies between 3% and 40% transmission	0.965", 1.25"	Standard thread; two polarizing layers in rotating cell
ScopeTronix			
Moon filter	14% transmission	1.25"	Standard thread
Sky Instruments			
Color filters	7 colors/shades	1.25"	Standard thread; also available in sets
Moon filter	13% transmission	1.25"	
VERNONscope			
Color filters	13 colors/shades	1.25"	Nonstandard thread[1]
Polarizer	30% transmission	1.25"	Nonstandard thread[1]

[1]Note that VERNONscope filters are threaded to fit VERNONscope Brandon eyepieces only. An adapter to fit these filters to other, standard-thread eyepieces is sold separately.

There are two ways to view the Sun safely: either by projecting it through a telescope or binoculars onto a white screen or piece of paper or by using a special filter. Sun filters come in a couple of different varieties. Some fit in front of the telescope, while others attach to the eyepiece. **NEVER** use the latter... that is, the eyepiece variety. They can easily crack under the intense heat of the Sun (focused by the telescope as is the Sun's image), leading tragically to blindness. Happily, I know of no new telescope that is supplied with an eyepiece solar filter, but many were in the past.

Safe solar filters are made from either glass or a polymer, such as Mylar, and fit securely *in front of* a telescope or binoculars. Figure 7.5 shows one example. The filter *must* be placed in front of the telescope so that it can filter out both the dangerously intense solar rays as well as the accompanying heat prior to entering the optical instrument. *Never place the filter between you and the eyepiece.* Be sure to use only specially designed Sun filters; *do not* use photographic neutral-density filters, smoked glass, overexposed photographic film, or other makeshift materials that may pass invisible ultraviolet or infrared light.

The most popular solar filters are sold by Baader Planetarium, Thousand Oaks Optical Company, Orion Telescopes, JMB, and Roger W. Tuthill. Of these, the oldest firm is Tuthill, who has been selling Solar Skreen filters since 1973. (Although Roger died in 2000, his legacy and company continue.) Then, as now, Solar Skreen filters consist of two tissue-paper thin pieces of aluminized Mylar material held together in a cell. While the raw material can be purchased separately, for this test, I used one of Tuthill's premounted filters, which comes in a cell made from PVC piping. Slip the filter cell over the front of the telescope, or in the case of a refractor, over the dew cap, and tighten three equidistant nylon thumbscrews.

Thousand Oaks Optical, another established name in the solar-filter field, offers both glass and nonglass filters. Their Type 2 Plus filters are made from float glass, the same stuff used in window glass. Many people have expressed doubts about the quality of the Sun's image through glass solar filters due to the possibility of optical distortions. But according to Pat Steele from Thousand Oaks, each piece of glass used for a Type 2 Plus filter is "hand selected" for its flatness, then coated with Solar II Plus, a mixture of chrome, stainless steel, and titanium. Thousand Oaks Type 3 Plus filters are made the same way, but they have a lighter density (Neutral Density 4) for photographic use. Type 3 Plus filters are **not** appropriate for visual observing.

Figure 7.5 *Full-aperture solar filter. Photo courtesy of Thousand Oaks Optical.*

Thousand Oaks Optical also sells a nonglass solar filter, the Polymer Plus, made from relaxed polymer film material that has also been coated with Solar II Plus. Polymer Plus filters are available in the same sizes as the Type 2 Plus. In addition, Thousand Oaks sells a Black Poly filter, which is designed for binoculars and finderscopes only (the film tends to blur high-power views, but is suitable for low-power viewing). Because their Black Poly filters are not designed for telescopic use, they were not included in this test.

JMB offers three different glass filters. The company's advertising claims that all of their filters are made of "machine-polished" float glass. JMB Identi-View Class A filters, coated with an alloy of nickel, chromium, and stainless steel, are also sold by the Orion Telescope Center. Premounted filters are available for telescopes ranging in aperture from 2 inches to 14 inches. The only difference between the Class A and less-expensive Class B is that the Class B's coating does not include stainless steel (and is, therefore, slightly less durable). Finally, like Thousand Oaks's Type 3 Plus filters, JMB's Class C filters have a lighter density that enable solar photographers to operate their cameras at faster shutter speeds when using slow, fine-grain films.

The Baader Planetarium in Germany is a newcomer to the field of solar filters. Its AstroSolar solar filter is also made from a polymer film that is specially darkened to reduce internal reflection. The material is heat-annealed like glass to reduce internal strains, then undergoes what is described as "an ion implantation and metallization" coating process. The Baader solar filter material is imported into the United States by Astro-Physics, which sells it only in raw sheets. In this case, the consumer must come up with his or her own cell to hold it in place, either by following the included instructions or in some other manner. Kendrick Astro-Instruments sells the Baader film in premade mountings, if you'd prefer, while Celestron also sells the material in cells made specifically for many of their instruments.

The most obvious difference between the filters is the color of the Sun's image. The JMB and Thousand Oaks Polymer Plus filters show the Sun as a yellowish disk, while it appears yellow-orange through the Thousand Oaks Type 2 Plus. The Baader filter turned the disk white, while the Tuthill Solar Skreen displayed a blue-white Sun. None of these shades represents the Sun's "true" color; instead, they are dependent on the type of coatings used.

The Baader AstroSolar filter impresses me both for its amazingly low price (as much as 60% lower than the others) as well as for showing the finest level of detail in sunspots. This was especially evident at higher powers when the other filters showed comparably dim images. The JMB Identi-View Class A and B also show a good amount of sunspot structure. For faculae and granulation, Tuthill's Solar Skreen outshines the others quite handily. But if you plan on showing the Sun to the public or school groups on a regular basis, I would recommend the Thousand Oaks Type 2+, since that produces a yellow-orange Sun, which is what everyone expects. Table 7.6 gives a summary of each filter.

All of the solar filters just discussed might be thought of as broadband filters, because they filter the entire visible spectrum (and you can't get much broader than that). There are also special narrowband solar filters that allow observers to see our star in a completely different light! These filters block all

Table 7.6 ***Solar Filters: A Comparison***

Company	Material	Image Color (solar disk/ sky background)	Diameters	Comments
Baader Planetarium				
AstroSolar Film	Polymer	White/gray	Unmounted sheets only	Brightest image of all; excellent sunspot detail and image contrast; faculae seen along limb, but washed out elsewhere due to image brightness; bargain price
JMB				
Identi-View Class A	Glass	Pale yellow/black	2" to 14"	Excellent sunspot detail and image contrast; cell held in place by friction with rubber strip—better than felt, which can rip and tear more easily
Identi-View Class B	Glass	Pale yellow/black	2" to 14"	Similar performance to Class A, but not as durable; cell held in place by friction with rubber strip
Identi-View Class C	Glass	Pale yellow/black	2" to 14"	Neutral Density (ND) 4, only suitable for photography, not visual use; cell held in place by friction with rubber strip
Orion Telescope Center				
Full-Aperture Filters	Glass	Pale yellow/black	3.56" to 12.45"	Made by JMB; see Identi-View Class A
Roger W. Tuthill				
Solar Skreen	Polymer	Blue-white/blue	4" to 14"	Mounted filters in cells; sizes to fit more than 50 specific telescope models; excellent sunspot detail, though slightly less clear than Baader; best on faculae and granulation; best, most secure cell
Thousand Oaks Optical				
Polymer Plus	Polymer	Yellow/black	2" to 8"	Also available unmounted; excellent sunspot detail, though slightly less clear than Baader; very good on faculae and granulation; cell held in place by friction with felt, which can rip and tear more easily than rubber lining

(*continued*)

Table 7.6 ***(continued)***

Company	**Material**	**Image Color (solar disk/ sky background)**	**Diameters**	**Comments**
Type 2 Plus	Glass	Yellow-orange/black	2" to 14"	Also available unmounted; very good sunspot detail; faculae visible, but not as contrasty as others; cell held in place by friction with felt
Type 3 Plus	Glass	Yellow-orange/black	2.4" to 14"	Also available unmounted; neutral Density (ND) 4, only suitable for photography, not visual use; cell held in place by friction with felt

of the light from the Sun with the exception of one distinctive wavelength: 656 nanometers, also known as hydrogen-alpha. Viewing the Sun at this wavelength with an H-alpha filter changes the view from a placid disk peppered with the occasional sunspot to a dynamic stellar inferno highlighted by ruby-red solar prominences, bright white filaments, and intricate surface granulation.

Hydrogen-alpha filters typically consist of two separate pieces: an energy-rejection filter (ERF) that fits over the front end of a telescope to prevent overheating and the hydrogen-alpha filter itself, which fits between the telescope and eyepiece. The filters are available in several different, extremely narrow bandwidths usually expressed in angstroms (1 angstrom equals 0.1 nanometer). The narrower the bandwidth, the higher the contrast but fainter the image. Bandwidths of 0.6 to 0.9 angstrom are the most popular because these offer the best compromise between brightness and contrast.

Unfortunately, H-alpha solar filters are expensive. DayStar Filter is one of the leading sources for these special accessories. Most expensive of the DayStar filters is the University model, with bandpasses from 0.4 to 0.8 angstrom available. As the name implies, it is geared toward the rigid requirements of professional institutions. The ATM filter, available in bandwidths from 0.5 to 0.95 angstrom, is designed with the serious amateur solar astronomer in mind. Both the University and ATM filters need to be heated in order to function properly and therefore require an external source of AC power. The least expensive H-alpha filter of all the DayStar models is the T-Scanner. Unlike the other two, the T-Scanner requires no external power supply and so is completely portable, making it even more attractive. Available in bandwidths from 0.5 to 0.8 angstrom, the T-Scanner is probably the most popular H-alpha filter available today.

Coronado solar filters are generally conceded to be the best. Although the company is based in Arizona, Coronado manufactures their filters on the Isle of Man. The least expensive brand, the ASP-60, has a bandwidth of less than 0.7 angstrom, providing excellent image contrast. Granulation, filaments, sunspot intricacies, and erupting prominences are all seen in vivid detail, creating a Sun that is much more alive than it appears with regular filters. The pricier Coronado AS1-90 has a slimmer, 0.6-angstrom bandwidth for better contrast with larger apertures. The ASP-60 is designed to thread directly onto a Tele Vue Pronto, while the AS1-90 fits the Tele Vue NP-101 or TV-102 refractors. Adapters are available for attaching either to a TMB Optics 4-inch refractor, Celestron 8, and 10- and 12-inch Meade LX200s. In addition, larger filters up to 140 mm in aperture are also available. Lastly, for those who are *really* serious about solar viewing, Coronado sells the Helios 1, a 2.8-inch f/5.7 refractor dedicated specifically for that task. The views produced by its permanently attached hydrogen-alpha filter are just amazing. There is always a long line by the Helios whenever Coronado has a display set up at astronomy conventions!

Lumicon also manufactures H-alpha solar filters for the amateur marketplace. The Lumicon Solar Prominence Filter works on the same no-heater principle as the T-Scanner. Its 1.5-angstrom bandpass, wider than the DayStar and Coronado filters, renders good views of prominences, but is not designed for viewing flares, filaments, and granulation. Complete with ERF prefilter and all required adapters and hardware, the Lumicon filter is significantly cheaper than the T-Scanner and the ASP-60 and is an excellent value for anyone living within a budget who wishes to get into this phase of solar observing.

Other Accessories

Collimation Tools

A telescope will deliver only poor-quality, lackluster images unless it is in proper optical alignment, or collimation. Several companies manufacture collimating tools for amateur telescopes, with some more useful for certain optical designs than others. Table 7.7 offers a summary. Typically, refractors should never have to be collimated, while reflectors and catadioptric telescopes should be checked regularly. Rich-field reflectors (that is, those with focal ratios less than f/6) should be checked each time they are set up.

The least expensive collimation tool sold today, at less than $10, is the Aline from Rigel Systems. The Aline is simply a black plastic cap with a centrally drilled peephole and a reflective surface on the cap's inside surface. Place the Aline into your telescope's 1.25-inch focuser and look through the hole. Adjust the telescope's diagonal and primary mirrors until both appear centered. The Aline is a handy tool for slower focal ratio reflectors, but faster instruments require some of the more sophisticated tools discussed below.

The three most popular collimation tools remain the sight tube, the Cheshire eyepiece, and the Autocollimator. The sight tube is helpful for approximating the collimation of just about any telescope. At one end lies a small peephole used by the observer to check collimation, while at the other, a pair of thin crosshairs serve as a reference for centering optical components. (Collimation procedures are outlined in detail in chapter 9.) Tectron Instruments,

Table 7.7 ***Collimation Tools***

Company	Product	Type/Comments
AstroSystems	Light Pipe	Sight tube; fits 1.25" focusers
	Autocollimator	Autocollimator; fits 1.25" focusers
Jim Fly	CatsEye	Similar to a Cheshire eyepiece; fits 1.25" focusers
Orion Telescope Center	Collimating eyepiece	Sight tube/Cheshire eyepiece combination; fits 1.25" focusers
Rigel Systems	Aline	Similar in principle to Autocollimator; fits 1.25" focusers
Tectron	Sight tube	Fits 1.25" focusers
	Cheshire eyepiece	Fits 1.25" focusers
	Autocollimator	Fits 1.25" focusers

AstroSystems, and Orion Telescopes all sell 1.25-inch-diameter sight tubes that will fit most amateur telescopes.

The Cheshire eyepiece (Figure 7.6), invented 80 years ago by Professor F.J. Cheshire at a British university, is intended primarily for in-the-field adjustments of fast Newtonians. Not an optical assembly, a Cheshire eyepiece is a variation on the sight tube with a part of one side cut out and a 45° mirrored surface inside. A light is shone through the cutout, reflects from the diagonal to the primary mirror, back to the diagonal, and out through a hole in the center of the eyepiece. The telescope is properly collimated when the reflected image of the Cheshire's center hole is centered in the primary mirror.

Figure 7.6 *A Cheshire eyepiece, a must for collimating fast Newtonian reflectors.*

Tectron, AstroSystems, and Orion Telescopes all sell Cheshire eyepieces. Those sold by Orion and AstroSystems include sight tube adapters that create handy all-in-one sight tube/Cheshire eyepiece combinations. Jim Fly sells a similar tool called the CatsEye, which uses a red reflective triangle to mark the center of the primary rather than the more typical white ring. Another difference is that the user must shine a light onto the mirror from the front of the tube, rather than through the cutout in the side of the tube, as with a traditional Cheshire eyepiece. Which is better? Fly's reflective triangle is a very good idea, but his system also makes it possible to drop the flashlight into the telescope tube. All things being equal, both a traditional Cheshire eyepiece and the CatsEye work about equally well.

The Autocollimator is the most sensitive of the three tools for collimating a fast Newtonian reflector. At first glance, it looks like a short sight tube; that is, a hollow tube capped with a piece of metal that has a tiny hole drilled in the center. The difference only becomes evident when you look inside the Autocollimator. There, encircling the peephole and perpendicular to the eyepiece mount's optical axis, a flat mirror serves to reflect all light from the primary mirror back down the telescope.

Are all three necessary? The sight tube is a *must* for anyone who owns a reflecting or catadioptric telescope and is concerned with optimizing its performance. I also consider the Cheshire eyepiece a requirement for owners of faster Newtonian reflectors. Don't leave home without it! The Autocollimator, although a handy tool for fine-tuning the alignment of fast Newtonians, is not absolutely necessary.

Recently, laser collimators have become all the rage in certain circles. Lasers emit precise, narrow beams of light that do not expand or elongate as they travel away from their source. Within the past few years, small, low-power units have been created and incorporated into many diverse applications, including telescopes.

But do they work as advertised? The short answer is yes, but with some qualification. Most of these accessories are not designed to be used with refractors or catadioptric instruments; they are only intended for deep-dish (less than f/6) Newtonian reflectors. If your Newtonian has a longer focal ratio than this, the collimation tools mentioned above are just fine.

The principle behind a laser collimator is simple enough. Place the laser into the eyepiece holder and turn on the beam. If the telescope's optics are properly collimated, the beam should bounce from the diagonal mirror to the primary mirror, then back exactly on itself. You then look through the front of the telescope tube at the bottom of the collimator protruding out of the focusing mount. If you see an off-center image of the reflected beam on the collimator, then the optics are not properly aligned. Table 7.8 lists current offerings.

Of those listed, some project simple dots, while others generate intricate grid patterns. Personally, after trying both types through my 18-inch f/4.5 reflector, I find the latter to be more precise. They seem to make it easier to determine which way the mirrors must be tilted to achieve optimal alignment. But they are also more expensive, which may be a consideration.

Table 7.8 ***Laser Collimators***

Company/Model	Pattern	Diameter	Power Source
APM			
Laser Collimator	Dot	1.25"	Three type AD-13 batteries
AstroSystems			
AstroBeam	Dot	1.25" and 2"	Two AA batteries
Helix Observing Accessories			
Laser collimator	Dot or several holographic projections	1.25"	Two LR44 batteries
Howie Glatter			
Laser Collimator	Dot or 9 × 9 square holographic grid	Either 1.25" or 2"	One 123A lithium cell
LaserMax			
TLC-202N	Holographic crosshairs with graduated concentric circles	1.25" and 2"	Three AAA batteries
Kendrick Astro Instruments			
2062	Dot	2"	Two AAA batteries
2063	Dot	1.25"	Two AAA batteries
Orion Telescope Center			
LaserMate	Dot	1.25"	Three SR44 batteries
Tech 2000			
Laser Collimator	Dot	1.25"	Three SR44 batteries

Cost is one of the drawbacks to laser collimators, although those from Orion Telescopes and Helix Observing Accessories are well within reach of most observers. The Helix collimator is especially noteworthy, because it is the only unit priced under $100 that can be outfitted with one of several holographic patterns at the time of purchase. It may also be retrofit after purchase, but only by returning the unit to the manufacturer.

Another, bigger concern is whether the laser collimator itself is collimated. Those listed in the table should be, since all are made from metal that has been machined to very tight tolerances. Some other laser collimators might be less expensive but may have plastic housings, which are far more prone to flexure and warping over time. Even among the collimators listed here, there is always the possibility that a defective unit may slip through quality-control measures. If you suspect that something is amiss with a laser collimator, contact the manufacturer about having it checked and cleaned.

A third problem is that the snout of the collimator may not protrude enough to make it easily visible in some focusing mounts. This is usually not a problem with the so-called low-profile focusers included with many Newtonian reflectors, especially open truss-tube models. I could not, however, use my laser collimator with the 8-inch f/6 Skywatcher reflector mentioned in chapter 5, since the focuser was too tall and the telescope's solid tube blocked the view from the side. If you can't see the bottom of the collimator through the focuser's tube, it can't be used.

This last dilemma is not applicable to the laser collimators from Kendrick Astro Instruments, which are perhaps the most versatile on the market. Although they only project single dots, they are long enough to extend out from the bottom of all but the longest Newtonian focusers. And unlike other laser collimators, these can also be used to collimate all Cassegrain-style telescopes. Kendrick has adapter tubes that have cutouts on one side through which the user can see the reflected laser beams (rather than having to look down the front of the telescope tube, as with other models). Although designed primarily for Cassegrains, this arrangement can also be used conveniently with Newtonians and Newtonian-catadioptrics.

Star Diagonals

Rather than make observers crane their necks to look straight through the eyepiece, most refractors and Cassegrain-style instruments come supplied with either a 45° or a 90° star diagonal that diverts the optical path to a more comfortable angle. While they certainly make life at the eyepiece much more enjoyable, some star diagonals work much better than others. Indeed, I have seen some supplied with beginners' telescopes that are so far out of alignment that they make focusing impossible. Others, because of relatively poor optics, dim images greatly. To owners of those telescopes, your best bet, if the telescope is acceptable otherwise, is to replace the original star diagonal.

There are two types of star diagonals, based on their angles: 45° and 90°. Traditionally, 45° prisms are associated with terrestrial rather than astronomical viewing. They are not nearly as easy to view through when a telescope is aimed at, or anywhere near, the zenith. Most 45° diagonals also generate upright images. To do this, the image must flip one additional time, dimming it further. My best advice is that if your telescope has such a star diagonal (perhaps it would be better called a terra-diagonal), you should consider replacing it with a real 90° star diagonal. But which one?

Some 90° diagonals use prisms to redirect the light, while others use flat mirrors. In general, assuming all other things are equal, a high-quality mirror diagonal will outperform a prism diagonal. The problem is that some light is dispersed and lost as it enters the prism, even if the prism is fully coated. More is lost as it exits. A mirror diagonal, however, has only one optical surface, resulting in a brighter, sharper image. The best mirror diagonals, like other telescopic mirrors, are coated with enhanced aluminizing for even brighter images.

Should you own a telescope that takes only subdiameter eyepieces, consider purchasing a hybrid star diagonal. These fit into 0.965-inch focusers but accept standard 1.25-inch eyepieces. Unless you are using some exceptional 0.965-inch eyepieces, the change will be quite an improvement.

Table 7.9 summarizes some of the most popular 90° star diagonals on the market today. Prices can vary widely from one model to the next, but in general they range from $30 for a plain-vanilla prism diagonal to $300 for a top-quality mirror diagonal.

Table 7.9 ***Star Diagonals***

Company/Model	Diameter	Type	Comments
Apogee			
Hybrid Diagonal	1.25"	Prism	Accepts 1.25" eyepieces in 0.965" focuser
Prism Diagonal	1.25"	Prism	
Mirror Diagonal	1.25"	Mirror	
Astro-Physics			
MaxBrite	2"	Mirror	99% reflectivity; threaded for 48-mm filters
Celestron			
#94106	1.25"	Mirror	Hybrid, accepts 1.25" eyepieces in 0.965" focuser
#94115-A	1.25"	Prism	Aluminum housing
#93519	2"	Mirror	Aluminum housing; screws onto standard Meade/Celestron threads
Earth and Sky Products			
Spectiva	1.25"	Mirror	
Intes			
Standard diagonal	2"	Mirror	
Deluxe diagonal	2"	Mirror	96% reflectivity
Lumicon			
Star diagonal	1.25"	Mirror	96% reflectivity
Star diagonal	2"	Mirror	96% reflectivity
Meade Instruments			
#915	0.965"	Prism	
#917	1.25"	Prism	Hybrid, accepts 1.25" eyepieces in 0.965" focuser
#918A	1.25"	Prism	Aluminum housing
#929	2"	Mirror	Aluminum housing; screws onto standard Meade/Celestron threads
#930	2"	Mirror	Aluminum housing
Orion Telescopes			
#8761	0.965"	Prism	Aluminum housing
#8763	1.25"	Prism	Aluminum housing
#14008	1.25"	Prism	Hybrid, accepts 1.25" eyepieces in 0.965" focuser
#8778	1.25"	Mirror	Aluminum housing
#8772	2"	Mirror	Aluminum housing
#8773	2"	Mirror	Aluminum housing; screws onto standard Meade/Celestron threads
Tele Vue			
Star diagonal	1.25"	Mirror	Aluminum housing
Star diagonal	2"	Mirror	Aluminum housing
EverBrite	1.25"	Mirror	Enhanced aluminizing; aluminum housing
EverBrite	2"	Mirror	Enhanced aluminizing; aluminum housing

Rubber Eyecups

To help shield an observer's eye from extraneous light, many eyepieces sold today come with collapsible rubber eyecups, and they certainly can make a big difference!

In the previous chapter, I tried to point out whenever an eyepiece came with a built-in eyecup. But just because an eyepiece is cupless doesn't mean it is not worth consideration. Most can be retrofitted with after-market eyecups that work just as well. Orion Telescopes, Edmund, Tele Vue, and several other companies sell rubber eyecups designed to fit both their own particular lines of oculars as well as others. Always match the proper size eyecup to your eyepiece before ordering, since a too-tight or too-loose eyecup isn't of much value.

Eyepiece Cases

If cared for, any of the equipment listed in this book should outlive you. Proper care is critical, especially when it comes to sensitive optical surfaces. While most of us treat our telescopes with kid gloves, eyepieces are often handled and stored far more haphazardly. Any eyepiece, whether a $20 Ramsden or $600 Nagler, should be protected from dust and dirt whenever not in use. To do that, several companies offer eyepiece carrying cases and individual storage packs. Orion Telescopes, for instance, sells several plastic and aluminum briefcase-style cases, each lined with cubed high-density foam that is easily customized to accommodate your accessories. Their smaller plastic cases are fine to start with, but if you really get into the hobby, you will need to expand to a larger aluminum unit as your eyepiece collection inevitably grows.

Tele Vue has adapted its case for the Pronto refractor into a nice padded eyepiece tote, complete with a handle, an adjustable over-the-shoulder strap, and compartments for up to 13 1.25- and 2-inch eyepieces. High-density foam protects the contents from shocks and keeps the gray nylon case from sagging. As we have come to expect from Tele Vue, their eyepiece carrying bag is very well made but expensive.

Some people prefer to store their eyepieces individually. For them, there is a series of inexpensive, translucent plastic cases from Absolute Astronomy. Each screws shut to seal out dirt quite effectively. Meade Instruments includes similar cases with many of their eyepieces.

Binocular Viewers

Research has shown that when it comes to viewing the night sky, binocular (two-eyed) vision is definitely better than monocular (one-eyed) vision. Our powers of resolution and ability to detect faint objects are dramatically improved by using both eyes. Some people enjoy up to a 40% increase in perception of faint, diffuse objects by viewing them with both eyes instead of just one. In addition, color perception and contrast enhancement also benefit.

While few experts will argue against the benefits of using binoculars over a same-size monocular telescope, the advantages of binocular viewing attachments for conventional telescopes are not as clear. Binocular viewers (Figure

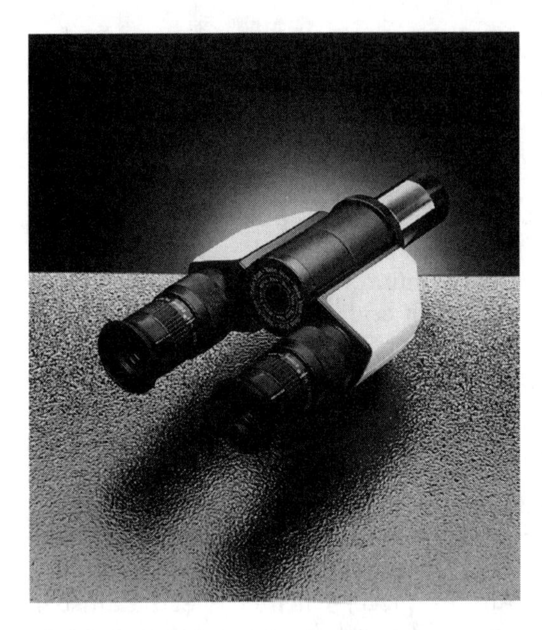

Figure 7.7 *Binocular viewing attachment for telescopes. Photo courtesy of Tele Vue Optics, Inc.*

7.7) customarily either screw onto a telescope (as with Schmidt-Cassegrain telescopes) or slide into a telescope's focusing mount. Inside, a beam splitter cuts the telescope's light into two equal-intensity paths, sending the light toward two eyepiece holders.

Are these rather costly accessories worth the price? Well, yes and no. First, they must be used with two absolutely identical eyepieces, raising the total investment even higher. Even then, however, they do not work as well as true binoculars. This should come as no surprise when you stop and think about it. In essence, binoculars are two independent telescopes strapped together. Binocular viewers, on the other hand, must rely on the light-gathering power of a single telescope. As such, the images perceived will be dimmer than through the same telescope outfitted with a single, equal-magnification eyepiece.

Several companies offer binocular viewers, but many use prisms that are too small for the intended purpose, causing field vignetting at lower magnifications. These so-called microscope-head accessories usually tilt their eyepieces at 45°, a very uncomfortable viewing angle through Newtonian reflectors (and not terribly convenient in other instruments either). Another problem with this style is the need to constantly refocus the view whenever the interpupillary distance is adjusted. This proves to be especially awkward if more than one person is viewing, since the distance will probably need to be readjusted back and forth. Microscope-head binocular viewers do have one advantage, that of price. They usually cost in the $500 to $700 range, less than the second type of viewers.

The second group are true binocular viewers. Though they cost as much as twice the price of microscope-head viewers, these superior units do not suffer from image vignetting or focusing shift when eyepiece spacing is adjusted, just like regular binoculars. Since they do not tilt the eyepieces, they are far easier to use with Newtonian reflectors, although pairs of star diagonals should be outfitted for viewing through refractors and Cassegrain-style instruments. Finally, images also seem brighter through these than through the microscope-head viewers.

Both types of binocular viewers are listed in Table 7.10.

Of those listed, the Tele Vue and Astro-Physics viewers get very high marks for quality by their owners. If money is tight (a relative term with binocular viewers, since they are all expensive to begin with), the Harry Siebert and LOMO/Gary Russell units are amazingly good values. For less than half the price of the other name-brand viewers, you get a well-made product that is quite usable in nearly any telescope (given the caveats about focusing that follow). But in side-by-side comparisons, the premium units will work better.

Note that not all telescopes (especially Newtonian reflectors) have enough range in their eyepiece focusing mounts to accommodate the long path that light must follow inside a binocular viewer, regardless of style. One owner described to me that the only way that he could get his 18-inch Newtonian to focus with a binocular viewer was to cut off 5 inches of the instrument's truss poles! He notes that "while that part was easy, it created yet another difficulty... with the diagonal mirror that much closer to the primary, the light cone at that point was larger than the diagonal mirror, causing an effective loss in aperture." He ultimately had to use a larger diagonal mirror built into a new upper tube / eyepiece assembly exclusively with the binocular viewer. Not a terribly convenient solution!

To avoid this situation, many insert a 2× Barlow lens in between their telescopes and binocular viewers. A Barlow extends the instrument's focal point, which cancels out the focusing problem but creates another. Just as when placed before a star diagonal, a Barlow in front of a binoviewer will mul-

Table 7.10 **Binocular Viewers**

Company/Model	Style	Comments
Celestron #93690	Microscope	Eyepieces tilted at 60°; adapters included for 1.25" and 2" focusers as well as threads for Celestron/Meade catadioptric telescopes.
Tele Vue BinoVue	True	Includes 2× compensator lens for use with telescopes that otherwise have insufficient focus travel.
Baader Planetarium	True	Sold in the United States by Astrophysics; can be outfitted with optional Convertible Photo-Visual Barlow to assist with focusing.
Lumicon	True	Focus "relay lens" sold separately.
Takahashi Twin View	True	Includes "doubler" focal-point extender.
LOMO/Gary Russell Optics	True	An excellent value for the money, though not up to the level of the more expensive Tele Vue or Astro-Physics binocular viewers.
TL Systems	Microscope	Sold in kit form only.
Harry Siebert	True	Covered with a unique vinyl bag; optional 1.7× Barlow lens available to extend telescope's focus point.

tiply the system's overall focal length not by 2×, but by approximately 4×. So a telescope / eyepiece combination that normally yields a nice, low 40× is suddenly increased to a comparatively high 160×.

Tele Vue has a better idea. Each BinoVue unit includes a special screw-on, parfocal Barlow lens (they call it a compensator) that overcomes the focusing problem while only increasing magnification by 2×. This is especially welcome for those who use fast-focal-ratio Newtonians, which excel in low-power viewing. For those customers using other instruments that can handle the 5.1 inches of internal travel within the BinoVue, Tele Vue also includes an empty extension tube to replace the compensator. The Baader Planetarium binocular viewer, sold in the United States by Astro-Physics, offers a similar solution with an optional Convertible Photo-Visual Barlow, which increases magnification by 2.4×.

Binocular viewers come into their own when viewing the brighter planets and, especially, the Moon. Our satellite seems to take on a three-dimensional effect (actually, an optical illusion) that cannot be duplicated with a traditional monocular telescope. In addition, an increase in planetary detail can also be expected with a binocular viewer thanks to the aforementioned improvement in subtle contrast and color perception. As one owner summed up his experiences: "When I have compared Saturn [through an Astro-Physics 180EDT apochromatic refractor] with and without the Tele Vue binocular viewer at similar magnifications, I find the difference in illumination almost unnoticeable. Markings on the disk of Saturn, however, are decidedly more apparent with the binocular viewer than without. Admittedly, however, some dimming of faint objects is detectable." To my eye, the dimming amounts to half a magnitude, give or take.

Books, Star Atlases, and Computer Software

Not long ago, I read that no pastime has more new books published about it each year than amateur astronomy. While that's bad news for us authors (too much competition!), it is good for the hobby. In fact, with so many excellent books and periodicals available, it is difficult to draw up a short list. Here is a *brief* listing of some of the better astronomy books, periodicals, and software available today.

Periodicals

Even though our electronic age affords an excellent opportunity to stay informed on all late-breaking discoveries and announcements in the astronomical world, there is still something nice about going out to the mailbox every month or so and finding a copy of an astronomy magazine waiting for you. The following magazines all cater to the amateur astronomer and come highly recommended. And for those readers living outside the United States, I have included several magazines that come from beyond. I would strongly recommend subscribing to one, if available, to stay in touch with your local universe.

Amateur Astronomy, 5450 NW 52 Court, Chiefland, FL 32626; www.amateurastronomy.com. Quarterly aimed toward observers and amateur telescope makers.

Astronomy, P.O. Box 1612, Waukesha, WI 53187; www.astronomy.com. Very good monthly general-purpose magazine, offering both technical and observational articles.

Astronomy and Space, Astronomy Ireland, P.O. Box 2888; Dublin 1, Ireland; www.astronomy.ie. Very good monthly general-purpose magazine, offering both technical and observational articles with an emphasis on events of interest to amateurs in Ireland.

Astronomy Now, 193 Uxbridge Road, London W12 9RA, United Kingdom; www.astronomynow.com. Very good monthly general-purpose magazine, offering both technical and observational articles with an emphasis on events of interest to amateurs in England.

Griffith Observer, Griffith Observatory, 2800 East Observatory Road, Los Angeles, CA 90027; www.griffithobs.org. Bimonthly geared more for the armchair astronomer.

Mercury, Astronomical Society of the Pacific, 390 Ashton Avenue, San Francisco, CA 94112; www.astrosociety.org. Bimonthly that features articles of general interest as well as basic articles on observing.

Sky & Telescope, P.O. Box 9111, Belmont, MA 02178; www.skypub.com. Very good monthly general-purpose magazine, offering both technical and observational articles.

Sky Calendar, Abrams Planetarium, Michigan State University, East Lansing, MI 48824; www.pa.msu.edu/abrams. Not a magazine per se, but rather a monthly flyer that details naked-eye events in a calendar format.

Sky News, Box 10, Yarker, Ontario K0K 3N0 Canada; www.skynews.ca. Very good bimonthly general-purpose magazine, offering both technical and observational articles with an emphasis on events of interest to amateurs in Canada.

Stardate, University of Texas, McDonald Observatory, Austin, TX 78712; www.stardate.utexas.edu. Bimonthly features articles of general interest as well as basic articles on observing.

Annual Publications

Although most of the periodicals listed above highlight monthly sky events and goings-on, it is often nice to know of things to come farther in advance. For this, there are several annual publications that focus primarily on the year's major happenings. These include:

Astronomical Calendar; Ottewell, G.; Universal Workshop

Exploring the Universe; from the editors of *Astronomy* magazine; Kalmbach Publishing

Observer's Handbook; Gupta, R., et al.; Royal Astronomical Society of Canada

Sky Watch; from the editors of *Sky & Telescope* magazine; Sky Publishing Corporation

Star Atlases

Just as we need road maps to plan our summer vacations, astronomers need star maps to find their way around the sky. The following star atlases are suitable for both telescopes and binoculars, but with certain caveats. First, note that each atlas's limiting magnitude (i.e., the faintest stars plotted) is listed, as is the number of deep-sky objects. If you are new to astronomy, and still learning your way around the sky, you would be best advised to stick with the simpler atlases that plot fewer objects. Those who are more accustomed to the sky should look toward the more sophisticated atlases.

Cambridge Star Atlas; Tirion, W.; Cambridge University Press, 2001. Stars to magnitude 6.5, 866 deep-sky objects. Recommended for beginners.

Edmund Mag 6 Star Atlas; Dickinson, T., Costanzo, V., and Chaple, G.; Edmund Scientific Company, Barrington, NJ 1982. Stars to magnitude 6.2, hundreds of deep-sky objects. Recommended for beginners.

Herald-Bobroff Astroatlas; Herald, D., and Bobroff, P.; HB2000 Publications, 1994. Three sets of black-stars-on-white-background charts: stars to magnitude 6.5, over 2,900 deep-sky objects; stars to magnitude 9.0, more than 13,000 deep-sky objects to magnitude 14; supplemental charts show overly crowded regions of the sky. These are augmented with a third set of charts that show stars as faint as magnitude 14 and deep-sky objects to magnitude 15. Highly recommended for all levels.

Millennium Star Atlas; Sinnott, R., and Perryman, M.; Sky Publishing, 1997. Three-volume set, plotting over 1 million stars to magnitude 11 and more than 10,000 deep-sky objects. Cumbersome to use at the telescope. For advanced amateurs only, who might consider a computer-based star atlas instead.

Norton's Star Atlas and Reference Handbook; Ridpath, I., et al.; Longman Scientific and Technical, 1998. Includes stars to magnitude 6.49 and 600 deep-sky objects, as well as extensive discussions on telescopes and observing techniques. Recommended for beginning and intermediate amateurs.

Sky Atlas 2000.0; Tirion, W.; Sky Publishing Corporation, second edition, 1998. Stars to magnitude 8, 2,700 deep-sky objects. The Sky Atlas 2000.0 is available in three editions: an unbound "field edition" (showing white stars on a black background), a "desk edition" (black stars on white), and a spiral-bound "deluxe color edition" (black stars, red galaxies, etc.). Highly recommended for intermediate and advanced amateurs.

Uranometria 2000.0; Tirion, W., Rappaport, B., and Lovi, G.; Willmann-Bell, 2001. This second edition includes 22 lower-detail finder charts to help orient the user, while the main charts show more than 280,000 stars to magnitude 9.75 as well as over 30,000 deep-sky objects. Sold in two volumes, which may be purchased separately. Volume 1 covers the northern sky from declination +90° to −6°, while the second volume includes +6° to −90° (the overlap is intentional). This atlas is recommended for intermediate and advanced amateurs only.

Introductory Books

Nightwatch; Dickinson, T.; Camden House, 1998. Highly recommended for beginners.

Peterson's Field Guide to the Stars and Planets; Pasachoff, J., and Menzel, D.; Houghton Mifflin, 1998. Recommended for beginners, although the small format makes reading the star charts difficult.

Sky Observer's Guide; Mayall, N., et al.; Western Publishing, 2000. Excellent guide to introduce the hobby of amateur astronomy to newcomers. Includes valuable observing tips and tricks.

365 Starry Nights; Raymo, C.; Fireside Press, 1992. Excellent night-by-night format to introduce the night sky to all.

Observing Guides

Note that while a few of these guides include finder charts, most require the use of a separate star atlas.

Burnham's Celestial Handbook volumes 1, 2, and 3; Burnham, R., Jr.; Dover, 1978. A classic reference book, listing just about every celestial object visible through small- and medium-aperture telescopes. Facts and figures are becoming a bit dated.

Deep-Sky Wonders; Houston, W., and O'Meara, S.; Sky Publishing, 1998. For nearly 50 years, Walter Scott Houston's "Deep-Sky Wonders" column was one of *Sky & Telescope* magazine's most popular features. O'Meara has taken the best, separated by month, and edited them together to create a read that is as timely today as when it was originally published.

Eclipse!; Harrington, P.; John Wiley & Sons, 1997. For those who, like me, are counting the minutes to the next solar or lunar eclipse.

Exploring the Moon through Binoculars and Small Telescopes; Cherrington, E.; Dover, 1984. An excellent guide to the lunar surface, although the maps are a bit cramped.

Night Sky Observer's Guide, volumes 1 and 2; Kepple, G., and Sanner, G.; Willmann-Bell, 1998. A monumental work that combines observations of 5,500 deep-sky objects, made by the authors as well as several contributors. Highly recommended for intermediate and advanced deep-sky observers.

Sky Atlas 2000 Companion; Strong, R., and Sinnott, R.; Sky Publishing, 2000. This second edition contains data on each of the 2,700 objects plotted on the Sky Atlas 2000.0. Objects are itemized twice, first alphabetically, then by chart number.

Touring the Universe through Binoculars; Harrington, P.; John Wiley and Sons, 1990. Are you an intermediate or advanced amateur who is looking to get the most out of your binoculars? This guide lists more than 1,100 binocular targets for both small and large glasses.

Turn Left at Orion; Cosmolmagno, G., and Davis, D.; Cambridge University Press, 2000. An excellent introductory guide to the night sky for beginners, with instructions for finding 100 sky objects.

Year-Round Messier Marathon Field Guide; Pennington, H.; Willmann-Bell, 1998. An excellent guide to the Messier objects, complete with maps and directions. Organized by area of sky rather than numerically.

Astrophotography

The Art and Science of CCD Astronomy; Ratledge, D., et al.; Springer Verlag, 1999. Excellent introduction to this ever-advancing aspect of astrophotography.

Astrophotography Basics (Publication P150); Eastman Kodak; Kodak, 1988. Call the Kodak Customer Service hotline at (800) 242-2424 and ask for a copy of Publication P150. You can't beat it for the price (single copies are free).

Astrophotography for the Amateur; Covington, M.; Cambridge University Press, second edition, 1999. The best general-purpose book on astrophotography in print today.

The New Astronomy; Wodaski, R.; self-published, www.newastro.com, 2001. A unique book offered in a unique way, either in traditional paper or online at the author's web site. Good, practical instruction.

Wide-Field Astrophotography; Reeves, R.; University of Alabama Press, 1999. Practical guide that will be a great help for beginners looking to start some basic astrophotography.

Telescope Making and Optics

Amateur Telescope Making (Books 1, 2, and 3); Ingalls, A., et al.; Willmann-Bell, 1996. *The* classic reference in the field.

Backyard Astronomer's Guide; Dickinson, T., and Dyer, A.; Camden House, 1991. Excellent guide to the aesthetics of backyard astronomy along with a good dose of practical advice.

Build Your Own Telescope; Berry, R; Willmann-Bell, 1993. Contains complete plans and photos for building several excellent telescopes at home.

Making and Enjoying Telescopes; Miller, R., and Wilson, K.; Sterling, 1995. Like the Berry book, contains complete plans and photos for building several telescopes.

Perspectives on Collimation; Menard, V., and D'Auria, T.; self-published, 1993. The most thorough treatment of this critical topic. Order directly from the authors (Vic Menard, 2311 23rd Avenue, West Bradenton, FL 34205, or Tippy D'Auria, 1051 NW 145th Street, Miami, FL 33168).

Scientific American: The Amateur Astronomer; Carlson, S., ed.; John Wiley & Sons, 2000. A compilation of the best amateur astronomy projects published over the years in *Scientific American* magazine. If you're a tinkerer, this is a must-have book.

Star Testing Astronomical Telescopes; Suiter, H.; Willmann-Bell, 1994. Although written at a fairly high level, the book offers a thorough treatment of this very revealing topic.

Telescope Optics: Evaluation and Design; Rutten, H., and Van Venrooij, A.; Willmann-Bell, 1988. In-depth coverage of optics and telescope design.

Miscellaneous Reading

Alvan Clark and Sons: Artists in Optics; Warner, D., and Ariail, R.; Willmann-Bell, 1995. The story of nineteenth-century America's first family of telescope makers.

Starlight Nights; Peltier, L.; Sky Publishing, 1999. The autobiography of America's most famous amateur astronomer. If anyone ever asks why you are interested in astronomy, tell them to read this book. No one has ever expressed the fascination we all feel so eloquently as Leslie Peltier.

Computer Software

Astronomical software can be divided into three basic categories based on their intent: observing programs (which include planetarium simulations and, often, telescope motion control), image-processing programs (typically used in conjunction with CCD cameras, but may also be used to enhance scanned-in conventional photographs), and special-purpose programs (covering a wide range of subjects, from optical design to predicting eclipses). Since most amateur astronomers are interested primarily in observing the sky, I have restricted the listing here to some of the best observing programs.

Many of the programs detailed below can be purchased directly from their companies via the Internet. Most also offer downloadable demonstration versions that give users a chance to try before they buy.

Deep Space (CD-ROM; DOS); David Chandler Company. Deep Space includes a huge database of more than 18 million stars, 10,000 deep-sky objects, and 10,000 comets and asteroids. Like most other programs of this type, *Deep Space* will print finder charts, but what makes these charts unique is the way a user can move titles and labels around to minimize overlapping. The only drawback is that since *Deep Space* is a DOS-based program rather than Windows, some users may find it a little less friendly to use.

Earth-Centered Universe (CD-ROM; Windows); Nova Astronomics. Graphically plots all objects in its database of over 15 million stars, the planets, the Sun, the Moon, comets, over 19,000 asteroids, and more than 10,000 deep-sky objects.

Deepsky 2000 (CD-ROM; Windows); Steven S. Tuma. An integrated planetarium program as well as searchable spreadsheet database listing more than 18 million stars and 426,000 deep-sky objects. Very helpful for organizing observations before and after observing sessions.

Guide (CD-ROM; Windows); Project Pluto. An excellent program for the deep-sky observer at a terrific price! Guide plots over 18 million stars, over 20,000 asteroids, hundreds of comets, and 75,000 deep-sky objects from the Messier, NGC, and IC lists as well as many comparatively unknown catalogs. *Guide*'s inexpensive price makes it a "best buy."

Megastar (CD-ROM; Windows); Willmann-Bell. The best program ever created for the diehard deep-sky observer. *Megastar* plots over 15 million stars to magnitude 15, an incredible 110,000 deep-sky objects, and more than 13,000 archived asteroids and comets; in short, more viewable objects than just about any other program currently available.

RealSky (CD-ROM; Windows or Macintosh), Astronomical Society of the Pacific. The Palomar Observatory Sky Survey, an invaluable photographic survey long held in high regard by professional astronomers, is now available in a nine-disk CD-ROM set for either Windows or Macintosh systems. *RealSky North* covers the northern sky to declination −15°, while *RealSky South* extends to the South Celestial Pole. Both display stars as faint as 19th magnitude, as well as innumerable deep-sky objects.

SkyMap Pro (CD-ROM, Windows); SkyMap Software. Many consider this one of the best planetarium-type software packages available. *SkyMap Pro* plots more than 15 million stars as faint as magnitude 15 and more than 200,000 deep-sky objects, as well as over 11,000 asteroids and comets. Updated asteroid and comet catalogs can be downloaded free of charge from the SkyMap web site.

Starry Night (CD-ROM; Windows or Macintosh), Space.com. One of the most widely acclaimed astronomy programs offered, *Starry Night* is available in three levels. The deluxe version accurately plots more than 19 million stars and deep-sky objects, displays the sky from any planet or moon in the Solar System at any time, prints customized star charts, and aims and controls computerized telescopes. And the graphics are terrific!

The Sky (CD-ROM; Windows or Macintosh); Software Bisque. The Sky is one of the finest sky-simulation and telescope-control programs available. With this program, you can show the Sun, the Moon, and the planets against myriad background stars for any time, any date, any place on Earth.

Touring the Universe through Binoculars Atlas (TUBA) (CD-ROM; Windows); Phil Harrington and Dean Williams. Warning: shameless commercialism ahead. TUBA displays stars to 9th magnitude and the more than 1,100 deep-sky objects listed in the book by the same name. We have received some very nice comments about it, but I'll spare you the details (which can be found on my home page).

Voyager II (CD-ROM; Windows or Macintosh); Carina Software. An outstanding sky simulation program, now available for both Windows and Macintosh users. *Voyager II* displays 259,000 stars; 50,000 deep-sky objects; and many comets, asteroids, and Earth-orbiting satellites. Animation compresses time to show eclipses or planets orbiting the Sun.

The Electronic Age

As with just about every other aspect of our lives, amateur astronomy is becoming more sophisticated thanks to tremendous advances in electronics and computerization. Here is a short review of some of the more popular electronic equipment on the market today.

Digital Setting Circles

One of the biggest complaints that amateur astronomers have had for years is that the setting circles that come with many equatorial mounts are useless. Because of their small size and gross calibration, they are little more than decoration. All this changed with the invention of electronic digital setting circles, which make it possible to aim a telescope accurately to within a small fraction of a degree. And forget polar aligning (although it is still usually required to use the clock drive); just aim the telescope at two of the many stars stored in memory, tell the unit which ones they are, and the built-in algorithms do the rest. This ease of operation means that the setting circles can be attached to just about any kind of telescope mount—either equatorial or alt-azimuth. Don't know what to look at? That's OK, since many of these units come with a whole catalog of thousands of sky objects from which to choose. Simply move the telescope until the LED prompt announces that you have hit the preselected target, look in the eyepiece, and there it is! Want one?

Each digital setting circle listed in Table 7.11 is a neat little unit that mounts unobtrusively to just about any telescope. Two encoders attach to the instrument's axes and are connected to the "brain" of the outfit by means of a pair of thin wires. Once properly secured and calibrated, the digital setting circles will automatically keep track of the passage of time as well as where the telescope is aimed. The LED typically reads to within an accuracy of 10 arc-minutes in declination and 1 minute of right ascension.

The biggest difference between the models lies in their onboard libraries of objects. The simplest units contain data on a little over 200 objects, while the most advanced include data on more than 12,000 sky sights. The latter can also be connected to a personal computer via a serial port, which allows the observer to link to various software packages like *Megastar, SkyMap Pro,* or *The Sky.*

Watch those object numbers, as they do not necessarily mean *unique* objects. Some manufacturers count the same object more than once if it belongs to more than one of the catalogs in their database. For instance, all but a couple of the Messier objects are also listed in the New General Catalog (NGC). It's not fair to count them twice, but some manufacturers do. Be sure to find out which catalogs are listed. At the very least, I would only recommend units that include all of the objects in the Messier and NGC lists.

All of these computerized telescope-aiming systems require that a pair of encoders be mounted to the telescope's axes. Some come with them, while others sell them separately. Keep that in mind when checking prices. Many also offer customized packages for attaching the encoders to fork-mounted Celestron or Meade Schmidt-Cassegrains simply by unscrewing and replacing one screw on both axes. Other telescopes, however, may not have it so easy, requiring some drilling and tapping in order to fit the encoders to their mountings. I strongly urge you to contact the manufacturer to find out what is required to attach the encoders to your particular telescope before you purchase any of them.

Another important question for readers considering the purchase of one of these units for their 8-inch Schmidt-Cassegrain telescope: Can the telescope swing into its storage position without hitting the right-ascension encoder?

Table 7.11 ***Digital Setting Circles***

Company/Model	Objects: Deep-sky objects[1]	Objects: User defined	Catalogs	RS-232 Port
Celestron International				
Advanced Astro Master	10,000	25	M, NGC, IC, more	Yes
Jim's Mobile, Inc. (JMI)				
NGC-microMAX	200	28	M, more	No
NGC-miniMAX	3,900	28	M, some NGC and IC, more	No
NGC-MAX	12,000	28	M, NGC, IC, more	Yes
Lumicon				
Sky Vector I	200	0	M, some NGC, more	No
Sky Vector II	3,500	25	M, most NGC, some IC, more	No
NGC Sky Vector	12,000	25	M, NGC, some IC, more	Yes
Meade Instruments				
Magellan I	12,000	125	M, NGC, some IC, more	Yes
Magellan II	12,000	125	M, NGC, some IC, more	Yes
Orion Telescopes				
Sky Wizard 2	2,100	27	M, some NGC and IC, more	No
Sky Wizard 3	10,000	27	M, NGC, some IC, more	Yes
Sky Engineering				
Sky Commander	9,000	120	M, NGC, IC, more	Yes
Vixen				
SkySensor 2000-PC	15,000	60	M, NGC, IC, more	Yes (GoTo)

1. Rounded to the nearest 100.

The encoders engage the right-ascension axis at its pivot point, smack dab in the middle of the fork. If the encoder sticks up too high, the front of the telescope tube will not clear. This may force the user to detach the encoder whenever the instrument is stored. Be sure to ask the manufacturer before you buy!

While most digital setting circles can be attached to nearly any telescope, either on an alt-azimuth or an equatorial mount, two are designed specifically for their own telescope models. These include the Magellan I and Magellan II systems from Meade Instruments. The Magellan I, for the Meade 8-inch LX-10 Schmidt-Cassegrain and Starfinder reflectors, includes a library of more than 12,000 objects and can be updated with more advanced data that may become available in the future. Each Magellan I kit comes with all necessary encoders, cables, and hardware. The Magellan II, made expressly for Meade's Starfinder equatorial reflectors, includes all of the features of the Magellan I and has a built-in drive corrector for photographic guiding. The Magellan II Starfinder kit also includes the required declination motor assembly and a new control panel that replaces the one originally supplied with the instrument. One caveat for owners of older AC-powered Starfinder equatorial telescopes: Sorry to say, but neither Magellan system will work with your instrument's drive.

Celestron's Advanced Astro Master, like the Meade, is designed expressly for Celestron-brand telescopes supported on either fork, CG-4, or CG-5 equatorial mounts, although it may be adapted to other brands and mounting styles with a drill (careful!) and a little creativity.

Finally, Vixen's SkySensor 2000-PC custom fits to its Great Polaris and GP-DX mounts as well as a few others. Motors and cables included with the SkySensor turn both mounts into computer-driven GoTo systems.

Speaking of GoTo systems, (also discussed in chapter 5), Astro-Physics offers voice recognition software so that now you don't even have to type numbers into a keypad. Instead, with its *DigitalSky Voice* software, you simply speak into a microphone and the telescope automatically points to whatever object you said. Say "M42" into the mike and next thing you know, the Orion Nebula is in the field of view (assuming it's above the horizon, wise guy). In fact, you can even ask "What's up?" and the computer will offer many of its own suggestions. A demonstration of its ability is quite convincing, although the early version I saw had some trouble with a few voice inflections. Usually, it was simply a matter of speaking up. The software and telescope respond instantly for most users, sweeping to whatever object the speaker had requested. Remember, *DigitalSky Voice* is strictly software and must be hooked up, via computer, to a GoTo-controlled telescope. It works with several GoTo systems marketed today, including Meade's LX200 and ETX-EC, Celestron's NexStar, the Vixen SkySensor 2000-PC, Astro-Physics GoTo mounts, and others using ACL Protocol.

Digital Setting Circles: A Commentary

Back in the 1980s, one manufacturer advertised that an observer could see "100 galaxies per hour" with its computer-aided digital circles. Sure, they are a great way to increase your productivity as an observer, if productivity is only measured in terms of sheer numbers. But since when are we out to break the land-speed record? The idea is not to whiz across the universe at warp speed, but rather to get to know the universe intimately. Without becoming too philosophical, I cannot recommend digital setting circles, especially if you are just starting out. It would be like giving a calculator to someone who does not know how to multiply. Sure, the calculator will give the right answer, but without it, the person is lost. Unless you are involved in an advanced observing program, such as variable-star observing or hunting for supernovae in other galaxies, run frequent public observing sessions, or are forced to observe from an area so polluted by light that you can't see enough stars to star-hop by, resist the impulse until you finish the section of chapter 10 entitled "Finding Your Way."

Smile . . . Say "Pleiades"

One of amateur astronomy's most popular pastimes is trying to capture the night sky on film. As most soon discover, however, there is a lot more to taking a good picture of the night sky than one might suspect at first. Patience is the most important requirement of the astrophotographer, followed closely by the right equipment. It's tough to bottle the former for sale, but there are lots of companies looking to sell you the latter!

Cameras

While successful astrophotographs can be taken with many different types of cameras, most amateurs prefer the 35-mm single-lens reflex (SLR) camera. Single-lens reflexes allow the photographer to look directly through the lens of the camera itself, a critical feature for aligning the image, especially when photographing through a telescope (with most other cameras, the photographer is viewing through a separate viewfinder). SLRs also offer the maximum flexibility in terms of film and lens availability, both discussed later in this chapter.

Not all 35-mm SLRs are suitable. For astrophotography, a camera must have a removable lens with manually adjustable focus, provisions for attaching a cable release to the camera and the camera to a tripod, a manually set, mechanical shutter with a "B" (bulb) setting, mirror lockup, and interchangeable focusing screens. Unfortunately, few of today's 35-mm SLRs fit this bill. In an attempt to attract more weekend photographers, most camera manufacturers offer cameras with automatic everything, from focus to exposure to flash control. All of these are nice for taking pictures of the family picnic, but are of no use to astrophotographers. Quite to the contrary, the long exposures required for astrophotos (usually measured in minutes, even hours) will quickly drain the power from expensive camera batteries. When that happens, the camera shuts down and becomes useless until a fresh set of batteries is inserted.

Which cameras are best for astrophotography? Table 7.12 lists several excellent alternatives, from both the past and the present.

Expensive does not necessarily mean better for astrophotography. All of these cameras will work well for wide-field constellation shots as well as through-the-telescope photos of the Moon and the Sun, but their differences will become more apparent when taking long-exposure telescopic shots. Here, the benefit of interchangeable focusing screens and mirror lockup will become apparent. Most subjects photographed through telescopes are very faint, making it difficult to line up and focus the shot when viewing through most standard focusing screens. A simple ground-glass screen will provide the brightest possible images, a great aid in focusing and composing (see further discussion under "Focusing Aids" later in this chapter). Mirror lockup is recommended for reducing *mirror slap,* which occurs every time the shutter is tripped and the camera mirror pivots out of the way. Swinging the mirror out of the way before the shutter is opened eliminates most vibration, reducing the chances for blurred images.

Lenses

Just as all cameras are not suitable for photographing the sky, neither are all lenses. But before an educated choice can be made, the photographer must

Table 7.12 ***Suggested Cameras for Astrophotography***

Model	Operates without Batteries	Manual Focus	Interchangeable Focus Screen	Mirror Lockup
Today's Best				
Leica R6.2	Y	Y	Y	Y
Nikon F3HP	Y	Y	Y	Y
Nikon FM2N	Y	Y	Y	N
Nikon FM10	Y	Y	N	N
Olympus OM-4TI	Y	Y	Y	N
Olympus OM-3TI	Y	Y	Y	N
Olympus OM 2000	Y	Y	N	N
Phoenix P2	Y	Y	N	N
Phoenix P1	Y	Y	N	N
Promaster 2000PK	Y	Y	N	N
Vivitar V4000	Y	Y	N	N
A Few of Yesterday's Best				
Canon F-1	Y	Y	Y	N
Canon FTb	Y	Y	N	Y
Minolta SRT-101	Y	Y	N	Y[1]
Miranda G	Y	Y	Y	Y
Nikon F2	Y	Y	Y	Y
Nikon F	Y	Y	Y	Y
Olympus OM-2	Y	Y	Y	Y
Olympus OM-1	Y	Y	Y	Y
Pentax LX	Y	Y	Y	Y
Pentax K1000	Y	Y	N	N

1. I thought that all Minolta SRT-101s had mirror lockup until I purchased a used one recently that did not. Be sure to check before purchasing.

first determine what he or she wants to photograph. For wide-field photography, either with the camera attached to a fixed tripod or guided with the stars, the standard lens may be all that is needed. Most 35-mm single-lens reflex cameras are supplied with lenses that have 50 mm to 55 mm of focal length; these cover an area of sky 28° × 40°. If a wider field is desired, then either a 28-mm or a 35-mm lens would be a good choice. They cover 50° × 74° and 40° × 59°, respectively. On the other hand, if a magnified view is what you want, try an 85-mm, a 135-mm, or a 200-mm telephoto lens, with 16° × 24°, 10° × 15°, and 6.9° × 10.3° fields of view, respectively.

In general, an astrophotographer wants to use as fast a lens as possible, because the faster the lens, the shorter the required exposure. Photographers refer to the speed of a lens just as astronomers talk about the focal ratio (the f-number) of a telescope. They mean the same thing.

Quite simply, the faster the lens, the lower the f-number. Holding the focal length constant, the only way to lower the focal ratio is to increase the lens's aperture. For instance, most 50-mm and 55-mm camera lenses have f-ratios between f/2 and f/1.2, resulting in apertures that range between about 1 and 2 inches. The larger the aperture, the greater the light-gathering ability of the lens, and, therefore, the shorter the required exposure.

Years ago, lens quality varied dramatically from one company to the next, but today, design and manufacturing procedures have been so perfected that most lenses will produce fine results (of course, every photographer thinks his or her brand is best). Flare and distortion, two of the biggest problems in older lenses, have been all but eliminated thanks to optical multicoatings. In general, most lenses made by reputable companies (all those listed above, as well as those by Vivitar, Sigma, and Tamron, to name a few) are fine for astrophotography. A *big* exception to all this is the zoom lens. Although extremely popular for everyday photography, zoom lenses almost always produce inferior results to their fixed-focal-length counterparts.

Film

The film industry has seen advances in the past two decades the likes of which have never been witnessed before. Without a doubt, the wide variety of films that are readily available today is a great boon to astrophotography. The selection is so vast, in fact, that it can leave the photographer confused. "Which film shall I use?" is a question often posed, even by veterans.

To help alleviate some of this bewilderment, Table 7.13 lists some currently popular films. The table is broken into three broad categories by film type (black-and-white, color print, and color slide) and then sorted by film speed, or ISO value. ISO (International Standards Organization) is the modern equivalent of the older ASA (American Standards Association) designation. Basically, the greater the numerical ISO value, the faster the film records light. For instance, an ISO 400 film will record the same amount of light in one-quarter of the time required by ISO 100 film. High-ISO-value films allow shorter exposures and, therefore, less chance of photographer error (accidentally kicking or hitting the telescope, tracking error due to polar misalignment, etc.).

Why, then, would anyone consider using slower films? If a frame of film is studied close up, it will be found to be made up of a pebbly surface called *grain*. It is one of those irrefutable laws of nature that fast film has larger grain structure than slower films. Larger grain means lower resolution and, therefore, poorer image quality.

Another reason to use slower film is something called *reciprocity failure*. This is one of those terms bounced around by most photographers but probably understood by few. As an illustration, think back to the example of ISO 100 film versus ISO 400 film. The films differ in ISO value by a ratio of 1 to 4. At the same time, ISO 100 film requires an exposure four times as long as ISO 400 film to record the same scene. Since these two ratios are reciprocals of each other, this concept is called reciprocity. But reciprocity does not remain constant as exposure times increase; instead, the film's ability to record light falters and finally ceases. After that, no further light buildup will occur regardless

Table 7.13 ***Film Comparisons***

Film (Manufacturer's Code)	ISO	Grain	
Black-and-White Film			
Kodak Tech Pan 2415 (TP-2415)	25–200[1,2]	Extra fine	
Kodak T-Max 100 (TMX)	100	Very Fine	
Fujinon Neopan 400[3]	400	Moderate	
Kodak T-Max 400 (TMY)	400	Moderate	
Kodak Tri-X (TX)	400	Moderate	
Fujinon Neopan 1600[3]	1600	Moderate	
Kodak T-Max P3200 (TMZ)	3200	Coarse	
Film (Manufacturer's Code)	**ISO**	**Grain**	
Color-Print Films			
Fujicolor Superia X-tra 400 (CH)	400	Fine	
Kodak Pro 400 (PPF)[1]	400	Fine	
Kodak Royal Gold 400	400	Fine	
Fujicolor Superia X-tra 800 (CZ)	800	Moderate	
Konica Centuria 800 (BD236)	800	Moderate	
Kodak Royal Gold 1000	1000	Moderate	
Fujicolor Super HG 1600 (CU)	1600	Coarse	
Film (Manufacturer's Code)	**ISO**	**Grain**	**Sky Color (background)**
Color-Slide Films			
Kodak Elite Chrome 100 (EB)	100	Very fine	Neutral
Fujichrome Provia 100F (RDP III)[1]	100	Fine	Green
Kodak Elite Chrome 200 (ED)	200	Very fine	Neutral
Fujichrome Sensia II 400 (RH)	400	Moderate	Green
Kodak Elite Chrome 400 (EL)	400	Fine	Neutral
Ektachrome P1600 Professional (EPH)[1]	1600[3]	Moderate	Neutral
Fujichrome Provia 1600 (RSP)	1600[3]	Moderate	Neutral

1. Also available hypersensitized.
2. Depends on developing time. See manufacturer's instructions for details.
3. May be processed at various speeds.

of length of exposure, hence the term *reciprocity failure*. Typically, the faster the film's speed, the quicker it fails.

There are ways to diminish a film's reciprocity failure while also increasing its ISO speed. The most popular technique is called *gas hypersensitizing*. The film is baked for days in a special oven containing a combination of nitrogen and hydrogen called *forming gas*. Though some amateurs prefer to cook their own, gas-hypersensitized film may be purchased from many mail-order sources, such as Lumicon. Users should note that not all films respond well to hypering, and that the hypered film has an effective shelf life of about a month. All hypered film should be sealed in an airtight package (such as the plastic film can it came in) and stored in a freezer before and after use.

Photographic Tripods

If you will be affixing your camera to a tripod (as opposed to shooting through a telescope), then pay close attention to the tripod you will be using. Many less-expensive tripods sold in department stores and other mass-market outlets are just not sturdy enough to support a camera steadily for any length of time. If the tripod is shaky, then the photographs will be hopelessly blurred. It makes no sense mounting a camera outfit costing hundreds, even thousands, of dollars on a cheap tripod!

Here are a few things to look for when purchasing a camera tripod. First, the legs should be extendable so that the camera may be raised to a comfortable height. Make certain, however, that the tripod remains steady when fully extended. Sturdier models feature braces that bridge the gap between the tripod's legs and the center elevator post. Next, take a look at the footpads. Better tripods have convertible pads that feature both a rubber pad for use on a solid surface as well as a spike for softer surfaces like grass or dirt. (A tip: When using a tripod on sand, place a plastic coffee can lid under each foot for added rigidity.)

Of all the tripods made, most photographers agree that the sturdiest are manufactured in Italy by Manfrotto and marketed in the United States under the Bogen name. For instance, the Bogen model 3036 is sturdy enough to hold a 4-inch refractor even with its legs fully extended; lesser tripods would collapse under such a load. Other brands worth considering are Tiltall, Star-D, Velbon, Vivitar and Slik.

Camera-to-Telescope Adapters

For prime-focus photography, the most common way to affix a camera to a telescope is a two-piece T-ring/adapter combination. The T-ring attaches to the camera in place of its lens, while an adapter attaches to the telescope. The ends of the adapter and T-ring are then screwed together to form a single unit.

Different cameras require different T-rings. For instance, a T-ring for a Minolta will not fit a Canon. Likewise, different adapters are required for different telescopes. In the case of most catadioptric telescopes, an adapter called a *T-adapter* screws onto the back of the instrument in place of the visual back that holds the star diagonal and the eyepiece. Most refractors and reflectors, on the other hand, use a different item called a *universal camera adapter,* which is inserted into the eyepiece holder.

Positive-projection astrophotography, commonly used when shooting the planets or lunar close-ups, requires that an eyepiece be inserted between the lensless camera and the telescope. Most camera adapters, such as the one shown in Figure 7.8, come with an extension tube for this purpose. The eyepiece is inserted into the tube, and the tube is then screwed in between the adapter and T-ring.

Celestron, Meade, Questar, and some other telescope manufacturers offer camera adapters that custom-fit onto their telescopes. Aftermarket brands, often less expensive, are also available. For instance, Orion Telescope Center sells several different adapters to fit most popular telescopes. None are supplied

Figure 7.8 *Camera-to-telescope adapter system. Photo courtesy of Orion Telescope Center.*

with camera T-rings, which must be purchased separately. Better photographic supply stores carry T-rings for most common single-lens reflex cameras, as do many astronomical mail-order companies.

Off-Axis Guiders

For long-exposure, through-the-telescope astrophotography, photographers have no choice but to visually monitor their telescopes' tracking. To do this, they must peer through either a side-mounted guide scope or an "off-axis" guider or use an autotracker, outlined later under "CCD Cameras." Mounting a guide scope onto the side of the main instrument can be both clumsy and expensive, leading many astrophotographers to choose an off-axis guider.

An off-axis guider looks like the letter *T,* with two hollow tubes attached to each other at a 90° angle. The main body of the guider fits between the telescope and camera, while the perpendicular leg contains a tiny prism that is used to divert a small amount of starlight toward an illuminated-reticle eyepiece that fits into the guider. To use the guider, one plugs or screws it into the telescope's eyepiece holder between the telescope and the camera and aligns the crosshairs of the eyepiece with a star. During the exposure, monitor the guide star through the reticle eyepiece to make sure it never leaves the crosshairs.

Off-axis guiders are available from a number of different sources. The simplest and, therefore, cheapest feature rigid prisms, while more expensive models come with prisms that may be rotated around the field. In practice, the latter are easier to use because the freedom of prism movement permits a much wider choice of guide stars. The Orion UltraGuider, the Celestron Radial Guider, and the Lumicon Easy-Guider are among the best off-axis guiders for the amateur astrophotographer.

Focusing Aids

Focusing a telescope for the eye alone is easy, but achieving a sharp focus when viewing through a camera viewfinder can be quite another matter.

Although standard focusing screens work well under bright conditions, they produce dim, ill-defined images when used at night. Many astrophotographers swap their camera's replaceable focusing screen for a clear ground-glass matte screen. Special viewfinder screens called Intenscreens are available from Beattie Systems and deliver images up to four times as bright as standard screens. (Some camera light meters that take readings off the focusing screen might need their ASA/ISO speed dials adjusted to compensate for the Intenscreen's extra brightness. While you aren't likely to use a light meter for astrophotography, it is something to keep in mind if you use the same camera for terrestrial photography.)

If you are interested in photographing bright objects like the Sun (remember your filter!), the Moon, the planets, or maybe the brightest deep-sky objects, then one of these focusing devices may not be needed. To get a sharp focus first time, every time, make a Hartmann mask for your telescope by cutting a circular piece of cardboard, foam rubber, or a similar material the same size as your telescope's front opening. Now cut two (some people prefer three) smaller circles in the mask directly opposite each other, making certain that they are not cut off by the telescope's secondary mirror (if a reflector or catadioptric). Two-inch diameters work well for 8-inch and larger telescopes. With the mask secured in front of the telescope (by using Velcro straps or some other type of fastener), attach the camera, aim at the target, and look through the viewfinder. If the telescope is out of focus, you will see two images. Turn the focusing in or out until the two slowly blend into one. When only a single image is seen, the telescope is properly focused. A commercial Hartmann mask, called the Kwik Focus, is marketed by Kendrick Astro Instruments.

CCD Cameras

The pursuit of astrophotography has changed dramatically in the past decade. The availability of supersensitive charge-coupled devices, or CCDs for short, now make it possible to take photographs of the Moon, the planets, and deep-sky objects using exposures many times shorter than those required with conventional film. A CCD camera uses a silicon chip made up of thousands of light-sensitive areas called *pixels* to convert light into electrons during an exposure. At the end of the exposure, each pixel converts its stored electrons into a digit and then downloads that result to a computer. The computer then converts the values received from each pixel into a shade of gray and combines them into an image displayed on the computer's monitor.

Exposures as short as 30 seconds with a CCD camera (Figure 7.9) will record the same detail as a half-hour exposure using conventional film. (Estimates indicate that these cameras have an equivalent ISO speed value in excess of 20,000 after computerized image processing!) In addition, the effects of light pollution can be eliminated by computer processing, making it possible to capture great deep-sky images from within cities. Fantastic!

But there is much more to this than just buying a CCD camera and hooking it up to your telescope. Besides having to buy a computer as well as a sturdy equatorial mounting, you will also need to have some idea of what kind

Figure 7.9 *The Celestron Fastar CCD imaging system, showing the SBIG ST-237 CCD camera at the CM-1400's f/2.1 focus. Photo courtesy of Kevin Dixon.*

of imaging you are interested in doing. The ideal setup for planetary work may not necessarily be appropriate for deep-sky imaging and vice versa. Planetary images are typically bright and small, requiring short exposures. The amount of sky coverage is usually not critical. Deep-sky objects, on the other hand, may cover wide expanses and are comparatively very dim, requiring larger, more expensive chips and longer exposures. Figure 7.10 graphically compares the area captured by some of the more common CCD chips in use today to that of a 35-mm negative. The comparison may surprise you!

Experts tell us that the ideal telescope-CCD combination depends on what you are imaging. The optimal deep-sky setup will have each pixel on the CCD chip covering between 1.5 and 2.5 arc-seconds of sky, while sky coverage between 0.4 and 0.7 arc-seconds produces the best planetary results. To find out just how much sky will be covered by each pixel, plug numbers into the formula

$$(P \div FL) \times 206$$

where P is the pixel size in microns and FL is the telescope's focal length in millimeters.

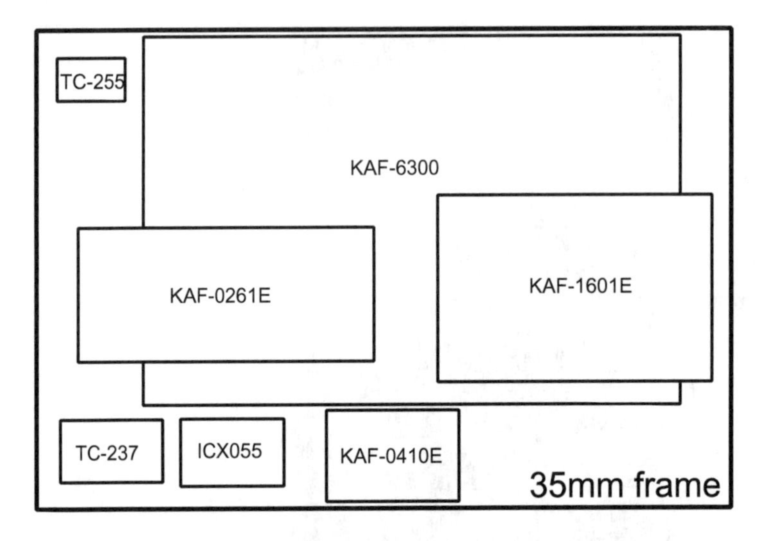

Figure 7.10 *A comparison of the sky coverage offered by today's most popular CCD chips to that of a standard frame of 35-mm film.*

For instance, an 8-inch f/10 SCT with a focal length of 2,032 mm and a CCD imager that uses the popular Texas Instruments TC255 chip (with pixels measuring 10 × 10 microns) would yield a sky coverage of 1 arc-second per pixel, which is really not appropriate for either application. To determine how much sky the entire CCD chip will record, simply multiply the arc-second values by the dimensions of the array (324 × 242 in this example), which works out to be 5.4 × 4 arc-minutes. A TC241 chip with this same telescope has a sky-per-pixel coverage of 2.3 × 2.7 arc-seconds and an overall field of 14.5 × 24.7 arc-minutes, which is better for deep-sky imaging. A good planetary CCD for this telescope would have pixels in the 5- to 7-micron range.

Because of their small fields, CCD imagers are notoriously hard to aim. One way is to aim your telescope at the object while viewing through an eyepiece first, then replacing the eyepiece with the imager. Of course, a lot can happen during that process; for example, you might knock the instrument off the target. An elegant way around this hurdle is to use a flip-mirror unit, which can be likened to the mirror in a single-lens reflex camera that diverts light to the camera's viewfinder. Flip mirrors are described later in this chapter.

As exposure grows, so does the background electronic noise, or *dark current,* generated by the CCD camera. While not a concern if exposures are restricted to the Moon and the planets, dark current does pose a problem with the long exposures required for deep-sky imaging. Dark current can be reduced by cooling the CCD chip, but some chips require more cooling than others to bring this noise level down to workable levels.

Some CCD chips are listed as being 8-bit, 12-bit, and 16-bit. These numbers refer to the analog-to-digital (abbreviated A/D) converter used for storing and processing images. Simply put, the larger the number, the greater the dynamic range in image brightness and, therefore, the better the image quality. But keep in mind that a 12-bit converter will require about 50% more memory space to store and is slower to process than an 8-bit converter, while a 16-bit requires twice the allotted memory.

Antiblooming is another important feature to consider when selecting a CCD. *Blooming* occurs when a bright object oversaturates the chip, causing an excess charge to bleed down a column of pixels. The result is a bright shaft of light above and below the offending star. Most chips nowadays feature antiblooming, although it can also reduce the chip's sensitivity and amplify dark current.

Table 7.14 lists many of today's most popular CCD chips, including their dimensions, the size of their pixels, and the imagers in which they can be found.

Not every telescope is CCD friendly. The focuser, for instance, is critical. It must be solid enough to support the CCD imager without changing its position.

Table 7.14 **Popular CCD Chips**

Chip	# of Pixels on Chip	Pixel Size (microns)	Imagers
Kodak KAF-0261E	512 × 512	10.2 × 10.2	Apogee KX260, SBIG ST-9E
Kodak KAF-0401E	765 × 510	9 × 9	SBIG ST-7E, Meade Pictor 416XTE, Apogee AP1E
Kodak KAF-1001E	1,024 × 1,024	24 × 24	Apogee AP6E, SBIG ST-1001E
Kodak KAF-1601E	1,530 × 1,024	9 × 9	Apogee AP2E
Kodak KAF-1602E	1,530 × 1,024	9 × 9	SBIG ST-8E, Meade Pictor 1616XTE
Kodak KAF-3200E	2,184 × 1,472	14.9 × 10	SBIG ST-10E
Kodak KAF-4200	2,048 × 2,048	9 × 9	Apogee AP4
Kodak KAF-6300	3,072 × 2,048	9 × 9	Apogee AP9E
Kodak KAF-16800	4,096 × 4,096	9 × 9	Apogee AP16E
SITe SIA003AB	1,024 × 1,024	24 × 24	Apogee AP8p
SITe SIA502AB	512 × 512	24 × 24	Apogee AP7p
Sony ICX055AL	500 × 290	9.8 × 12.6	Starlight Xpress MX5
Sony ICX055CL	500 × 290	9.8 × 12.6	Starlight Xpress MX5C
Sony ICX083AL	752 × 580	11.6 × 11.2	Starlight Xpress MX916
Sony ICX084AL	660 × 494	7.4 × 7.4	Apogee LISÄÄ, Apogee LISÄÄ Guider, Starlight Xpress HX516
Sony ICX084AK	660 × 494	7.4 × 7.4	Apogee LISÄÄ Color
Sony ICX085AL	1,300 × 1,030	6.7 × 6.7	Apogee LISÄÄ Megapixel, Starlight Xpress HX916
Sony ICX249AK	752 × 582	8.6 × 8.3	Starlight Xpress MX7C
Texas Instruments TC-211	192 × 164	13.75 × 16	Cookbook 211
Texas Instruments TC-237	640 × 480	7.4 × 7.4	SBIG ST-237A, SBIG ST-V
Texas Instruments TC-245	252 × 242	25.5 × 19.8	Cookbook 245
Texas Instruments TC-255	320 × 240	10 × 10	Meade Pictor 201XT, Meade Pictor 208XT, Meade Pictor 216XT, SBIG ST-5C
Thomson THX7899M	2,048 × 2,048	14 × 14	Apogee AP10

Ideally, it should have a locking thumbscrew, just to make sure that everything stays put. Also, it must not shift when that thumbscrew is locked down. Even a slight movement will blur the image, leading to wasted time and ever-growing frustration.

Even with the finest focuser, some focus shift can result as the night wears on and the temperature drops. This problem can be especially noticeable in catadioptric telescopes and multiple-element apochromatic refractors, in which different optics contract at different rates.

The telescope's mount must also be extremely sturdy and outfitted with an accurate clock drive (one that has very little periodic error). Remember, any inaccuracies that you can detect visually, either due to your telescope or its mounting, will only be magnified by the CCD.

The CCD cameras listed in Table 7.14 image in black-and-white only (save for two—the MX5C and the MX7C—from Starlight Xpress), yet color CCD photography is possible due to a variation on the tricolor photographic technique introduced by James Maxwell in 1861. Three separate exposures are made with red, green, and blue filters over the camera; these are then combined electronically to produce a true-color image. Some CCD cameras have optional built-in color filter wheels, while others require a separate attachment.

Several companies market CCD cameras for today's amateur astronomer. Some of the most highly regarded are manufactured by Santa Barbara Instrument Group, or SBIG for short. Built around Texas Instruments' popular TC255 CCD, the SBIG ST-5C includes a feature that SBIG calls "Track and Accumulate." This allows the camera to spend most of its time imaging, but once in a while, it switches briefly to autotrack mode to make sure the telescope is tracking properly, thereby freeing the observer from the tedious task of manually checking that the instrument is still on target. This is a great advance in CCD design, because it eliminates the need either to track the sky manually or to purchase a separate autoguider.

The ST-237A's small pixel size produces very-high-resolution images when used in Fastar mode with a Celestron Schmidt-Cassegrain telescope (see chapter 5 for more details). Using SBIG's Camera Lens Adapter, the ST-237A can also be attached to standard camera lenses with focal lengths from 50 to 400 mm for additional versatility. Like the ST-5C, the ST-237 features Track and Accumulate autoguiding. The ST-237A also has an internal mechanical shutter as well as an electronic shutter for easier exposure control, a feature usually not found on cameras in this price range. An optional color filter wheel can also be mounted inside the CCD head for tricolor imaging.

As impressive as the ST-5C and ST-237 are, SBIG's ST-7 through ST-10 imagers are in a whole different league. Built around much larger CCD chips from Kodak, all four include a separate TC-211 CCD chip that provides full-time autoguiding. Although the Track and Accumulate flip-flop method works fine, it can cause a *skipping* (jerking) motion if the telescope is slightly off in polar alignment. But with a separate chip devoted only to autoguiding on a field star, that problem is eliminated entirely. The ST-5C through ST-10E models all connect to a Windows-based computer through the PC's parallel port, while Macintosh computers require an additional SCSI / Parallel adapter (note that the ST-10E does not support use with a Macintosh). All require a separate power source.

Finally, SBIG sells the ST-1001E, which uses Kodak's megapixel KAF-1001E CCD chip. Although this affords a much larger imaging area, it also has a few drawbacks. Among those are no autoguiding (SBIG recommends using their ST-V imager for autoguiding; see "Video Cameras," later) and lack of a tricolor filter wheel, although the latter is in development and will probably be released by the time this book hits the streets.

Meade sells a number of CCD products. The least expensive is the Pictor 201XT autoguider. This is not an imaging device; rather, the 201XT is inserted into a telescope's focuser and connected to the dual-axis clock drive to follow a centered star automatically. An autoguider eliminates the need to adjust the instrument's tracking rate manually during a conventional photographic exposure. No personal computer is needed. Several owners report trouble trying to lock the 201XT onto a star, but once one is acquired, the autotracker works properly. Personally, however, I would recommend that prospective consumers should invest a little more money and purchase a CCD imager/autotracker, like the Meade 208XT.

The 208XT is one of four CCD imagers offered by Meade. Both it and the 216XT use the Texas Instruments TC255 chip. The difference between them? Sophistication. The 208XT uses an 8-bit A/D converter, while the 216XT enjoys a wider dynamic image range thanks to its selectable 12- or 16-bit imaging, plus will also attach to the #616 color filter system for tricolor imaging. Both may also be used as separate autoguiders or in a sophisticated *shift-and-combine* mode where the imager pauses momentarily during an exposure to autoguide, like the SBIG ST-5C.

The two most advanced CCD cameras in the Meade line are the 416XTE and 1616XTE, which are based on Kodak's KAF-0401E and KAF-1602E CCD chips, respectively. Meade recommends coupling these cameras to computers with an SCSI interface for fastest image downloading, which is available on very few computers these days. (Most other imagers nowadays have either series, parallel, or USB connectors. SCSI connectors require either a separate controller or a dedicated section on the motherboard, which few computers have. USB ports have pretty much made SCSI a thing of the past.) With an SCSI interface, image download time is 1 second for the 416XTE and 4 seconds for the 1616XTE; but with the more common serial interface, download times climb to 65 seconds and 260 seconds, respectively!

Apogee Instruments sells a number of top-of-the-line CCD imagers for the amateur astronomy market, all using CCD chips from Kodak, SITe, or Thomson. Its least expensive unit, the AP1E, is built around the KAF-0401E, the same chip used in the SBIG ST-7E and the Meade 416XTE, while its most expensive, the AP16E, uses Kodak's KAF-16800, currently the largest CCD chip suitable for amateur astronomical imaging. All yield excellent results thanks to low noise and precise chip-cooling control. Although more expensive than some competing models, none are capable of autoguiding. Instead, Apogee sells a universal autoguider called the LISÄÄ, described below. Note also that, except for the AP7p and AP260Ep, which have parallel-port interfaces, all

Apogee AP imagers require the installation of a separate ISA-bus interface board into the host computer.

LISÄÄ stands for "Low-cost Imaging System for Amateur Astronomy." Several models make up the LISÄÄ line, including an autoguider, a monochromatic imager, and a color imager, the latter two also available in Megapixel versions. The LISÄÄ system consists of one or two separate components. The LISÄÄ Guider can be used alone, like any autoguider, or can be coupled to nearly any CCD imager (though it may overlap larger CCD chips, such as Apogee's AP10 and AP16E). Apogee tells us that the autoguider's performance will not be interfered with when used with its own filter wheel, because it is placed in front of the color wheel, but it may with others. Of course, this also makes it necessary to purchase several separate pieces of equipment, rather than a single system. The LISÄÄ Guider can also be attached directly to a conventional 35-mm single-lens reflex camera by using separate adapters. Other autoguiders require either separate guidescopes or off-axis guiders.

From England, Starlight Xpress sells some of the smallest CCD imagers on the market. Its least expensive unit, the original MX5, measures only 2 inches in diameter by 4 inches long. This lets it fit directly into a 2-inch focusing mount, reaching focal planes in some telescopes that might otherwise be inaccessible (a 1.25-inch adapter is also included). The heart of the MX5 is the Sony ICX055AL CCD chip, which measures 500 × 290 pixels, or about twice the area of the Texas Instruments TC255.

The MX-5C is a unique "color" CCD imager that takes exposures through different color filters simultaneously. The trick is that yellow, magenta, cyan, and green microscopic filters are distributed over the pixel array. The software supplied with the MX-5C decodes the intensity data from the pixels and reconstructs a full-color image in one step. Pretty clever, although image scale is very limited.

To answer the demand for a larger imaging area, Starlight Xpress came out with the MX-7C imager. The same physical size as the MX-5 series, the MX-7C also uses a four-filter internal system to generate some very impressive results. Inside, the Sony ICX249AK CCD measures 6.47 × 4.83 mm, as compared to 4.9 × 3.6 mm in the MX-5C, and has smaller pixels, making the MX-7C suitable for fast-focal-ratio instruments.

The final MX imager, the MX916, is best used with longer-focal-length telescopes and is ideal for imaging faint galaxies or conducting asteroid searches. Its aluminum barrel, measuring 63 mm in diameter and 95 mm long, comes supplied with adapters for both 1.25- and 2-inch focusers.

To help keep things on track, Starlight Xpress sells autoguiding software for use with their MX-series imagers. Called S.T.A.R., which is short for "Simultaneous Track and Record," this Windows-based software relies on the imager's CCD chip for tracking information, while image-processing software gathers data for the final photograph. To use the software, an interface box is placed between the serial port of the computer and the autoguider port of the telescope mount using an RJ11 connector.

Designed for small telescopes, Starlight Xpress's HX516 can also be attached to telephoto lenses (although results through the latter leave something to be desired, in my opinion). A megapixel version of the same imager, known as the HX916, has also been released that promises some outstanding results. Both cameras are constructed using the same body as the MX916, the latter supplied with a high-speed USB interface to give a very fast image download speed. Note, however, that neither the HX516 nor the HX916 are compatible with Starlight Xpress's Simultaneous Track and Record (S.T.A.R.) tracking software.

Perhaps your needs are more basic. If you are looking for an inexpensive CCD imager, just to test the waters, then perhaps the SAC II or SAC IV from Sonfest is for you. Like the homemade CCD imager detailed in the next chapter, these two imagers are very basic units designed solely for lunar and planetary targets. Both of these color cameras come housed in plastic bodies and are ready to attach to a computer's parallel port (in the case of the SAC II) or USB port (SAC IV). The SAC II includes the 320 × 240 pixel Sharp LZ2313H5 CMOS (short for Complementary Metal-Oxide Semiconductor) chip, with individual pixels measuring 12 microns square. CMOS chips were developed before CCDs came along and are still quite popular in some low-end applications because of their low price and low-power needs. Although the latest CMOS chips generate surprisingly high-quality images, they are not quite up to CCD-quality—yet. The SAC IV has a 640 × 480 CCD chip from Micoteck. The basic SAC II-b fits directly into a 1.25-inch focuser, permitting only prime-focus photography, while the SAC II-c and SAC IV both have standard camera T-mounts, so they can also be used for projection photography. While they can't compete with any of those above, the SAC imagers are certainly worth more than a passing glance like if you want to enter the CCD world on the cheap.

Looking for a challenging project? Consider making your own CCD camera. The *CCD Camera Cookbook* by Richard Berry, Veikko Kanto, and John Munger outlines how anyone who is knowledgeable in electronics can make not one but two different CCD cameras. The simpler of the two is based on the Texas Instruments TC211 chip, while the other uses the larger TC245 chip. Both work quite well when assembled correctly but have very small fields.

While still more expensive than 35-mm SLR cameras, CCD cameras continue to drop in price just as CCD-based home video camcorder prices have fallen as the technology becomes cheaper to produce. Of course, all imagers require a personal computer and a high-resolution monitor, adding further to the start-up cost, yet many amateurs think this is a small price to pay for the outstanding quality that can be achieved with CCDs.

Unfortunately, the area of CCD photography is far too complex to address adequately in the small space provided here. While the discussion here may be enough to whet your appetite, if you are new to this burgeoning field, I recommend that you consult one of the books listed earlier in this chapter before purchasing *anything*.

Video Cameras

Adirondack Video Astronomy sells five terrific little video cameras that are ideal for mating with a telescope. Each of their AVA Astrovid CCD video cameras is small enough to fit into a jacket pocket yet powerful enough to get some terrific views of the Moon and the brighter planets through just about any telescope on any mounting. All come complete with power supply, cables, a T-C adapter (needed to couple the camera to a telescope), and instructions. All you need to add is a telescope and a video recorder (for field use, it might be best to attach the video camera to a battery-powered camcorder rather than a conventional VCR).

Weighing only 6.6 ounces (190 grams), the color Astrovid Planetcam features a sensitivity of 1 lux (a unit of illuminence or field brightness) and a horizontal resolution of 480 lines, which is fine for lunar and planetary shots. A deluxe version is also available that offers complete image control with the supplied software. Both use Sony's ICX0208AK CCD chip, which measures 4.5 mm × 3.8 mm (the pixel array measures 768 × 494; individual pixels cover 4.75 × 5.55 microns). The result: sharp images across a very small field of view.

The monochromatic Astrovid 2000 weighs 10.6 ounces (300 grams). The Astrovid 2000 offers higher resolution than the Planetcam, with 600 lines of horizontal resolution, and enjoys an increase in sensitivity to 0.01 lux. Although these cameras are designed primarily for lunar and planetary photography, the manufacturer states that both are sensitive enough to record 9th-magnitude stars through an 8-inch telescope.

Recently, SBIG introduced a completely different kind of CCD imager that combines the best of an autoguider with the convenience of a supersensitive black-and-white video camera. The result is the first totally self-contained CCD imager, called the ST-V. Unlike the CCD imagers from SBIG or any of the others discussed earlier, the ST-V does *not* require a separate computer. Instead, the ST-V consists of a camera head and a control box that are connected to one another by a 15-foot cable. The system is designed to run on 12 volts DC. A standard version includes both the imager and the control box, but it needs to be attached to a video monitor to be able to monitor exactly what the imager is seeing. The control box supplied with the deluxe ST-V includes a 5-inch LCD display, making a separate monitor unnecessary.

How about using your family's camcorder with your telescope? Absolutely! Astrovideography is practiced by surprisingly few amateurs, yet a regular camcorder is perfect for capturing the Moon, the Sun (again, using appropriate filters), and the brighter planets. You'll have to shoot afocally (that is, with the camera lens and telescope eyepiece both in place), but the results can be quite surprising. Best of all, if they don't work out, you can always record your favorite soap opera right over it!

Camcorders are significantly heavier than CCD imagers or the astro-videocams mentioned above, so how can they be aimed precisely at the telescope? Plans for a camcorder bracket were published in the second edition of this book, and a company called Stratton Video Brackets in Simpsonville, South Carolina, makes suitable brackets for securing both video and digital cameras to many of Celestron's and Meade's catadioptric telescopes. The brackets are sturdily made from black anodized aluminum and are easily adjustable for precise camera / eyepiece alignment. Bear in mind that the camera will undoubtedly hit the fork mount's base when the instrument is tilted toward the zenith, limiting the aim somewhat. Stratton brackets are pricey, however, with the cheapest costing a little under $300.

Flip Mirrors

One of the best things about a single-lens reflex camera is that the photographer is actually looking directly through the camera lens when peering through the viewfinder, enabling the shot to be composed exactly as it will appear in the final photograph. As the shutter is depressed, the mirror that has been directing light to the viewfinder flips out of the way, letting the light pass to the film frame that lies behind the shutter. How do you compose a shot using a CCD imager? Some people rely on the dim, ill-defined image appearing on their computer screen, although trying to get the target centered and in sharp focus can be difficult, if not impossible.

Happily, there is a better way; it's called a *flip mirror.* In effect, a flip mirror combines the mirror of a single-lens reflex camera into a star diagonal. With the mirror down, it redirects light to an eyepiece for focusing and centering. Flip it up, and the mirror swings out of the way so that light passes onto the CCD chip.

Several companies offer flip mirrors, including Meade Instruments, Murnaghan Instruments, True Technology, Taurus Technologies, and Van Slyke Engineering. Make sure you select one that is compatible with both your CCD imager and your telescope. In addition, some companies sell beam splitters, which let users view through the eyepiece and image using the CCD at the same time. In general, these result in dimmed images that are often inferior due to optical aberrations introduced by the beam splitter itself.

Star Wear

An area that few manuals of amateur astronomy address is the environment around the observer. Sure, most books complain about excessive light pollution and the need for good sky conditions, but there is so much more to enjoying the night sky than just the sky.

Baby, It's Cold Outside

The old saying that clothing makes the man (excuse me, person) is certainly true in astronomy. Nothing can take the enjoyment out of observing faster than weather-related discomfort. Although this is usually not a problem during summer, it certainly can be at other times of the year. Even the sturdiest telescope mount will wobble if the observer using it is shivering!

It goes without saying that the clearest nights occur after a high-pressure weather front sweeps the atmosphere of clouds, haze, and smog. Unfortunately, the clear atmosphere also causes Earth to lose a great deal of the heat that it has built up during the day. Many amateurs decide to sit these cold nights out, but by doing so they are missing some of the clearest skies of the season. Others try to brave the cold wearing their usual overcoat and a thin pair of gloves, but they soon return inside, teeth chattering and fingers numb. Is this any way to enjoy the wonder of the universe?

Most hardy souls agree that layering clothes works best. For temperatures above about 20°F, wear (from the inside out) a T-shirt, flannel shirt, sweater, and parka above the waist, while underwear, long underwear, and heavy pants should be worn below. In colder temperatures, or when the wind is howling, replace the sweater with a one-piece worksuit, snowmobile suit, or insulated coveralls. These provide a good, windproof barrier between you and the cold, cruel world.

These items should keep you warm enough in moderately cold conditions, but they also can make you stiff as a board! The multilayered look can make it difficult to pick up that pencil that was dropped on the ground or even to bend for a peek through the eyepiece. There must be a better way! Happily, today there is.

Thanks to modern synthetic fabrics, it is now possible to stay outside even in subzero temperatures in relative comfort and with full freedom of movement. Increasing numbers of observers are joining other outdoor enthusiasts who wear clothing made of advanced materials such as Dupont's Thermax and Thinsulate. Both have amazing heat-retention properties, yet they are thin and light enough to permit the wearer to bend with ease.

The best selections of cold-weather apparel are found at either local sporting goods retailers or in national mail-order catalogs such as L. L. Bean and Campmor. Unlikely as it may sound, I have gotten much of my cold-weather clothing either from local bike shops or from two mail-order outlets: Bike Nashbar and Performance Bicycle Shop (I'm an avid bicyclist by day).

Keeping the extremities warm is the most critical part of your cold-weather regimen. In less extreme temperatures, a pair of thick socks and work boots for the feet, a hat for the head, and a pair of gloves for the hands should do the trick. Under colder conditions, the head is best protected with a silk or wool balaclava. Looking like a full-face ski mask, a balaclava is thin enough to hear through yet warm enough that it may make a heavy hood or hat unnecessary.

For the hands, try a pair of ski mittens stuffed with Thinsulate or a similar material. Unfortunately, though they are warmer than gloves, mittens can make it difficult even to focus the telescope. Some mittens come with a thin insulating glove that may be worn separately, which is handy when changing eyepieces but still a problem when taking notes or making drawings. One other possibility is to wear a separate thin glove liner inside each mitten. Once again, many outdoor outlets sell mittens and glove liners that work quite well.

Nothing is more painful than frozen feet. I have seen some people walk out in 10°F weather wearing a heavy parka, a down vest, long underwear, a thick hood, heavy gloves, and a pair of sneakers! They didn't last long.

Work boots and so-called sorels and moon boots have excellent heat-retention qualities but only when used with a good pair of socks. The two-pair strategy usually works best. Wear a thin pair on the inside and a thicker, thermal pair on the outside. For truly frigid weather, even the best insulated gloves and boots may not do the trick. While some outdoorsfolk use heated hand warmers that run on cigarette lighter fluid, I get nervous keeping an incendiary device near clothing. (Do you smell something burning?)

No doubt about it, battery-powered socks will keep your feet warm, but they fail on several different fronts. First, all of those that I have tried tend to warm my feet like a microwave oven warms a bowl of leftover spaghetti—unevenly! Invariably, part of my foot was too hot, while another part was too cold! Then, too, you must decide what to do with the battery pack. In most cases, it must be strapped onto your leg, adding to the discomfort. Some people love electric socks, which is fine, but I prefer another approach.

A safer means for fighting off the cold are nontoxic chemical hand- and feet-warming pads, such as those manufactured by Mycoal Warmers Company in Japan, and sold in North America by Grabber of Grand Rapids, Michigan. Once removed from their packaging and exposed to the air, these nontoxic heaters maintain a temperature of about 140°F for several hours. (I have found the advertised claim that they last "seven hours or more" to be a little optimistic. From actual use, 4 or 5 hours is probably a better estimate.) Afterward, simply discard the pad. It is important to note, however, that the warmer must be wrapped in a cloth or other protective material before use. Burns can result if these heaters are left in direct contact with the skin for long. Other companies make similar heat pads, but I have found those by Mycoal to be the best.

Finally, those of us who wear glasses must face the problem of lens fogging when we go from a warm room out into the cold night. While we can fit our telescopes with dew shields and heating elements, it's tough to walk around at night with a dew cap on your head. But leave it to modern science to offer a solution. An eyeglass antifogging agent called Visaclean Spray Eyeglass Cleaner/Anti-Fog is available from Visine, a division of Pfizer Consumer Health Care. You'll find it at pharmacies and eyeglass stores.

Don't Bug Me!

All that talk about cold weather makes me long for spring and warmer nights. But, of course, with warmer weather come things that go buzz in the night. Mosquitoes, gnats, and blackflies can prove more annoying than the cold. Can anything be done to ward off these nighttime pesties? Different solutions, ranging from voodoolike rituals to toxic chemical brews, have been advanced over the years with varying degrees of success.

The best way to avoid insect bites is a combination of long-sleeved clothing, an observing site that is high and dry, and a good insect repellent. Studies show that the most effective brands use N, N-diethyl-3-methylbenzamide, better known as DEET. Most commercial repellents specify a DEET concentration of no more than 25% to 30% because higher amounts are potentially harmful. Common repellents that use DEET include Cutter, Off!, Deep Woods Off!, and 6-12 Plus.

Exercise caution whenever applying an insect repellent. DEET may be applied directly to the skin and clothing, but be sure to use it far from your telescope or other equipment. Although it works well at warding off insects, DEET also acts as a wonderful solvent when sprayed onto vinyl, plastic, painted surfaces, and optical coatings! DEET should not be used on infants or young children.

A repellent growing rapidly in popularity is actually not a repellent at all. For years, Avon has been selling Skin-So-Soft lotion and bath oil as a way to promote youthful skin; according to modern-day folklore, it works as a mosquito repellent, too. Tests in *Consumer Reports* magazine and by others have found that it works marginally, if at all. Maybe that's why so many astronomers look so young!

Still More Paraphernalia

Flashlights

Every astronomer has an opinion of what makes a flashlight astronomically worthy. Some prefer pocketable penlights, others like focusable halogen models, and a few favor dual-bulb models (which provide a built-in replacement just in case one bulb burns out). Most agree that the best are small enough to fit into a pocket but large enough not to get lost at night. White or brightly colored housings are also preferred to black or dark models, because they are easier to find if dropped.

Regardless of the style or design, a flashlight must be covered with a red filter to lessen its blinding impact on an observer's night vision. There are many different ways of turning a white light red. Some of the more common methods include painting the bulb with red fingernail polish or using red tissue paper or transparent red cellophane from stationery and party supply stores. These tend to chip, tear, or fade with time, forcing repeated filter renewal or replacement. A more permanent solution is to use red gelatin filter material sold in art supply or camera stores. One or two layers of Wratten gelatin filter #25 (yes, the same classification system as eyepiece filters) works especially well. But perhaps the most versatile red filter material is sold by auto parts stores as repair tape for car taillights. Sold in rolls of several feet, its adhesive backing makes it ideal for sticking onto a flashlight.

Far better are flashlights that use a light-emitting diode (LED) instead of a conventional bulb. These are growing in popularity thanks to their deep, pure red color and low power drain. Many astronomical suppliers sell LED "astronomy flashlights," with one of my favorites being the Starlite by Rigel Systems. The Starlite features an adjustable-brightness LED and runs on a single 9-volt battery. This and other astronomer's flashlights are listed in Table 7.15.

Dews and Don'ts

One of the most frustrating things that we, as amateur astronomers, are forced to deal with is dew. The formation of dew on a telescope objective, finder, or eyepiece can end an observing session as abruptly as the onset of a cloud bank.

Dew forms on any surface whenever that surface becomes colder than the *dew point* temperature, which varies greatly with both air temperature and humidity. To illustrate this, think of a can of cold soda. In the refrigerator, the can's exterior is dry because its surface temperature is above the dew point of

Table 7.15 ***Astronomer's Flashlights***

Company/Model	Bulb or LED	Batteries	Comments
Astro-Lite			
Astro-Lite II	LED	Two AA	Swivel head, clip; 5" long; 3.2 ounces
Astro-Lite III	LED	Two AAA	Swivel head, clip; 4.6" long; 2.1 ounces
AstroSystems			
ChartMaster	LED	Two AA	6" long
Celestron International			
#93590 Red Astro Lite	Bulb	n/a	Disposable squeeze-type flashlight
#93588 Night Vision	LED	One 9-volt	Adjustable brightness
Helix Observing Accessories			
Swivel Head	LED	Two AA	Swivel head, clip; 4.25" long; 3.5 ounces
Lumicon			
Lumilamp	LED	Two AA	Pocket clip
Lumilite	LED	Two AA	
Orion Telescope Center			
RedBeam	LED	Two AA	Yellow plastic body; 6.25" long; 2.6 ounces
Nitewriter	Bulb	Two AAA	Combination pen and flashlight
Illuminated magnifyer	Bulb	Two AA	3.25" magnifying glass
Gooseneck flashlight	Bulb	Two AAA	Flexible 9" neck; focusable beam
Rigel Systems			
Skylite	LED	One 9-volt	Adjustable brightness; switchable between red and white
Starlite	LED	One 9-volt	Adjustable brightness

n/a: not applicable.

the surrounding refrigerated air. Now take the can out of the refrigerator. Almost immediately, its surface becomes laden with moisture because the can is now colder than the warmer air's dew point. Under a clear sky, objects radiate heat away into space and soon become colder than the surrounding air.

Nothing, neither telescopes, binoculars, eyepieces, star atlases, nor cameras, is impervious to the assault of dew, but there are ways to slow the whole process down (no, one way is *not* to give up and go inside—although there have been times that I was tempted!). One option is to install a dew cap on the telescope. A *dew cap* is a tube extension that protrudes in front of a telescope tube to shield the optics from wide exposure to the cold air, thus slowing the radiational cooling process. Binoculars, refractors, and catadioptric telescopes stand to gain the most from dew caps because their objectives and corrector plates lie so near the front of the telescope tube. Reflectors usually do not need a dew cap because their primary mirrors lie at the bottom of the tube (which itself acts as a dew cap). The only exceptions to this are if the secondary mirror dews over (only in exceptionally damp conditions) or if the reflector has an open-truss design. The former situation can be slowed by installing battery-powered heater elements around the secondary, a technique that will be described later in this section. For the latter, many amateurs wrap the truss with cloth, effectively shielding the mirror from radiational cooling.

Dew caps merely slow the cooling of an objective lens or corrector plate and, therefore, only delay dew from forming. Depending on the humidity, this may be enough, but to be effective, a dew cap must extend in front of the objective or corrector at least 1.5 times (preferably 2 to 3 times) the telescope's diameter. While most refractors built this century have been supplied with dew caps, most binoculars and catadioptrics are not (two exceptions being the Questar and Orion Argonaut Maksutovs). Owners must therefore either purchase or construct dew caps.

Most dew caps are made of molded plastic and designed to slip on and off the telescope as needed. For instance, Orion Telescope Center supplies Flexi-Shield dew caps. Made of thin, durable ABS plastic, FlexiShields are sold as flat rectangles. They are formed into tubes by wrapping the ends around and sealing the full-width, permanently attached Velcro closure. After the observing session, simply pull the Velcro seal apart and the FlexiShield lies flat for easy storage. FlexiShields come in sizes to accommodate telescopes from 3.5 to 14 inches in diameter. Another company, Astrozap, sells very similar dew caps, also made from ABS plastic. Astrozap dew caps are lined with felt, which is better at absorbing both moisture and extraneous light than plain plastic.

DewGuard also sells flexible dew caps, but these are made from a thicker, fabric material that includes a black velvet interior and a reflective exterior. Like the Orion FlexShield, DewGuard caps are wrapped around the front end of the telescope and then held together by sewn-on Velcro. Sizes range up to 16 inches in diameter, although prices are typically higher than the Orion products.

Roger W. Tuthill, Inc., also offers a complete line of No-Du dew caps (Figure 7.11). The No-Du caps are rigid, plastic cylinders, making them more difficult to store than the Orion caps. Each is lined with black felt, which is claimed to increase the cap's effectiveness. Tuthill's larger dew caps also feature built-in cylindrical heating elements. By warming the corrector plate and the air inside the cap *ever so slightly* by convection, dewing can be effectively prevented even in high humidity. The cap's heating element emits between 10 and 20 watts of heat, depending on telescope size, and requires a 12-volt DC power source, such as a car battery, a DC power supply (useful only if a 110-volt AC outlet is nearby), or a rechargeable battery with at least 5 (preferably 10 or more) ampere-hour capacity. No-Du caps are available for telescopes ranging from 3.5 to 14 inches in diameter, but heated versions are sold only for 8-inch and larger scopes.

Jim Kendrick Astro Instruments sells dew caps that seem to be a combination of Orion FlexiShields and Tuthill unheated No-Du caps. Kendrick's models all have sewn-in Velcro strips, felt linings, and they unfold flat for easy storage. They are available for most popular-size telescopes.

For 8- through 11-inch Schmidt-Cassegrains, Orion Telescope Center also offers a different style of contact heating element called the Dew Zapper, which wraps around the front end of the telescope tube to heat the corrector plate by conduction to just above the air's dew point. Although it may be used independent of a dew cap, the Dew Zapper is more effective with a cap in place. Both 110-volt AC and 12-volt DC versions are available; the former delivers up to 25 watts of heat while the latter supplies about 12 watts at 1 amp.

Figure 7.11 *Tuthill No-Du Cap.*

Perhaps the most flexible dew-prevention system is sold by Kendrick Astro Instruments. The Kendrick Dew Remover works equally well on all types of telescopes as well as eyepieces, finderscopes, camera lenses, and just about any other optical surface likely to be turned skyward by today's stargazers. The Kendrick system consists of a small control box that is powered by a 12-volt DC source, such as a car battery. Kendrick recommends using, at a minimum, a 12 amp-hour battery. The system may also be run off house current by plugging the controller into an AC / DC converter of adequate amperage.

The Kendrick heater elements are wrapped in black nylon strips of various lengths and widths, each with elastic straps and Velcro pads for fastening. They come in various shapes and sizes, some for wrapping around tubes and others for laying under primary and secondary mirrors. Depending on just how damp the conditions are, power to the heating elements can be set at either low, medium, or high and will cycle off automatically to prevent overheating.

Some find the Kendrick system, the Orion Dew Zapper, and the Tuthill No-Du cap inconvenient to use because they all require external power sources, which means each comes with a long electric cord. Cords are easy to snag and trip over, especially at night, but because of the required power draw, there seems no way around this.

Right about now, if you listen carefully, you can hear the traditionalists in the crowd jumping up and down, screaming that all this is blasphemy. Many amateurs believe that imparting *any* heat to a telescope will cause optical distortion. True enough, heat will upset the delicate figures of optics, which is why telescopes must be left outside for an hour or more before observing to let the optics adjust to the outdoor temperature. In practice, however, the disturbance from contact heaters will be minimal as long as the heating is done in moderation. By definition, the dew point can only be less than or equal to the air temperature, never greater (under most clear-sky conditions, the dew point is going to be much less than the air temperature). The purpose of a contact heater is to raise the telescope's temperature not above that of the air but only above the dew point. Therefore, the telescope should never feel warm to the touch, and overheating should never become a problem.

A third way to wage war against dew is to use a low-power hair dryer or heat gun. By blowing a steady stream of warm air across an optical surface, dew may be done away with, albeit temporarily. If an AC outlet is within reach, a small portable hair dryer at its lowest setting makes a good dew remover, but if you are out in the bush away from such amenities, use a heat gun designed to run off a car battery (DC heat guns draw too much power to use with rechargeable batteries). A wide variety of sources sell basically the same 12-volt heat gun at a wide variety of prices for a wide variety of uses. Although Orion Telescope Center sells it as a "dew remover gun," auto parts stores call it a "windshield defroster" and camping equipment outlets offer it as a "mobile hair dryer." Call them what you will, all are pretty much the same. The gun puts out about 150 watts of heat—not exactly enough to dry hair or defrost a windshield but plenty to remove dew from an eyepiece.

Although I never go observing without one, heat guns do have some drawbacks. First, although they are fine for undewing finderscopes and eyepieces, their small size limits their effectiveness for objectives and corrector plates. (If the lens or mirror is much larger than about 3 inches, it is likely that the entire surface will not be cleared before a portion becomes fogged again.) A second shortcoming is that, sadly, the dew will return as soon as the surface cools below the dew point, making it necessary to halt whatever you are doing and undew the optics all over again. If dew is a big problem from where you observe, the Orion Dew Zapper and, especially, the Kendrick system are hard to beat; in drier environs, you can probably make do with a dew cap and a heat gun. But regardless of where you are observing from, never wait for dew to form. Before you begin observing, always turn the heater on, put the dew cap in place (remember the finderscope, too), and have the heat gun at the ready to clear any fog that may form on eyepieces. By following this three-step program, optical fogging should be minimized, if not eliminated.

Table 7.16 summarizes antidew accessories.

Light Baffles

As mentioned earlier in this chapter under "Light-Pollution Reduction (LPR) Filters," one of the biggest problems facing suburban and urban stargazers is the issue of local light pollution, which can come from just about any source, including a streetlight, a porch light, an illuminated sign, and so on. Not only do these light sources continually damage our eyes' dark adaptation, their light can also enter a telescope from the side of its aperture. This unwanted light will wash out image contrast, which is exactly what we are interested in enhancing. LPR filters offer little against the onslaught of *side light,* so we must look elsewhere.

First, let's baffle your telescope. By *baffle,* I don't mean *confuse,* although you may find your telescope baffling. Rather, baffling a telescope means adding a partition of some sort to block side light from entering the telescope's tube. An effective light baffle should extend at least one tube diameter

Table 7.16 ***Antidew Products***

Company/Model	Comments
Clear Sky Products	
TeleWrap	Dew cap for ETX-90
DewGuard	Foldable, unheated dew cap in 7 sizes to fit telescopes between 90 mm and 16 inches in diameter
Just-Cheney Enterprises	
Insulated Dew Shields	Foldable dew caps available in standard and custom sizes for telescopes between 70 mm and 14 inches in diameter
Kendrick Astro Instruments	
Dew Remover System	Wraparound heating elements; 16 sizes to fit telescope objectives, primary and secondary mirrors, corrector plates, binoculars, and eyepieces
Flexible Dew Shields	Foldable, unheated dew cap in 9 sizes to fit telescopes between 90 mm and 14 inches in diameter
Orion Telescope Center	
FlexiShield	Foldable, unheated dew cap in 11 sizes to fit telescopes between 90 mm and 14 inches in diameter
Dew Remover Gun	12-volt heat gun with cigarette lighter plug
Dew Zapper	Flexible heating element wrap; 3 sizes to fit 6-inch, 8-inch, and 9.25-inch telescopes; 12-volt DC and 110-volt AC models available
Roger W. Tuthill, Inc.	
No-Du caps	Cylindrical caps fit over front of telescope; available in unheated models for 3.5-inch to 5-inch telescopes; heatable models for 7-inch to 14-inch telescopes
Virgo Astronomics	
Dew/guards	Dew caps for binoculars and spotting scopes from 40 mm to 80 mm aperture

in front of the telescope. Most of the dew caps described earlier also serve as light baffles.

Just about every refractor sold today comes with a dew cap that doubles as a light baffle, although it would not hurt to lengthen the baffle, if light pollution is severe. Newtonian reflectors and catadioptric instruments usually do not come properly baffled, yet they are more prone to light intrusion than other instruments because of how close the eyepiece is to the tube's opening. For these, a light baffle can make a big difference in image quality. I already mentioned in chapter 5 that Discovery Telescopes offers removable light baffles for their Newtonians. There is really nothing proprietary about their design, so they may also be used equally well with other instruments. An independent company, DarkSide Light Baffles, offers a similar product for telescopes that ranges in size from 6 to 25 inches. Either can make a world of difference in your telescope's performance.

In the discussion of Newtonian reflectors in Chapter 5, it was noted that an open mirror cell is preferred over a closed cell because of the increased air circulation. At the same time, however, an open cell can also let light enter from behind the primary. Opaque light shrouds of black cloth, similar to the material used for tube shrouds, can help eliminate that problem, as well. You will have to get out your sewing machine to make one, however, since I know of no company that manufactures them. If one pops up, however, you will read about it on the Star Ware Home Page, (http://www.philharrington.net).

Finally, to help protect your eyes from assault, try using what I call a *cloaking device*. I recommend using a black turtleneck shirt turned inside out and placed over your head to act like a light cloth used by the first photographers. Orion Telescope Center also sells an Observing Canopy for those who prefer a flat cloth.

Observing Chairs

It is a proven fact that faint objects and subtle detail will be missed if an observer is fatigued or uncomfortable, yet many amateurs spend hours outside at their telescopes without ever sitting down. Observing is supposed to be fun, not a marathon of agony!

Observing chairs help relieve the stress and strain associated with hours of concentrated effort at the telescope. The best have padded seats and may be raised or lowered with the eyepiece (not always possible with long-focus refractors or large-aperture Newtonians).

Musician's stools or drafting-table chairs make very good observing chairs. Check the offerings at local music shops and drafting / art supplies stores.

For the astronomical market, Orion Telescope Center offers three observing seats. Two of these differ little from piano stools; the standard model adjusts between 19 and 25 inches, while the deluxe chair rises between 21 and 27 inches above the ground. The third chair in Orion's stock is an innovative design that is also sold as the Starbound Viewing Chair. The padded seat can be set at any height between 9 and 32 inches by sliding it along the chair's framework of black metal tubes, making it usable with most amateur telescopes. Although much more flexible than stools, the Orion Starbound chair is also considerably more expensive.

The Tele Vue Air Chair is very similar to Orion's Heavy-Duty Observer's Chair, although it has a slightly greater range in height and includes a very handy pneumatic height adjustment. This makes setting the height in the dark much easier than with the Orion chair; with that model you have to get off the seat, flip a small lever, and then pull or push the seat up or down.

If you prefer wood to metal, consider the all-oak Catsperch chair from Jim Fly. Similar in design to the Starbound chair, the Catsperch can be adjusted in increments from 7.5 up to 42.5 inches off the ground. Although not quite as flexible as the infinitely adjustable Starbound chair, the Catsperch's seat is less likely to slip inadvertently. Like the Starbound chair, the Catsperch folds flat for storage. The Catsperch is available in three versions, depending on how much "sweat equity" you want to put into it: in kit form, with everything precut and predrilled; as an assembled chair ready for painting or staining; or completely assembled and coated with polyurethane. A seat cushion is available optionally for those long viewing sessions.

Kendrick Astro Instruments also sells two very nicely crafted wooden observing chairs. Called the Ultimate Observing Chair, each is constructed of Baltic birch plywood and comes completely assembled and finished with clear coat. Ultimate I comes as a chair alone, while Ultimate II includes a small folding shelf that has holes drilled to accommodate two 2-inch eyepieces and three 1.25-inch eyepieces and enough extra space to hold a book or a clipboard. Like the others, both Kendrick chairs fold flat for easy storage.

Finally, StarMaster Telescopes offers the unique StarStep chair / stepladder combination (see Figure 7.12). Two aluminum side rails, each standing 38 inches tall and placed at right angles to each other, support a triangular platform of oak, which in turn holds the side rails together. The platform can be locked in four height positions; the standard chair sits the observer as high as 29 inches. Unlike the others described here, the StarStep chair may also be used as a short stepladder, although the manufacturer warns that this should only be done with the platform set in either of its two lowest positions. The entire chair disassembles for flat storage.

Table 7.17 lists some facts and figures for each of these observing chairs.

Binocular Mounts

After purchasing a pair of giant binoculars, most people suddenly come to the realization that the binoculars are too heavy to hold by hand and must be attached to some sort of external support. The favorite choice is the trusty camera tripod, but not all tripods are sturdy enough to do the job. All of the brands mentioned earlier in this chapter are strong enough to support binoculars, but most are too short to permit the user to view anywhere near the zenith comfortably (the bigger Bogen / Manfrotto models being an exception to this). What are the alternatives? Basically only one: Use a special mount designed specifically for the purpose.

No fewer than four companies offer parallelogram binocular mounts: Blaho Company, T&T Binocular Mounts, Universal Astronomics, and Virgo

Figure 7.12 *The StarMaster observing chair / stepstool. Photo courtesy of StarMaster Telescopes.*

Table 7.17 ***Observing Chairs***

Company/Model	Material	Seat Height off Ground
Jim Fly		
Catsperch Observing Chair	Wood	7.5" to 42.5"
Kendrick Astro Instruments		
Ultimate I Observing Chair	Wood	9" to 30"
Ultimate II Observing Chair	Wood	9" to 30"
Orion Telescope Center		
Standard Observer's Chair	Metal frame, padded seat	21" to 27"
Heavy-Duty Observer's Chair	Metal frame, padded seat	21" to 27"
Deluxe Adjustable Chair	Metal frame, padded seat	9" to 32"
Star Bound		
Viewing Chair	Metal frame, padded seat	9" to 32"
StarMaster Telescopes		
StarStep	Wood	6" to 29"
Tele Vue		
Air Chair	Metal frame, padded seat	20.5" to 28"

Astronomics. All range in price from $175 to $300, with very quick delivery times.

The Blaho Company offers two mounts, the Stedi-Vu and Stedi-Vu Junior, for different size binoculars. The Stedi-Vu is designed to support up to 80-mm giant binoculars, while Junior is good up to 50 mm. Both offer sleek, black anodized finishes, the nicest of any binocular mount sold today. While the others are made from aluminum square tubing, Blaho uses solid aluminum bar stock. A neat feature of the Stedi-Vu is the integrated counterweight used to balance the binoculars. The others mount their counterweights on long shafts.

The only drawback to the Stedi-Vue is that binoculars can only be aimed directly in line with the parallelogram, meaning that the mount and tripod are always in front of the observer. This makes it less convenient to view while sitting down, because the observer's legs can easily hit those of the tripod. Several of the other mounts reviewed here let the binoculars pivot to either side of the parallelogram, thereby placing the tripod to the side of the observer, away from any confrontation.

Virgo Astronomics sells a variety of binocular accessories, including two mounts: the Sky/mount (also sold by Orion Telescopes and Binoculars) and the Nova/mount. Each is essentially the same unit, made from 1-inch-square aluminum tubing nicely painted a semigloss black. A 5-pound counterweight comes standard, which is adequate for balancing binoculars up to 80 mm in size. In tests, however, I found that both should be limited to 70-mm and lighter binoculars. While perfectly adequate for my 50-mm binoculars, my 11 × 80s caused both to wobble. The problem was traced to the small size of each mount's base that attaches to the tripod. Tightening the pivot bolt helped some, but it did not eliminate the flexure.

The Sky/mount provides three axes of motion: azimuth (left/right), altitude (up/down), and binocular elevation. A large knob on the altitude axis lets users apply just enough pressure so that the binoculars can be aimed and released without constantly having to be loosened and tightened—a nice convenience. Motions in all three axes were very smooth.

The Nova/mount comes equipped with a very nice *cradle-within-a-cradle* fork bracket that lets binoculars move in both altitude and azimuth, independent of the parallelogram mounting itself. This feature greatly enhances the mount's versatility, especially if you are viewing from a seated position. Unfortunately, the fork's locking altitude knobs need to be loosened, then retightened, when aiming the binoculars, rather than having enough "stiction" to hold the binoculars when released.

T&T Binocular Mounts offers a greater variety of binocular mounts than any other company. The aluminum T&T mounts are neither painted nor anodized, giving them a more homemade look than those from Blaho or Virgo. This does not detract from their performance, however. T&T's wooden mounts, including the Four-Arm Mount and the Kid's Mount, show very nice woodworkmanship.

T&T's most popular mount, the ArtiMount (short for "Articulated Mount"), swings around four different axes of motion: altitude, azimuth, binocular elevation, and a pivoting head. This latter capability, also found on the Virgo Nova/mount and Universal UniMount, lets the user swing the binoculars left or right perpendicular to the mounting and tripod, making it much easier to enjoy binocular stargazing sitting down. Balance is easy to achieve, thanks to the counterweight and the mount's long counterweight shaft, the longest of the group. The shaft slides into the bottom parallelogram tube and is held in place with a captive pin.

One of the nicest features of most T&T mounts is that binoculars can be aimed at any angle and then released without having to lock them in place manually. By adjusting a large side knob before use, just enough pressure can be applied to let them stay put when released. On the downside, the aluminum side plates that hold the parallel beams in place seem a little thin, possibly leading to some flexure. But all in all, these mounts are excellent for the price.

Universal Astronomics also offers several binocular mounts that I consider the most complicated but also the most versatile. Their most popular mount, the UniMount (Figure 7.13), is a real swinger! With the optional Ultra Swing Mounting System that was included with this test, binoculars can move in any of five different ways: left to right in two axes, up and down in two axes, and rotating about their central pivot or hinge.

Adaptability is certainly a word that comes to mind with all of Universal Astronomics' mounts. Their mounting systems typically consist of several plate assemblies that slide together and can be adjusted in any direction to achieve a perfect balance. Because of this flexible design, no locking knobs are required to hold the binoculars in place; just mount the glasses, slide the plates back and forth, and adjust the counterweight until balanced. Nice!

Table 7.18 gives some vital statistics for the binocular mounts mentioned above.

Figure 7.13 *Universal Astronomics' UniMount binocular mount.*

Table 7.18 **Binocular Mounts**

Manufacturer/Model	Height Range	Weight Limit (pounds)	Degrees of Freedom
Blaho			
Stedi-Vu Junior	22"	3	2
Stedi-Vu	24"	5	2
T&T			
Kid's Mount	17"	4	3
Spring Mount	18"	3	3
Sidesaddle Mount	31"	8	3
ArtiMount	31"	8	4
Three-Arm Mount	25"	10	4
Three-Arm Mount Mark II	25"	12	4
Four-Arm Mount	27"	20	4
Garden Mount	30"	20	4
Wheelchair Mount	31"	8	4
Universal Astronomics			
UniMount Light	34"	8	3 (Basic model) 5 (Deluxe model)
UniMount	34"	12	5
Millennium Mount	34"	30	6
Virgo Astronomics			
Sky/mount	30"	5	3
Nova/mount	30"	5	4

Although the most expensive of the bunch, Universal Astronomics mounts deserve the title of best in class because of its well-thought-out designs and smooth motions in all directions. The T&T ArtiMount is a close runner-up in my opinion, and it costs less.

What about those binocular mounts that aim the glasses downward toward a mirror that reflects starlight into the objective lenses? I am sure that there are many readers out there who own and love them. Perhaps you are one of them. If so, great! I encourage anyone to use anything that enhances his or her enjoyment of the night sky. To my way of thinking, however, these binocular mounts are compromises at best. For openers, all commercial mirror-style binocular mounts I know of use only standard aluminizing, which reflects between 85% and 88% of the light striking them, effectively reducing the binoculars' aperture. Another problem is that the mirrors used in these contraptions are usually made of float glass, which may appear flat to the eye but is far from optical quality. Although distortions may not be apparent through low-power binoculars, image quality is likely to suffer as magnification grows. These accessories also lack the freedom of movement of parallelogram mounts. Finally, these also flip the images around, negating one of the greatest appeals of binocular astronomy: that the binoculars act as direct extensions of our eyes. Therefore, I recommend that you shouldn't use them.

Vibration Dampers

These make great stocking stuffers for any amateur who owns a tripod-mounted telescope. One of the biggest problems facing amateurs using such telescopes is how to combat vibration. Vibration is caused by a wide range of sources; anything from nearby automobile traffic or another person moving about to the wind can be responsible.

While nothing can guarantee shake-free viewing, Celestron has introduced a simple little gadget that can make a big difference. Its vibration damper is made of a hard outer resin and soft inner rubber polymer, isolated from each other by an aluminum ring. By simply placing one under each tripod leg, any vibration will be absorbed by the inner polymer before it can be transmitted to the telescope. Such a simple yet effective idea—why didn't someone think of this before?

Observatories

Most amateurs dream of one day owning their own observatory. Apart from a backyard observatory making a statement that you are a serious astronomer, they offer many practical advantages. A permanent structure makes setting up and breaking down a telescope a snap, and it also shelters us from earthly distractions, including light pollution and wind.

Observatories are usually broken into two broad categories, depending on their construction. Most professional observatories use large domes that open a thin slit to the night sky. Domed observatories are also quite popular among amateurs, although not as much as they were, say, a few decades ago.

One reason is that most domed observatories have relatively high walls, which means their telescopes must be on tall-standing mounts. Refractors and Cassegrain-style telescopes are fine for these, but Newtonians, especially those on low-riding Dobsonian mounts, are better housed in roll-off roof observatories, the second category. Unfortunately, this latter style does not offer the same protection against wind and light pollution that is afforded by domes. They do, however, give the observer a much more panoramic view of the entire sky at once.

To my knowledge, there are no commercially sold roll-off roof observatories. In constructing one of those, most amateurs modify a wooden garden shed, slicing off the roof and placing it on four or more wheels that ride on a pair of tracks that extend behind the building. Another approach, the Lookum Observatory, is shown in chapter 8.

Some of today's most popular domed observatories for the amateur are made by Boyd Observatories. Owner Garry Boyd sells four complete models or the domes alone for do-it-yourselfers. All models include fiberglass domes and walls that require very little maintenance. Designed primarily for pier-mounted Cassegrain-style instruments, the Boyd model 5500 is 7 feet in diameter, which is barely large enough to hold the telescope, the observer, and perhaps a small table. A small bump-out area that protrudes outside for a computer is available, adding a little more room for occupants. Still, because of the tight quarters, the telescope must be offset toward the south side of the observatory, which makes access to the northern part of the sky difficult in practice.

Although it is several hundred dollars more expensive, the Boyd model 6000 gives users a little more elbow room, thanks to a larger square base. Optional bump-out areas for a computer and a settee improve comfort even more. In either case, the supplied dome has two hatches, although a more secure single hatch is also available at no extra charge.

The most unique Boyd observatory is the model 5000C. Unlike the others, both of which are intended to house both telescope and observer(s), the 5000 looks a little like a domed doghouse. It is intended solely to house a remotely controlled robotic telescope, such as a CCD-equipped, GoTo Schmidt-Cassegrain. Cables from the telescope would run to another structure (presumably a home or a school), where the observer would use the telescope via computer control. A trailer-mounted version is also available for astronomers on the go. If this is your preferred modus operandi, then the Boyd 5000C should be on your shopping list.

Home Domes and Pro Domes, both from Technical Innovations of Barnesville, Maryland, are also quite popular. Both domes are white on the outside and deep blue on the inside to dampen stray light from damaging the observer's dark adaptation. Shutters and slot openings overlap to prevent leaks.

The smallest of the Home Domes (6 feet in diameter) are the HD-6S and HD-6T (for "short" and "tall," respectively). The former is best suited for attaching to an existing building, while the latter is a freestanding structure with 45-inch walls. The larger HD-10 (10 feet in diameter) is also meant to be added on to existing walls.

Pro Domes come in 10- and 15-foot versions. Both have the unique ability to stack fiberglass wall rings, which let the customer make as tall an observatory as desired. Finally, there is the Robo-Dome, a small, robotic observatory that, like the Boyd 5000C, is only large enough for a remotely controlled telescope, not an observer.

All of the Technical Innovations domes can be controlled automatically with two optional software/hardware systems. The most sophisticated, called Digital Dome Works, controls both the dome's rotation as well as the opening and closing of the slot. Two (or preferably three) people are needed to assemble any of these products, although the manufacturer states that no special hoisting equipment is needed (though, obviously, it may be if the dome is being placed on very high walls or on a roof). Motors to rotate the domes are sold separately. While not necessarily required for the 6-foot models, they are recommended for the larger units.

Virgo Astronomics, known primarily for binocular mounts, offers the Sky 360 domed observatory. Measuring 10 feet in diameter, which many consider to be the smallest size for a useful observatory, the Sky 360 includes corrugated steel walls and a fiberglass dome. The dome features a single 36-inch-wide slit that is raised by hand (it slides up along a track) and opens past the zenith. Once assembled, the Sky 360 is a single unit, with the dome and walls joined together with stainless steel hardware. To rotate the dome, the entire structure must be turned by hand, riding on a galvanized steel ring around the stationary, pressure-treated wood floor. That floor, incidentally, has a hole in its center for a pier up to 10 inches in diameter. Framing under the floor also has predrilled holes for electrical wiring. Although the manufacturer states that the observatory can be placed on level ground without a concrete foundation, I would recommend raising it on concrete blocks and placing a vapor barrier underneath (such as a simple plastic tarp) to prevent invasion of moisture and unwanted visitors.

Finally, Astro Haven makes fiberglass, clamshell-style domes that combine attributes of conventional domes as well as some of the benefits of a roll-off roof observatory. Available in 7-, 12-, and 16-foot diameters, all are split at the zenith and open up fully in sections. This has the immediate benefit of making it possible to see the entire sky without needing to rotate the dome. All models are normally made with white gel coat exteriors and black interiors, but other light colors are available. The 7-foot model comes preassembled, so it's a simple matter (simple if you have a fork truck or a lot of friends, since it weighs 600 pounds) to bolt the dome to a concrete foundation; once that's done, it's ready to go. The larger domes require a fair amount of assembly.

Over the past few years, several companies have introduced portable observatories geared for observers who must flee their homes in search of dark skies. Modeled after or adapted from tents, portable observatories can give observers the best of both worlds, affording both shelter during and after a long night under the stars as well as the ability to remain nomadic.

The portable observatories listed here seem to fall naturally into three categories, based on their design and function. Both the Kendrick Shelter-Dome and Clear Sky TeleDome are intended to be on-the-fly shelters, and as such share many common characteristics. Each is designed to be set up and taken down quickly and is light enough for easy transport. Neither has a sewn-in floor.

The TeleDome, looking like the amputated leg of a giant ant bent slightly at the knee when stored, is made from steel, aluminum, and black Cordura fabric, all sewn together in a single assembly. It requires no assembly at all; rather, it just needs to be unfolded for use. After setting up and placing your equipment inside, zip the door flap shut and unzip the T-shaped half opening in the opposite wall for a waist-high view covering a horizontal span of about 60°. Observers with lower-riding reflectors will do better peering through the doorway itself. The TeleDome's Cordura fabric is completely opaque to light pollution, making it ideal for combating offending photons from neighborhood lights.

To rotate the TeleDome, the user must pick up the entire structure in the center (at the peak) and spin it by hand, which proves inconvenient (and quite a stretch for shorter users). The TeleDome's tall profile also makes it less effective against the wind. Staking it to the ground will hold it in place in windy conditions but also limits its usefulness, since it can't be rotated. The TeleDome is currently offered in two models. Both are octagonal in shape, measuring 7 feet 6 inches at their widest. The smaller TeleDome measures 6 feet high, while the taller version is 6 feet 6 inches at its peak.

While the TeleDome may be likened to a domed observatory, the Shelter-Dome is philosophically more akin to a roll-off roof observatory in that it can give the observer a nearly unobstructed view of the entire night sky at once. The Shelter-Dome measures 7 feet × 10 feet × 6.5 feet high and is made from fabric-reinforced vinyl. Users can elect to keep one or more of the completely opaque sides zipped up either partially or entirely to block wind and light pollution. The Kendrick Shelter-Dome, like the TeleDome, is also intended to be a fair-weather observatory, although a fly is sold separately for inclement weather.

The second category of portable observatories are adaptations of camping tents. The Kendrick Observatory Tent (Figure 7.14) was the first of this kind on the market and still shows an ingenious design. It is easy for one person to set up quickly; it took this veteran camper only 8 minutes the first time, without instructions. Once erected, the Observatory Tent has two rooms separated by a polyester wall. The smaller area (9 feet × 6 feet) is for sleeping and storing supplies, while the larger area (9 feet × 9 feet) is devoted to telescope space. The roof peaks at 6.5 feet but slopes off toward either end.

Actually, floor space is not the limiting factor for the size of telescope that the Observatory Tent will accommodate. That is set by the size of the D-shaped door that leads into the telescope area, which measures only 30 inches wide × 50 inches high. I would guess that the largest telescope you could squeeze through there comfortably would be a 10-inch reflector and possibly a 12-inch Schmidt-Cassegrain. The Observatory Tent's floor, which runs across both the telescope area as well as the sleeping section, is made from a plasticized rip-stop polyester for extra strength. The floor also gives extra sleeping room should a few extra observers decide to spend the night, but it could also cause some lighter telescopes to shake as people move about. Speaking of the walls, they diffuse light pollution but are not opaque enough to block it entirely.

Figure 7.14 *The Kendrick Astro Instruments Observatory Tent.*

With the help of the Royal Astronomical Society of Canada, North Peace International has modified two square dome tents to serve as portable observatories: the Galileo (measuring 96 inches × 96 inches × 76 inches high in the center) and the Orion (84 inches × 84 inches × 60 inches high in the center). North Peace certainly did its homework when designing the tent frame, as it is just about the simplest that this camper of more than 25 years has ever seen. Unlike the Kendrick tent, which zips open across the tent's top, the single-room North Peace tents are designed so that all four sides zip open fully. Unfortunately, even with all four doors open, the tent's crisscrossing frame blocks the zenith as well as four equally spaced vertical slices of sky. Owners of larger instruments will find this a greater hindrance than those who have smaller telescopes because the latter can be repositioned more readily.

For those who want a semipermanent observatory but don't want the zoning hassles that frequently accompany permanent buildings, a Sky Tent may be your salvation. The Sky Tent is designed to weather just about anything Mother Nature has on her agenda while protecting your valuable astronomical equipment. Although the largest observatory in this collection, the Sky Tent is surprisingly easy to put together. Simply snap together the framework of PVC pipes, then overlay the PVC-coated polyester fabric. The Sky Tent's base is staked to the ground, while the PVC-framed dome rotates manually 360° by riding smoothly along hook-shaped guides made of nylon. The dome's flap is zippered closed and has a generous overlap that is also sealed with Velcro strips. But despite everything being battened down, the Sky Tent can leak a little after a heavy rain.

Table 7.19 **Must-Have Accessories for the Well-Groomed Astronomer**

Commercial

- ☐ Red flashlight
- ☐ Star atlas (Sky Atlas 2000.0 recommended)
- ☐ Narrowband LPR filter
- ☐ Straight-through finderscope
- ☐ Adjustable observer's chair
- ☐ Warm clothes
- ☐ Hand and foot warmers
- ☐ Bug repellent (Cutter, Deep Woods Off!)

Homemade

- ☐ Eyepiece and accessory tray or table
- ☐ Two old briefcases—one for charts, maps, and flashlights, the other lined with high-density foam for eyepieces, filters, etc.
- ☐ Observation record forms
- ☐ Cloaking device (see chapter 10)

The only other fault I can find with Sky Tents are their prices. Their base model measures 10 feet in diameter × 7 feet high and has 48-inch high walls, but it costs a little less than $2,000. A larger version that measures 10 feet in diameter × 8.5 feet high and has 66-inch-high walls costs a few hundred dollars more. Dobsonian instruments, which can get very low to the ground when viewing objects near the horizon, would do better with either the 12-foot × 8.5-foot Sky Tent, which features 24-inch-high walls or the 12-foot × 10-foot Sky Tent, both in the $3,000 range. Be forewarned that delivery is quite slow, usually measured in several months.

The Well-Groomed Astronomer

The list of available accessories for today's amateur astronomer could run on for pages and pages, but I must cut it off somewhere. To help put all of this in perspective, Table 7.19 lists what the well-groomed amateur astronomer is wearing these days. Some of the items may be readily purchased from any of a number of different suppliers; others can be made at home. A few may even be lying in your basement, attic, or garage right now.

The most important accessory that an amateur astronomer can have is someone else with whom to share the universe. Although many observers prefer to go it alone, there is something special about observing with friends; even though you may be looking at completely different things, it is always nice to share the experience with someone else. If you have not already done so, seek out and join a local astronomy club. For the names and addresses of clubs near you, contact a local museum or planetarium, the Astronomical League, or, if you have access to the Internet, check out the World Wide Web; you will find addresses in Appendix E. If you have an astronomy buddy, then you have the most important accessory of all, and one that money cannot buy!

8

The Homemade Astronomer

Amateur astronomers are an innovative lot. Although manufacturers offer a tremendous variety of telescopes and accessories for sale, many hobbyists prefer to build much of their equipment themselves. Indeed, some of the finest and most useful equipment is not even available commercially, making it necessary for the amateur to go it alone.

Many books and magazines have published plans for building complete telescopes. Rather than reinvent that wheel here, I thought it might be fun to include plans for a variety of useful astronomical accessories. For this third edition, here are 11 new projects, ranging from the very simple and inexpensive to the advanced and costly. Their common thread is that they were created by amateurs to enhance their enjoyment of the universe. These projects are just a few samples of the genius of the amateur astronomer.

An Inexpensive Introduction to CCD Imaging

Many amateurs dream of owning advanced charge-coupled device (CCD) imagers, such as those described in chapter 7, that will allow them to take electronic photographs of celestial objects, store them in their computers, and later adjust image brightness and contrast for the best possible results. But to enjoy this technology usually requires investing thousands of dollars in the CCD imager, to say nothing of the portable computer system behind it (most use laptop computers). While I can't offer much help with the computer, here are step-by-step instructions on how to adapt a common CCD camera to short-exposure astrophotography. Best of all, it will not cost a fortune. In fact, I built the one outlined here for about $35 in one afternoon and was out shooting the Moon that evening! Although the results, such as in Figure 8.1, cannot compare to an expensive CCD astro-imager, they are still pretty impressive.

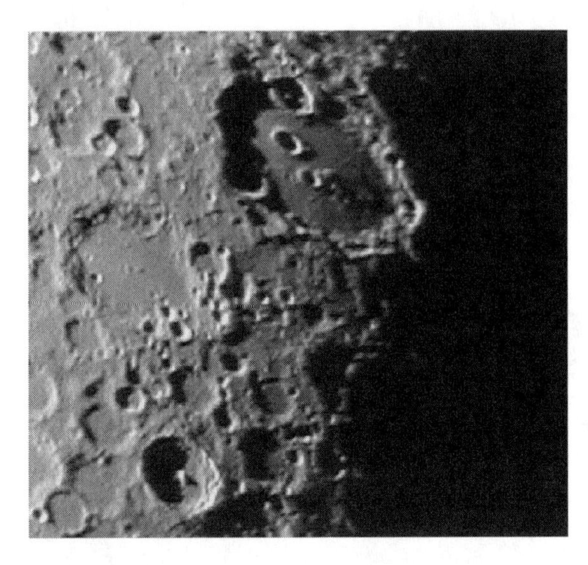

Figure 8.1 *CCD image of the lunar crater Clavius and vicinity taken with the QuickCam astro-imager at the prime focus of the author's 4-inch f/9.8 Vixen refractor.*

The homemade CCD imager shown in Figure 8.2 is based around the Logitech QuickCam from Connectix. The Logitech QuickCam series are designed as *web cams,* and are aimed primarily at the market for Internet-based conferences and informal images of family and friends. I got mine, a color QuickCam VC, for $30 on the Internet auction site eBay (www.ebay.com).

Table 8.1 lists the parts you will need to make your own QuickCam astro-imager.

Although you can use a QuickCam for afocal astrophotography right out of the box (that is, with the camera's lens and telescope eyepiece both in place), most prefer to adapt the QuickCam to prime-focus astrophotography by removing the lens and placing the imager's circuit board in a new housing. To

Figure 8.2 *The QuickCam astro-imager in its new housing. The original QuickCam ball is seen in the background.*

Table 8.1 ***QuickCam Astro-Imager Parts List***

Quantity	Description
1	QuickCam (exact model is up to you)
1	Plastic box (Radio Shack part # 270-1805 or equivalent)
1	35-mm film can (Kodak recommended)
—	Silicone adhesive
—	Foam (enough to fill box)

do that, the camera's ball-shaped housing must be opened. Now, we can do that in a couple of different ways. Because the camera body will not be reused, there is always the brute-force approach (that is, using a hammer). This approach, though tempting, will probably damage the sensitive CCD chip and printed circuit board, so it is not recommended.

How can the ball's two hemispheres be opened delicately? There's a secret! Look along the seam where the two halves meet for a tiny hole about the diameter of a paper clip. Depending on your camera model, it might be hidden behind a little sticker, which can be peeled back easily. Insert a straightened paper clip or small jeweler's screwdriver into the hole and gently press it in until you feel a snap. You have just unlocked one of three retaining clips that hold the ball together. Pry the ball apart using a screwdriver, again taking care not to damage the internal components.

For the next several steps, you should be properly grounded to prevent the buildup of static electricity, which can damage the camera's sensitive electronics. Most computer supply stores sell tiny grounding straps that tie you to a pipe or some other grounded structure.

Inside the QuickCam you will find several pieces, including the lens assembly, the metal spacer adjacent to the lens, and a metal counterweight that doubles as a tripod socket. Remove all three. Carefully extract the round circuit board and you will see the CCD chip through what looks like a reddish window—the infrared filter. The filter is held in place by two tiny Allen-head cap screws. Unscrew them slowly, remove the filter, and set the board aside while you prepare its new home.

You can use any of a number of different enclosures for the camera's circuit board. In fact, some people simply put the board back into the QuickCam ball, but I chose to remount it in a 3-inch × 6-inch × 2-inch-deep plastic box from Radio Shack. First, drill a 1.25-inch hole in the center of the box's top, using either a hole saw or a drill bit. Next, take a Kodak plastic 35-mm film can, remove the cap, and cut out the bottom with either a fine-tooth coping saw or a hacksaw. I recommend Kodak film cans, since they have a protruding lip around the open end, whereas those from Fuji and most other film manufacturers do not. Insert the hollow film can through the hole in the box top from the inside and glue it in place with a flexible adhesive, such as RTV. (A better idea might be to use a T-mount from a 35-mm camera in lieu of the 1.25-inch plastic film canister. This way, you can screw the QuickCam onto any mating camera lens as well as to your telescope.)

After this has dried, place the circuit board in the box so that the CCD chip is centered and faces out through the 1.25-inch hole, as shown in Figure 8.3. I am sure there are more elegant ways to hold the circuit board in place, but I simply taped it down using electrical tape, and then I glued it in place using more RTV adhesive, squeezing the RTV between the board and the plastic box. Just be careful not to get any on the CCD chip!

Next, drill or cut a slot in the side of the box for the cable that runs from the camera to your computer. Finally, pack the box with high-density packing foam to keep the board from moving or coming loose. Screw the box's top in place, and you are done.

QuickCams attach to the keyboard and mouse jacks of most computers or plug directly into a computer's USB port, depending on the model. Regardless, you will need to install the camera's imaging software before using the camera. Follow the instructions that come with the camera.

Once everything is set, insert your QuickCam astro-imager into your telescope's eyepiece holder just as you would an eyepiece or regular camera. Aim the telescope at a bright target (the Moon is ideal for first light) and look for its

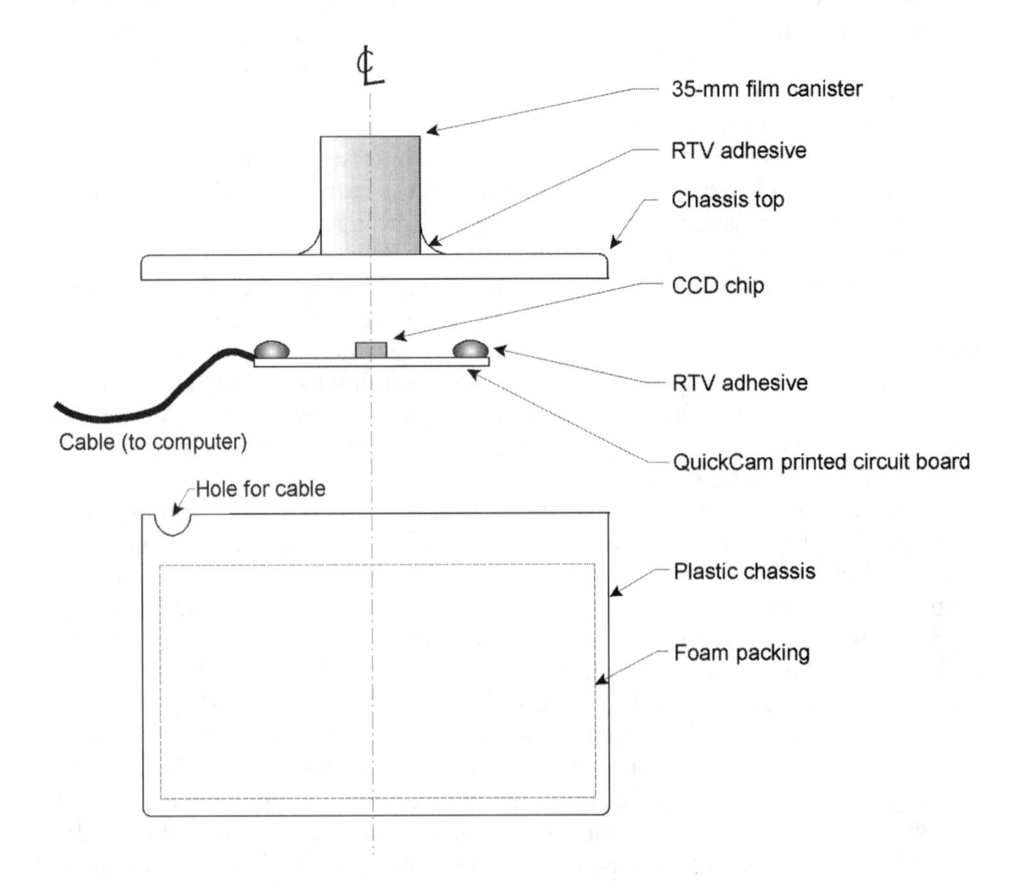

Figure 8.3 *Assembly drawing for the QuickCam astro-imager.*

image on the computer screen. The QuickCam will produce real-time images, meaning that you can adjust the aim of your telescope and see the results on the computer screen at the same time. With the Moon centered, turn the focus until the image is sharpest. If it is not evenly sharp across the field, that means that the CCD is tilted a little. Open up the camera and shim one side of the circuit board to even things out, then try the Moon again.

It's a simple matter to take a photograph with the astro-imager: You simply click the QuickCam's program screen with your computer's mouse button. Once a photo is taken, save it so that you can adjust its quality with the QuickCam's software or another software package, such as *Adobe Photoshop*. You will be amazed at how sharp the results can be from a $35 camera! And since the camera weighs so little, even owners of Dobsonian telescopes can use it without affecting the balance of their instruments.

A Backlit Chart Box

One of the most common complaints heard from amateur astronomers is that the views through their telescopes do not agree with the orientation of their star charts. Astronomical instruments tend to flip things upside down, left to right, or both. While there is no up or down in space, there is here on Earth, and trying to get charts to agree with what's seen through the eyepiece can be tough. This is especially true through refractors and Cassegrain-style instruments, which usually produce mirror images due to their star diagonals.

Dave Kratz of Poquoson, Virginia, has come up with an effective yet simple way of addressing the problem by creating a backlit chart box (Figure 8.4).

Figure 8.4 *Dave Kratz's chart box.*

The design centers around a wooden box and two 15.5-inch × 10-inch pieces of clear acrylic, which are available from any well-stocked hardware or plastics supply store (check the Yellow Pages). Table 8.2 lists everything you need. All hardware and materials for the box should be available at a well-stocked lumberyard or home improvement center, while all of the electronics can be found at Radio Shack.

Begin by making the box. Cut the 4-foot length of 1 × 6 lumber into two 14-inch lengths and two 10-inch lengths. Using a router, cut a 0.25-inch-wide × 0.25-inch-deep channel into each of the 14-inch-long pieces 3 inches from one edge. These will hold a piece of plywood that will be used to prevent light from sneaking out of the box. Kratz also routed channels lengthwise along each 14-inch-long piece to serve as paths for the wires that attach between LEDs. This is not absolutely necessary, but it makes for a neater end result.

Decide which of the 10-inch pieces will be the box's control panel. Drill a hole through the wood, offset to one side as shown in Figure 8.5, for the rheostat.

Assemble the four sides to the box as shown by gluing and then either nailing or screwing the four pieces together. Be careful not to split the wood during assembly (predrilling pilot holes might be a good idea). Simple butt joints are fine, or you can try a fancier approach, such as a rabbet or a miter joint (although overall dimensions may vary). The end result should be a wooden rectangle that measures 10.0 inches by 15.5 inches.

Next, cut the bottom panel from the 0.25-inch plywood, sized to match the dimensions of the box. Glue the bottom panel to the box using wood glue and

Table 8.2 ***Kratz Light Box Parts List***

Quantity	Description
4	Red LED (Radio Shack # 276-307 or equivalent)
1	Rheostat, 5,000 ohm (Radio Shack # 271-1714 or equivalent)
1	Knob for rheostat (Radio Shack # 900-2546 or equivalent)
1	SPST potentiometer switch (Radio Shack # 271-1740 or equivalent)
1	C or D size battery holder for two batteries (Radio Shack # 270-385 or equivalent; don't forget to buy the batteries)
1	Length of wire to connect LEDs, resistor, battery holder, and rheostat (Radio Shack # 278-1215 or equivalent)
1	4' length of 1 × 6 wood (pine will work, but a hardwood, such as maple or oak, will look better and last longer)
1	15.5" × 10" sheet of 0.25" plywood
1	5.5" × 8.5" sheet of 0.25" plywood
2	15.5" × 10" plexiglass sheets
1	Handle for the top plexiglass sheet
1	Set of Velcro to attach one of the plexiglass sheets to the top of the box and to form a hinge between the two plexiglass sheets
1	Can of spray paint (red or white)
—	Silicone adhesive
—	Assorted hardware (nails, screws, etc.)

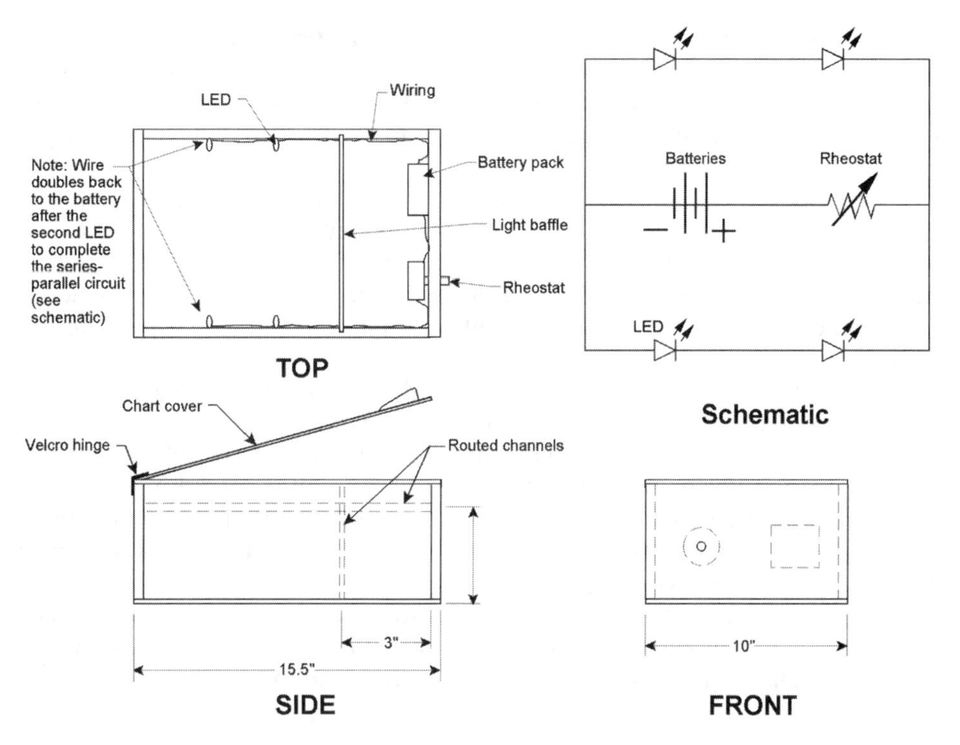

Figure 8.5 *Plans for the Kratz chart box.*

fasten it in place with either small nails or screws. From the same plywood stock, cut the box's light baffle panel to suit (5.5 inches × 8.5 inches in the original plans); when the baffle is inserted into the routed channels, it will limit the light to the 11-inch × 8.5-inch illuminated area. Note that the partition has to be removable so that you can easily replace the batteries.

With the box now complete, paint the inside with either flat white or red paint. Glossy paint might cause hot spots and should be avoided. You can also paint the outside of the box, although Kratz recommends staining and polyurethaning because the end result is much nicer. Set everything aside to dry.

While the paint is drying, fashion the top of the box and chart cover from the pieces of acrylic, again taking care to match the dimensions of the box. Decide which will be the chart cover and attach the wooden handle, as shown on the drawing. Be very careful when drilling the two pilot holes for the handle's screws to avoid cracking the plastic. Don't fasten the top to the box just yet, however. Instead, assemble the electrical components first.

The battery, rheostat, potentiometer switch, and four LEDs are wired together in a series-parallel circuit, as shown schematically in the drawing. Follow the diagram carefully when soldering the components together, making sure to use enough wire between the LEDs so they can be spaced evenly in the box. Take extra care when soldering the LEDs, because they can be damaged if accidentally overheated. Insulate the LED leads from each other with electrical tape or shrink sleeving to prevent electrical shorting. Once the circuit is assembled, insert the batteries and turn on the rheostat. Do the LEDs light? If none of them work, first make sure the battery is good. If it is, then recheck all the solder connections. If only one or two light, the faulty LED may have been damaged during soldering and must be replaced.

Now place the electronics into the box. Start by inserting the rheostat through the hole in the box's control panel and screw the battery holder to the box wall as shown. Next, feed the LEDs and wires through the channels along the inside of the box. Secure the wires and LEDs in place with a silicone adhesive, tape, or other household glue. Finally, slide the light baffle panel into the channels.

The clear acrylic top can be mounted to the box in any of a number of different ways. Just make sure that whichever method you choose, you can remove the cover to replace the batteries when needed! Kratz elected to use adhesive-backed Velcro pads in the four corners to hold everything in place. He also made hinges from Velcro straps on the outside of the chart cover. Metal hinges can be substituted, if desired, though care must be taken during assembly to avoid cracking the plastic.

The chart box makes turning and twisting charts to match the eyepiece view a snap. And you will especially appreciate it on those windy nights when loose charts and papers tend to sprout wings and take to the sky!

An Alt-Azimuth Mount and Tripod for Small Refractors

Okay, you did it! You bought a telescope before you read this book. Worse yet, you bought a telescope from a large discount department store. It sure did sound impressive, didn't it, promising 675× views of Jupiter, Saturn, the Moon, and the galaxies. Surely the Hubble Space Telescope itself couldn't do better.

But that dream was shattered when you took the telescope out of its box, set it up, and took that first, breathless look. Aiming it carefully at the Moon or a bright planet, you peered through the eyepiece. And what did you see? Celestial Jell-O quivering before your eyes. No matter how hard you tried, you just couldn't focus the view or get it to stand still.

Did you just waste $200 or more? Not necessarily. Many of these so-called Christmas Trash Telescopes (CTTs) can be saved from their inevitable fate—storage in the basement before Easter—but it will require some additional investment of time and money.

First, the eyepieces. The eyepieces that come with the typical CTT are almost always either Huygens or Ramsdens, two old designs that can be summed up in one word: terrible. Stop right here and read chapter 6, which includes everything you ever wanted to know about eyepieces but didn't even know to ask. A new set of Plössl eyepieces will certainly help clear things up, but there are still other problems to conquer.

What else can be done to improve these telescopes? Certainly that mount has to go! Most CTTs include mountings and tripods that fail the all-important rap test. The mountings and especially the aluminum tripods are just too flimsily made to support the telescope steadily. And as has already been mentioned elsewhere in this book, even the finest telescope will be reduced to a useless waste of money if the mount shakes when the instrument is focused or nudged.

Department-store telescopes are not necessarily the only ones that need a decent mounting. Many of the finest apochromatic refractors are sold without mountings at all. Owners of these instruments have no choice but to spend a goodly sum on a commercially available mount. Is there no hope?

Here is your salvation in either case. Do as many amateurs have done: Remount your refractor on a homemade mounting fashioned from wood. Not only is wood easy to work with, it is also great at squelching vibrations. And while wooden equatorial mounts are certainly possible, the project presented here is a simple alt-azimuth mount that anyone can make with commonly available tools. Ghyslain Loyer of Quebec, Canada, made the nicely designed alt-azimuth mount and tripod shown in Figure 8.6. The entire project cost him less than $60 and holds his 12-pound telescope quite steadily. First, take a look at the parts list in Table 8.3.

Referencing the layout drawing (Figure 8.7), cut the three tripod legs (labeled B) from the 2 × 4 studs (note the 15° bevel) and all other pieces from 8-inch-wide by 0.75-inch-thick solid birch or maple. Take care to lay everything out beforehand in order to conserve wood. One possibility is that parts A, D, E, G, and I can be cut from a single 3-foot-long piece, while C, F, and H can be cut from a second 3-foot-long section. Drill holes where indicated in the

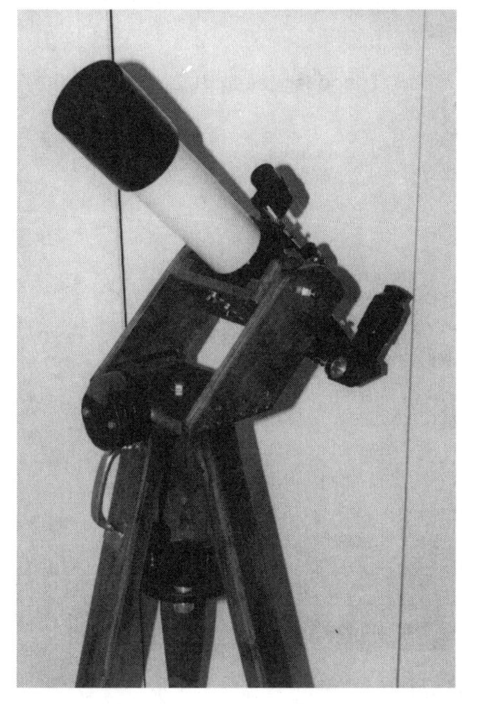

Figure 8.6 *Ghyslain Loyer's alt-azimuth mount and tripod, holding his 4-inch Tele Vue refractor. Photo courtesy of Ghyslain Loyer.*

Table 8.3 ***Alt-Azimuth Mount and Tripod Parts List***

Quantity	Description
3	Legs, 2" × 4" × 48" studs (spruce recommended)
6 linear feet	Birch or maple stock, 8" wide × 0.75" thick
3	Hinges
1	Handle
3	Bolt, 0.5" in diameter × 2.25" long
2	Wing nuts, 0.5" in diameter
1	Lock nut, 0.5" in diameter
6	Washers, 0.5" in diameter
1	Weight (barbell), approximately 10 pounds
1	6-inch chair bearing (lazy Susan bearing)
2	Plastic margarine, milk, or juice containers
50 (approx.)	Wood screws, #10 × 1.5" long
18 (approx.)	Wood screws, #10 × 0.5" long
—	Wood glue
—	Stain or paint

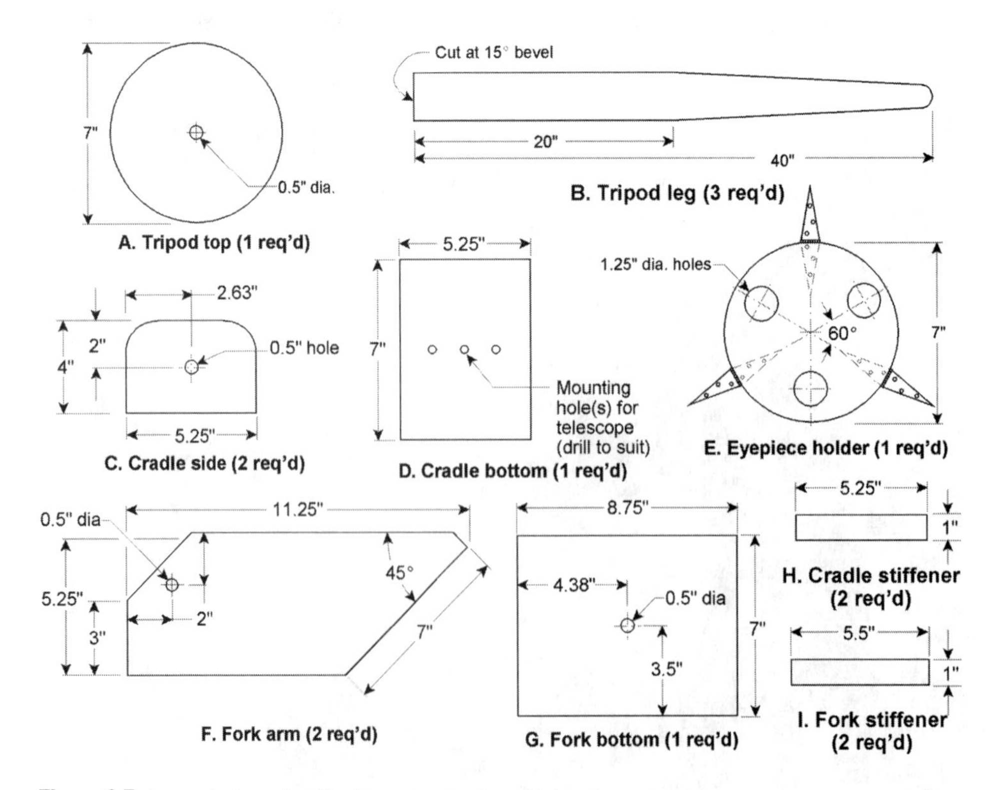

Figure 8.7 *Layout plans for the Loyer mount and tripod.*

drawing. Note that the eyepiece holder has three 1.25-inch holes. If your telescope uses different diameter eyepieces (either 0.96- or 2.00-inch), change these accordingly. That same advice goes for all of the dimensions. The plans here are designed to match Loyer's telescope. As they say in advertising, your actual numbers may vary.

Using a router or sandpaper, round off all edges for a nice, finished look. Last, stain or paint everything prior to assembly.

Once all of the parts are dry, it's time to put it all together. Begin by screwing the three hinges onto the eyepiece holder, spacing them evenly as shown on the drawing. Next, lay out the tripod top so that all three legs are evenly spaced and attach each with two or more 1.5-inch wood screws. With the tripod upside down and the legs raised in the air, screw the eyepiece shelf to the insides of the three legs, taking care that it is level. Turn the tripod right-side up and set it aside for the moment. (Note that the tripod's legs will not fold in this design. Although this is a little less convenient, it makes the mount much more rigid.)

Now glue and screw the telescope cradle together. Loyer used six 1.5-inch wood screws to attach each fork side to the fork bottom, as shown in the construction drawing (Figure 8.8). Attach the two fork stiffeners between the fork arms and bottom where shown to increase the mount's rigidity. Again, glue

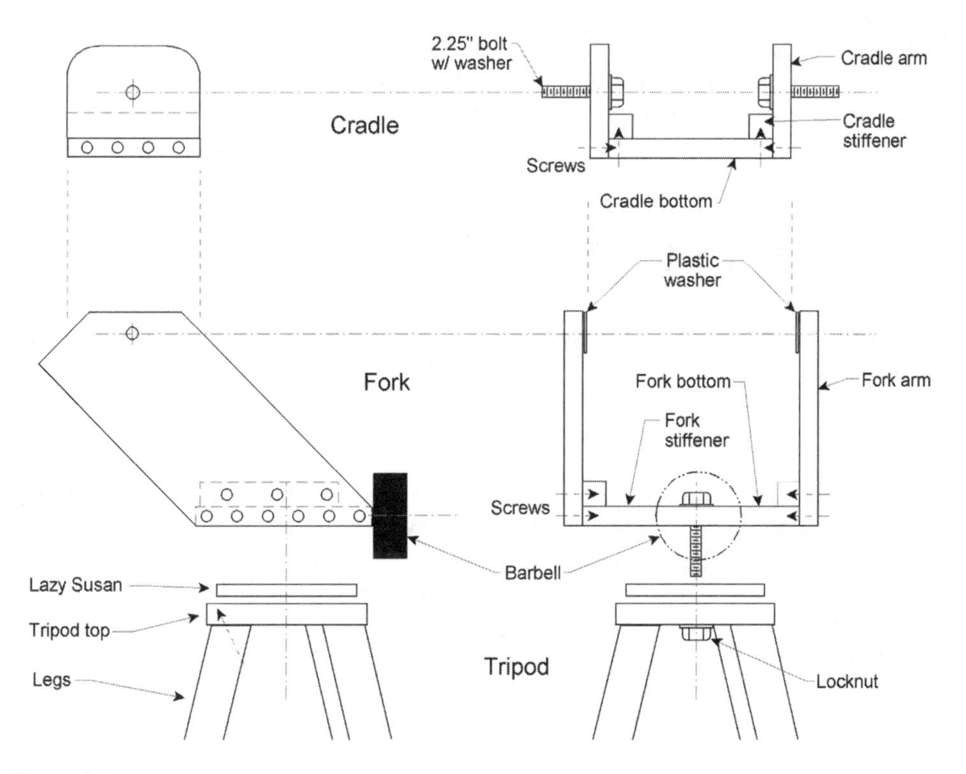

Figure 8.8 *Assembly plans for the Loyer mount and tripod.*

and screw the stiffeners into place using 1.5-inch wood screws, using three per surface. Similarly, assemble the parts of the telescope cradle.

The cradle is held to the fork arms using two 0.5-inch × 2.25-inch-long bolts. Make sure that the bolt heads are on the inside of the telescope cradle, letting the bolts protrude out through the fork arms. Place a 3-inch plastic washer between each fork arm and cradle arm. These act as friction pads, smoothing the mount's altitude motion. Your local hardware store may not have these washers, but your local supermarket does! Loyer happened to choose soft margarine containers, but gallon milk or juice containers will also do the job.

With the fork and cradle assemblies now together, it's time to make the azimuth axis and put it all together. Loyer recommends a 6-inch-diameter chair bearing or lazy Susan ball-bearing ring, available at most hardware stores. Screw the bearing onto the tripod top plate, center the cradle assembly over the hole in the tripod top, and pass through the third 2.25-inch bolt. Don't forget the washers on either side to protect the wood. After assembly, adjust the friction in azimuth by loosening or tightening the 0.5-inch locknut that holds the cradle to the tripod. Finally, attach the barbell (or other suitable counterweight) to the front of the cradle bottom, using long wood screws. (Alternatively, threaded inserts can be set into the cradle bottom, and long machine screws can be used to hold the weight in place.)

The final step is to attach the carrying handle to the outside of one of the legs. Its exact location depends on your telescope. Attach the telescope to the fork bottom, and pick up the entire assembly. Where is the center of gravity? Wherever that is (probably close to the top of the tripod), that's where the handle should go. This way, everything will be balanced as you carry it out for a night under the stars.

A Pair of Binocular Mounts

Binoculars play a big role in my astronomical enjoyment, but one of their most frustrating aspects is trying to support them by hand. Not only do they shake, it's also annoying that I have to put them down every time I look back at my star chart, only to have to retrace my celestial footsteps to get back to where I was in the first place. And even those impressive image-stabilized pairs get heavy after a while.

To help ease those inevitable aches and pains, many binocularists support their glasses on camera tripods. Not all camera tripods are suitable for supporting binoculars, however; indeed, most are not. The flimsy legs and weak heads of less expensive tripods are little better than just hand-holding the binoculars, especially when used with heavier giant models. Several companies make binocular mounts, as already detailed in chapter 7, but there is no reason why you can't make your own. Kurt Maurer of League City, Texas, did just that, basing his design (Figure 8.9) on plans published in the Summer 1999 issue of *Amateur Astronomy* magazine. If you look back at the previous project in this chapter, the alt-azimuth mount for small refractors, you will see

Figure 8.9 *Kurt Maurer's binocular mount. Photo courtesy of Kurt Maurer.*

an immediate resemblance. If it works so well with telescopes, why shouldn't an adaptation do equally well with binoculars?

Maurer's design is made from 0.5-inch-thick Baltic birch plywood. He writes that "the only fancy ingredient is the 0.25-inch aluminum bottom plate." You can substitute another piece of wood and a common 0.25-20 tee nut, if you prefer. But before we look at the details, let's list the parts needed for the mount (Table 8.4).

Table 8.4 ***Binocular Mount Parts List***

Quantity	Description
1	Plywood (Baltic birch preferred), 13" × 36", 0.5" thick
1	Aluminum plate, 6" × 3", 0.25" thick
1	Carriage bolt, 0.25-20 thread × 1.5" long
1	PVC pipe, 2.5" diameter × 4" long (minimum)
1	Formica sheet (or equivalent), 7" in diameter
3	Teflon pads, 1" square
1	Thumbscrew, 0.25-20 × 1" long
3	Fender washers, 0.25" inside diameter
1	C-clip, 0.25" inside diameter
—	Drywall screws
—	Wood glue

Note that once again you may have to vary the dimensions to suit your particular binoculars. Maurer's design (Figure 8.10) is sized for his 20 × 80 binoculars, which is probably a good recommendation even if you only own a pair of 7 × 35 binoculars at the present time. You never know, but if you get hooked on binocular astronomy, you just might move up to a larger pair some day. If so, your mount will be there at the ready.

Lay out all of the parts for the mount on a 4-foot-square piece of 0.5-inch plywood. While any plywood will work, don't cut corners. I always recommend that readers choose high-quality wood, such as Baltic birch in this case, for a more refined finished product. Once everything is laid out (check all dimensions twice), cut out each piece with a jigsaw or a band saw and assemble as shown in the drawing. A router will come in handy when creating the semicircular cutouts that will eventually support the altitude bearings.

Assemble the binocular cradle as shown in the drawing. You may use aluminum angle or wood blocks for corner reinforcing, but Maurer has found that glue-and-screw joints are sufficiently strong for his 20 × 80 binoculars. For the two side altitude bearings, trim a pair of 2-inch × 2.5-inch-diameter PVC bushings to about 0.5 inches wide. Be sure to place them at or very near the centerline of the binoculars to avoid balance problems (at the risk of repeating myself, measure your binoculars before cutting *anything* and adjust the cradle dimensions accordingly). When everything is aligned correctly, drill

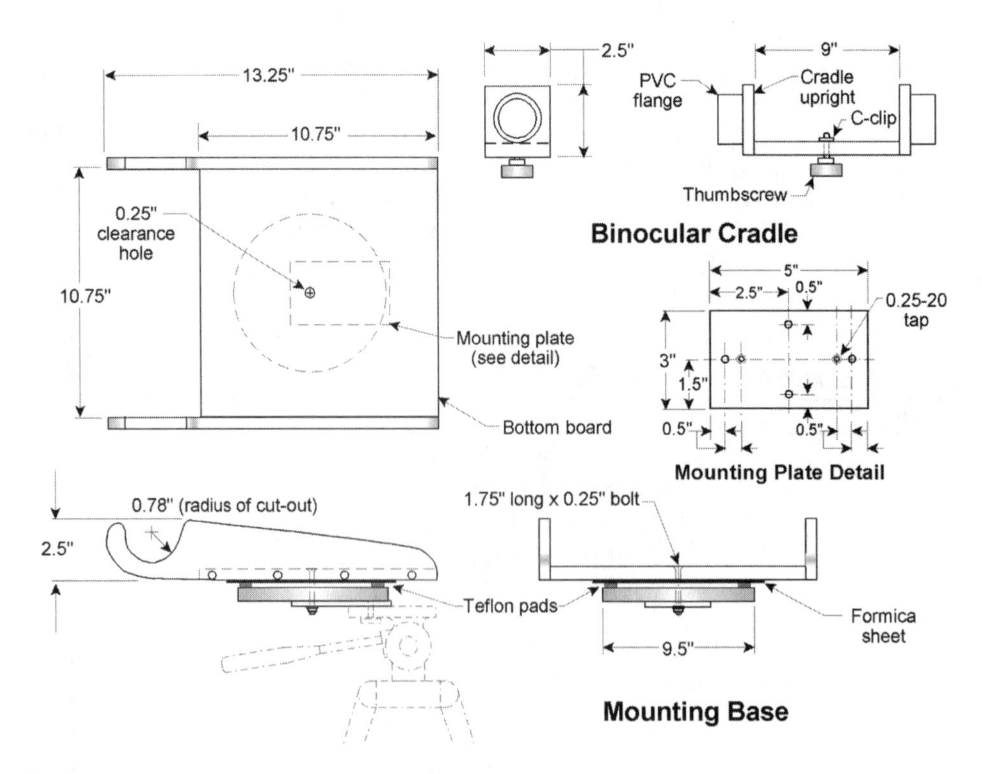

Figure 8.10 *Plans for the Maurer binocular mount.*

a 0.25-inch-diameter hole centrally in the base of the cradle. This will be used for a thumbscrew that will hold the binoculars' L-shaped tripod adapter, which can be purchased from nearly any astronomical or photography retailer. Use a 0.25-inch C-clip to hold the thumbscrew captive. Be sure to drill a recess with an oversize bit for the clip to nest in; this will eliminate teetering. And should your binoculars twist around on their mounting bolt on the cradle as you move in azimuth, Maurer recommends attaching a wood guide along either side of the L-bracket.

Just as the binocular cradle should be wide enough to accommodate your binoculars, the mounting base should be just wide enough to accept the cradle. Cut the square bottom board and two sides approximately as shown, again changing dimensions to suit your needs. Glue a sheet of Formica or a similar countertop material to the underside of the bottom board to act as the bearing surface. As many amateur telescope makers are aware, the best material to use has a rough, pebbly surface. Maurer also suggests that an old 45-rpm record (remember those?) makes a good bearing surface. Drill a 0.25-inch clearance hole through the center of the bottom board, then glue and screw the three pieces together using wood glue and drywall screws, as shown.

Drill a second 0.25-inch clearance hole through the base's circular ground board. Attach three Teflon pads, each about 1 inch square, spaced equally toward the board's outer edge. If you can't find Teflon pads, nylon furniture skid pads, often used under chair and sofa legs, will suffice.

For the mounting plate, which attaches between the base and circular ground board, Maurer recommends a piece of 0.25-inch aluminum plate, large enough so that a portion of it will sit flatly on the tripod head for maximum stability. Drill and tap two 0.25-20 holes approximately as shown in the drawing. One is for the bolt that will tie the aluminum plate to the mounting base, while the other will be used to screw the entire mount to your tripod. Be sure to allow enough room in between the two so that they clear the tripod head. If, after everything is set up, you find that the binocular mount loosens from the tripod, you may need to go back and drill a second hole to pin both together.

To assemble everything, pass a 1.5-inch-long, 0.25-20 bolt through the center of the mounting base and the circular ground board, then thread it into the aluminum mounting plate. Use fender washers on either side of the bolt to protect the surfaces. After adjusting for correct tension, secure it with a locknut.

Glen Warchol of Salt Lake City, Utah, has taken a slightly different approach, fashioning a sidesaddle flexible-parallelogram binocular mount out of a pair of old crutches (Figure 8.11). What a great idea! Table 8.5 has the simple parts list.

Certainly the most unusual aspect of the Warchol mount has to be the crutches. If you prefer the conventional approach, a pair can be purchased from a pharmaceutical supply store for between $40 and $160. If that's a bit expensive for your tastes, scavenge from friends and family for a pair that are (hopefully) no longer needed. Other possible sources include yard sales, especially in the spring right after the ski season, and your local Goodwill store. Warchol proudly says that he purchased his pair for only $1.

Figure 8.11 *Glen Warchol's binocular crutch mount. Photo by Glen Warchol.*

Table 8.5 **Binocular Crutch Mount Parts List**

Quantity	Description
1	Binocular L-bracket
2	Aluminum crutches (longest available)
1	Plywood, 9" × 16" × 0.5" thick
1	2 × 4 wood, 12" long
1	Pine, 1" × 2" × 36" long
1	Threaded rod, 28" long × 0.25-20 thread
16	Nylon washers, 0.25" inside diameter
1	Nylon or Teflon washer, 1" outside diameter with a 0.25" centered hole
1	Fender washer, 0.25" inside diameter
1	Carriage bolt, 0.25-20 thread × 5" long
9	Wing nuts, 0.25-20 thread
1	Hardwood dowel, 4" long by 0.5" diameter
1	Hardwood dowel, 4" long, diameter depends on the inside diameter of the crutches
1–2	Barbell weight, 2.5 pounds
2	Cotter pins
—	Wood screws
—	Wood glue

Once you find the crutches, measure their inside span. Use this dimension when laying out and cutting the four pieces of plywood for the two wooden boxes (the design in Figure 8.12 is based on a 4-inch width). Cut the length of the 1 × 2 pine stud into four 9-inch-long pieces. Assemble each 4-inch-wide × 9-inch-tall × 3-inch-deep box with wood screws, four per edge, but do *not* glue the boxes together just yet.

Once the boxes have been sized and assembled together correctly, take each apart, remembering which pieces go with which. Drill a 0.25-inch hole through each 1 × 2 pine board, about 7.5 inches apart. Be sure to space them as uniformly as possible.

Decide which of the two boxes will mount at the top of the crutches. Drill a 0.25-inch hole through the center of the box's 4-inch side. Run a 5-inch carriage bolt through the hole so that it sticks out the far side of the box. Slide on a nylon or Teflon washer that measures about 1 inch in diameter. If you have trouble finding the material, a couple of washers made from 1-gallon plastic milk bottles should work adequately.

Here the plans deviate a bit from the original design in Figure 8.11. Rather than use PVC piping as illustrated, which works fine as well, I suggest using a foot-long piece of 2 × 4 wood to support the binoculars. Drill a 0.25-inch hole through the 2 × 4 about 4.5 inches up from one end and slide it over the 5-inch carriage bolt so that it matches Detail A in Figure 8.12. Tighten everything

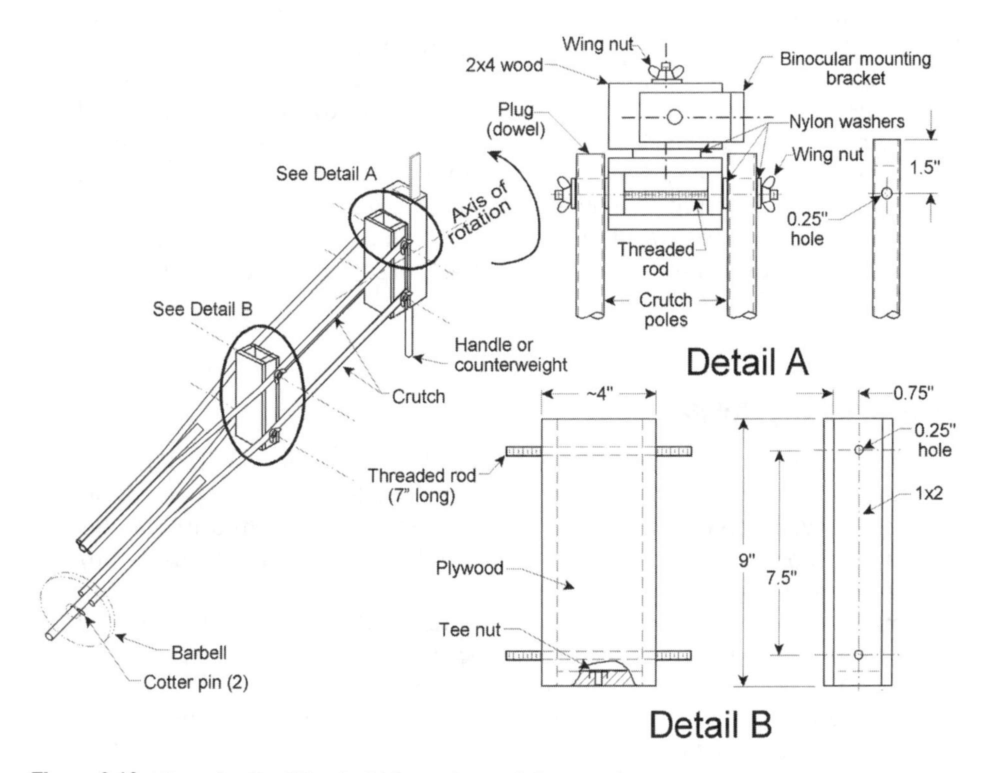

Figure 8.12 *Plans for the Warchol binocular crutch mount.*

together with a nylon washer, a fender washer, and a wing nut. Attach the 90° binocular L-bracket to the end of the 2 × 4 with a suitable wood screw. Drill a 0.5-inch hole in the other end of the 2 × 4 for the 0.5-inch hardwood dowel, which can simply act as a handle or as a counterweight shaft, as illustrated in Figure 8.11. Glue the dowel in place with a good-quality wood glue.

The other small, wooden box will hold the mount to your tripod. Because the mount must support the cantilevered binoculars steadily, it is best to use a sturdy tripod, such as those described in chapter 7. Once again, however, I have seen some simple homemade tripods that are the equal of commercial units. Some of the best, ironically, use crutches for legs! Back to the yard sales.

Cut four 0.5-inch-long slices from the wooden dowel. Plug them into the top ends of the crutches as shown; they will prevent the crutches from crushing as you drill 0.25-inch holes through the outside edges of each, about 1.5 inches from their ends. Drill four more 0.25-inch holes about 4 inches up from where each crutch begins to curve toward the bottom.

Now cut the threaded rod into four 7-inch-long pieces. Line up the top box with the holes at the top of one crutch and slide one of the rods through. Be sure to place two nylon washers on each side, one between the box and the crutch and another between the crutch and flat washer, to smooth motions in altitude. Hold everything together with two wing nuts, one on each side. Repeat the same process for the bottom crutch, then do it again for the lower box, which will attach to the tripod.

Cut the foot extension off the top crutch but leave the bottom crutch as is. The extra length will be used to hold the counterweight.

At long last, put it all together, including the binoculars, and see how everything balances. Slide the weight back and forth until you find a good balance point and make a mark on the pipe. Drill a hole on either side of the counterweight, then insert and bend the two cotter pins to hold the weight in place to complete the assembly. Use the tripod to move left and right in azimuth and pivot the binoculars in altitude by grabbing the handle. Adjust all of the wing nuts until the mount moves smoothly in all directions. You're now ready to tour the universe on crutches!

Three-Filter Slide

If you have ever used either color or light-pollution reduction (LPR) eyepiece filters, then you know what a hassle they can be to change, especially when wearing gloves in cold weather. Ed Stewart, who was featured in *Star Ware*'s second edition for the telescope sled that he uses to move his Schmidt-Cassegrain telescope, offers a different sort of transportation system in this edition. He has created a simple-to-make filter holder that will slide several filters directly under the eyepiece of a reflecting telescope. Stewart's filter slide (Figure 8.13) makes changing filters a thing of the past—and it makes comparing filters a snap. Best of all, unlike some designs that require extensive metalworking, Stewart's design is made from strips of plywood.

Figure 8.13 *Ed Stewart's filter slide. Note that the slide rails have been highlighted with white tape to show up better in the picture; normally, they are painted flat black to dampen stray light. Photo courtesy of Ed Stewart.*

The first step is to draw an arc of your telescope tube's inside diameter on a piece of paper. The full diameter is not needed; instead, draw an arc representing the area directly below and adjacent to the focuser. Locate the exact center of the arc, which will mark the center of your focuser's drawtube. Extend a 3-inch line (more or less, depending on your telescope's size) perpendicular to the focuser's centerline so that its ends are about 0.38 inches below the tube's arc. Figure 8.14 should clarify what your drawing ought to look like. Note that this latter dimension might have to be increased if your focuser protrudes into the tube (to prevent it from hitting the slide).

With the layout complete, let's review the list of parts needed for the project (Table 8.6).

Stewart advises that the thin plywood is available from hobby shops, while the other items should be found at most hardware stores or home-improvement centers. Also note that the slide's 12-inch dimension may need to be changed, depending on your telescope. To determine the length for your instrument, consider that you need to allow about 0.5 inches between filters and that there must be enough room at one end of the filter slide for a grasping hole. If you have 2-inch filters or want the slide to hold more than three filters, increase the length accordingly. Stewart notes that his telescope "is an open tube so that the slide can be reached from either side of the secondary cage."

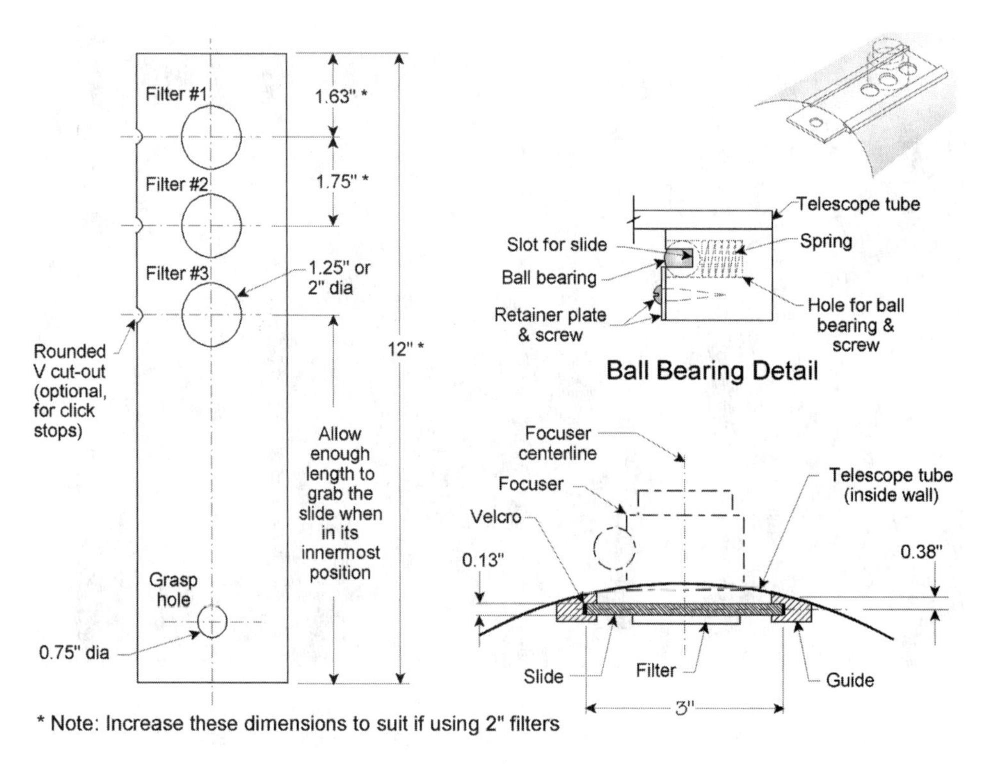

Figure 8.14 *Plans for the Stewart filter slide.*

After the slide is cut to size, position and cut the filter holes. Although a hand drill can be used for this part of the operation, Stewart recommends the steadiness of a drill press. He also reminds us that an adjustable-beam circle cutter is needed, because the hole diameters must be cut precisely. Why? It's not enough to cut 1.25-inch holes for 1.25-inch filters, or 2-inch holes for 2-inch filters. Rather, each hole must be just slightly *smaller* so that its filter can be forcibly threaded into the plywood, which is so hard that threads can actually be cut into it. Be sure to use a scrap piece of wood first to find the *exact* diameter that works. Then, and only then, cut the holes into the slide itself. Also drill a finger grasp hole, perhaps 0.75 inches in diameter, at the end of the slide, as shown in the drawing. When all the holes are cut, use masking tape to cover about 0.25 inches of the slide's edge (which will fit into the two guides)

Table 8.6 **Stewart Filter Slide Parts List**

Quantity	Description
1	Plywood, 3" wide × 12" long, 0.09" (3/32") or 3 mm thick
2	Pine strips, 0.75" wide × 0.5" tall × 12" long
2	Velcro strips (fuzzy side only), 0.25" wide × 12" long
1	Flat-black spray paint

as well as the inside edges of the filter holes, then spray-paint both sides of the slide with flat black paint.

The two filter guides can be cut from ordinary pine or any other wood that happens to be handy. You might even use a 1-foot length of a 0.75-inch square dowel, then sand one corner to match the curve of your telescope tube. (Note to owners of large Dobsonians: The slide can often be mounted to the underside of the flat focuser board so that shaping the rails to conform to the inside radius of the tube is unnecessary. When this is the case, the entire slide assembly may be fabricated independently of the telescope by connecting the two rails with two aluminum mounting plates. The assembly can then be mounted using stacked washers to keep the two from hitting when the focuser is turned in all the way.) Paint both guides with flat black paint and set them aside to dry.

Once the paint is dry, cut a 0.13-inch groove into each guide, as shown. Here, a table saw is almost a must. Stewart notes that this happens to be the thickness of his table-saw blade, a handy coincidence. Be sure not to cut the groove too deeply, or the track will split right in half. And watch those fingers!

It's decision time. At this point, you might want to add click stops for each filter position to make it easier to center each filter exactly under the focuser. This modification comes from Kurt Maurer, who also constructed one of the binocular mounts described earlier in this chapter. All that is required is a small ball bearing, a small spring, a thin piece of aluminum (cut from a soda can, for instance), two small wood screws, and a drill. See "Ball Bearing Detail" in the drawing.

Begin by adding rounded V-shaped notches along one side of the filter slide, as shown in the drawing. These must be *exactly* in line with each filter station. Drill a hole a little larger than the ball bearing in one of the slide's slots, *precisely* in line with the focuser. Now place the tiny spring in the hole, followed by the ball bearing. Cut an aluminum strip that is just wide enough to cover part of the hole and trap the bearing, letting a bit of it protrude. Use two small wood screws to attach the aluminum to the guide. When assembled, the spring will constantly push the ball bearing against the filter slide. As the slide is moved along the guides, the ball will engage the notches; remember, for this to work, the alignment must be precise.

Once the grooves are cut in the guides, line each with a strip of Velcro (the fuzzy side only), which will prevent the slide from sliding too much, especially when the telescope is aimed toward the zenith. (If you added the Maurer click stops, the Velcro might not be necessary. To play it safe, try it first before adding the Velcro.) Lay everything on a flat surface and test the slide's movement, striking a balance between smoothness and friction. Drill slightly oversize pilot holes through the two guides so that they can be adjusted later, and then screw them to the inside of the telescope tube. Try to position the guides so that they are as parallel as possible, and be very careful not to split the wood.

Place the slide into the guides and move it back and forth while looking through the focuser. Do the filter holes align correctly? If not, loosen the guides' mounting screws and adjust as necessary. When the guides and slide line up with the focuser, remove the slide and screw each of your filters into the mounting holes. Be sure not to touch the glass itself or strip the filters' fine threads. It might be best to cover the filters with optical cleaning tissue before manhandling them. If the holes prove to be a little too large, hold each filter in place with three equally spaced dabs of a silicone adhesive (not a bad idea even if they do fit, just as a safeguard). Finally, if the slide doesn't move smoothly, try waxing its edges with a little candle wax or bar soap.

Accessories like the Stewart filter slide can sell for more than $100, but for less than 1/10 the price, you can build one in a day and take it for a spin that night.

Planet Blocker

There is no denying that astrophotography is one of the most enticing aspects of amateur astronomy. Sure it's easy to pick up a book or magazine or visit any of a number of web sites that have some outstanding celestial photos, but there is something special about actually taking an outstanding astrophoto *yourself!* Many of us dabble with the sport, first trying our luck with the Moon, then perhaps graduating to the planets, but rarely do the photos convey what the eye sees. That's because the human eye has an amazing dynamic range. We can see objects that are both bright and dim simultaneously, such as the planet Jupiter and its four Galilean moons. Photographic film is far more limited, requiring an exposure that is going to show one or the other but not both. Darkroom techniques, such as dodging, can be used to bring out latent details, but this usually requires a good deal of experience. More recently, computer processing has become popular in certain circles; again, this requires some advanced equipment.

Chris Flynn from Fort Smith, Arkansas, attacked this problem in a different way. Instead of waiting until after a photograph has been taken, he uses a simple *blocker* (Figure 8.15) to alternately cover and uncover bright objects while also recording dim objects on the same frame of film at the same time.

Figure 8.15 *Chris Flynn's blocker. Photo courtesy of Chris Flynn.*

Table 8.7 ***Planet Blocker Parts List***

Quantity	Description
1	Wire coat hanger
1	Thin piece of aluminum
1	Can of spray paint (flat black)
—	Tape (duct or masking) *or*
—	Epoxy
1	Eyepiece projection extension tube (see chapter 7)

Flynn's blocker couldn't be simpler to make, as you can see from the parts list in Table 8.7.

Using a wire cutter, cut a 3-inch straight length from the coat hanger. Using tin snips, cut a piece of thin aluminum, such as from a soda can, that measures approximately 0.75 inch long × 0.19 inch wide. Center the wire over the aluminum as shown in Figure 8.16. Using either duct tape, masking tape, or if you want to get a little fancier, epoxy, fasten the two together. Paint the pair flat black and set them aside to dry.

While the blocker is drying, drill a 0.25-inch hole into the eyepiece projection extension tube about 4.5 inches in from where it sticks into the focuser. This dimension may vary depending on the length of the tube, so use this as a

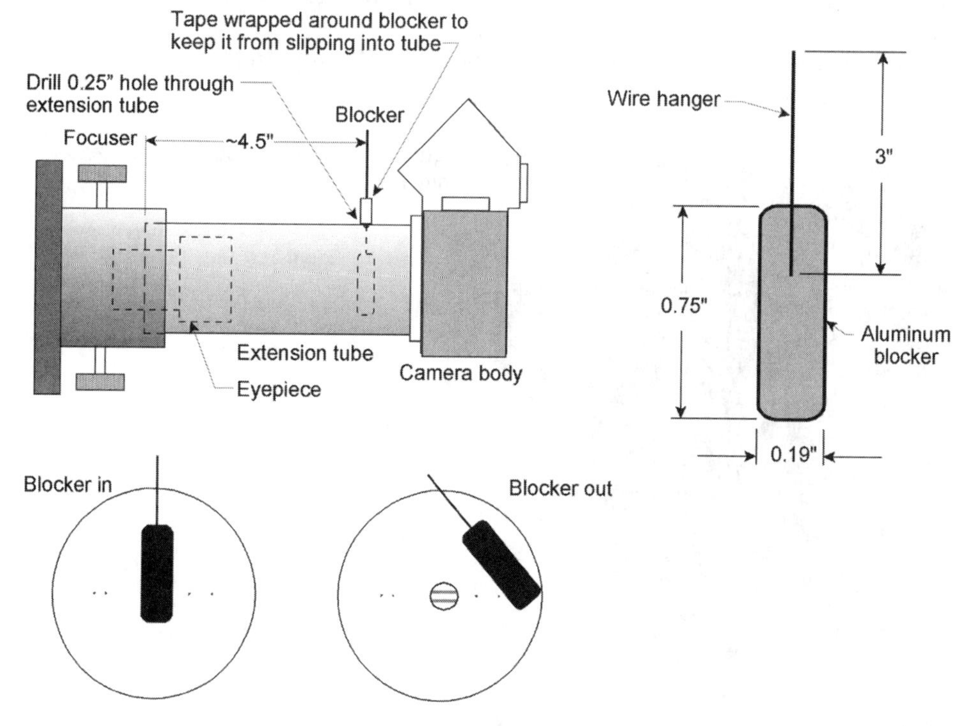

Figure 8.16 *Assembly drawing for the Flynn blocker.*

suggestion only. The idea is that you want the hole close to the camera body, but not so close that it interferes.

When the paint is dry, simply stick the blocker through the hole in the extension tube and wrap a piece of tape around the wire to keep it from falling out, as shown in the drawing. That's all there is to it!

To use the blocker on, say, Jupiter or the crescent Moon, Flynn offers the following suggestions: "All you do is position the objects as you want them to appear on the film. Using one hand, line up the blocker so that it covers the bright object or section. Then, using the other hand, carefully open the camera shutter using a cable release. I prefer an air-bulb release, as it is less prone to vibration. Wait a few seconds so that the dim object(s) records, then swing the blocker to the side and stop the exposure. There's a bit of timing here, but it becomes easier to do with a little practice."

Ezekiel Unity Finder

Certainly one of the most oft-recommended accessories for any telescope has to be a 1×, or unity, finder. You're probably already familiar with some of the commercially sold versions that are available, such as the Telrad, the Rigel QuikFinder, the Orion EZPoint, and others. But here is a design from Randall McClelland that offers all of the benefits of the others and adds one or two more. The Ezekiel unity finder, seen in Figure 8.17, features a rotatable reticle pattern that takes the guesswork out of lining up stars and estimating distances between reference points in the sky.

Table 8.8 lists the parts needed for the Ezekiel unity finder.

Figure 8.17 *Randall McClelland's Ezekiel unity finder.*

Table 8.8 ***Ezekiel Unity Finder Parts List***

Quantity	Description
1	Plywood, 6" × 12" × .09" thick, cut as follows:
2 pieces	2.25" × 6" (left and right sides)
2 pieces	2" × 4" (front and back)
2 pieces	2" × 2" (lens mount and baffle, each with 1.38" holes centered)
1 piece	2" × 2" (bottom, with centered hole to match diameter of medicine bottle)
1 piece	2" × 2.25" (top)
1 piece	2" × 0.25" (cross member to hold front screw)
1	Window, 2" × 2.5", glass or plastic
1	Fresnel lens card, 3× magnification, Edmund Scientific part # T30384-56
1	Reticle (copied 1:1 from Figure 8.18)
1	Medicine bottle, plastic, 1.5" diameter, with two caps (a 35-mm film canister can also be used)
1	Tee nut, 0.25-20 thread
1	Red light-emitting diode (Radio Shack part # 276-307)
1	Battery holder (Radio Shack part # 270-430)
1	Battery, 3-volt wafer type (CR 2025 or similar)
1	Variable resistor, 10k ohm (Radio Shack part # 271-282)
1	On-off switch, miniature (Radio Shack part # 275-409A)
—	Wire
1	Wood screw, 0.38" long

The plywood is best obtained from a hobby shop, the 3× Fresnel lens magnifier from Edmund Scientific, the window from a piece of scrap (often available from hardware stores), and the electrical components from Radio Shack or a similar outlet. The reticle can simply be a photocopy of the reticle in Figure 8.18, which also shows the assembly plans.

Cut the plywood to the recommended dimensions shown, although the only dimension that is actually critical is the 3.62-inch spacing between the reticle and the Fresnel lens. This is because of one of the unique features of the Ezekiel finder. Not only does it act as a unity finder, it also serves as a rotatable sky protractor for scaling distances in the sky. The tick marks along the outer circumference of the reticle pattern are marked in 10° intervals. These are useful for measuring angles and directions. The outer circle measures 24° across, while the central circle measures 2° in diameter. The neat part is that if you enlarge the pattern accordingly onto a piece of acetate (transparency material), you can make a to-scale overlay for any popular star atlas. For instance, to create an overlay to the scale of the *Sky Atlas 2000.0,* the outer circle should measure 6.75 inches.

Drill a 1.38-inch hole centrally through two of the 2-inch × 2-inch squares. Next, carefully measure the exact outside diameter of the medicine bottle and drill an equal diameter hole through the third 2-inch × 2-inch square. This third hole must be measured and drilled with care, as it must hold the bottle without play but still allow it to be turned easily. McClelland suggests using a

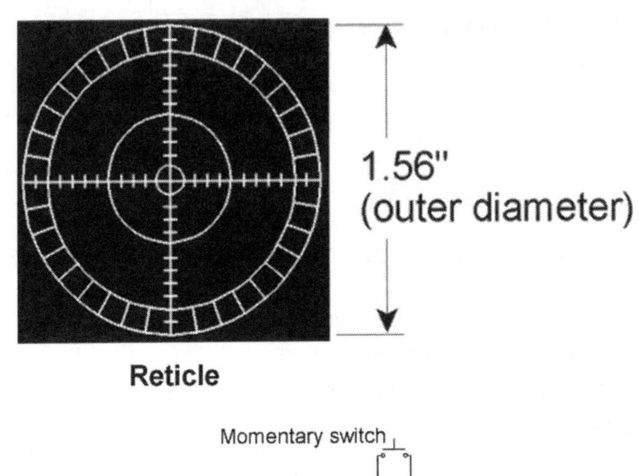

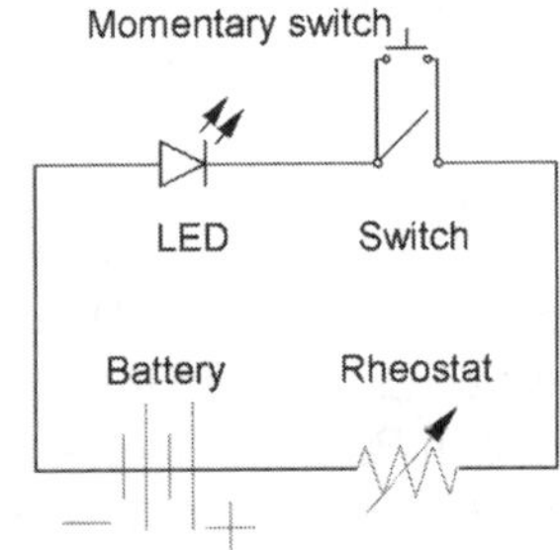

Figure 8.18 *Plans for the McClelland Ezekiel unity finder.*

Dremel tool for the cut. Next, cut the Fresnel magnifier to size and glue it to one of the diaphragm boards, keeping the smooth side up (that is, facing toward where the glass window will go).

Attach and glue one of the medicine bottle caps to the bottle. Stick the bottle and cap through the hole of the bottom square and cut the bottle so that about 0.25 inch of it protrudes.

To make the reticle, first photocopy the pattern shown in the figure. Carefully cut it out so that its diameter matches that of the other medicine bottle cap and then glue the two together. Finding the exact center of the bottle cap might be tricky, unless the manufacturer left a central mark. To position the reticle, stick a straight pin through its center and then through the center of the cap. Apply some glue to the cap and then smoothly press the reticle onto it evenly, leaving the pin in place until the glue has dried. Finally, glue the second bottle cap to the bottle and cap in the bottom square.

To make the LED light, solder together the components, as shown. Note that although it is possible to use standard AA or AAA batteries as the power source, McClelland prefers a flat lithium battery, such as those used in watches. Allow enough wire between the battery, the resistor, and the LED so that the LED can hang fairly close to the reticle. Then, once the wiring is complete, mount the battery holder, the variable resistor, and the on-off switch to the front panel. Loop the wire to the LED through a slot in the front corner of the baffle so that the panel can be removed, complete with wires and LED, if needed. When everything is assembled, turn the switch on to check that the LED lights correctly.

Assemble the box using wood glue and tiny nails, tapping each in place gently so as not to crack the wood. Put away the side carrying the switch and battery for the moment. Glue the glass window in place, checking that it makes about a 45° angle with the box, as shown. Finally, attach the panel with the LED, but rather than gluing and nailing it in place, use small wood screws so that it can be removed when needed.

To mount the Ezekiel finder to his telescope, McClelland attached a block with an embedded 0.25-20 tee nut, as shown. You may prefer to use a different approach, but whatever method you choose, make sure that the finder can be tilted left to right and up or down so that it may be aligned with the telescope.

Adjustable Observing Chair

Anyone who has ever spent more than, perhaps, half an hour at the eyepiece of a telescope knows that it can be a painful experience. The constant bending over to look through the eyepiece can put a tremendous strain on the neck, the back, and the legs. To help alleviate some of the discomfort, many amateur astronomers use commercially made observing chairs or stools. James Crombie of Prince Edward Island, Canada, decided to design and build his own observing chair, shown in Figure 8.19. The benefits were twofold. Not only did this let him customize the dimensions to his own needs, he also saved some money.

To make it easy to store, transport, and set up, the chair is designed to fold flat. Simply open the chair as you would a small ladder, set it down on the

Figure 8.19 *James Crombie's observing chair. Photo courtesy of James Crombie.*

ground, and it's ready for use. To change the height of the seat, lift the front of the seat board, slide it up or down as desired, and gently release it. What could be simpler?

Table 8.9 is an itemized list of materials for the Crombie chair.

Cut the 1 × 2 hardwood lumber to the appropriate lengths, listed above. Round off the tops of the two leg sides and bevel one side of each seat crosspiece, as shown in Figure 8.20. Drill 0.38-inch holes centrally in each of the 18 spacers as well as where shown in the three main uprights. Counterbore the 0.38-inch holes in two of the uprights as well as in the tops of the leg side pieces for the washers and nuts that will hold the chair together.

Cut the threaded rod so that you have two pieces that are each 8.875 inches long and one piece with an 11-inch length. Spread glue on all mating faces of the uprights and spacers and slip the rods through each, as well as through the two rear leg sides. Slip a flat washer between each rear leg side and the uprights. Assemble the pieces as shown and clamp everything together using the threaded rods, washers, and nuts. Check that all pieces are aligned and square before final tightening.

After the glue has dried, measure the width of the two rear leg sides and transfer this dimension to the leg base. Center the leg base as shown and drill pilot holes for the four #10 × 2-inch wood screws that hold the base to the leg sides. Spread some wood glue on the bottom of the legs' sides and screw everything together. The chair's frame is now complete.

Make the frame for the seat from the four remaining pieces of 1 × 2 lumber. Double-check that the 11.06-inch width of the seat frame is not too large or too small, since the seat needs to be able to slide up and down on the upright frame smoothly, without binding. If the seat frame is too narrow, it won't fit on

Table 8.9 ***Crombie Observing Chair Parts List***

Quantity	Description	
1	Plywood, 14" × 10", 0.5" thick	
1	Threaded rod, 0.31-18 thread × 36" long	
4	Carriage bolts, 0.31-18 × 2.5" long	
12	Flat washers, 0.31"	
10	Nuts, 0.31-18	
4	Eyehooks	
4	Wood screws, #8 × 2"	
4	Wood screws, #6 × 0.75"	
1	Plywood, 14" × 10", 0.25" thick	
1	Foam, 14" × 10", 2" thick	
1	Vinyl or other suitable material for seat cover, 18" × 14"	
1	Bungee cord	
1	Chain, cable, or rope; length to suit	
24 linear feet	1 × 2 hardwood (maple, oak, etc.), cut as follows:	
Main upright	3 pieces	36" long each
Spacer	18 pieces	2" long each
Leg side	2 pieces	18" long each
Leg base	1 piece	16" long
Seat side	2 pieces	14" long each
Seat crosspiece	2 pieces	11.06" long each

the frame; if it's too wide, it will twist when moved. Assemble the seat frame using the four 0.31 × 2.5-inch carriage bolts, drilling and counterboring as shown. Now cut the seat itself from the 10-inch × 14-inch piece of 0.5-inch plywood. Round and smooth the corners for a more refined look as well as to prevent splintering. Position the plywood on the seat frame and then glue and nail it into place.

Apply cork or rubber to the beveled edge of the two crosspieces. The cork provides enough friction to keep the seat from slipping when sat upon.

With everything now together, slide the seat over the upright frame. Spread the frame until the seat is level (approximately 30°). Measure the distance between the rear legs and the back of the upright frame and cut a piece of chain, cable, or rope to this length. Connect the chain between the legs and the frame as shown, using two eyehooks. After you have determined that everything assembles as it should, temporarily remove the eyehooks and chain.

Use a block plane, sanding block, or router, if available, to round all of the chair's edges. Sand all surfaces smooth. Apply several coats of exterior varnish, allowing each to dry thoroughly. Be careful not to coat the rubber on the seat crosspieces. After the last coat has dried, reinstall the eyehooks and chain.

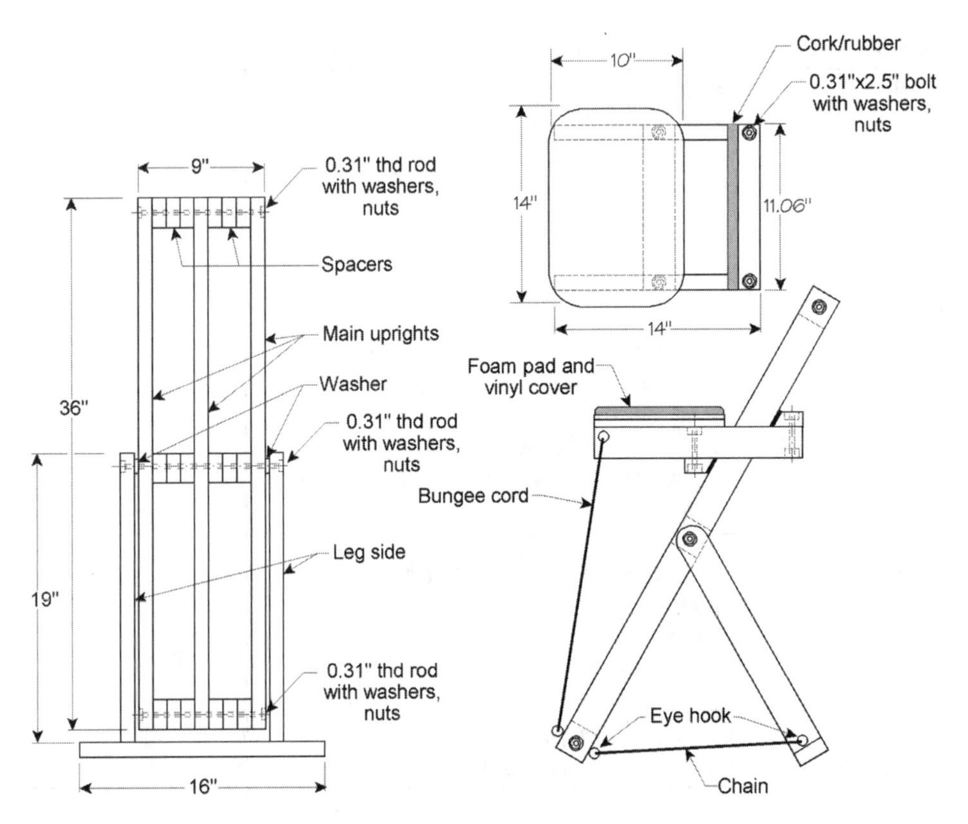

Figure 8.20 *Assembly drawing for the Crombie observing chair. (In the drawing,* thd *is the abbreviation for "threaded.")*

Finally, make the padded seat using the foam, vinyl, and 10-inch × 14-inch piece of 0.25-inch plywood. Lay the foam on the plywood, cover it with the vinyl, and then use a heavy-duty stapler to fix the vinyl to the bottom of the plywood. Attach the seat to the 0.5-inch plywood using four #6 wood screws drilled through the underside. Crombie also suggests stretching an elastic bungee cord between the bottom of the seat and the leg to keep the seat from slipping when you get up.

Whether you choose to buy one or make your own, an observing chair is one of the most important accessories for the amateur astronomer. Being seated in a relaxed position is certain to make your time at the eyepiece more productive and much more enjoyable.

Mini Observatory

I have to admit that I am guilty of missing many clear nights of observing simply because I don't have the ambition to drag my telescope outside after a long day at work. You, too? Aren't those the nights that you wished you had everything in an observatory? Rather than having to trudge it all outside and then back in again afterward, it's all right there, waiting for you. But if you are like me, perhaps the job of building a full-fledged observatory is beyond your abilities as a carpenter.

Dave Trott of Englewood, Colorado, has come up with a solution that even someone like me could build. Rather than constructing a full-size building, his mini observatory (shown in Figures 8.21a and 8.21b) is just large enough to shelter his telescope from the elements. Although it offers no shelter from the wind or light pollution, it does give the advantage of having the telescope ready to go in a matter of seconds. Just flip the top open; there's no assembly and no cool-down time. Best of all, the simple design is made of 0.25-inch plywood and some common hardware, and it requires only an electric jigsaw and basic hand tools to put it all together.

Take the parts list in Table 8.10 with you when you go shopping.

First, pick out the best location in your yard for the observatory, compromising between horizons, blockage against local light pollution, and any other factors that may hamper its usability. Trott's original design, presented here in Figure 8.22, requires that a permanent pier be sunk into the ground for the telescope, but you may prefer to adapt it to your particular situation.

In any case, you'll need to sink some sort of footing into the ground to hold the observatory in place. Trott recommends setting a 6-inch steel pier in concrete at the bottom of a hole about 3 feet deep and surrounded by sand (for vibration damping and to prevent heaving in cold weather). Once set, fill the pipe with more concrete. If your telescope will simply be set up on the ground,

Figure 8.21 *Dave Trott's mini observatory photo, (a) closed and (b) open.*

Table 8.10 ***Trott Mini Observatory Parts List***

Quantity	Description
2	Plywood sheets, 4' × 8', 0.5" thick
4	Lumber, 2" × 2" × 8' long
40–50	Corner braces, 1" × 1", steel
6	Heavy-duty door hinges
1	Hasp with padlock
4	Screened soffit vents
Approx. 8 feet	Rubber gasketing, 0.5" wide
2	Screen-door handles
—	Drywall screws, 0.5" long
—	Drywall screws, 1.5" long
—	Nails, 16d
150	Carriage bolts, 0.25" × 1" long, with washers and nuts
—	Rolled roofing
1	Tube of roofing adhesive
1	Steel pipe, 6" diameter, length to suit
—	Steel plate, 0.5" thick, cut to suit
2+/–	Bags of concrete
2+/–	Bags of sand
—	Miscellaneous hardware

it is probably best to place a 2- or 3-mil-thick plastic sheet underneath to act as a moisture barrier.

As can be seen from the illustrations, the mini observatory is actually two separate sections hinged together. The bottom section is simply a rectangular box built from 2-inch × 2-inch studs and 0.5-inch-thick plywood, screwed together with drywall screws. The box's top and bottom are intended to embrace the telescope's pier base as securely as possible. Adapt each section to match the dimensions of your telescope and pier. Make certain that the observatory's interior will be large enough for your telescope in its stowed position. There is nothing worse than completing the job only to find out that the telescope is a few inches higher or wider than the shelter! The plans here were designed for Trott's 7-inch Meade Maksutov and should be sufficient for similar size instruments, including 8-inch Schmidt-Cassegrains.

Once the bottom section is complete, slide it over the pier. If needed, secure them together with four 90° corner braces and sheet-metal screws spaced evenly around. (The top and bottom of the original observatory base gripped the pier so tightly that further attachment proved unnecessary.)

With the pier and the bottom half of the observatory in place, it's time to make the adapter plate that will connect the telescope to the pier. This can be done in any of a number of different ways, depending on the instrument. How does it currently attach to its tripod? Perhaps it uses a large, central bolt. If so, sink a similarly threaded bolt into the concrete before it sets, making certain

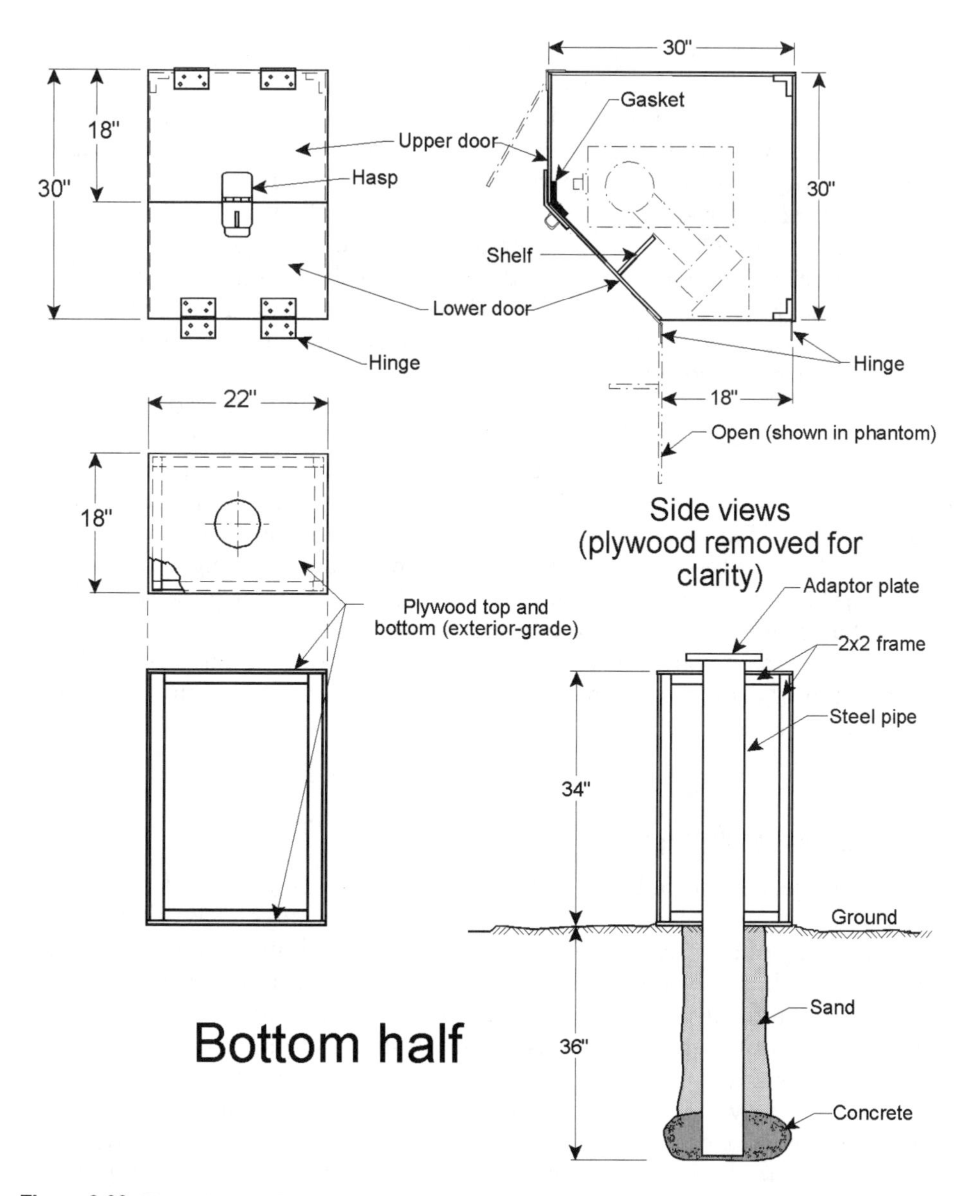

Figure 8.22 *Plans for the Trott mini observatory.*

that it is vertical. You may instead prefer to cut the steel pipe to match your latitude angle, then attach a metal plate either by welding it in place or securing it with brackets and corner braces.

The top half of the observatory is also built from 0.5-inch plywood, but instead of 2 × 2 studs, Trott used 1-inch × 1-inch corner braces spaced evenly around the inside in an effort to slow vandals from breaking through. Use 0.25-inch × 1-inch-long carriage bolts to hold everything together. Although they are not as powerful against the forces of evil, 2 × 2 studs will work just as well structurally. Just to reemphasize, be sure to check the length, width, and breadth of your telescope and mount *twice* before laying out and cutting the plywood. Use drywall screws and/or machine screws, washers, and nuts to secure everything together. Also place two exterior soffit vents in both sides of the observatory so that air can pass through the structure even when closed, thereby preventing mold and mildew. Finally, install two handles on both sides; these are used to open the observatory for use.

Use exterior-grade hardware for the two doors that open on the top half of the observatory. To keep rain out, put rubber gasketing along the inside edges of the door panels and beads of silicone caulking around all of the other inside joints and seams on both halves.

To make his observatory a little more user friendly, Trott added a small shelf that pops out when the observatory doors are opened. This is strictly optional, but if you do decide to include it, just make sure it isn't so large that it hits the telescope when everything is folded shut. You might also consider drilling some 1.25-inch or 2-inch holes for eyepieces in the shelf.

Now place the top section on the bottom and attach the hinges and latches to secure the two halves together. Once again, place rubber gasketing where the two halves meet. Finally, add a hasp and padlock.

While not as elaborate as the observatory described below, Dave Trott's mini observatory is simple to make and will certainly increase your impromptu observing sessions.

Maxi Observatory

The final project in this edition of *Star Ware* comes in the form of the unique Lookum Observatory (Figure 8.23) created by P. J. Anway of Munising, Michigan. Like many of us, Anway often found that "the last thing I want to do is haul scope, tripod, charts, eyepieces, etc. back indoors until the next observing night."

Unfortunately, a project as daunting as an observatory is impossible to describe in step-by-step instructions in a book such as this. Instead, it is hoped that the details that follow are enough to inspire some readers to plan observatories of their own based on this design. Fortunately, in this day and age of instant, global electronic communication, I can offer some additional help. By visiting the Star Ware section of my Star Ware (and more!) home page at http://www.philharrington.net, you will find a link to Anway's own web site, which offers additional photos and construction details as well as an e-mail address to which you can send specific questions.

Having built an earlier observatory, Anway found that the design should be based on two important factors: the demands of the observing site and the observer's preferences. In his case, the observing site's northern horizon is solidly covered by a tree line. Because that is also the direction of the prevailing wind during the cold winter, he decided to forgo any northerly view in

Figure 8.23 *P.J. Anway's Lookum Observatory.*

favor of a solid wall. His observatory is also built on steeply sloping land, necessitating that the door had to open into the observatory, where the footing was level. Anway's CCD imaging with his Celestron 11 Schmidt-Cassegrain telescope also required that he have enough room for a computer as well as access to power.

After investigating different observatory designs, including the popular roll-off roof and traditional dome approaches, he chose a unique system of hinging the roof in two sections. The building is oriented such that the roof opening is toward the south but is wide enough to take in the east and west as well.

Where does a project like this begin? Before the first nail can be driven, take a trip to your local town hall to check your municipality's building code. Are there any restrictions on building a garden shed on private property? Fill out the required paperwork and submit it to the proper office.

Next, draw up plans for your observatory similar to Figure 8.24. Decide just how large you want the observatory to be. The dimensions shown here are suggestions only and are by no means set in stone. If your telescope does not require that much room, then shrink the numbers as you see fit (but not too much). If more room is required, let the observatory grow as needed. Should you foresee the possibility of someday upgrading your telescope to something larger, take this into consideration when planning the building. The important thing is to make the observatory fit your needs, not the other way around.

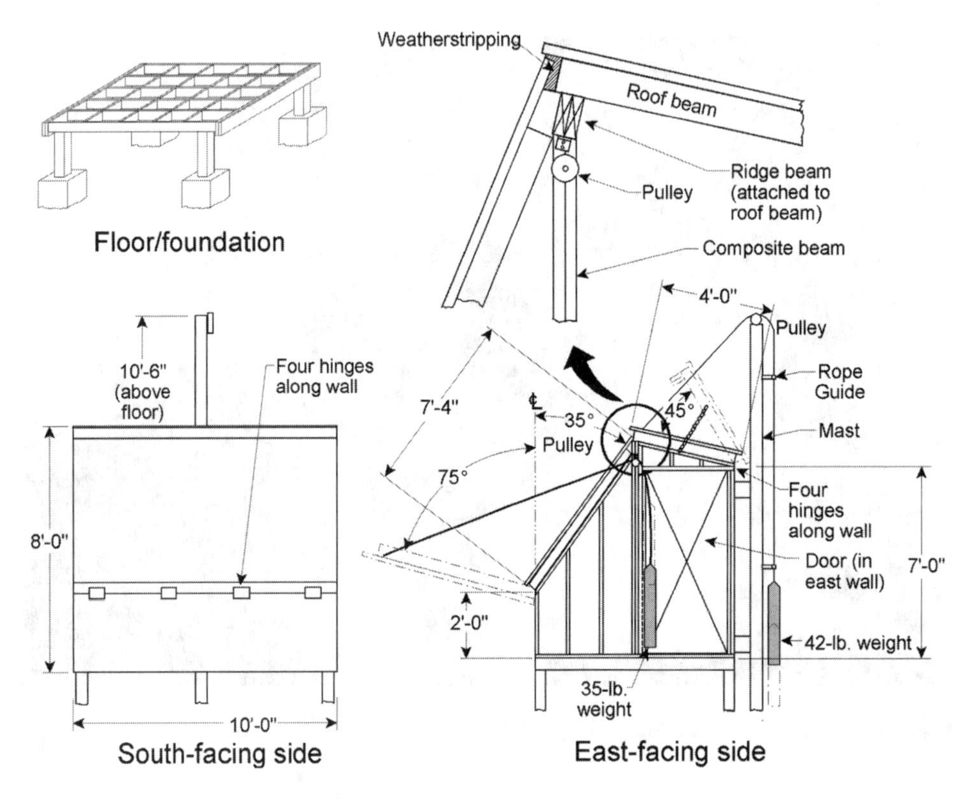

Figure 8.24 *Plans for the Lookum Observatory.*

Table 8.11 lists the basic parts for Lookum Observatory, but keep in mind that in a project of this magnitude it's not possible to list every single item that might be needed during construction. Instead, use this inventory as a guide only.

Start by staking out the area, making sure that everything remains square throughout the project. If you want to sink a permanent pier into the ground to support your telescope, it is best to do that now. Anway sank a 6-inch steel pipe into the ground in much the same way as Dave Trott did for his mini observatory, described earlier in this chapter. Place the pier more or less in the center of the observatory, making certain that it extends below your locale's average frost line to prevent heaving. Also double-check that the pier rises far enough above the floor for the telescope. The pier in the Lookum Observatory tops off at 30 inches above floor level.

Due to the steep slope of his land, Anway bypassed a solid concrete pad and foundation in favor of a simpler, less-expensive post-and-beam construction. Each of the four 4-inch × 4-inch corner posts are attached to 12-inch × 12-inch concrete footings using 3-inch × 90° angle iron. Each footing, in turn, sinks 42 inches below ground level.

To these four posts, nail an 8-foot × 10-foot framework of 2-inch × 10-inch × 10-foot beams and 2-inch × 8-inch × 8-foot floor joists located 16 inches on center. The floor itself is made from 5/8-inch plywood. As the floor's

Table 8.11 ***Lookum Observatory Parts List***

Quantity	Description
Floor	
4	4" × 4" × 6' treated posts
6	Bags of ready-mix concrete
2	2" × 10" × 10' joists (beams)
8	2" × 8" × 8' floor joists
16	Joist hangers
3	4' × 8' × 5/8" plywood flooring
Walls	
33	2" × 4" × 8' studs and plates
4	2" × 4" × 10' plates
8	4' × 8' × 0.5" plywood sheeting
1	Door, 32" wide × 76" (prehung in frame)
Roof	
2	2" × 8" × 10' roof joists (beams)
10	2" × 4" × 8' roof joists
9	1" × 4" × 12' furring strips
8	Commercial-grade stainless hinges
4	38" × 12' sheets of metal roofing
12'	Heavy foam rubber weather stripping
Mast	
1	4" × 4" × 12' treated post
Landing and Steps	
2	4" × 4" × 6' treated posts
2	Bags of ready-mix concrete
6	5/4" × 6" × 8' treated decking
2	2" × 12" × 4' treated stringers
Hardware	
4	Pulleys
4	35-lb. weights
1	45-lb. weights
—	Assorted fasteners

framework takes shape, frame around the pier, leaving about an inch clearance on all sides so that any foot movement won't disturb the telescope.

Next, build the four walls from 2-inch × 4-inch framing and 0.5-inch plywood sheathing. The rear wall is 7 feet tall, the front only 2 feet. The peaks of the two sides are each 8 feet high. Two composite columns, each made from three 2 × 4s, rise to the peaks of the two side walls. Install a prehung exterior doorway in the rear wall for access into the observatory, along with at least one screened, weatherproof air vent in each wall to promote airflow.

Construct the two roof sections from frames of 2-inch × 4-inch studs. The small upper roof section also includes a longitudinal 2-inch × 8-inch ridge beam that forms the structural peak of the roof. This beam is critical, because it serves to tie the whole structure together. When the roof is closed, the beam is supported at its ends by angle brackets attached to the two central 4 × 6 composite columns at the peaks of the side walls. The front roof section also leans against this beam when closed. Place a piece of rubber weather stripping along that section's upper edge to seal the observatory against weather. Anway chose to cover the roof's plywood sheathing with metal roofing, although you may prefer to use asphalt roofing shingles.

Attach each roof section to the observatory with four heavy-duty hinges, such as those used on commercial-grade steel doors. One is placed near each end of the roof sections with the other two spaced evenly between.

The two roof sections are far too heavy to open by hand, so they are opened by means of pulleys, ropes, and counterweights. The lower roof section is raised and lowered with a counterweighted rope that runs through a pulley attached to the top of one of the 4 × 6 composite side columns. The upper roof section is also opened by means of a pulley attached to the top of a 4 × 4 mast that is nailed to the back of the observatory but sunk separately into the ground. Make certain to attach chains to each roof section as shown; this is to limit how much each can swing.

To batten down the roof after each use, position a 0.25-inch threaded stud protruding about 4 inches out from the upper edge of the lower front roof section. Close both sections and see where the pin hits the upper section's ridge beam. Open the roof and drill an oversize hole through the ridge beam, such that when the sections are closed, the threaded stud will stick through the hole. To close the observatory for the night, put a fender washer and a wing nut onto the stud to clamp the roof sections together.

Top off the observatory with a coat of paint and some decorative trim, and you are ready for the dedication ceremony.

9

Till Death Do You Part

If this were a book about how to run a business, then this chapter might be entitled "Standard Operating Procedures" because it details methods and strategies that although not part of the business's primary product or service, help to enhance the company's operational efficiency. This chapter contains lots of little tidbits to help you get the most out of your telescope. It addresses a wide variety of topics, ranging from care and maintenance to traveling with a telescope.

Love Thy Telescope As Thyself

Unlike so many other products in our throwaway society, in which planned obsolescence seems the rule, telescopes are designed to outlast their owners. They require very little care and attention, cost nothing to keep, and eat very little. With a little common sense on the part of its master, a telescope will return a lifetime of fascination and adventure. But if neglected or abused, a telescope may not make it to the next New Moon. Are you a telescope abuser?

Storing Your Telescope

Nothing affects a telescope's life span more than how and where it is stored when *not* in use. *How* to store a telescope will be addressed farther along in this chapter, but first let's consider *where* the best places are to keep an idle instrument. The choice should be based on a number of different factors that at first glance might appear unrelated. A good storage place should be dry, dust-free, secure, and large enough to get the telescope in and out easily. Ideally, a telescope should always be kept at or near the outside ambient air temperature; doing so reduces the cool-down time required whenever the telescope is first set up at night. The quicker the cool-down time, the sooner the telescope will be ready to use.

Without a doubt, the best place to keep a telescope is in an observatory. It offers a controlled environment and easy access to the night sky. Of course, not everyone can afford to build a dedicated observatory, and neither is an observatory always warranted. Clearly, an observatory is pointless if the nearest good observing site is an hour's drive away. In cases such as these, a few compromises must be struck.

If an observatory is not in the stars (so to speak), other good places to store telescopes include a vented, walled-off corner of an unheated garage or a wooden tool shed. I keep my telescopes in a corner of my garage that is completely walled off to protect the optics from the inevitable dust and dirt that accumulates in garages as well as from any automotive exhaust that could damage delicate optical surfaces. A pair of louvered vents were installed in the outside wall of the garage to let air move freely in and out of the telescope room, reducing the risk of mildew. Wooden tool or garden sheds share many of the advantages of observatories and garages for telescope storage, but again I recommend installing one or two louvered vents in the shed's walls for air circulation. Metal sheds are not as good, as they can build up a lot of interior heat on sunny days.

Many amateurs choose to store their equipment in basements, which are certainly secure enough and large enough to qualify. Furthermore, they offer easy access provided there is a door leading directly to the outside. Their cool temperatures also keep the optics closer to that of the outside air. While all of these considerations weigh in their favor, most basements fail when it comes to being dry and free of dust. If a basement is your only alternative, invest in a dehumidifier. Clothes closets, another favorite place to hide smaller telescopes, also fall short because clothes act as dust magnets. Remember, unless a spot meets all of the criteria, continue the search.

Regardless of where a telescope is kept, seal the optics from dust and other pollutants when it is not in use. Usually this is simply a matter of putting a dust cap over the front of the tube. Most manufacturers supply their telescopes with a custom-fit dust cap just for this purpose. Use it diligently. If the telescope did not come with a dust cap or if it has been lost over time, then a plastic shower cap makes a great substitute. If you are into a more sophisticated look, some of the companies listed in appendix C sell dust caps made of rubber in a wide variety of sizes. Further, if the telescope or binoculars came with a case, use it. Not only will a case add a second seal against dust, but it also will protect the instrument against any accidental knocks or bumps.

A dark, damp telescope tube is the perfect breeding ground for mold and mildew. To avoid the risk of turning your telescope into an expensive petri dish, be sure that all of its parts are dry before sealing it up for the night. Tilt the tube horizontally to prevent water from puddling on the objective lens, primary mirror, or corrector plate. No matter how careful you are, optics are bound to become contaminated with dust eventually. A moderate amount of dust has, surprisingly, little effect on a telescope's performance. But if there's a great deal of it, or if the optics have become coated with a film or mildew, the observer sees dimmer, hazier views that lack clarity.

Having a telescope that is a little dusty is cause not for panic but for cautious action. "Clean a telescope?!" you ask. "Isn't that a job that should be left to professionals?" Not at all. While to the uninitiated it might seem a formidable task, it's actually not, as you are about to discover.

An optic should be cleaned only when dust or stains are apparent to the eye; otherwise, leave well enough alone. *Never* clean a telescope lens or mirror just for the sake of cleaning it, because every time an optic is touched, there is always the risk of damaging it. Remember this rule: If it ain't broke, don't fix it.

The methods described here are for cleaning *outer* optical surfaces only. Unless you really know what you are doing, I strongly urge against dismantling sealed telescopes (such as refractors and catadioptrics), binoculars, and eyepieces. Dirt and dust will never enter a sealed tube if it is properly stored and protected. Nevertheless, if an interior lens or mirror surface in a sealed telescope becomes tainted by film or mildew, it should be disassembled and cleaned only by a qualified professional. Contact the instrument's manufacturer for recommendations on how to do so. If you don't, you may discover that the telescope is much easier to take apart than it is to put back together! I know someone who once decided to take apart his 4-inch achromatic objective lens to give it a thorough cleaning. All was going well until it came time to put the whole thing back together. Seems he tightened a retaining ring in the lens holder just a little too much and... CRACK! The edges of the crown element fractured. Don't make the same mistake, y'hear?

Never start to clean a telescope if time is short. For instance, it is not time to decide that your telescope is absolutely filthy as the Sun is setting in a crystal-clear sky. Instead, check the optics well beforehand so there are no surprises. To help you along the way, I have divided the cleaning process into two parts, one for lenses and corrector plates and another for mirrors.

Cleaning Lenses and Corrector Plates

Begin the cleaning process by removing all abrasive particles that have found their way onto the lens or corrector plate. This does NOT mean blowing across the lens with your mouth; you'll only spit all over it. Instead, use either a soft camel-hair brush (Figure 9.1a) or a can of compressed air, both available from photographic supply stores. Some brushes come with air bulbs to allow blowing and sweeping at the same time. If the brush is your choice, lightly whisk the surface of the lens in one direction only, flicking the brush free of any accumulated dust particles at the end of each stroke.

Many amateurs prefer to use a can of compressed air instead of a brush for dusting optical surfaces, as this way the lens is never physically touched. Hold the can perfectly upright, with the nozzle away from the lens *at least* as far as recommended by the manufacturer. If the can is too close or tilted, some of the spray propellant may strike the glass surface and stain it. Also, it is best to use several short spurts of air instead of one long gust.

With the dust removed, use a gentle cleaning solution for fingerprints, skin oils, stains, and other residue. Don't use a window glass–cleaning spray or other household cleaners. They could damage the lens' delicate coatings

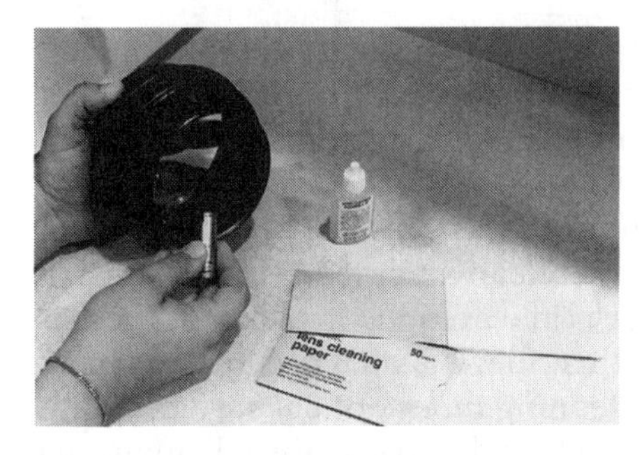

Figure 9.1 *The proper way to clean a lens or corrector plate: (a) Lightly brush the surface with a soft camel-hair brush and then (b) gently wipe the lens with lens-cleaning solution.*

beyond repair. Photographic lens-cleaning fluid can be used but can occasionally leave a filmy residue behind. One of the best lens-cleaning solutions can be brewed right at home. In a clean container, mix three cups of distilled water with a half cup of pure-grade isopropyl alcohol. Add two or three drops of a mild liquid dishwashing soap (you know, the kind that claims it will not chap hands after repeated use).

Dampen a piece of sterile surgical cotton (not artificial so-called cotton balls) or lens tissue with the solution. Do not use bathroom tissue or facial tissue; they are impregnated with perfumes and dyes. Squeeze the cotton or lens tissue until it is only damp, not dripping, and gently blot the lens (Figure 9.1b). Avoid the urge to use a little elbow grease to get out a stubborn stain. The only pressure should be from the weight of the cotton ball or lens tissue. Once done, use a dry piece of lens tissue or cotton to blot up any moisture.

The operation for cleaning the corrector plate of a catadioptric telescope is pretty much the same as just detailed. The only difference is in the blotting direction; in this case, begin with the damp cotton or tissue at the secondary mirror holder in the center of the corrector and move out toward the edge. Follow a spokelike pattern around the plate, using a new piece of cotton or tissue with each pass. As you stroke the glass, turn the cotton or tissue in a backward-rolling motion to carry any grit up and away from the surface before it has a chance to be rubbed against the optical surface. Overlap the strokes until the entire surface is clean. Again, gently blot dry.

Mirror Cleaning

Cleaning a mirror is much like cleaning a lens in that special care must be exercised to ensure against damaging the fine optical surface. In fact, a mirror is even more susceptible to scratches than a lens. The mirror's thin aluminized coating is extremely soft, especially when compared to abrasive dirt, and so is easily gouged. This is not meant to scare you out of cleaning your mirror if it really needs it but only to heighten your awareness.

The operation for cleaning a telescope's primary or secondary mirror requires it to be removed from the telescope and the cell that holds it in place.

Consult your owner's manual for more details on mirror removal. With the naked mirror lying on a table, blow compressed air across its surface to rid it of any large dust and dirt particles. (Remember, keep the can of compressed air vertical.) Do not use a brush for this step, as even the softest bristles can damage a mirror's coating.

Next, inspect the mirror's coating for pinholes and scratches. A good aluminum coating should last at least 10 years, even longer if the mirror has been well cared for. To check its condition, hold the mirror, its reflective side toward you, in front of a bright light. It is not unusual to see a faint bluish image of the light source through the mirror if the source is especially bright, but its image should appear the same across the entire mirror. If not, there may be thin, uneven spots in the coating. Any scratches or pinholes in the coating also will become immediately obvious as well. A few small scratches or pinholes, while not desirable, can be lived with. But if scratches or pinholes abound or if an uneven coating is detected, then the mirror should be sent out for realuminizing. Appendix C lists several companies that realuminize mirrors; consult any or all of them for prices and shipping details. It is also a good idea to have the secondary mirror realuminized at the same time, because it probably suffers from the same problems as the primary. In fact, most realuminizing companies will work on the secondary at no additional cost.

If the coating is acceptable, bring the mirror to a sink. Be sure to clean the sink first and lay a folded towel in the sink as a cushion, just in case—OOPS—the mirror slips. Run lukewarm tap water across the mirror's reflective surface. This should lift off any stubborn dirt particles that refused to dislodge themselves under the compressed air. End with a rinse of distilled water. Tilt the mirror on its side next to the sink on a soft, dry towel and let the water drain off the surface. Examine the mirror carefully. Is it clean? If so, quit.

If you want to go further, thoroughly clean the sink to remove any gritty particles that may not have washed down the drain. Next, fill the sink with enough tepid tap water to immerse the mirror fully and add to it a few drops of gentle liquid dish soap (the same as used in the lens-cleaning solution). As shown in Figure 9.2a, carefully lower the mirror into the soapy water and let it sit for a minute or two. With a big, clean wad of surgical cotton, sweep across the mirror's surface ever so gently with the same backward-rolling motion, being careful not to bear down. Now is not the time to act macho. After you've rolled the cotton a half-turn backward, discard it and use a new piece. If stains still exist after this step is completed, let the mirror soak in the water for five to ten minutes and repeat the sweeping with more new cotton.

With the surface cleaned to your satisfaction, drain the sink once again. Run tepid tap water on the mirror for a while to rinse away all soap. Then turn off the tap and pour room-temperature distilled water across the surface for a final rinse.

Finally, rest the mirror on edge on a towel where it may be left to air dry in safety. I usually rest it against a pillow on my bed (Figure 9.2b). Tilt the mirror at a fairly steep angle (greater than 45°), its edge resting on the soft towel, to let any remaining water droplets roll off without leaving spots. Close the door behind you to prevent any nonastronomers from touching the mirror.

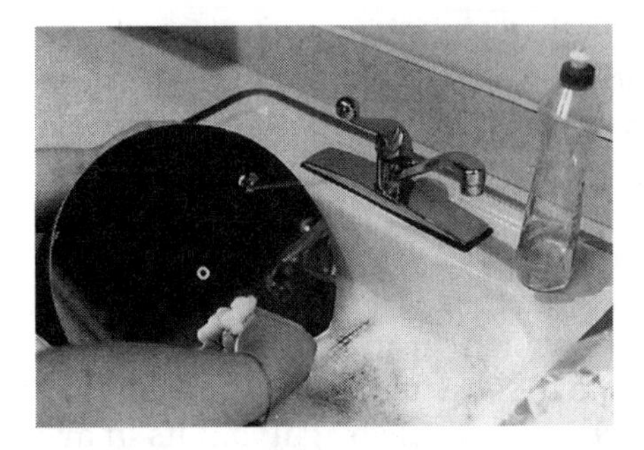

Figure 9.2 *The proper way to clean a mirror: (a) After using compressed air or a camel-hair brush to remove any loose contaminants, wash the mirror in the sink using a gentle liquid dishwashing detergent. (b) After rinsing the mirror, tilt the mirror nearly vertically to prevent water from puddling.*

When the mirror is completely dry, reassemble the telescope. Recollimate the optics using the procedure described later in this chapter, and you're done!

Other Tips

Other telescope parts require occasional attention as well. For instance, rack-and-pinion focusing mounts will sometimes bind if not lubricated occasionally. To prevent this from happening, spray a little silicone-based lubricant (such as WD-40) on the mount's small driving gear. To do this, remove the screws (typically two) that hold the small plate onto the side of the focuser housing, taking care not to lose anything along the way. With the plate out of the way, take a look inside at the small pinion gear that meshes with and drives the focusing mount's tube up and down. Squirt a little (*very* little) lubricant on the pinion teeth, reaffix the cover plate, and wipe off any drips as required.

If your telescope has a cardboard tube, make sure its ends do not fray and begin to unravel. Tilt the tube horizontally, remove or otherwise protect the optics, and brush on a thin layer of varnish to seal the ends. Checking the tube's condition once every now and then can add years to its life.

If a metal telescope mounting begins to move roughly or starts to bind, put a drop or two of lubricating oil on the axes' bearing points. This will keep the telescope moving freely and evenly, rather than binding and grabbing. Some manufacturers recommend this be done at specific intervals, while others make no mention of it at all. If nothing is said in the owner's manual, then do it once a year or so.

The typical wood-Formica-Teflon construction of Dobsonian mounts requires little in the way of maintenance. However, if your Dobsonian does not move freely in altitude or azimuth, take the mount apart and spray a little furniture polish on the contact surfaces. Buff the polish as you would a coffee table until it shines with a luster. Put the mounting back together and take it for a test spin. The difference should be immediately noticeable.

Some clock drives also need an occasional check to keep them happy. Carefully remove the drive's protective cover plate or housing and put a little thin grease between the two meshing gears. While the drive is open, put a drop or two of thin oil on the motor's shaft as well. Finally, reassemble the drive and turn it on. Listen for any noises. Most clock drives hum as they slowly turn. If unusually loud, angry, grinding noises are coming from it, turn the drive off immediately and contact the manufacturer for recommendations.

Get It Straight!

Have you ever gotten as frustrated with your telescope as the character depicted in Figure 9.3? (Hopefully you stopped short of whacking it with an axe.) There is nothing more vexing or disappointing to an amateur astronomer than to own a telescope that doesn't work as expected. "That lousy company," you think to yourself. "I spend over a thousand dollars on a telescope, and what do I get? A big, expensive, piece of junk!" All too often we are quick to blame the poor images produced by our telescopes on faulty optics and poor workmanship. Yet this is not necessarily the case. Although some telescopes are truly lacking in quality, most work acceptably *if they are in proper tune.* Even with the finest optics, a telescope will show nothing but blurry, ill-formed images if those optics are not in proper alignment. Correct optical alignment, or collimation, is a must if we expect our telescopes to work to perfection.

Figure 9.3 *From* Russell W. Porter *by Berton Willard (Bond Wheelwright Company, 1976). Reprinted with kind permission of the author.*

What exactly is meant by *collimation*? A telescope is said to be collimated if all of its optics are properly aligned to one another. Refractors, for instance, are collimated when the objective lens and eyepiece are perpendicular to a line connecting their centers (of course, this is assuming the lack of a star diagonal).

A Newtonian reflector is collimated when the optical axis of the primary mirror passes through the centers of the diagonal mirror and the eyepiece. Furthermore, the eyepiece focusing mount must be at a right angle to the optical axis. For our purposes here, all Cassegrain reflectors and all Cassegrain-based catadioptric telescopes will be lumped together. To be precisely collimated, their primary and secondary mirrors must be parallel to one another, with the center of the secondary lying on the optical axis of the primary. The eyepiece holder must be perpendicular to the mirrors, with the telescope's optical axis passing through its center. Refer to Figure 2.4 for clarification if these descriptions are unfamiliar.

The optics in some types of telescopes are apt to go out of alignment more easily than others. For example, it is rare to find a refractor that is not properly collimated unless its tube is bent or warped, usually as the result of abuse and mishandling. On the other hand, most Schmidt-Cassegrain telescopes will experience some collimation difficulties in their lives, whereas Newtonian and Cassegrain reflectors are notoriously easy to knock out of alignment.

It's easy to tell if a telescope is collimated properly. On the next clear night, take your telescope outside and center it on a bright star using high magnification. Place the star slightly out of focus and take note of the diffraction rings that result. If the telescope is properly collimated, then the rings should appear concentric around the star's center. If, however, the rings appear oval or lopsided—and remain oriented in the same direction when you turn the focuser both inside and outside of best focus—then the instrument is in need of adjustment.

Collimating a Newtonian

In general, the faster the telescope (that is, the lower the f-number), the more critical its collimation. Though slower Newtonians may be collimated adequately by eye, I have chosen to lump them together with deep-dish scopes in the interest of brevity. Throughout the discussion, I will make it clear when a step is required for both and when it is for deep-dish reflectors only.

Begin by purchasing or making a *sight tube,* which is essentially a long, empty tube with a pair of crosshairs at one end and a precisely centered pinhole at the other. The outer diameter of the sight tube must match the inside diameter of the telescope's eyepiece holder (typically, 1.25 inches). Tectron, the Telescope and Binocular Center, and AstroSystems make excellent sight tubes. Or you can make one yourself (see chapter 8).

As an aid for collimation, many Newtonians (especially those that are f/6 or less) come with small, black dots right in the middle of their primary mirrors. (Although at first it might appear that these dots will impede mirror performance, in reality they have no effect because they lie in the shadow of the diagonal mirror.) If your telescope mirror does not have a centered dot, now is the time to put one there. Remove the mirror from the telescope and place it on a large piece of paper. Trace its diameter with a pencil and then move the mirror to a safe place.

Now to find the *exact* center of the mirror tracing: Cut out the tracing precisely along its edge. Once you've done that, fold the paper circle in half and then in half again so that you end up with a quarter-circle, 90° wedge. Snip off the tip of the wedge with a pair of scissors and unfold the paper. The resulting hole marks the center of the mirror tracing.

Return with the mirror and align (gently, please) the paper so that the tracing matches the mirror's edge. Using the tracing as a template, make a tiny mark at the center of the mirror with a permanent felt-tip marker. Carefully check the spot's accuracy with the ruler; if acceptable, then enlarge the spot to about one-eighth of an inch (0.25-inch for mirrors of 12 inches and larger). Finally, stick an adhesive-backed hole reinforcer (the kind used to strengthen loose-leaf paper) on the mirror around the black dot.

Aim the telescope toward a bright scene (the daytime sky works well). Rack the focusing mount in as far as it will go, insert the sight tube, and take a look. The scene will probably look like one of those shown in Figure 9.4. Ideally, you

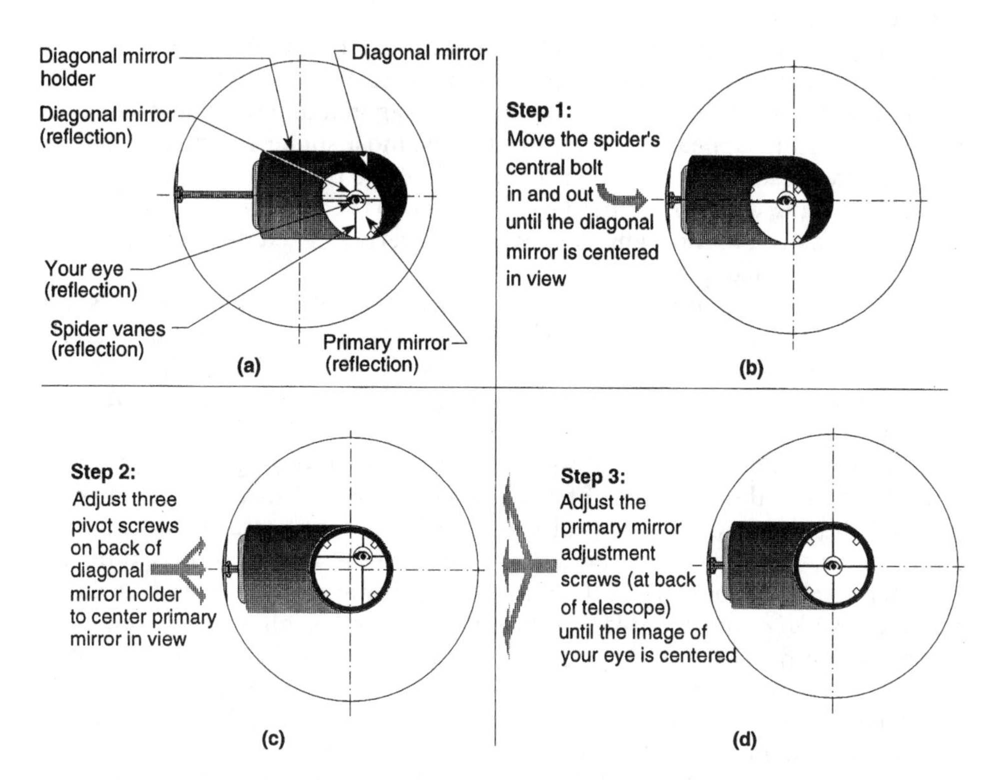

Figure 9.4 *Collimating a Newtonian reflector is as easy as 1-2-3. (a) The view through an uncollimated telescope. (b) Adjust the diagonal mirror's central post until the diagonal is centered under the eyepiece tube. (c) Turn the diagonal mirror's three adjustment screws behind the mirror until the reflection of the primary mirror is centered. (d) Finally, adjust the primary mirror's cell until the mirror image of your eye is centered in view.*

should see the diagonal mirror centered in the sight tube. If it is, skip to the next step; if not, then the diagonal must be adjusted. Move to the front of the telescope tube and look at the back of the diagonal mirror holder. Most Newtonians use a four-vane spider mount to hold and position the diagonal in place. Spider mounts typically grasp the diagonal in a holder supported on a central bolt. Loosen the nut(s) holding the bolt in place and move the diagonal in and out along the optical axis until its outer diameter is centered in the sight tube. Before tightening the nut, check to make sure that the diagonal is not turned away from the eyepiece, because it can also rotate as it is moved. When done with this step, the view through the sight tube should look like Figure 9.4b.

Some less-expensive telescopes use single rods attached to the eyepiece mounts to hold their diagonals in place. These are much more prone to being knocked out of adjustment and are unfortunately much harder to aim accurately. Loosen the setscrew that holds the rod in place, being certain to hold on to the diagonal or it will drop into the tube. By rotating the rod or moving it up and down, recenter the diagonal in the sight tube. Be careful not to bend the rod in the process. (Unless the rod is bent, the primary should appear centered in the diagonal at the end of this step. If so, skip the next step; if not, then repeat this process.)

With the diagonal centered in the sight tube, look through the tube at the reflection of the primary mirror. You ought to see at least part of the primary and the far end of the telescope tube. For the diagonal to be adjusted correctly, the primary must appear centered. To do this, most spider mounts have three equally spaced screws that, when turned, pivot the diagonal's angle. By alternately loosening and tightening the screws, move the reflection of the primary until it is centered in the sight tube. When properly aligned, the view after this step should look like Figure 9.4c.

Aiming the primary mirror will go much faster if you have an assistant. First, look at the back of the primary mirror's cell. You should see three (sometimes six) screws facing out. These adjust the tilt of the primary. Turn one or more of the adjustment screws until the diagonal's silhouette is centered in the sight tube.

With your helper at the adjustment screws, look through the sight tube. You should see the reflection of the primary centered in the crosshairs. In bright light, the primary's central spot should also be seen. Following your instructions, have your assistant turn one or more of the screws until the reflected image of your eye in the diagonal mirror appears centered as shown in Figure 9.4d. (Primary mirror mounts with six screws follow pretty much the same procedure. Three of the screws [usually the inner three] adjust the mirror, while the others prevent the mirror from rocking. If your telescope uses this type of mirror mount, the three outer screws must be loosened slightly before an adjustment can be made.)

If you own a shallow-dish Newtonian, the mirrors should now be collimated adequately. If, however, you own a deep-dish reflector, then some fine-tuning will be needed. For this, it is strongly recommended that you use a Cheshire eyepiece. Here's how it works. Shine a flashlight beam into the eyepiece's side opening and look through the peephole. Centered in the dark silhouette of the diagonal will be a bright donut of light—the reflection of the eyepiece's mirrored surface. The dark center is actually the hole in that surface. Adjust the primary mirror until its black reference spot is centered in the Cheshire eyepiece's donut. That's it.

The tell-all test is the *star test*. Take the telescope outside and let the optics fully adjust to the outside air temperature. Aim toward a moderately bright star. Unless your telescope is polar-aligned and the clock drive is turned on, use Polaris for the star test. Unlike all other stars, Polaris has the distinct advantage of not moving (at least, not much) in the sky, which makes the test a little easier to perform. Using a medium-power eyepiece, place the star in the center of the field. Move it slightly out of focus, transforming it from a point into a tiny disk surrounded by bright and dark rings. If the mirrors are properly collimated, then the rings should be concentric, like a bull's-eye. If not, one or both of the mirrors may need a little fine-tweaking. Be sure to recenter the star in the field every time an adjustment is made. (Defects in the mirror's curve can also cause the rings to appear irregular, but more about this later.)

Collimating a Schmidt-Cassegrain

Unlike the Newtonian, in which both the primary and secondary mirrors can be readily accessed for collimation, commercially made Schmidt-Cassegrain telescopes have their primary mirrors set and sealed at the factory. As such, an owner cannot adjust the primary if misalignment ever occurs. Fortunately, SCTs are rugged enough to put up with the minor bumps that might occur during setup without affecting collimation.

This leaves only the secondary mirror to adjust, as shown in Figure 9.5. Take a look at the secondary-mirror mount centered in the front corrector plate. There you will see the heads of three adjustment screws spaced 120° apart. (On some models, a plastic disk covers the screw heads. If so, it must be removed—carefully—to expose the adjustment screws.) In addition, some telescopes have a fourth, large screw or nut in the center of the secondary holder. If so, do not touch it! Loosening that central screw will release the secondary from its cell and drop it into the tube. Talk about a quick way to ruin your day!

Remove the star diagonal and insert a sight tube into the eyepiece holder. Take a look. Ideally, you should see your eye centered in the secondary mirror. If the view is more like Figure 9.5a, turn one or more of the adjusting screws until everything lines up as in Figure 9.5b.

To check your success, move outside on a clear night. Set up the telescope as you would for an observing session, giving it adequate time to cool to the night air. With the telescope acclimated to the outdoor temperature, remove the star diagonal, if so equipped, and insert a medium-power eyepiece into the telescope. Center the instrument on a very bright star. Turn the focusing knob until the star moves out of focus and its disk fills about a third of the eyepiece field. Take a look at the dark spot on the out-of-focus disk. That's the dark silhouette of the secondary mirror. For the telescope to be properly collimated, the secondary's image must be centered on the star, creating a donutlike illusion. If the donut is asymmetric, then the secondary must be adjusted.

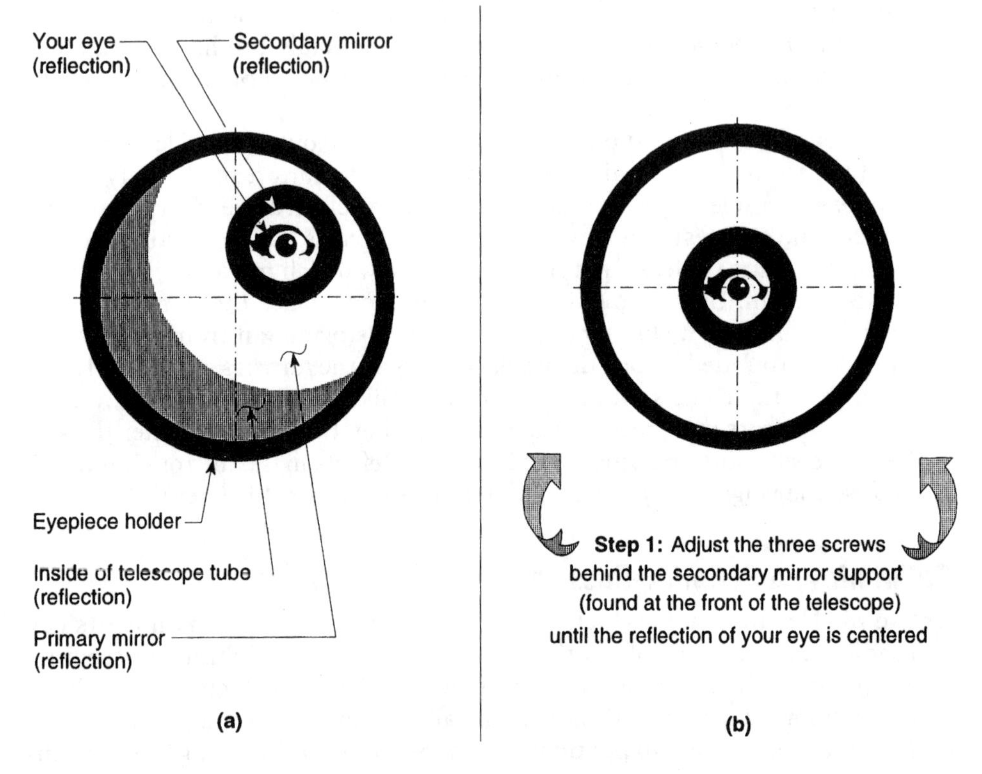

Figure 9.5 *Collimating a Schmidt-Cassegrain involves adjusting the secondary mirror (a) until its reflection appears centered in the primary (b).*

Make a mental note of which direction the silhouette favors, go to the front of the telescope, and turn the adjustment screw that most closely coincides to that direction *ever so slightly*. Now, return to the eyepiece, recenter the star in view, and look at the dark spot again. Is it better or worse? If it's worse, turn the same screw the opposite way; if it's off in a different direction, turn one of the other screws and see what happens. Continue going back and forth between eyepiece and adjustment screws until the dark spot is perfectly centered in the star blob.

To double-check the adjustment, switch to a high-power eyepiece. Place the star back in the center of the field and defocus its image only slightly. If the secondary's dark outline still appears uniformly centered, then collimation was a success and the telescope is set to perform at its best. If not, repeat the procedure, but with finer adjustments.

If the image is still not correctly aligned even after repeated attempts, then there is a distinct possibility that the primary is not square to the secondary. Focus on a rich star field. If any coma (ellipticity) is evident around the stars at the center of view, then chances are good that the primary is angled incorrectly. In this case, your only alternative is to contact either the dealer from which the telescope was purchased or the manufacturer.

To learn more about the fine art of telescope collimation, consult the book *Perspectives on Collimation* by Vic Menard and Tippy D'Auria. The capsule review in chapter 7 gives further information.

Test Your Telescope

There is no way to guarantee that every telescope made, even those from the finest manufacturers, will work equally well. Although companies have quality-control measures in place to weed out the bad from the good, a lemon is bound to slip through every now and then. That is why you should always check an instrument right after it is purchased. Examine the telescope for any overt signs of damage. Next, look at the optics. They should be free of obvious dirt and scratches. The mounting should be solid and move smoothly. If any problems are detected, immediately contact the outlet from which the telescope was purchased so that the problem can be rectified.

It is easy to distinguish clean, scratch-free optics from those that are not, but a poor-quality lens or mirror may not be so obvious. Fortunately, an elaborate, fully equipped optical laboratory is not required to check the accuracy of a telescope's optics. Instead, all that you need are a well-collimated telescope, a moderate-power eyepiece, and a clear sky. With these ingredients, any amateur astronomer can perform one of the most sensitive and telling optical tests available: the star test. A fourth ingredient (this book) will help you interpret the star test's results.

But before you can run a valid star test, you must make certain that the optics are properly collimated. It is impossible to get accurate results from the star test if a telescope's optics are misaligned. Now might be a good time to review that procedure (found earlier in this chapter) if you are uncertain.

Once everything is in alignment, select a star with which to run the star test. Which star? That depends on your telescope. First, if your scope does not have a clock drive or an equatorial mount or is not polar aligned, it is best to aim toward a bright star near the celestial pole because these move more slowly in the sky than, say, stars near the celestial equator. Which individual star depends on the telescope's aperture. For 4-inch and smaller instruments, select a first- or second-magnitude star; 6- to 10-inch telescopes are best checked with a third- or fourth-magnitude star, and larger instruments may be best tested with a fifth-magnitude star. Use a moderately high power eyepiece for the test. Suiter[1] recommends a magnification equal to 25× per inch (10× per centimeter) of aperture. So for a 6-inch telescope, magnification should equal 150×, and so on.

Focus the star precisely and examine its image closely. You'll recall from chapter 1 that at higher magnifications, a star will look like a bull's-eye, that is, a bright central disk (the Airy disk) surrounded by a couple of faint concentric rings (diffraction rings). By definition, any telescope that claims to have "diffraction-limited" optics must show this pattern. Is that what you see?

[1]Suiter, H.; *Star Testing Astronomical Telescopes;* Willmann-Bell, Inc., 1994.

Probably not, or at least not at first glance. Many factors affect the visibility of diffraction rings, such as telescope collimation, warm-air currents inside the telescope tube, the steadiness of the atmosphere, and the focal ratio of the telescope. Poor atmospheric steadiness, or "seeing" as it is commonly known, causes star images to "boil," making it impossible to detect fine detail. Diffraction rings can be seen only under steady seeing conditions (see the section in chapter 9 entitled "Evaluating Sky Conditions" for more on this). The aperture of the telescope also plays a big role in seeing diffraction rings. The larger the aperture, the smaller the diffraction pattern and, thus, the more difficult it is to see.

Slowly rack the image *slightly* out of focus. Just how far out to go is a matter of focal ratio. Most people foil the star test simply by being too aggressive when defocusing the star's image. Suiter recommends that for an f/10 instrument, the eyepiece should be moved no more than 0.20 inch off focus, while an f/4.5 telescope requires the eyepiece be moved a mere 0.06 inch! In either case, the star's light should enlarge evenly in all directions, like ripples expanding after a small stone is tossed into a calm body of water.

Begin by moving the eyepiece inside of focus. Examine the out-of-focus star image. It should look like one of the patterns illustrated in Figure 9.6. Then reverse the focusing knob, bringing the star slightly outside of focus. Examine the ring pattern again. If the telescope has first-rate optics (that is,

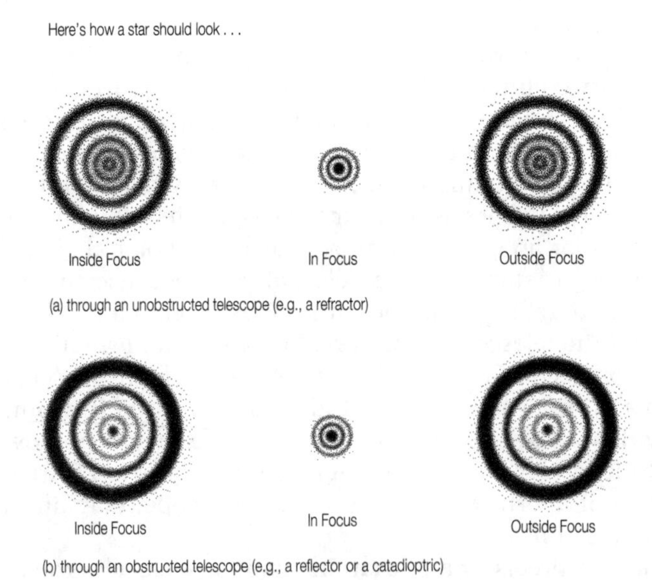

Figure 9.6 *What should a star look like through the perfect telescope? Ideally all light should come to a common focus, causing a star to appear as a tiny disk (the Airy disk) surrounded by concentric diffraction rings. The star should expand to look like one of the illustrations here when seen out of focus, the image being identical on either side of focus: (a) through a refractor (unobstructed telescope) and (b) through a refractor on catadioptric (obstructed telescope).*

substantially better than merely diffraction-limited optics), both extrafocal images should appear identical.

What if they don't? Compare the exact shapes of the inside-focus and outside-focus patterns with those shown in Figure 9.7. Are the patterns at least circular? No? What geometric pattern do they resemble? An oval? If the oval shapes look the same both inside and outside of best focus, the optics are doubtless out of collimation. If the images are dancing wildly, then either the telescope optics are not yet acclimated to the outside air temperature or the atmosphere is too turbulent to perform the test. What if the images are either

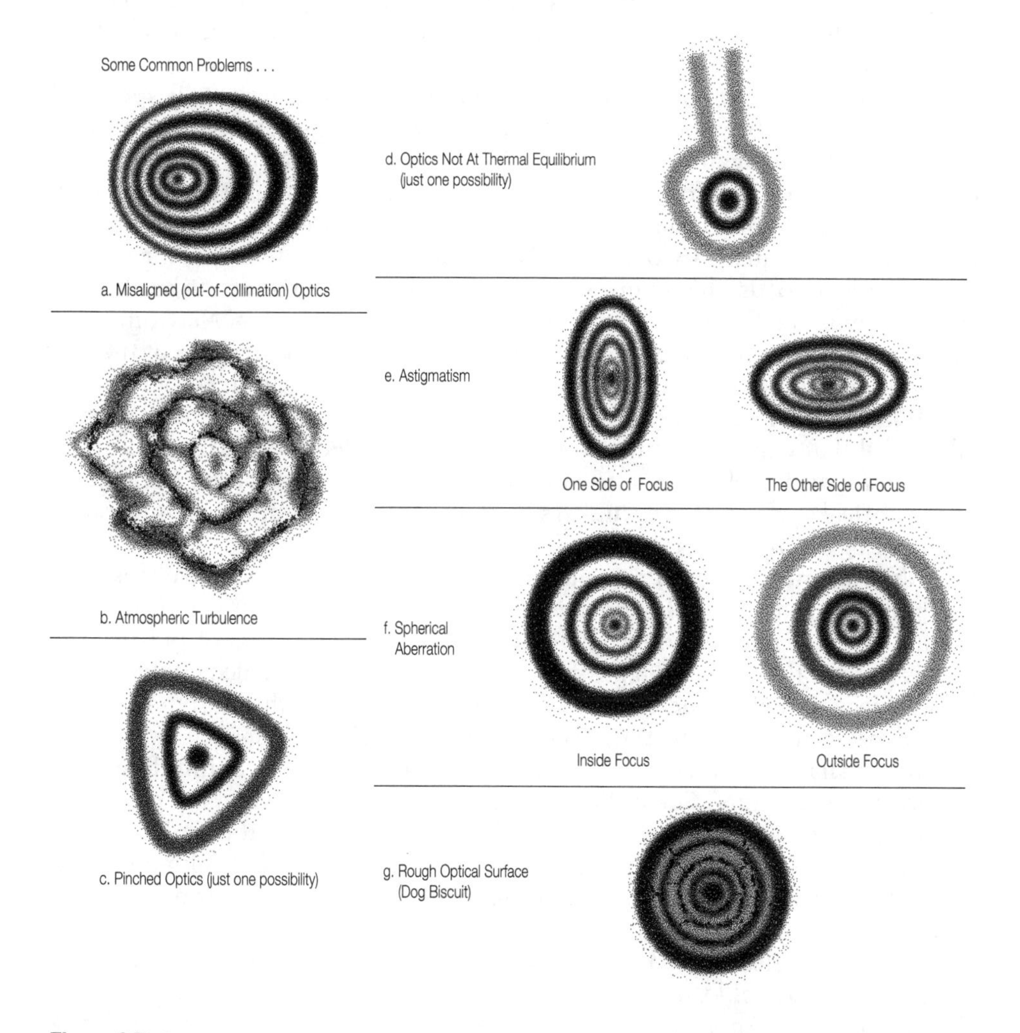

Figure 9.7 *Something is amiss if the star test yields any of these results: (a) optics are out of collimation; (b) turbulent atmospheric conditions are present; (c) optics are being pinched or squeezed by their mounting; (d) optics have not acclimated to the outside air temperature; optics suffer from (e) astigmatism, (f) spherical aberration, and (g) rough optical surface.*

triangular or hexagonal? If the patterns have one or more sharp corners, then the optics are probably being pinched, or distorted, by their mounts. Pinched optics are especially common in Newtonian reflectors when the clips holding the primary in the mirror cell are too tight.

"My telescope never focuses stars sharply. I just did the star test, but the out-of-focus patterns appear oval. When the eyepiece is racked from one side of focus to the other, the ovals flip orientation 90°." If that is your telescope, then it suffers from astigmatism. Astigmatism may be caused by poorly figured optics, but it may also be caused by pinched optics, uncollimated optics, or by the cooling process after the scope is taken outdoors. In Newtonian reflectors, a slightly convex or concave secondary mirror is also a common cause of astigmatism. If the axis of the star oval is parallel to the telescope tube, suspect the secondary mirror.

The most common optical defect found in amateur telescopes is spherical aberration, which becomes evident when a mirror or lens has not been ground and polished to its required curvature. As a result, light from around the edge of the optic comes to a focus at a different distance than light from the center. Spherical aberration comes in two varieties: one caused by undercorrected optics, and one caused by overcorrected optics. Both produce similar effects: On one side of focus, the outermost part of the bull's-eye pattern is brighter or sharper than on the other side of focus. In the case of a Newtonian or Cassegrain design, the shadow of the secondary mirror in the center of the star disk will look larger on one side of best focus and smaller on the other.

What if, even when a star is brought out of focus, rings cannot be seen at all and all that can be seen instead is a round, mottled blob? This condition, more common in reflectors and catadioptrics than in refractors, indicates a rough optical surface. Mirror makers have an especially appropriate nickname for this: *dog biscuit*.

Although not as accurate as the star test, another simple way to check optical quality of reflecting telescopes is the Ronchi test, which uses a diffraction grating made of many thin parallel lines printed on clear glass or plastic. The idea is to hold the diffraction grating up to the reflector's eyepiece holder, with the eyepiece removed, and examine a star's blurred image through the grating.

To perform a Ronchi test, you will need a 100-line-per-inch diffraction grating, available from the Telescope and Binocular Center, Schmidling Productions, and Edmund Scientific Company, among others. With the grating in hand, set up your telescope outside and allow it to cool to the ambient temperature. While you are waiting for the optics to acclimate, check and adjust the instrument's collimation. The Ronchi test will not work properly with poorly aligned optics.

Center the telescope on a bright star, remove the eyepiece (and star diagonal, if used), and hold the diffraction grating up to the empty eyepiece holder. Turn the focusing knob slowly in one direction until you see an enlarged disk with four superimposed black lines. Are the lines straight, or are they curved or crooked? If the latter, in what direction are they bent? Repeat the test by turning the focusing knob in the opposite direction. You'll see the star shrink in size and then enlarge again. Continue defocusing until the star's disk expands and four dark lines are visible again. As before, examine their curvature. Note how this time the lines curve in the opposite direction.

Depending on how much and in what manner the lines curve, your optics may have one of several problems or no problem at all. Figure 9.8 interprets some of the common results. If the lines are straight, then your telescope *probably* enjoys a properly corrected mirror or lens (Even if it passes the Ronchi test, be sure to double-check the instrument with the more telling star test. Some instruments will pass the Ronchi test but fail the star test miserably.) If, however, the lines are bent or crooked, then the telescope suffers from one of the imperfections shown in the figure. Be forewarned that as in the star test, an improperly collimated or uncooled telescope may show the same result. Before you accuse your instrument of having poor optics, double-check their alignment, let the optics cool, and then repeat the test.

Both the star test and Ronchi test examine all of the optical components collectively; they can ascertain if there is a problem but cannot immediately distinguish *which* optical component is at fault. To point the finger of blame at a reflector, try rotating the primary mirror 90° and retesting. If the defect in the pattern turns with the mirror, then the primary is at fault. If it does not, then the secondary mirror is the culprit.

The moral to all this? Simple: There is no substitute for good optics. Pay close attention to manufacturers' claims and challenge those that appear too good to be true. If you are considering a telescope made by the XYZ Company, write or call to inquire about its optics. Who makes them? What tests are performed to evaluate optical quality? And finally, what kind of guarantee is offered? Remember, if you want absolute perfection, you will probably have to pay a premium for it.

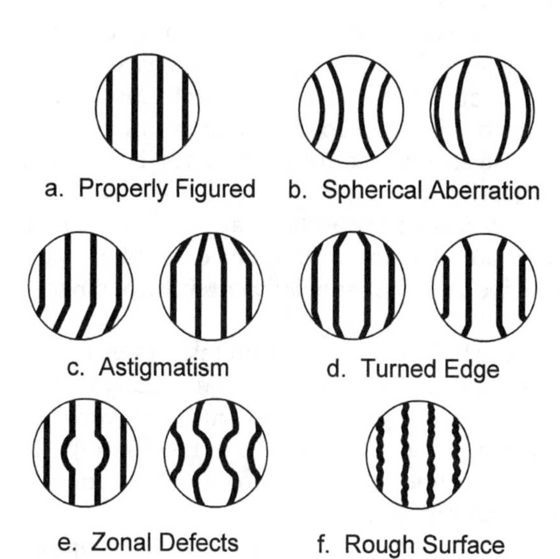

Figure 9.8 *Some possible results from the Ronchi test and their causes. As explained in the text, ideally you should see straight lines during the test.*

Have Telescope, Will Travel

Not too long ago, the most traveling a telescope would ever do was the trip from the house to the backyard, but not anymore. With the increased popularity of large regional or national star parties, as well as the ever-worsening problem of light pollution, many amateur telescopes routinely travel tens or hundreds of miles in search of dark skies.

Whenever a telescope is picked up and moved, the risk of damage is present. That is why some hobbyists tend to shy away from driving around with such delicate instruments. With a little forethought and common sense, though, this risk can be minimized.

Let's first examine traveling with a telescope by car. Begin by examining the situation at hand. If you own, say, a subcompact sedan and an 18-inch Newtonian, you have a big problem with only one solution: Buy another car. (That is exactly what a friend of mine did after he purchased an NGT-18 from JMI a few years ago! He traded in his sports car for a minivan.)

If your telescope and automobile are already compatibly sized, then it is just a matter of storing the instrument safely for transport. Take apart as little of the telescope as necessary. If there is room in the car to lay down the entire instrument, mounting and all, then by all means do so. Take advantage of anything that can minimize the time required to set up and tear down.

If the telescope tube must be separated from its mounting, it is usually best to place the tube into the car first. Be sure to seal the optics from possible dust and dirt contamination, which usually means simply leaving the dust caps in place, just as when the telescope is stored at home. If the telescope does not come with a storage case, protect the tube from bumps by wrapping it in a clean blanket, quilt, or sleeping bag. Strategically placed pillows and pieces of foam rubber can also help minimize screw-loosening vibrations. If possible, strap the telescope in place using the car's seatbelt.

Next comes the mounting. Carefully place it into the car, making sure that it does not rub against anything that may damage it or that it may damage. Wrap everything with a clean blanket for added protection. Again, use either the car's seatbelts or elastic shock cords (bungee cords) to keep things from moving around during a sharp turn. Be sure to secure counterweights, which can become dangerous airborne projectiles if the car's brakes are hit hard.

Transporting telescopes by air presents many additional problems. Their large dimensions usually make it impossible to carry them on a plane and store them under the seat! Therefore, we must be especially careful when packing these delicate instruments.

Some owners of 8-inch and smaller Schmidt-Cassegrain telescopes prefer to wrap their instruments in foam rubber and place them in large canvas duffel bags and carry them onto the plane, placing the scopes into the overhead compartments. While this method usually works (some overhead compartments are just not large enough), the chance of damaging the telescope is great. This practice, therefore, is *not* recommended.

Scott Roberts of Meade Instruments recommends placing the telescope in its carrying case and then placing them in a double-walled cardboard shipping carton. Use the original padding that came with the telescope from the factory, if available. He also recommends, for Meade 10-inch and larger instruments, reinstalling the Mirror Lock Shipping Bolt that originally came with the telescope, as well as releasing the RA and Dec locks to prevent gear damage should the telescope be dropped in shipment. Celestron says that their deluxe hard cases (Celestron part number 302070), made by Penguin, well known in the photo industry for their excellent camera cases, are strong enough to take the jostling that air cargo can go through. These cases are packed with cubed high-density foam rubber so that they may be customized to hold any comparably sized instrument. Their other cases, as well as Meade's soft cases, are not appropriate for airline travel.

Even if you will be using one of these high-impact-resistant cases, take a few additional precautions before sending your baby off onto the loading ramp. Begin by completely enclosing the telescope in protective bubble wrap, available in larger post offices and stationery stores. Bubble wrap is a two-layer plastic sheet impregnated with air to form a multitude of cushioning bubbles. The bubbles are available in many different sizes—select the larger variety.

Owners of other types of telescopes face even greater challenges. First and foremost, the optics must be removed and placed into a suitable case to be carried on the plane. The empty tube may then be bubble-wrapped, surrounded by styrofoam pellets (both inside and outside the tube), and placed into a strong wooden crate. Make certain that the shipping carton can be used again on the return trip and bring along a roll of packing tape or duct tape just in case an emergency repair is needed.

Due to their weight, tripods and mountings pose special problems. I have traveled by air with a large tripod by first wrapping it in two thick sleeping bags and then packing everything in a large tent carrying case. Heavy equatorial mountings, on the other hand, must be packed professionally. Seek a local crating company for help.

No industry-wide policy exists regarding the transport of telescopes by air. Some airlines permit telescopes to be checked as luggage provided they do not exceed size and weight restrictions. (The purchase of optional luggage insurance is strongly recommended.) Other airlines will not accept telescopes as check-in items at all. In these instances, you must ship the instrument separately ahead of time via an air cargo carrier. Owing to the amount of paperwork involved, especially on international shipments, air cargo services usually require that advance arrangements be made. Contact your airline well ahead of departure to find out exact details and damage insurance options.

Make sure that each piece of luggage has both a destination ticket and an identification tag and that both are clearly visible on the outside. Information on the ID tag should include your name, complete address, and telephone number. Although permanent plastic-faced identification tags are preferred, most check-in points provide paper tags that may be filled out on the spot. I always make it a habit to include a second identification label inside my luggage as well, just in case the outside tag is torn off.

Whatever you do, don't forget to bring along all the tools needed to reassemble the telescope once you arrive. It is best to keep the tools in a piece of checked baggage. While returning from Mexico after the July 1991 solar eclipse, a friend was stopped from boarding his flight because he was carrying a screwdriver. He ultimately had to check the screwdriver as a separate piece of luggage because all of his bags had already been boarded. It was quite a sight at the baggage claim area when his screwdriver came down the ramp among all these suitcases!

Finally, compile a thorough inventory of all equipment that you plan to bring. Include a complete description of each item, such as its dimensions, color, serial number, manufacturer, and approximate value. U.S. Customs requires owners to register cameras and accessories with them on a "Certificate of Registration for Personal Effects Taken Abroad" form before departure. Contact your nearest Customs office for further information. Keep a copy of the list with you at all times while traveling, just in case any item is lost or stolen. Carriers will be able to find the missing piece more quickly if they know what to look for.

10

A Few Tricks of the Trade

A telescope alone does not an astronomer make. Sure, telescopes, binoculars, eyepieces, and other assorted contraptions are all important ingredients for the successful amateur astronomer, but there is a lot more to it than that. If a stargazer lacks the knowledge and skills to use this equipment, then it is doomed to spend more time indoors gathering dust than outdoors gathering starlight. Here is a look at some techniques and tricks used by amateur astronomers when viewing the night sky.

Evaluating Sky Conditions

Clearly nothing affects our viewing pleasure more than the clarity of the night sky. However, just because the weather forecast calls for clear skies does not necessarily mean that it's time to get out the telescope. As amateur astronomers are quick to discover, "clear" is in the eye of the beholder. To most people a clear sky simply means an absence of obvious clouds, but to a stargazer it is much more.

To an astronomer, sky conditions may be broken down into two separate categories: *transparency* and *seeing*. Transparency is the measure of how clear the sky is, or in other words, how faint a star can be seen. Many different factors, such as clouds, haze, and humidity, contribute to the sky's transparency. The presence of air pollutants, both natural and otherwise, also adversely affects sky transparency. Artificial pollutants include smog and other particulate exhaust, whereas volcanic aerosols and smoke from large fires (for example, forest fires) are forms of natural air pollution. Still, the greatest threat to sky transparency comes not from nature but from ourselves. We are the enemy, and the weapon is uncontrolled, badly designed nighttime lighting.

Today's amateur astronomers live in a paradoxical world. On one hand, we are truly fortunate to live in a time when modern technology makes it possible for hobbyists to own advanced equipment once in the realm of the professional only. On the other, we hardly find ourselves in an astronomical Garden of Eden. Although technology continues to serve the astronomer, it is also proving to be a powerful adversary. The night sky is under attack by a force so powerful that unless drastic action is taken soon, our children may never know the joy and beauty of a supremely dark sky. No matter where you look, lights are everywhere: buildings, gigantic billboards, highway signs, roadways, parking lots, houses, shopping centers, and malls. Most of these lights are supposed to cast their light downward to illuminate their earthly surroundings. Unfortunately, many fixtures are so poorly designed that much of their light is directed horizontally and skyward. The result: light pollution, the bane of the modern astronomer.

Have you ever driven down a dark country road toward a big city? Long before you get to the city line, a distinctive glow emerges from over the horizon. Growing brighter and brighter with each passing mile, this monster slowly but surely devours the stars; first the faint ones surrender, but eventually nearly all succumb. Several miles from the city itself, the sky has metamorphosed from a jewel-bespangled wonderland to a milky, orange-gray barren desert. Although I know of no astronomer advocating the total and complete annihilation of all nighttime illumination, we must take a critical look at how it can be made less obtrusive.

Responsible lighting must take the place of haphazard lighting. But let's face it—not many people will be interested in light conservation if the point is debated from an astronomical perspective only. To win them over, they must be convinced that more efficient lighting is good for *them.* The public must be educated on how a well-designed fixture can provide the same amount of illumination over the target area as a poorly designed one but without extraneous light scattered toward the sky and with a lower operating cost. That latter phrase, *lower operating cost,* is the key to the argument. The cost of operating the light will be lower because all of its potential is specifically directed where it will do the most good. The wattage of the bulb may now be lowered for the same effect, resulting in a lower cost. Everybody wins—taxpayers, consumers, and, yes, even the astronomers!

Can one person effect a change? That was the dream of David Crawford, the person behind the International Dark-Sky Association. The nonprofit IDA has successfully spearheaded anti-light-pollution campaigns in Tucson, San Diego, and many other towns and cities. It provides essential facts, strategies, and resources to light-pollution activists worldwide. For more information on how you can join the fight against light pollution, contact the IDA at 3545 North Stewart, Tucson, Arizona 85716, or on the Internet at www.darksky.org.

In mid-northern latitudes, the clearest nights usually take place immediately after the crossing of an arctic cold front. After the front rushes through, cool, dry air usually dominates the weather for 24 to 48 hours, wiping the atmosphere clean of smog, haze, and pollutants. Such nights are characterized by crisp temperatures, high barometric pressure, and low relative humidity.

To help judge exactly how clear the night sky actually is, many amateurs living in the Northern Hemisphere use the stars of Ursa Minor (the Little Dipper) as a reference because they are visible every hour of every clear night in the year. Figure 10.1a shows the major stars of Ursa Minor, while Figure 10.1b shows those in the Southern Cross (Crux), a favorite check for observers in the Southern Hemisphere. The numbers next to several of the stars represent their visual magnitudes. Note that in each case the decimal point has been eliminated to avoid confusing it with another star. Therefore, the 20 next to Polaris indicates it to be magnitude 2.0, and so on. You may find that all of the Dipper stars are visible only on nights of good clarity, while other readers may be able to see them on nearly every night. Still others, observing from light-polluted environs, may never see them all.

The night sky is also judged in terms of seeing, which refers not to the clarity but rather the sharpness and steadiness of telescopic images. Frequently, on clear, dark nights of exceptional transparency, the twinkling of the stars almost seems to make the sky come alive in dance. To many, twinkling adds a certain romantic feeling to the heavens, but to astronomers, it only detracts from the resolving power of telescopes and binoculars.

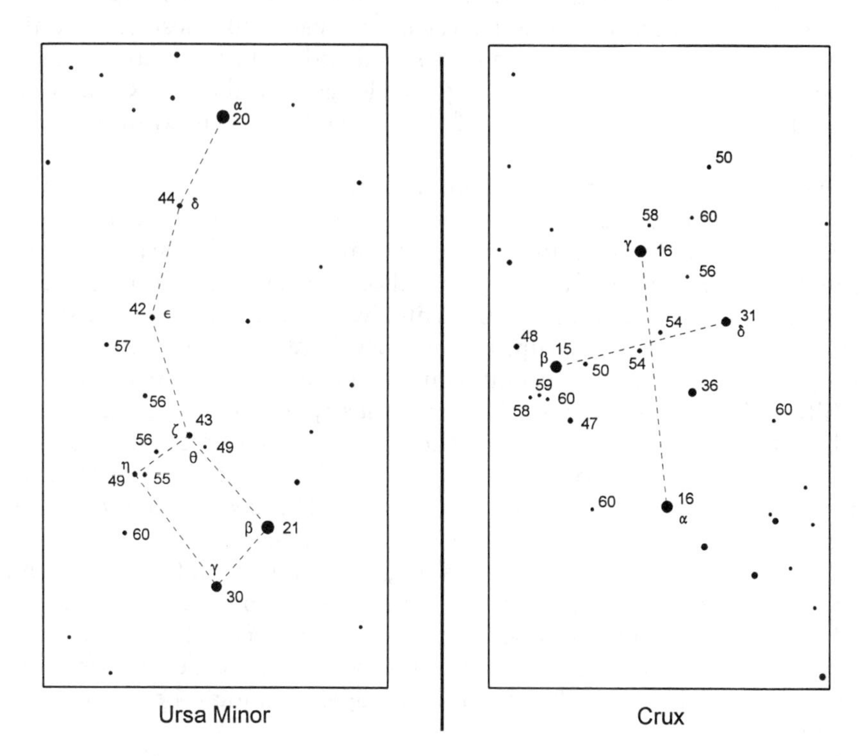

Figure 10.1 *How clear is the sky tonight? Many amateurs in the Northern Hemisphere use the visibility of the Little Dipper (a, Ursa Minor) as a gauge, while amateurs south of the equator use the Southern Cross (b, Crux). In both figures, decimals have been omitted to avoid confusing them with stars. Therefore, a magnitude 2.0 star is shown as "20" on the chart and so on.*

The twinkling effect, called *scintillation,* is caused by turbulence in our atmosphere. Density differences between warm and cold layers refract or bend the light passing through, causing the stars to flicker. Ironically, the air seems steadiest when a slight haze is present. Although any cloudiness may make faint objects invisible, the presence of thin clouds can actually enhance subtle details in brighter celestial sights such as double stars and the planets.

Although the Earth's atmosphere greatly influences seeing conditions, image steadiness also can be adversely affected by conditions inside and immediately surrounding the telescope. If you are like most amateur astronomers, you probably store your telescope somewhere inside your house or apartment . . . your *warm* house or apartment. Moving the telescope from a heated room out into the cool night air immediately sets up swirling heat currents as the instrument and its optics begin the cooling process toward thermal equilibrium. Peering through the eyepiece of a warm telescope on a cool night is like looking through a kaleidoscope, with the stars writhing in strange ritualistic dances. As the telescope tube and its optics reach equilibrium with the outside air, the images will begin to settle down. The night's observing may then begin in earnest.

How long it takes for these heat currents to subside depends on the telescope's size and type as well as the local weather. Newtonian and Cassegrain reflectors seem to acclimate themselves the fastest. Still, these can require at least one hour in the spring, summer, and fall and up to two hours in the winter (the greater the temperature change, the longer it will take). Refractors and catadioptric instruments, because of their sealed tubes, need up to twice as long!

Several steps can be taken to minimize the time required for instrument cooling. For openers, find a cool and dry place to store the telescope when not in use. If a dedicated observatory is impractical, then good alternatives include a vented wooden garden shed or a sealed-off corner of an unheated garage. This way, the telescope's temperature will always be close to the outside temperature even before it is set up. If you must drive with your telescope to an observing site, try to travel without using the heater on the way there.

Tilting the open end of a reflector vertically like a smokestack can help speed air-temperature equalization because warm air rises. In the case of sealed tubes, where a lens or corrector prevents the smokestack effect, it is best to turn the instrument broadside to the wind. The cooling breeze will help wick away heat as it is radiated by the telescope tube.

If the telescope is of your own making, select a material other than metal for the tube. Wood, cardboard, and fiberglass have a lower heat capacity than metal and, therefore, will have less heat to radiate once set outside. Unfortunately, nonmetal tubes can also retain the warm air inside the tube, but in practice the benefit outweighs the disadvantage. Metallic tubes also dew over more quickly.

A variation on this theme is not to use a solid tube at all but rather an open-truss framework to support the optics. Such an idea has been popular among observatory-installed telescopes for a century or more, as it speeds thermal equalization while cutting down the instrument's weight. Many Dobsonian-style reflectors utilize this same principle for the same reasons, as well as the resulting ability to break down a large instrument into a relatively small package for transport. Unfortunately, while the speed of temperature adaptation is increased, so are the chances of stray light intruding into the final image. Open tubes are also more susceptible to interference from the observer's body heat, turbulence caused by crosswinds, and possible dewing of the optics. A sleeve made from black cloth and wrapped around the truss acts to slow these interferences but will not eliminate them.

As noted in chapter 5, some manufacturers build cooling vents into their instruments' designs—a good idea. Some, such as JMI's NGT instruments, Obsession and Starsplitter Newtonians, and Meade's Maksutovs, even offer small, flat, so-called muffin fans to help rid their telescopes of trapped pockets of warm air. Amateur telescope makers would do well to consider including some type of venting system in their instruments as well, for they go a long way in speeding up a telescope's reaching temperature equilibrium. Just be sure to seal the vents against dust infiltration when the telescope is not in use.

A telescope is not the only piece of equipment that needs to adjust to the change in temperature after being brought outdoors for the first time at night; the Earth must as well. The ground you place the instrument on, having been exposed to the comparatively warm temperatures of daytime, also has to adapt to the cool of night. Different surfaces absorb heat better than others. Concrete and blacktop are the worst offenders because they readily absorb and retain heat. Grass, though also requiring a cool-down period, is better because it does not retain as much heat.

Your Observing Site

This brings up another hot topic among amateur astronomers today: where to view the sky. Choosing a good observing site is becoming increasingly difficult. The ideal location should be far from all sources of light pollution and civilization in general. In addition, it should be as high above sea level as possible to avoid low-lying haze and fog, it must be safe from social ne'er-do-wells and other possibly harmful trespassers, and it should allow for an obstacle-free view of the horizon in all directions. Wouldn't it be nice if this was a description of your backyard? We can all dream of finding such a Shangri-la, but a few compromises usually must be made.

Over the years I have used several different observing sites with varying degrees of success. Many national, state, county, and local parks and beaches offer excellent areas, but their accessibility may be limited to daytime hours only or by residency. Ask your local park office if special access is available. The local authority that oversees the state parks near my home offers a stargazing permit that allows after-hours access to more than half a dozen parks for a small annual fee. The parks not only have much better horizons than most observing sites but also offer the added benefit of round-the-clock security patrols, an important consideration in these times. (Unfortunately, the patrol cars are outfitted with more lights than a small city, but I guess you have to take the good with the bad.)

Other good alternatives include both private and public golf courses. They have wide open expanses but may also suffer from restrictions and excessive security lighting around the clubhouse. If the owner of the course is apprehensive at first, why not offer to run a free observing session for club members in return for nighttime access? Although they may not be bona fide amateur astronomers, most people jump at the chance to see celestial wonders such as the rings of Saturn and the Moon. Flat farmland can also provide a secluded view, but unless the land is your own, be sure to secure permission from the owner beforehand. The last thing you want is to be chased by a gun-wielding farmer at two in the morning!

Where are the best observing sites? Here's my top-ten list:

10. A beach (watch out for fog and salt spray!)
9. A flat rooftop (given a while to cool down after sunset)
8. A town, county, state, or national park
7. An open field or farmland
6. A club observatory
5. Your yard
4. A *daytime-only* airport or landing strip
3. The desert
2. A golf course
1. A hill or mountain

No matter where it is, a good observing site must be easy to reach. I know many urban and suburban amateurs who never see starlight during the week; instead, they restrict their observing time to weekends only, because the closest dark-sky site is more than an hour's drive away. Isn't that a pity? First, these amateurs may spend more time commuting to the stars than they do actually looking at them. Second, odds are they are forsaking many clear nights each month just because they believe the sky conditions closer to home are unusable. You really have to ask yourself if the local sky conditions are truly that bad. Remember, a telescope will show the Moon, Sun, and the five naked-eye planets as well from the center of a large city as it will from the darkest spots on Earth. Hundreds, if not thousands, of double and variable stars are also observable through even the most dismal of sky conditions. True, there is something extraspecial about observing under a star-filled sky, but never forgo a clear night just because the ambience is less than ideal. As the old saying goes, where there's a will, there's a way.

Star Parties and Astronomy Conventions

Perhaps as a reflection of the increasing interference of light pollution, the past few decades have seen a tremendous growth in regional, national, and international star parties and astronomy conventions (Figure 10.2). At these, hundreds, even thousands, of amateur astronomers travel to remote spots to hear top-notch speakers, set up their homemade and commercially purchased tele-

Figure 10.2 *The granddaddy of all amateur astronomy conventions, Stellafane hosts thousands of hobbyists each summer atop Breezy Hill in Springfield, Vermont.*

scopes, and enjoy dark-sky conditions far superior to the skies back home. Table 10.1 lists some of the world's largest and oldest.

Dozens, even hundreds, of smaller conventions are held across the country and around the world throughout the year. To help spread the word about when and where they will occur, astronomy magazines contain monthly listings giving information and addresses for further information, while the Star Ware home page at www.philharrington.net offers links to listings on the Internet. Take a look to see if there is an upcoming event near you. If so, by all means, try to attend. Astronomy conventions and star parties are great ways to meet new friends, learn a lot about your hobby and science, and get a terrific view of the night sky.

Finding Your Way

Once a site is selected, it is time to depart on your personal tour of the sky. Although most novice enthusiasts begin with the Moon and brighter planets, most objects of interest in the sky are not visible to the unaided eye or even through a side-mounted finderscope. How can a telescope be aimed their way if the observer cannot see the target in the first place? That is where observing technique comes into play. To locate these heavenly bodies, one of two different methods must be used. But before any of these systems can be discussed, it is best to become fluent in the way astronomers specify the location of objects in the sky.

Table 10.1 ***Annual Astronomy Conventions and Star Parties***

Alberta Star Party (Caroline, Alberta; August)
4612 17th Avenue NW, Calgary AB T3B 0P3, Canada
Web site: www.syz.com/rasc

Apollo Rendezvous (Dayton, Ohio; June)
Miami Valley Astronomical Society, 2600 Deweese Parkway, Dayton, OH 45414
Web site: www.mvas.org

Astrofest (Kankakee, Illinois; September)
Chicago Astronomical Society, P.O. Box 30287, Chicago, IL 60630
Web site: www.chicagoastro.org

Enchanted Skies Star Party (Socorro, New Mexico; October)
P.O. Box 743-I, Socorro, NM 87801
Web site: www.nmt.edu/~astro

Eta Aquilae Star Party (Rucava, Latvia; August)
Latvian Astronomical Society, Raina Boulevard 19 Riga, LV-1586 Latvia
Web site: www.astr.lu.lv/aquila

Grand Canyon Star Party (Grand Canyon National Park, Arizona; June)
1122 E. Greenlee Place, Tucson, AZ 85719
Web site: www.tucsonastronomy.org

Mason-Dixon Star Party (York, Pennsylvania; May)
York County Parks, 400 Mundis Race Road, York, PA 17402
Web site: www.masondixonstarparty.org

Mount Kobau Star Party (Osoyoos, British Columbia; August)
P.O. Box 20119 TCM, Kelowna, BC V1Y 9H2, Canada
Web site: www.bcinternet.com/~mksp

Nebraska Star Party (Valentine, Nebraska; August)
P.O. Box 540307, Omaha, NE 68154–0307
Web site: www.nebraskastarparty.org

Northeast Astronomy Forum (Suffern, New York; April or May)
Rockland Astronomy Club, 73 Haring Street, Closter, NJ 07624–1709
Web site: www.rocklandastronomy.com

Nova East Star Party (Fundy National Park, Nova Scotia; August)
11 Hickman Drive, Truro, NS B2N 2Z2, Canada
Web site: halifax.rasc.ca/ne

Okie-Tex Star Party (Kenton, Oklahoma; October)
P.O. Box 128, Mustang, OK 73064
Web site: www.okie-tex.com

Table 10.1 ***(continued)***

Oregon Star Party (Ochoco National Forest, Oregon; August)
P.O. Box 91416, Portland, OR 97291
Web site: oregonstarparty.org

Peach Tree Star Gaze (Jackson, Georgia; September)
1741 Bruckner Court, Snellville, GA 30078–2784
Web site: www.atlantaastronomy.org

Queensland Astrofest (Linville, Queensland; August)
145 Ardoyne Road, Oxley QLD 4075, Australia
Web site: www.ozemail.com.au/~mhorn2/afest.html

Riverside Telescope Maker's Conference (Big Bear Lake, California; May)
8300 Utica Avenue, Suite 105, Rancho Cucamonga, CA 91730
Web site: www.rtmc-inc.org

Rocky Mountain Star Stare (near Colorado Springs, Colorado; June/July)
Colorado Springs Astronomical Society, P.O. Box 62022, Colorado Springs, CO 80962–2022
Web site: www.rmss.org

South Pacific Star Party (Ilford, New South Wales, Australia; March)
Astronomical Society of NSW, GPO Box 1123, Sydney, NSW 2001, Australia
Web site: www.asnsw.com

Starfest (Mount Forest, Ontario; August)
North York Astronomical Association, 26 Chryessa Avenue, Toronto, ON M6N 4T5
Web site: www.nyaa-starfest.com

Stellafane (Springfield, Vermont; August)
P.O. Box 601, Springfield, VT 05156
Web site: www.stellafane.com

Swiss Star Party (Gurnigel, Switzerland; August)
Schaufelweg 109, CH-3098 Schliern bei Köniz, Switzerland
Web site: www.starparty.ch/english.html

Table Mountain Star Party (Ellensburg, Washington; August)
P.O. Box 785, Puyallup, WA 98371
Web site: www.tmspa.com

Texas Star Party (Fort Davis, Texas; May)
TSP Registrar, 4812 Twin Valley Drive, Austin TX 78731–3539
Web site: www.metronet.com/~tsp

Winter Star Party (West Summerland Key, Florida; February)
1051 N.W. 145th Street, Miami, FL 33168
Web site: www.scas.org

Celestial Coordinates

Like the Earth's spherical surface, the celestial sphere has been divided up by a coordinate system. On Earth, the location of every spot can be pinpointed by its unique longitude and latitude coordinates. Likewise, the position of every star in the sky may be defined by *right-ascension* and *declination* coordinates.

Let's look at declination first. Just as latitude is the measure of angular distance north or south of the Earth's equator, declination (abbreviated *Dec.*) specifies the angular distance north or south of the celestial equator. The celestial equator is the projection of the Earth's equator up into the sky. If we were positioned at 0° latitude on Earth, we would see 0° declination pass directly through the zenith, while 90° north declination (the North Celestial Pole) would be overhead from the Earth's North Pole. From our South Pole, 90° south declination (the South Celestial Pole) is at the zenith.

As with any angular measurement, the accuracy of a star's declination position may be increased by expressing it to within a small fraction of a degree. We know there are 360° in a circle. Each of those degrees may be broken into 60 equal parts called *minutes of arc*. Further, every minute of arc may be broken up to 60 equal *seconds of arc*. In other words:

$$\text{one degree } (1^\circ) = 60 \text{ minutes of arc } (60') \\ = 3{,}600 \text{ seconds of arc } (3{,}600'')$$

When minutes of arc and seconds of arc are spoken of, an angular measurement is being referred to, not the passage of time.

Right ascension (abbreviated *R.A.*) is the sky's equivalent of longitude. The big difference is that while longitude is expressed in degrees, right ascension divides the sky into 24 equal east–west slices called *hours*. Quite arbitrarily, astronomers chose as the beginning, or zero-mark, of right ascension the point in the sky where the Sun crosses the celestial equator on the first day of the Northern Hemisphere's spring. A line drawn from the North Celestial Pole through this point (the Vernal Equinox) to the South Celestial Pole represents 0 hours right ascension. Therefore, any star that falls exactly on that line has a right ascension coordinate of 0 hours. Values of right ascension increase toward the east by one hour for every 15° of sky crossed at the celestial equator.

To increase precision, each hour of right ascension may be subdivided into 60 minutes and each minute into 60 seconds. A second equality statement summarizes this:

$$\text{one hour R.A. } (1 \text{ h}) = 60 \text{ minutes R.A. } (60 \text{ m}) \\ = 3{,}600 \text{ seconds R.A. } (3{,}600 \text{ s})$$

Unlike declination, where a minute of arc does not equal a minute of time, a minute of R.A. does.

The stars' coordinates do not remain fixed. Due to a 26,000-year wobble of the Earth's axis called *precession,* the celestial poles actually trace circles on the sky. Right now, the North Celestial Pole happens to be aimed almost exactly at Polaris, the North Star. But in 13,000 years it will have shifted away from Polaris and will instead point toward Vega in the constellation Lyra. The passage of another 13,000 years will find the pole aligned with Polaris once again.

Throughout the cycle, the entire sky shifts behind the celestial coordinate grid. Although this shifting is insignificant from one year to the next, astronomers find it necessary to update the stars' positions every 50 years or so. That is why you will notice that the right ascension and declination coordinates are referred to as "epoch 2000.0" in this book and most other contemporary volumes. These indicate their exact locations at the beginning of the year 2000 but are accurate enough for most purposes for several decades on either side.

Star-Hopping

The simplest method for finding faint sky objects, and also the one preferred by most amateur astronomers, is called *star-hopping*. Star-hopping is a great way to learn your way around the sky while developing your skills as an observer. Before a telescope can be used to star-hop to a desired target, its finderscope must be aligned with the main instrument. Take a look at the telescope's finder. Chances are it is held by six thumbscrews in a set of mounting rings. Begin the process by aiming the telescope toward a distant identifiable object. Although the Moon, a bright star, or a bright planet may be used, I suggest using instead a terrestrial object such as a distant light pole or mailbox. The reason for this is quite simple: They don't move. Celestial objects appear to move because of the Earth's rotation, making constant realignment of the telescope necessary just to keep up. Center the target in the telescope's field and lock the mounting's axes. To alter the finder's aim, adjust the front three thumbscrews until the finder's tube is centered in the front mounting ring. Now loosen the back three screws. Move the finder by hand until the target is centered in the finder's crosshairs. Once set, tighten all of the adjustment screws and check to see that the finder or telescope did not shift in the process.

Some finderscopes are held by only three adjustment screws and are adjusted in much the same way (though they are much more prone to misalignment). Aim at a terrestrial target as previously suggested and lock the telescope's axes. Look through the finder to see which direction is out of alignment. Loosen the two opposite screws and move the finder by hand until the target is centered. Tighten all thumbscrews, making a final check to see that the finder is aimed correctly.

With the finder correctly aligned to the telescope, the fun can begin. After deciding on which object you want to find, pinpoint its position on a detailed star atlas. Scan the atlas page for a star near the target that is bright enough to be seen with the naked eye. Once a suitable star is located on the atlas, turn your telescope (or binoculars) toward it in the night sky. Looking back at the atlas, try to find little geometric patterns among the fainter suns that lie between the naked-eye star and the target. You might see a small triangle, an arc, or perhaps a parallelogram. Move the telescope to this pattern, center it in the finderscope, and return to the atlas. By switching back and forth between the finderscope and the atlas, hop from one star (or star pattern) to the next across the gap toward the intended target. Repeat this process as many times as it takes to get to the area of your destination. Finally, make a geometric pattern among the stars and the object itself. For instance, you might say to yourself, "My object lies halfway between and just south of a line connecting star A and star B." Then locate star A and star B in the finder, shift the view to the point between and south of those two stars, and your target should be in (or at least near) the telescope's field of view. Don't worry if you get lost along the way; breathe a deep sigh and return to the starting point.

Let's imagine that we want to find M31, the Andromeda Galaxy. (M31 is this galaxy's catalog number in the famous Messier listing of deep-sky objects.) Begin by finding it in Figure 10.3. M31's celestial address is right ascension 0 hours 42.7 minutes (written 00^h $42^m.7$), declination +41°16′. Notice that it is located northwest of the naked-eye stars Beta (β), Mu (μ), and Nu (ν) Andromedae. Find these stars in the sky and center your telescope's aim on them. Hop from Beta to Mu and then on to Nu. M31 lies a little over a degree to the west-northwest of this last star and should be visible through the finderscope (indeed, it is visible to the naked eye under moderately dark skies).

Easy, right? Now let's tackle something a little more challenging. Many observers consider M33 (seen back in Figure 6.3), the Great Spiral Galaxy in Triangulum, to be one of the most difficult objects in the sky to find. Most references note it as magnitude 5.9, which means that if the galaxy could somehow be squeezed down to a point, that would be its magnitude. However, M33 measures a full degree across, so its *surface brightness,* or brightness per unit area, is very low.

M33 and its home constellation are also plotted on Figure 10.3. How can we star-hop to this point? One way is to locate the naked-eye stars Alpha Trianguli and our old friend Beta Andromedae. Aim at the former with the finderscope and move slightly toward the latter. About a quarter of the way between Alpha Trianguli and Beta Andromedae and a little to the south lies a 6th-magnitude star. Spot it in your finderscope. Connect an imaginary line between it and Alpha and then extend that line an equal and opposite distance from the 6th-magnitude star. There you should see the galaxy's large, hazy glow. Nevertheless, when you peer through your telescope at that point, nothing is there. Recheck the map. No, that's not it; the telescope is aimed in the right direction. It's got to be there! What could be wrong?

At frustrating times such as this, it is best to pull back from your search and take a breather. Then go back, but this time use the averted-vision technique mentioned later in the section of this chapter entitled "Eye Spy." Look a little to one side of where the galaxy should be. By glancing at it with peripheral vision, its feeble light will fall on a more sensitive area of the eye's retina. Suddenly, there it is! You will wonder how you ever missed it before.

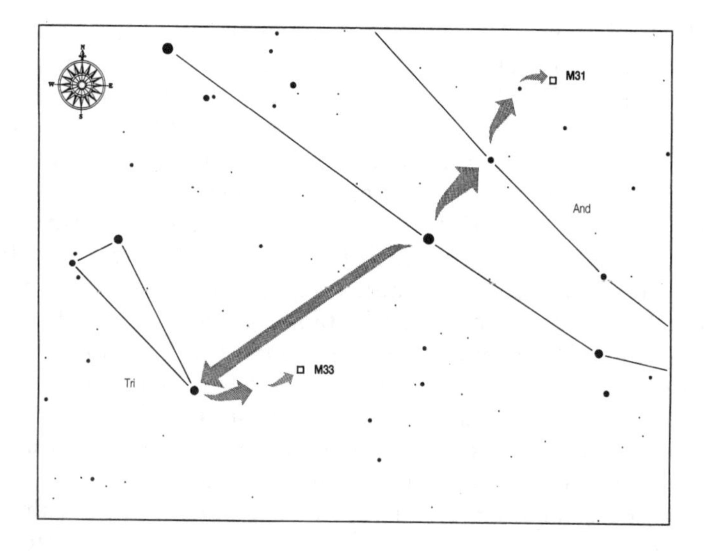

Figure 10.3 *Star-hopping is the only way to find deep-sky objects, according to some observers. Here are suggested plans of attack for locating M31, the Andromeda Galaxy, and M33, the Great Spiral in Triangulum. Both are in the autumn sky.*

Setting Circles

This brings us to the second method that amateur astronomers use to find faint objects in the sky. It sounds easiest in theory, because you don't have to know the sky. But in fact it's no shortcut—as the next few pages of (complicated) instructions should make clear.

Nowadays, most telescope mountings can be retrofitted with digital setting circles, which are, in effect, tiny computers that help the observer aim a telescope toward a preselected target. These were examined at some length in chapter 7, but the discussion here deals with old-fashioned, mechanical-style setting circles. Mechanical setting circles—round, graduated scales—are only found on equatorial mounts, one on each axis (Figure 10.4). The circle on the

Figure 10.4 *Setting circles allow an observer to take aim at a sky object by knowing its celestial coordinates.*

polar axis is divided into 24 equal segments, with each segment equal to one hour of right ascension. The setting circle attached to the declination axis is divided into degrees of declination, from 90° north to 90° south. With an equatorial mounting accurately aligned to the celestial pole (or any mounting with the properly encoded alt-azimuth digital setting circles), a sky object may be located by dialing in its pair of celestial coordinates.

Is that all there is to it? Unfortunately, no. First, to use setting circles, the mounting's polar axis must be aligned to the celestial pole. Begin by leveling the telescope. This may be done by adjusting the length of each tripod leg (if so equipped) or by placing some sort of a block (a piece of wood, a brick, etc.) under one or more corner footpads of the mount. Don't spend a lot of time trying to make the mount perfectly level; close is good enough. In fact, to be completely correct, it is not necessary to level the mount at all. The only thing that matters is that the polar axis be aimed at the celestial pole. In practice, however, it is easier to polar-align a mount that is level than one that is not, so take the time to do so. As an aid, many instruments come outfitted with bubble levels on their mountings.

Next, check to make sure that your finderscope is aligned to the telescope and that the entire instrument is parallel to its polar axis. This latter step is usually accomplished simply by swinging the telescope around until the declination circle reads 90°. Most declination circles are preset at the factory, although some are adjustable; others may have slipped over the years. If you believe that the telescope is not parallel to the polar axis when the declination circle reads 90°, consult your telescope manual for advice on correcting the reading. If your manual says nothing, or worse yet you cannot find it (it must be here somewhere, you think), try this test. Align the telescope to the polar axis as best you can by eye and lock the declination axis. Using only the horizontal and vertical motion (azimuth and altitude, respectively) of the mounting, center some distant object, such as a treetop or a star, in your finderscope's view. Now, rotate the telescope about the polar axis *only.* If the telescope/finderscope combination is parallel to the polar axis, then the object will remain fixed in the center of the view; in fact, the entire field will appear to pivot around it.

If, however, the object moved, then the finder and the polar axis are not parallel to one another. Try it again, but this time pay close attention to the direction in which the object shifted. If it moved side to side, shift the entire mounting in azimuth (horizontally) exactly half the *horizontal* distance it moved. If it moved up and down, then the mounting's altitude (vertical) pivot is not set at the correct angle. Loosen the pivot and move the entire instrument one-half the *vertical* distance that the object moved. Because, in all likelihood, it shifted diagonally, this will turn into a two-step procedure. Take it one step at a time, first eliminating its horizontal motion, then the vertical.

With the polar axis and telescope now parallel, it is time to set the polar axis parallel to Earth's axis. Some equatorial mounts come with a polar-alignment finderscope built right into the polar axis. These come with special clear reticles surrounding the celestial pole. Consult the telescope's manual for specific instructions.

Polar-alignment finderscopes are certainly handy, but for those of us without such luxuries, the following method should work quite well. With the telescope level and the declination axis locked at +90°, point the right-ascension axis by eye approximately toward Polaris. Release the locks on both axes and swing the instrument toward a star near the celestial equator. Once aimed at this star, spin the right-ascension circle (taking care not to touch the declination circle) until it reads the star's right ascension. Table 10.2 suggests several suitable

Table 10.2 **Suitable Stars for Setting-Circle Calibration**

Star	Epoch	Right Ascension		Declination	
		h	m	°	′
Alpheratz (Alpha Andromedae)	(2000)	00	08.5	+29	05
	(2005)	00	08.7	+29	07
	(2010)	00	09.0	+29	09
Hamal (Alpha Arietis)	(2000)	02	07.5	+25	28
	(2005)	02	07.9	+25	29
	(2010)	02	08.0	+25	31
Aldebaran (Alpha Tauri)	(2000)	04	35.9	+16	30
	(2005)	04	36.2	+16	31
	(2010)	04	36.4	+16	31
Procyon (Alpha Canis Minoris)	(2000)	07	39.3	+05	14
	(2005)	07	39.6	+05	13
	(2010)	07	39.9	+05	12
Regulus (Alpha Leonis)	(2000)	10	08.3	+11	58
	(2005)	10	08.6	+11	57
	(2010)	10	08.9	+11	55
Arcturus (Alpha Boötis)	(2000)	14	15.7	+19	11
	(2005)	14	15.9	+19	10
	(2010)	14	16.1	+19	08
Altair (Alpha Aquilae)	(2000)	19	50.8	+08	53
	(2005)	19	51.0	+08	53
	(2010)	19	51.3	+08	54
Polaris (Alpha Ursae Minoris)	(2000)	02	31.8	+89	16
	(2005)	02	37.6	+89	17
	(2010)	02	43.7	+89	18
South Pole Star (Sigma Octantis)	(2000)	21	08.6	−88	57
	(2005)	21	13.1	−88	56
	(2010)	21	17.2	−88	55

stars for this activity. Note that the stars' positions are given at five-year intervals. Choose the pair closest to your actual date. (Although this slight shift is not of much concern when aligning a telescope to use setting circles, it is of great consequence for long-exposure, through-the-telescope astrophotography.)

Swing the telescope back toward the celestial pole, stopping when the setting circles read the position of Polaris (also given in Table 10.2). If the telescope is properly aligned with the pole, then Polaris should be centered in view. If not, lock the axes and shift the entire mounting horizontally and vertically until Polaris is in view. Repeat the procedure again until Polaris is in view when the circles are set at its coordinates. Owing to the coarse scale of most circles supplied on amateur telescopes, a polar alignment within roughly 0.5 to 1° of the celestial pole is usually the best you can get. (Digital setting circles are much more accurate, but still the alignment need not be overly precise to be useful.)

Once the mounting is adjusted to the pole and the right-ascension circle (also known as the *hour circle*) is calibrated with the coordinates of a known star, the mount's clock drive must be turned on. Many equatorial mounts have a direct link that turns the hour circle in time with the telescope. This way, as the sky and its coordinate system shift relative to the horizon, the hour circle will move along with them. Some less-sophisticated equatorial mounts, however, do not have driven hour circles. In these instances, the hour circle must be recalibrated not just once a night but before each use—inconvenient, to say the least. The easiest way to do this is to reset it on a reference star immediately before swinging the telescope toward a new target.

Incidentally, though the previous paragraphs describe the procedure for aligning to the North Celestial Pole, the actions are the same for observers in the Southern Hemisphere, except the mounting must be aligned to the South Celestial Pole. Everywhere you see "Polaris" or "North Star," substitute in "Sigma Octantis" or "South Star." Sigma Octantis, a 5.5-magnitude sun, is located about 1° from the South Celestial Pole. Its celestial coordinates are also given in Table 10.2. Although not as obvious as its northern counterpart, Sigma works well for the task at hand.

With the mounting polar-aligned, the setting circles calibrated, and the clock drive switched on, it is now a relatively simple matter to swing the telescope around until the circles read the coordinates of the desired target. If all was done correctly beforehand, the target should appear in, or at least very near, the eyepiece's field of view. Be sure to use your eyepiece with the widest field when first looking for a target's field before moving up to a higher-power ocular.

What if it's not there? Don't give up the ship immediately. Instead, try recalibrating the setting circles on a known star near the intended target's location. This technique, called *offsetting,* is a lot more accurate than trusting the setting circles to give the correct reading by themselves.

How about digital setting circles and GoTo computer-driven telescopes? Both are much easier to get up and running. In fact, neither requires any polar alignment at all, except for needing to aim the telescope in the general direction of north (or south, in the Southern Hemisphere) to get going. You don't even need to be able to see—or, for that matter, even know—Polaris. Instead, a compass is more than adequate for initial aiming.

Begin by turning on the power to the setting circles or onboard computer. While both are likely to be battery powered, the latter can really draw quite a bit of power, draining a set of batteries very quickly. If at all possible, plug the GoTo control into an AC wall outlet or use an extension cord. Let the computer initialize. GoTo telescopes usually prompt you to enter a variety of information, such as the date, time, your location (which may be selected from the unit's built-in database), and possibly the telescope model. Manually driven digital setting circles do not. Instead, they require that you initialize the telescope by aiming it at two of the stars in the onboard database.

Once everything is initialized, both work pretty much in the same way. Select a target from the object database. In the case of GoTo telescopes, press go and the telescope does the rest. If you want the same effect with manual digital circles, say go, then move the telescope by hand. Watch the indicator on the hand control box until the location reads "00," which means the target is, in theory, centered in view. Some fine adjustment is inevitable, but if you set everything up correctly, the accuracy should be quite good.

Now that the technique for using setting circles is familiar, here is why they should *not* be used, especially by beginners who are unfamiliar with the sky. To my way of thinking (and you are free to disagree), using setting circles or a computer to help aim a telescope reduces the observer to little more than a couch-potato sports spectator flipping television channels between football games on a Sunday afternoon. Where is the challenge in that? Observational astronomy is not meant to be a spectator's sport; it is an activity that is best appreciated by doing. There is something very satisfying in knowing the sky well enough to be able to pick out an object such as a faint galaxy or an attractive double star using just a telescope, a finder, and a star chart. Even if you get your setting circles perfectly aligned and master their use, you will be missing out on the thrill of the hunt. Stalking an elusive sky object is much like searching for buried treasure: you never know what else you are going to uncover along the way. As you set your attention toward one target, you might turn up other objects that you have never seen before. Of course, there is also that slight possibility of striking astronomical pay dirt by stumbling onto an undiscovered comet. Imagine that thrill!

All right, so maybe I have been hard on setting circles, maybe even a bit unjustifiably. Make no mistake; setting circles are very useful tools *when used for the right reasons*. There is no question that they are required to aim the large, cumbersome telescopes in professional observatories where viewing time is at a premium. They also serve a very real purpose for advanced amateurs who are involved in sophisticated research programs, such as searching for supernovae in distant galaxies or estimating the brightness of variable stars. Both of these activities involve rapid, repetitive checking of a specific list of objects. Setting circles can also come in very handy when light pollution makes star hopping difficult. However, they should never be used as a crutch. Most amateur astronomers will do much better by looking up and learning how to read the night sky rather than looking down at the setting circles.

Eye Spy

Although we may think of our binoculars and telescopes as wonderful optical instruments, there is no optical device as marvelous or versatile as the human eye. Experts estimate that 90% of the information processed by our brains is received by our eyes. There is no denying it; we live in a visual cosmos. To better understand how we perceive our surroundings, both earthly and heavenly, let us pause a moment to ponder the workings of our eyes.

The human eye (Figure 10.5) measures about an inch in diameter and is surrounded by a two-part protective layer: the transparent, colorless *cornea* and the white, opaque *sclera*. (Remember, don't shoot until you see the whites of their scleras.) The cornea acts as a window to the eye and lies in front of a pocket of clear fluid called the *aqueous humor* and the *iris*. Besides giving the eye its characteristic color (be it blue, brown, or any of a wide variety of other hues) the iris regulates the amount of light entering the eye and, more important, varies its focal ratio. Under low-light conditions, the iris relaxes, dilating the *pupil* (the circular opening in the center of the iris) to about 7 mm. Under the brightest lighting, the iris contracts the pupil to 2.5 mm or so across, increasing the focal ratio and masking lens aberrations to produce sharper views. From the pupil, light passes through the eye's *lens* and across the eyeball's interior, the latter being filled with fluid called the *vitreous humor.* Both the lens and cornea act to focus the image onto the *retina,* which is composed of 10 layers of nerve cells, including photosensitive receptors called *rods* and *cones.* Cones are concerned with brightly lit scenes, color vision, and resolution; rods are low-level light receptors that cannot distinguish color. There are more cones toward the *fovea centralis* (the center of the retina and our perceived view), while rods are more numerous toward the edges. There are neither rods nor cones at the junction with the optic nerve (the eye's *blind spot*).

From an astronomical point of view, we are most interested in the eye's performance under dimly lit conditions, our so-called night vision. Try this test

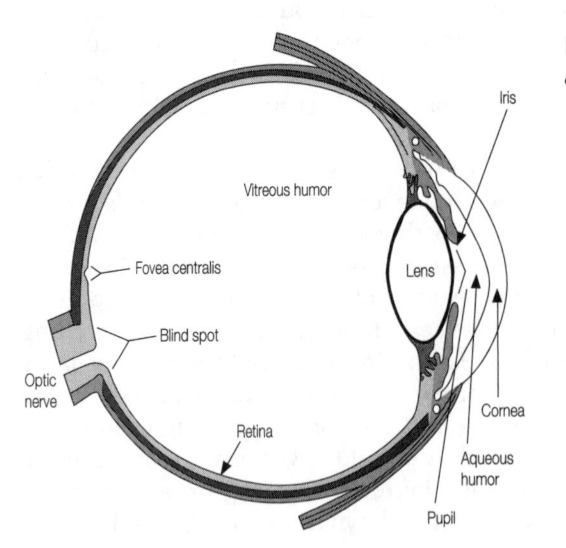

Figure 10.5 *The human eye, the astronomer's favorite tool.*

on the next clear, moonless night. Go outdoors from a well-lit room and look toward the sky. Chances are you will be able to pick out a few of the brightest stars shining in what looks like an ink-black background. Face away from the sky, wait three or four minutes, and then look up again. You will immediately notice that many more stars seem to have appeared in that short time, an indication that the eye has begun to adapt to its new, darker environment.

Complete the exercise by turning away from the sky a second time, making certain to block your eyes from any stray lights. Wait another 15 to 20 minutes and then look skyward once again. This time there will be even more stars than before. Over the ensuing time, the eye has become fully adapted to the dark. Not only has the pupil dilated, but a shift in the eye's chemical balance has also occurred. The buildup of a chemical substance called *rhodopsin* (also known as *visual purple*) has increased the sensitivity of the rods. Most people's eyes become adjusted to the dark in 20 to 30 minutes, though some require as little as 10 minutes or as long as one hour. This is enough to begin a night's observing, but *complete* dark adaptation takes up to another hour to occur.

Although the eye's sensitivity to dim lighting increases dramatically during the dark-adaptation process, it loses most of its sensitivity to color. As a result, visual observers can never hope to see the wide range of hues and tints in sky objects that appear so vivid in astronomical photographs. Nebulae are good examples of this. Unlike stars, which shine across a broad spectrum, nebulae are only visible at specific, narrow wavelengths. Different types of nebulae shine at different wavelengths. For instance, emission nebulae, those clouds that are excited into fluorescence by the energy from young stars buried deep within them, shine with a characteristic reddish color. Red, toward the long end of the visible spectrum, is all but invisible to the human eye under dim illumination. As a result, emission nebulae are among the toughest objects for the visual astronomer to spy.

The eye is best at perceiving color among the brighter planets, such as Mars, Jupiter, and Saturn, as well as some double stars where the color contrast between the suns can be striking. Most extended deep-sky objects display little color, apart from the greenish and bluish tints of some brighter planetary nebulae and the golden and reddish-orange tinges of some star clusters.

While the eye's blind spot does not adversely affect our night vision, the fact that the fovea centralis is populated only by cones does. What this means, quite simply, is that the center of our view is *not* the most sensitive area of the eye to dim light, especially when it comes to diffuse targets such as comets and most deep-sky objects. Instead, to aid in the detection of targets at the threshold of visibility, astronomers use a technique called *averted vision*. Rather than staring directly toward a faint object, look a little to one side or the other of where the object lies. By averting your vision in this manner, you direct the target's dim light away from the cone-rich fovea centralis and onto the peripheral area of the retina, where the light-sensitive rods stand the best chance of revealing the faint target.

Another way to detect difficult objects is to tap the side of the telescope tube very lightly. Your peripheral vision is very sensitive to motion, so a slight back-and-forth motion to the field of view will frequently cause faint objects to reveal themselves. I know it sounds a bit strange, but try it. It works, but be gentle.

Here's an aside for all readers who wear eyeglasses. If you suffer from either nearsightedness or farsightedness, it is best to remove your glasses before looking through the eyepiece of a telescope or binoculars, refocusing the image until everything is sharp and clear. On the other hand, if you suffer from astigmatism or require eyeglasses with thick, curved lenses, then it is best to leave the glasses on. Of course, if you use contact lenses, leave them in place when observing as you would doing any other activity.

Frequently, localized light pollution will also mask a faint object. The distraction caused by glare seen out of the corners of your eyes from nearby porch lights, streetlights, and so on can be enough to cause a faint celestial object to be missed. To help shield our eyes from extraneous light, many eyepieces and binoculars sold today come with built-in rubber eyecups. Although they prove adequate under most conditions, eyecups may not block out all peripheral light. Here are a couple of tricks to try if eyecups alone prove inadequate.

This first idea was already mentioned in my book *Touring the Universe through Binoculars,* but I think it merits repeating here. Buy a cheap pair of ordinary rubber underwater goggles, the kind you can find at just about any toy or sporting goods store. Cut out half of the goggles' front window. If you prefer to use one eye over the other for looking through your telescope, make certain to cut out the correct side. Of course, both windows would need to be cut out if used with binoculars. Spray the goggles with flat black paint, and they're done. To test your creation, put the goggles on (after the paint has dried, please!) and go out under the stars. The blackened goggles should provide enough added baffling to keep stray light from creeping around the eyepiece's edge and into your eyes.

Here another approach to the same situation that works well for viewing through telescopes but is not really applicable to binocular use. This involves wearing a dark turtleneck shirt or sweater, but not in the way you are used to. Instead of slipping it on from bottom to top, stick your head into the shirt through the neck opening. Let the shirt rest on your shoulders. Whenever you look through the eyepiece, simply pull the shirt up and over your head to act as a cloak against surrounding lights. This idea was first used by photographers a century ago, and it still works today. The only drawback, apart from looking a little strange to civilians, is that the eyepiece may dew over more quickly because of trapped body heat. Still, I can sometimes see up to a half-magnitude fainter just by using this cloaking device. Give it a try.

Record Keeping and Sketching

One of the best habits an amateur astronomer can develop is keeping a logbook of everything he or she sees in the sky. Recording observations serves the dual purpose of both chronicling what you have seen as well as how you have developed as an observer. It's also a great way to relive past triumphs on cloudy nights.

Although you are free to develop your own system, I prefer to record each object on a separate sheet of paper, including a few descriptive notes and a drawing. See Figure 10.6 for an example of a generic observation form. Most of the entries should be self-explanatory. *Transparency* rates sky clarity on a scale of 0, indicating complete overcast, to 10, which is perfect. *Seeing* refers to the steadiness of the atmosphere, from 0 (rampant scintillation) to 10 (very steady, with no twinkling even at high power). Because both of these are subjective judgments, as previously noted, I like to include the magnitude of the faintest star in Ursa Minor visible to the naked eye. This helps put the other two values in perspective. (The only exception to this is if the sky near the object under observation is noticeably different than that near Ursa Minor. In this case, record the faintest star visible near the object itself.)

It is difficult, if not impossible, to convey the visual impact of subtle heavenly sights with words alone, which is why including a drawing with all written

Observation Record

Object: ________________ Constellation: ____________

Date: ____________________ Time: ____________________

Observing Site: ____________________________________

Sky Trans'y: _____ Naked-eye Limit: _____ Seeing: _____

Telescope: ____

Eyepiece: ________________ Filter: __________________

Notes: ____

____

____

____

____

Figure 10.6 *Suggested observation form for recording celestial sights.*

observations is so important. The drawing should convey the perceived image as accurately as possible. Right now, some readers might be thinking, "I can't even draw a straight line." That's all right, because few objects in space are straight. Actually, sketching celestial objects is not as difficult as might be thought initially. It just takes a little practice.

Although astrophotographers require elaborate, expensive equipment to practice their trade, the astro-artist can enjoy his or her craft with minimal apparati. Besides a telescope or tripod-mounted binoculars, all you need to begin are a clipboard, a pad of paper, and a few pencils. All of these supplies are available from almost any art or stationery supply store for about twenty dollars. Here are a few specific things to look for. The delicate textures of celestial bodies are best rendered using an artist's pencil with soft lead such as H or HB or with sketching charcoal. (Just be careful with the charcoal, as it tends to get away from those unfamiliar with it.) As for surface media, most astro-artists prefer smooth white paper as opposed to rag bond typing paper. The grain of rag paper tends to overwhelm the fine shading of most astronomical sketches. Most sketching pads have a fine surface grain ideal for the activity, but even computer printer paper will suffice. Lastly, the paper must be kept from blowing away while the artist moves between eyepiece and sketch. The simplest approach is a clipboard with a dim red flashlight clipped on or otherwise held in place.

Sketching a sky object is a multistep process, one that requires patience and close attention to detail. First, examine the target at a wide range of magnifications. Select the one that gives the best overall view as the basis for the sketch. Begin the drawing by lightly marking the positions of the brighter field stars surrounding the target object. Next, go back to each star and change its appearance to match its perceived brightness. Convention has it that brighter stars are drawn larger than fainter ones. Once the field is drawn accurately, lightly depict the location and shape of the target itself. The last step is to shade in the target to match its visual impression. Using your finger, a smudging tool, or a soft eraser to smudge the lead, recreate the delicate shadows and brighter regions. Remember, the drawing is, in effect, a negative image of the object. As such, the brightest areas should appear the darkest on the sketch, and vice versa. Finally, examine the target again with several different eyepieces, penciling in any detail that previously went unseen.

Here are some final tips for beginning astro-artists. First, never try to hurry a drawing, even if it is bitter cold and your hands are turning blue. Better to put the pencil down, take a five- or ten-minute warm-up break, and then go back. Second, do not try to create a Rembrandt at the telescope. Instead, make only a rough (but accurate) sketch at the telescope itself; the final drawing may be made later indoors at your leisure. Last, avoid the urge to add a little stylistic license, such as putting spikes around the brighter stars or drawing in detail that isn't quite visible but that you think is there from looking at photographs. Remember, a good astro-artist is an impartial reporter.

Once filled out, the observation record may be filed in a large loose-leaf notebook. It is handiest to separate observations by category. Individual headings might include the Moon, planets, variable stars, deep-sky objects, and so on. Interstellar objects may be further broken down first by type and then by increasing right ascension beginning at 0 hours. Members of the Solar System might be separated by object and then filed chronologically.

The eye of the experienced observer can detect much more tenuous detail in sky objects than a beginner can spot. Does this mean that the veteran has better eyesight? Probably not. Like most things in life, talent is not inborn. It has to be nurtured and developed with time. That's why it is important to keep notes. You will be amazed at how far your observing skills have come when you look back at your early entries a few years later.

Observing versus Peeking (A Commentary)

Are you an astronomical observer or an astronomical peeker? There is a difference... a big one! A peeker flits from one object to the next, barely looking at each before... *swish*... it's off to another. He or she never writes down what was seen, let alone makes a simple drawing. Whenever asked what he or she saw during an observing session, the peeker will only say, "Oh, I don't know, just some stuff." The fact of the matter is that peekers usually cannot remember what they have seen and what they have not.

An observer takes a slow, methodical approach to the study of the night sky. Most compile long lists of objects they want to see before going outside in an effort to use each moment under the stars as effectively as possible. Unlike our friend the peeker, the true observer is not out to break any land-speed records. He or she prefers to take it a little more slowly, savoring each photon that reaches the eye.

Perhaps it is a sign of the hectic times in which we live, but more and more amateur astronomers seem to be peekers. If you are one of them, I have to ask you this: What's your rush? Take a deep breath and relax. Resist the urge to race impulsively across the sky. By taking a slower, more deliberate tour, you will see the heavens in a new and exciting light. Become an astronomical observer, a connoisseur of the universe.

Epilogue

The universe holds a lifetime of interesting sights for each of us to enjoy. In fact, there are so many sights that many observers simply do not know where to begin. My advice? Start with the easy stuff. Take a tour of the Moon, and learn its more striking features in the same way that you have learned your own neighborhood. Many amateurs tire of the Moon, probably because they consider it too easy, yet getting to know its many features intimately can take years of careful study. Next, if there is a bright planet visible in the sky, be sure to pay it a visit. Once again, avoid the peeker syndrome mentioned in the last chapter. Instead, savor the view slowly. Drink it in like a fine wine, rather than guzzling it like cheap Ripple. Finally, step beyond our solar system into the deep sky. There are many guidebooks available that will take you there, including a sequel to this book, due out in 2003.

Yes, indeed, we live at a remarkable time. In many ways, this is the golden age of amateur astronomy. Never before have we enjoyed such an amazing selection of equipment. Huge Newtonian reflectors, exquisite refractors, and astonishingly sophisticated catadioptric instruments are all at our beck and call. The personal computer has also served to revolutionize the hobby, just as it has influenced just about every other aspect of our daily lives.

I must reiterate one final thought. Please don't end this book by thinking that astronomy is only for the rich. That may be an easy impression to leave with, given some of the huge price tags that some telescopes and accessories carry. But I must impress upon you that it is *not* important how much money you have to spend or how much equipment you own. What *does* matter is that, be it a small, inexpensive instrument, or a pricey, advanced scientific device, the best telescope in the world is the one that *you* use often and enjoy.

Appendix A
Specs at a Glance

A. Binoculars

Brand/Model	Magnification	Aperture (mm)	Field of View (degrees)	Type of Focusing[1]	Exit Pupil (mm)	Eye Relief (mm)	Weight (oz.)	Tripod Adaptable?	Type of Prisms	Coatings[2]
Price Range: Under $100										
Celestron Enduro	7	35	9.2	C	5	11	22	Yes	BK-7	FC
Celestron Enduro	7	50	6.8	C	7.1	13	28	Yes	BK-7	FC
Celestron Enduro	8	40	8.2	C	5	12	25	Yes	BK-7	FC
Celestron Enduro	8	56	6	C	7	21	36	Yes	BK-7	FC
Celestron Enduro	10	50	7	C	5	13	28	Yes	BK-7	FC
Celestron WaterSport	7	50	7	C	7.1	19	39	No	BaK-4	FC
Edmund Scientific	7	50	7	C	7.1	17	29	No	BK-7	FC
Meade Infinity	7	35	9.2	C	5	11	24	No	BK-7	FC
Meade Infinity	8	40	8.2	C	5	11	26	No	BK-7	MC
Meade Infinity	10	50	7	C	5	13	30	No	BK-7	FC
Meade Mirage	7	35	8	C	5	14	21	No	BK-7	C
Meade Mirage	10	50	5.7	C	5	13	25	No	BK-7	C
Meade Travel View	7	35	9.2	C	5	12	23	No	BK-7	FC
Meade Travel View	10	50	7	C	5	13	29	No	BK-7	FC
Meade Travel View	12	50	5.4	C	4.2	13	30	No	BK-7	FC
Minolta Classic II	7	35	9.3	C	5	12	20	Yes	BK-7	C
Minolta Classic II	7	50	7	C	7.1	19	27	Yes	BK-7	C
Minolta Classic II	8	40	8.2	C	5	12	23	Yes	BK-7	C
Minolta Classic II	10	50	6.5	C	5	12	27	Yes	BK-7	C
Orion Scenix	8	40	9	C	5	14	24	Yes	BaK-4	FC
Orion Scenix	7	50	7.1	C	7.1	17	28	Yes	BaK-4	FC
Orion Scenix	10	50	7	C	5	14	28	Yes	BaK-4	FC
Orion Scenix	12	50	6	C	4.2		28	Yes	BaK-4	FC
Swift Aerolite	7	35	6.8	C	5	14	20	No	BK-7	FC

Brand/Model	Magnification	Aperture (mm)	Field of View (degrees)	Type of Focusing[1]	Exit Pupil (mm)	Eye Relief (mm)	Weight (oz.)	Tripod Adaptable?	Type of Prisms	Coatings[2]
Swift Aerolite	7	35	9.5	C	5	9	24	No	BK-7	FC
Swift Aerolite	7	50	7.1	C	7.1	14	39	No	BK-7	FC
Swift Aerolite	8	40	9	C	5	16	33	No	BK-7	FC
Swift Aerolite	10	50	5.2	C	5	13	37	No	BK-7	FC
Price Range: $100 to $200										
Bausch & Lomb Legacy	7	35	11	C	5	9	22	Yes	BaK-4	MC
Bausch & Lomb Legacy	8	40	8.5	C	5	9	24	Yes	BaK-4	MC
Bausch & Lomb Legacy	10	50	7.5	C	5	9	28	Yes	BaK-4	MC
Brunton Lite-Tech	7	50	7	C	7.1	24	39	Yes	BaK-4	FC
Brunton Lite-Tech	10	50	6.5	C	5	19	39	Yes	BaK-4	FC
Bushnell Marine	7	50	7.0	I	7.1	n/s	37	Yes	BaK-4	FC
Bushnell Natureview	8	42	8	C	5.3	19	26	Yes	BK-7	FC
Bushnell Natureview	10	42	6	C	4.2	19	25	Yes	BK-7	FC
Celestron Bird Watcher	7	35	10	C	5	14	20	Yes	BaK-4	MC
Celestron Bird Watcher	8	40	9	C	5	14	21	Yes	BaK-4	MC
Celestron Ultima	8	40	6.6	C	5	19	21	Yes	BaK-4	FMC
Celestron Bird Watcher	10	50	6.5	C	5	13	26	Yes	BaK-4	MC
Celestron Watersport	10	50	6.5	C	5	19	40	No	BaK-4	FC
Eagle Optics Voyager	8	42	6.6	C	5.3	18	21	Yes	BaK-4	FMC
Kowa ZCF	7	40	7.5	C	5.7	23	28	No	BaK-4	FMC
Kowa ZCF	10	40	6.3	C	4.2	17	28	No	BaK-4	FMC
Meade Safari Pro	7	36	9.3	C	5.1	19	24	Yes	BaK-4	MC
Meade Safari Pro	8	42	8.2	C	5.3	19	26	Yes	BaK-4	MC
Meade Safari Pro	10	42	6.1	C	4.2	16	24	Yes	BaK-4	MC
Meade Safari Pro	10	50	6.5	C	5	16	32	Yes	BaK-4	MC
Minolta Activa	7	35	9.3	C	5	19	24	Yes	BaK-4	FMC
Minolta Activa	7	50	6.3	C	7.1	24	31	Yes	BaK-4	FMC
Minolta Activa	8	40	8.2	C	5	18	26	Yes	BaK-4	FMC
Minolta Activa	10	50	6.5	C	5	19	31	Yes	BaK-4	FMC
Minolta Activa	12	50	5.5	C	4.2	14	31	Yes	BaK-4	FMC
Nikon Action Naturalist IV	7	35	9.3	C	5	12	23	Yes	BaK-4	MC
Nikon Action Shoreline	7	50	6.4	C	7.1	20	32	Yes	BaK-4	MC
Nikon Action Egret II	8	40	8.2	C	5	12	25	Yes	BaK-4	MC
Nikon Action Lookout IV	10	50	6.5	C	5	20	32	Yes	BaK-4	MC
Nikon Action Fieldmaster	12	50	5.5	C	4.2	9	32	Yes	BaK-4	MC
Oberwerk 8 × 60	8	60	5.5	C	7.2	18	37	Yes	BaK-4	FC
Oberwerk 9 × 60	9	60	5.5	C	6.5	16	37	Yes	BaK-4	FC
Oberwerk 11 × 56	11	56	6	C	5.1	19	36	Yes	BaK-4	FMC
Oberwerk 12 × 60	12	60	5	C	4.6	14	37	Yes	BaK-4	FC
Oberwerk 15 × 60	15	60	4.1	C	3.8	12	37	Yes	BaK-4	FC

Brand/Model	Magnification	Aperture (mm)	Field of View (degrees)	Type of Focusing[1]	Exit Pupil (mm)	Eye Relief (mm)	Weight (oz.)	Tripod Adaptable?	Type of Prisms	Coatings[2]
Oberwerk 15 × 70	15	70	4.3	C	4.6	15	53	Yes	BaK-4	FMC
Oberwerk 20 × 60	20	60	3	C	3	10	37	Yes	BaK-4	FC
Orion UltraView	7	50	6.5	C	7.1	22	32	Yes	BaK-4	FMC
Orion Vista	7	50	6	C	7.1	22	28	Yes	BaK-4	FMC
Orion UltraView	8	42	8.2	C	5.3	22	27	Yes	BaK-4	FMC
Orion Vista	8	42	6.5	C	5.25	18	22	Yes	BaK-4	FMC
Orion Mini Giant	8	56	5.8	C	7	18	32	Yes	BaK-4	FMC
Orion Mini Giant	9	63	5	C	7	18	36	Yes	BaK-4	FMC
Orion UltraView	10	50	6.5	C	5	22	32	Yes	BaK-4	FMC
Orion Vista	10	50	5.3	C	5	16	28	Yes	BaK-4	FMC
Orion Intrepid	10	60	5.3	C	6	20	39	Yes	BaK-4	FC
Parks GR	7	50	6.8	C	7.1	20	24	Yes	BaK-4	FMC
Parks ZWCF	8	42	8.2	C	5.25	22	24	Yes	BaK-4	FMC
Parks GR	10	50	5.3	C	5	18	25	Yes	BaK-4	FMC
Parks ZWCF	10	52	6.5	C	5.2	22	28	Yes	BaK-4	FMC
Parks ZWCF	12	50	6	C	4.2	7	26	Yes	BK-7	FC
Pentax PCF V	7	50	6.2	C	7.1	20	34	Yes	BaK-4	MC
Pentax PCF V	8	40	6.3	C	5	20	27	Yes	BaK-4	MC
Pentax PCF V	10	40	5	C	4	17	27	Yes	BaK-4	MC
Pentax PCF V	10	50	5	C	5	20	33	Yes	BaK-4	MC
Pentax PCF V	12	50	4.2	C	4.2	20	33	Yes	BaK-4	MC
Pro-Optic Provis	7	50	6	C	7.1	23	28	Yes	BaK-4	FMC
Pro-Optic Provis	8	42	6.5	C	5.3	16	22	Yes	BaK-4	FMC
Pro-Optic Provis	10	50	5.3	C	5	16	28	Yes	BaK-4	FMC
Pro-Optic Giant	11	70	4	C	6.4	21	45	Yes	BaK-4	FMC
Pro-Optic Giant	16	70	3.6	C	4.4	11	45	Yes	BaK-4	FMC
Swift Plover	8	40	9	C	5	14	25	Yes	BaK-4	FC
Swift Cougar	10	50	7	C	5	14	29	Yes	BaK-4	FC
Price Range: $200 to $300										
Bausch & Lomb Custom	8	36	6.5	C	4.5	19	22	Yes	BaK-4	MC
Bausch & Lomb Custom	10	40	5.2	C	4	19	29	Yes	BaK-4	MC
Celestron Ultima	7	50	7	C	7.1	20	27	Yes	BaK-4	FMC
Celestron Ultima	8	56	6.1	C	7	21	31	Yes	BaK-4	FMC
Celestron Ultima	10	50	5	C	5	21	27	Yes	BaK-4	FMC
Eagle Optics Voyager ED	9.5	44	6	C	4.6	14	24	Yes	BaK-4	FMC
Nikon 7 × 50 Sports and Marine	7	50	7.5	C	7.1	18	41	Yes	BaK-4	MC
Orion Little Giant II	11	70	4.5	C	6.4	18	47	Yes	BaK-4	FMC
Orion Mini Giant	12	63	4.7	C	5.3	18	40	Yes	BaK-4	FMC
Orion Mini Giant	15	63	3.7	C	4.2	19	40	Yes	BaK-4	FMC
Orion Little Giant II	15	70	4	C	4.7	8	47	Yes	BaK-4	FMC
Orion Little Giant II	20	70	3	C	3.5	8	47	Yes	BaK-4	FMC
Parks Deluxe Giant	10	70	5	C	7	20	60	Yes	BaK-4	FMC
Pentax PCF V	16	60	2.8	C	3.8	20	43	Yes	BaK-4	MC

Brand/Model	Magnification	Aperture (mm)	Field of View (degrees)	Type of Focusing[1]	Exit Pupil (mm)	Eye Relief (mm)	Weight (oz.)	Tripod Adaptable?	Type of Prisms	Coatings[2]
Pentax PCF V	20	60	2.5	C	3	21	43	Yes	BaK-4	MC
Pro-Optic Giant	11	80	4.2	C	7.3	14	84	Yes	BaK-4	FMC
Pro-Optic Giant	20	80	3.3	C	4	14	84	Yes	BaK-4	FMC
Swift UltraLite	7	42	7	C	6	25	21	Yes	BaK-4	FMC
Swift Sea Hawk	7	50	7.5	I	7.1	19	42	Yes	BaK-4	FC
Swift UltraLite	8	42	6.6	C	5.3	22	21	Yes	BaK-4	FMC
Swift Audubon	8.5	44	8.2	C	5.2	15	29	No	BaK-4	FMC
Swift UltraLite	10	42	6.6	C	4.2	13	21	Yes	BaK-4	FMC
Swift Kestrel	10	50	7	C	5	15	32	Yes	BaK-4	FMC
Swift Sea Wolf	10	50	7.1	C	5	19	33	Yes	BK-7	FC
Swift Vanguard	15	60	4.2	C	4	10	40	Yes	BK-7	MC
Price Range: $300 to $500										
Canon IS (Image Stabilizer)	10	30	6	C	3	15	21	No	Roof	FMC
Celestron Ultima	9	63	5.4	C	7	21	35	Yes	BaK-4	FMC
Celestron Deluxe Giant	20	80	3.5	C	4	16	89	Yes	BaK-4	MC
Fujinon FMT-SX	7	50	7.5	I	7.1	23	50	Yes	BaK-4	FMC
Fujinon MT-SX	7	50	7.5	I	7.1	12	45	Yes	BaK-4	FMC
Fujinon MT-SX	10	70	5.3	I	7	12	76	Yes	BaK-4	FMC
Leupold Wind River	8	42	6.5	C	5.3	16	25	No	BaK-4	MC
Leupold Wind River	10	50	5.3	C	5	16	32	No	BaK-4	MC
Orion Giant	11	80	4.5	C	7.3	16	77	Yes	BaK-4	FMC
Orion MegaView	15	80	3.5	C	5.3	20	84	Yes	BaK-4	FMC
Orion Giant	16	80	3.5	C	5	16	77	Yes	BaK-4	FMC
Orion Giant	20	80	3.5	C	4	15	77	Yes	BaK-4	FMC
Orion MegaView	20	80	3.5	C	4	16	84	Yes	BaK-4	FMC
Parks Deluxe Giant	11	80	4.5	C	7.3	20	85	Yes	BaK-4	FMC
Parks Deluxe Giant	15	80	3.5	C	5.3	18	85	Yes	BaK-4	FMC
Parks Deluxe Giant	20	80	3.5	C	4	16	86	Yes	BaK-4	FMC
Steiner Military/Marine	7	50	7	I	7.1	22	37	Yes	BaK-4	FMC
Steiner Military/Marine	9	40	5.7	I	4.4	20	26	Yes	BaK-4	FMC
Steiner Military/Marine	10	50	6.2	I	5	17	37	Yes	BaK-4	FMC
Swarovski Habicht	7	42	6.4	C	6	14	24	No	BaK-4	FMC
Swift UltraLite ED	8	44	6.5	C	5.5	21	25	Yes	BaK-4	FMC
Swift UltraLite	9	63	5.4	C	7	21	25	Yes	BaK-4	FMC
Swift Observer	11	80	4.5	C	7.3	16	78	Yes	BaK-4	FC
Swift Satellite	20	80	3.5	C	4	16	78	Yes	BaK-4	FC
Price Range: $500 to $1,000										
Canon IS (Image Stabilizer)	12	36	5.6	C	3	15	31	No	Roof	FMC
Fujinon FMT-SX	10	70	5.3	I	7	23	76	Yes	BaK-4	FMC
Fujinon FMT-SX	16	70	4	I	4.4	16	76	Yes	BaK-4	FMC
Fujinon MT-SX	16	70	4	I	4.4	12	76	Yes	BaK-4	FMC
Nikon IF SP (Prostars)	7	50	7.3	I	7.1	15	48	Yes[3]	BaK-4	FMC
Nikon Superior E	10	42	6	C	4.2	17	24	Yes[3]	BaK-4	FMC

Brand/Model	Magnification	Aperture (mm)	Field of View (degrees)	Type of Focusing[1]	Exit Pupil (mm)	Eye Relief (mm)	Weight (oz.)	Tripod Adaptable?	Type of Prisms	Coatings[2]
Nikon Superior E	12	50	5	C	4.2	17	30	Yes[3]	BaK-4	FMC
Orion MegaView	30	80	2.3	C	2.7	14	84	Yes	BaK-4	FMC
Pro-Optic Ultra Giant	14	100	3.3	C	7.1	14	136	Yes	BaK-4	MC
Pro-Optic Ultra Giant	25	100	2.5	C	4	10	136	Yes	BaK-4	MC
Steiner Admiral Gold	7	50	7.3	I	7.1	22	37	Yes	BaK-4	FMC
Steiner Nighthunter	7	50	7.1	I	7.1	21	33	No	BaK-4	FMC
Steiner Nighthunter	8	56	6.3	I	7	21	41	No	BaK-4	FMC
Steiner Nighthunter	10	50	6	I	5	21	35	No	BaK-4	FMC
Steiner Nighthunter	12	56	4.9	I	4.7	22	41	No	BaK-4	FMC
Swarovski Habicht	10	40	6.2	C	4	13	24	No	BaK-4	FMC
Swift Audubon ED	8.5	44	8.2	C	5.2	15	29	No	BaK-4	FMC
Vixen BT80M-A	36	80	1.1	I	2.2	n/s	11 lbs.	Yes	BaK-4	MC
Zeiss B/GA T* ClassiC	7	42	9.4	C	6	19	28	Yes	Roof	FMC
Zeiss B/GA T* ClassiC	10	40	6.3	C	4	15	27	Yes	Roof	FMC
Price Range: $1,000 and up										
Canon IS	15	45	4.5	C	3	15	36	No	Roof	FMC
Canon IS	15	50	4.5	C	3.3	15	42	No	Roof	FMC
Canon IS	18	50	3.7	C	2.8	15	42	No	Roof	FMC
Fujinon Techno-Stabi	14	40	4	C	2.9	10	43	No	Roof	FMC
Fujinon MT-SX	15	80	4	I	5.3	16	16 lbs.	Yes	BaK-4	FMC
Fujinon ED-SX	25	150	2.7	I	6	19	41 lbs.	Yes	BaK-4	FMC
Fujinon EM-SX	25	150	2.7	I	6	19	43 lbs.	Yes	BaK-4	FMC
Fujinon MT-SX	25	150	2.7	I	6	19	41 lbs.	Yes	BaK-4	FMC
Fujinon ED-SX	40	150	1.7	I	3.8	15	41 lbs.	Yes	BaK-4	FMC
Miyauchi BS-77	20	77	2.5	I	3.9	20	6.6 lbs.	Yes	BaK-4	FMC
Miyauchi BJ100iB	20	100	2.5	I	5	25	10 lbs.	Yes	BaK-4	FMC
Miyauchi BJ100iBF	20	100	2.5	I	5	25	13.2 lbs.	Yes	BaK-4	FMC
Miyauchi BR-141	25	141	2.6	I	n/s	n/s	26 lbs.	Yes	BaK-4	FMC
Nikon IF SP (Astroluxe)	10	70	5.1	I	7	15	70	Yes[3]	BaK-4	FMC
Nikon IF SP	18	70	4	I	3.9	15	68	Yes[3]	BaK-4	FMC
Nikon 20 × 120 III	20	120	3	I	6	21	34 lbs.	Yes[3]	BaK-4	FMC
Oberwerk Military Binoculars	25/40	100	2.5/1.5	I	4/2.5	14/8	26.5 lbs.	Yes	BaK-4	MC
Orion SuperGiant	25	100	2.6	C	4	10	7.5 lbs.	Yes	BaK-4	FMC
Parks Deluxe Giant	25	100	2.6	C	4	10	7.6 lbs.	Yes	BaK-4	FMC
Steiner Senator	15	80	3.7	I	5.3	13	56	Yes	BaK-4	FMC
Steiner Senator	20	80	3.7	I	4	13	56	Yes	BaK-4	FMC
Takahashi Astronomer	22	60	2.1	I	2.7	18	76	Yes	BaK-4	FMC
Vixen HFT-A	20	125	3	I	6.2	20	24 lbs.	Yes	BaK-4	MC
Vixen HFT-A	25–75	125	1.6–0.8	I	4.2–1.7	20	24 lbs.	Yes	BaK-4	MC
Vixen HFT-A	30	125	1.6	I	4.2	20	24 lbs.	Yes	BaK-4	MC
Zeiss B/GA T* ClassiC	7	50	8.2	I	7.1	18	41	Yes	BaK-4	FMC
Zeiss Victory B T	8	40	8.5	C	5	16	26	Yes	Roof	FMC
Zeiss B/GA T* ClassiC	8	56	6.9	C	7	15	36	Yes	Roof	FMC
Zeiss Victory BT	8	56	8.3	C	7	18	41	Yes	Roof	FMC
Zeiss Victory BT	10	40	6.3	C	4	15	26	Yes	Roof	FMC
Zeiss Victory BT	10	56	6.9	C	5.6	16	42	Yes	Roof	FMC

Brand/Model	Magnification	Aperture (mm)	Field of View (degrees)	Type of Focusing[1]	Exit Pupil (mm)	Eye Relief (mm)	Weight (oz.)	Tripod Adaptable?	Type of Prisms	Coatings[2]
Zeiss B/GA T* ClassiC	15	60	4.3	C	4	15	56	Yes	BaK-4	FMC
Zeiss BS/GA-T*	20	60	3.1	C	3	14	57	Yes	BaK-4	FMC

1. *Type of focusing: C = center focus; I = individual focus.*
2. *Coatings: The optical coatings applied to the binocular's elements. C = coated (e.g., single-layer magnesium fluoride, likely only on the outer surfaces of the objectives and eyepieces); FC = fully coated (e.g., single coating on all optical surfaces); MC = multicoated (e.g., multiple coatings on some optical surfaces); FMC = fully multicoated (e.g., multiple coatings on all optical surfaces).*
3. *A special adapter from the manufacturer is required to attach the binoculars to a tripod.*

n/s = not supplied by manufacturer.

B. Achromatic Refractors

Brand/Model	Aperture	Focal Ratio	Tube Material[1]	Mount[2]	Total Weight (heaviest component)[3]
Price Range: Under $300					
Apogee G185	3.1	5	Metal	Alt-az	n/s
Celestron 80 Wide View	3.1	5	Metal	Sold separately	4
Helios Startravel 80	3.1	5	Metal	GEM	n/s
Konus Vista 80	3.1	5	Metal	Sold separately	n/s
Murnaghan ValueScope 90	3.5	11	Metal	Alt-az or GEM	n/s
Orion Telescopes ShortTube 80	3.1	5	Metal	None or GEM	16 (12)
Pacific Telescope Co. Skywatcher 804	3.1	5	Metal	Sold separately	3
Price Range: $300 to $600					
Apogee 80mm f/7	3.1	7	Metal	Sold separately	n/s
Apogee Widestar 127	5	6.5	Metal	GEM	n/s
Celestron Firstscope 80 AZ	3.1	11	Metal	Alt-az	18
Celestron Firstscope 80 EQ	3.1	11	Metal	GEM	24
Celestron NexStar 80GT	3.1	4	Metal	Alt-az	11
Celestron 102 Wide View	4	5	Metal	Sold separately	5
Celestron C102-AZ	4	5	Metal	Alt-az	18
Celestron C102-HD	4	10	Metal	GEM	36
Discovery Telescopes System 90	3.5	5	Metal	GEM	10 (5)
Helios Evostar 80	3.1	11.3	Metal	GEM	n/s
Helios Evostar 90	3.5	10	Metal	GEM	n/s
Helios Evostar 102	4	9.8	Metal	GEM	n/s
Helios Startravel 102	4	5	Metal	GEM	n/s
Konuspace 910	3.5	10	Metal	GEM	n/s
Meade ETX-70AT	2.8	5	Plastic	Fork	7
Meade DS-2070AT	2.8	10	Metal	Alt-az	n/s
Meade DS-80EC	3.1	11.3	Metal	Alt-az	14

Brand/Model	Aperture	Focal Ratio	Tube Material[1]	Mount[2]	Total Weight (heaviest component)[3]
Meade DS-90EC	3.5	11.1	Metal	Alt-az	15
Murnaghan Astron 90	3.5	11	Metal	GEM	n/s
Murnaghan Astron 102	4	9.8	Metal	GEM	n/s
Orion Telescopes ShortTube 90 EQ	3.5	5.6	Metal	GEM	n/s
Orion Telescopes Explorer 90	3.5	10	Metal	Alt-az	16
Orion Telescopes AstroView 90 EQ	3.5	10	Metal	GEM	24
Pacific Telescope Skywatcher 909	3.5	10	Metal	Alt-az or GEM	26
Pacific Telescope Skywatcher 1025	4	5	Metal	Alt-az	31
Pacific Telescope Skywatcher 1021	4	9.8	Metal	GEM	37
Pacific Telescope Skywatcher 1206	4.7	5	Metal	GEM	n/s
Stellarvue AT1010	3.1	6	Metal	None or GEM	5 (OTA)
Stellarvue AT80/9D	3.1	9.4	Metal	None or GEM	4.5 (OTA)
Stellarvue 102D	4	6.9	Metal	None or GEM	10 (OTA)
Vixen Custom 80M	3.1	11.4	Metal	Alt-az	42
Price Range: $600 to $1,000					
Apogee Widestar	4	6.4	Metal	Sold separately	n/s
Borg Series 80 76-mm	3	6.6	Metal	Alt-az	n/s
Helios Evostar 120	4.7	8.3	Metal	GEM	n/s
Konuspace 102	4	9.8	Metal	GEM	n/s
Konusuper 120	4.7	8.3	Metal	GEM	n/s
Meade 102ACHR/300	4	9	Metal	GEM	43
Meade 102ACHR/500	4	9	Metal	GEM	47
Meade LXD55	5	9	Metal	GEM	n/s
Meade LXD55	6	8	Metal	GEM	n/s
Orion Telescopes AstroView 120ST	4.7	5	Metal	GEM	36 (28)
Orion Telescopes AstroView 120	4.7	8.3	Metal	GEM	39 (28)
Pacific Telescope Skywatcher 1201	4.7	8.3	Metal	GEM	40
Pacific Telescope Skywatcher 15075	6	5	Metal	GEM	n/s
Pacific Telescope Skywatcher 150/1200	6	8	Metal	GEM	46
Photon Instruments Photon 102	4	9.8	Metal	GEM	n/s
Photon Instruments Photon 127	5	9	Metal	GEM	n/s
Vixen Custom 90	3.5	11.1	Metal	Alt-az	42
Price Range: $1,000 to $2,000					
Borg Series 80 100 mm	4	6.4	Metal	Alt-az	n/s
Borg Series 115 100 mm	4	4 or 6.4	Metal	Alt-az	n/s
Celestron CR150-HD	6	8	Metal	GEM	30
D&G Optical	5	12 or 15	Metal	Sold separately	19–20
D&G Optical	6	12 or 15	Metal	Sold separately	26–28
Helios Evostar 150	6	8	Metal	GEM	n/s
Unitron 3 Alt-Az	3	16	Metal	Alt-az	23
Unitron 3 EQ	3	16	Metal	GEM	46
Vixen GP 102M	4	9.8	Metal	GEM	48

Brand/Model	Aperture	Focal Ratio	Tube Material[1]	Mount[2]	Total Weight (heaviest component)[3]
Price Range: Over $2,000					
D&G Optical	8	12 or 15	Metal	Sold separately	37–39
D&G Optical	10	12 or 15	Metal	Sold separately	75–80
Konus Konusky Evolution 150	6	8	Metal	GEM	n/s
Vixen GP NA120SS	4.7	6.7	Metal	GEM	48
Vixen GP NA130SS	5.1	6.2	Metal	GEM	n/s
Vixen GP NA140SS	5.5	5.7	Metal	GEM	n/s

C. Apochromatic Refractors

Brand/Model	Aperture	Focal Ratio	Tube Material[1]	Mount[2]	Total Weight (heaviest component)[3]
Price Range: Under $2,000					
Borg Series 80 76 mm	3	6.6	Metal	Alt-az	
Meade 102APO/500	4	9	Metal	GEM	55
Stellarvue 102EDT	4	6.8	Metal	None or GEM	8.4
Stellarvue 102APO	4	6.1	Metal	None or GEM	8.4
Takahashi FS-60	2.4	5.9	Metal	GEM (Teegul)	4.4
Takahashi FS-78	3	8	Metal	GEM	7 (OTA)
Tele Vue Ranger	2.8	6.8	Metal	Sold separately	6
Tele Vue Pronto	2.8	6.8	Metal	Sold separately	6
Tele Vue TV-76	3	6.3	Metal	Sold separately	6
TMB Apo-80	3.1	6 or 7.5	Metal	Sold separately	6
Vixen GP FL-80S	3.1	8	Metal	Sold separately	n/s
Vixen 80-ED	3.1	9	Metal	GEM	40
Vixen GP 102-ED	4	6.5	Metal	GEM	40
William Optics Megrez 80	3.1	6	Metal	Sold separately	4
William Optics Megrez 102	4	6.9	Metal	Sold separately	7
Price Range: $2,000 to $4,000					
Astro-Physics 105EDFS (Traveler)	4.1	6	Metal	Sold separately	9
Borg Series 80 100 mm	4	6.4	Metal	Alt-az	n/s
Borg Series 115 100 mm	4	4 or 6.4	Metal	Alt-az	n/s
Borg Series 115 125 mm	4.9	4	Metal	Sold separately	n/s
Borg Series 115 125 mm Astrograph	4.9	2.8	Metal	Sold separately	n/s
Meade Apochromatic 102ED	4	9	Metal	GEM	69 (23)
Meade Apochromatic 127ED	5	9	Metal	GEM	78 (23)
Takahashi FCL-90 (Sky 90)	3.5	5.6	Metal	OTA or GEM	n/s
Takahashi FS-102	4	8	Metal	GEM	11 (OTA)
Takahashi FSQ-106	4.2	5	Metal	GEM	n/s
Tele Vue Bizarro	3.3	7	Metal	Sold separately	8
Tele Vue TV-85	3.3	7	Metal	Sold separately	8
Tele Vue NP-101	4	5.4	Metal	Sold separately	10
Tele Vue TV-102	4	8.6	Metal	Sold separately	9
TMB Apo-100	4	8	Metal	Sold separately	10
TMB Apo-105	4.1	6.2	Metal	Sold separately	13

Brand/Model	Aperture	Focal Ratio	Tube Material[1]	Mount[2]	Total Weight (heaviest component)[3]
Vixen GP FL-102S	4	9	Metal	GEM	46
Vixen GP 114-EDSS	4.5	5.3	Metal	GEM	n/s
William Optics Fluoro-Star 108	4.3	6.5	Metal	Sold separately	n/s
Price Range: Over $4,000					
Aries 6" f/7.9	6	7.9	Metal	Sold separately	35
Astro-Physics 130EDFS	5.1	6	Metal	Sold separately	15
Astro-Physics 155EDFS	6.1	7	Metal	Sold separately	23
Astro-Physics 155EDF	6.1	7	Metal	Sold separately	27
Borg Series 140 150 mm	6	4.2 or 6.7	Metal	Sold separately	n/s
Meade Apochromatic 152ED	6	9	Metal	GEM	153 (55)
Meade Apochromatic 178ED	7	9	Metal	GEM	176 (55)
Takahashi FS-128	5	8.1	Metal	GEM	17 (OTA)
TMB Apo-130	5.1	6	Metal	Sold separately	24
TMB Apo-152	6	7.9	Metal	Sold separately	39
Vixen GP-DX 130-EDSS	5.1	6.6	Metal	GEM	n/s
William Optics Fluoro-Star 140	5.5	7	Metal	Sold separately	n/s
Price Range: Way over $4,000					
Aries 7" f/7.9	7	7.9	Metal	Sold separately	46
Aries 8" f /9	8	9	Metal	Sold separately	57
Takahashi FCT-150	6	7	Metal	GEM	53 (OTA)
Takahashi FS-152	6	8.1	Metal	GEM	24 (OTA)
Takahashi FCT-200	8	10	Metal	GEM	n/s
TMB Apo-175	6.9	8	Metal	Sold separately	48
TMB Apo-180	7.1	9	Metal	Sold separately	46
TMB Apo-203	8	9	Metal	Sold separately	62
TMB Apo-228	9	9	Metal	Sold separately	79
TMB Apo-254	10	9	Metal	Sold separately	100
William Optics Fluoro-Star 160F8	6.3	8	Metal	Sold separately	n/s
William Optics Fluoro-Star 160F6	6.3	6.3 or 8	Metal	Sold separately	n/s
William Optics Fluoro-Star 180	7.1	8	Metal	Sold separately	n/s
William Optics Fluoro-Star 202	8	8	Metal	Sold separately	n/s

D. Newtonian Reflectors

Brand/Model	Aperture	Focal Ratio	Tube Material	Mount	Total Weight
Price Range: Under $300					
Bushnell Voyager 100 × 4.5"	4.5	4.4	Plastic	Ball	12
Celestron Firstscope 114 Short	4.5	8.8	Metal	GEM	16
Helios Skyhawk-114	4.5	8.8	Metal	GEM	n/s
Konus Konusmotor 500	4.5	4.4	Metal	GEM	n/s
Murnaghan ValueScope 4.5	4.5	8	Metal	GEM	n/s
Novosibirsk Instruments TAL-M	3.1	6.6	Metal	Alt-az	26
Orion Telescopes SkyQuest XT4.5	4.5	8	Metal	Dob	18

Brand/Model	Aperture	Focal Ratio	Tube Material[1]	Mount[2]	Total Weight (heaviest component)[3]
Orion Telescopes ShortTube 4.5	4.5	8.8	Metal	GEM	n/s
Orion Telescopes SpaceProbe 130	5.1	6.9	Metal	GEM	32
Pacific Telescope Skywatcher 1145	4.5	4.4	Metal	GEM	n/s
Pacific Telescope Skywatcher 1149	4.5	8	Metal	GEM	n/s
Pacific Telescope Skywatcher 1141	4.5	9	Metal	GEM	n/s
Pacific Telescope Skywatcher 13065P	5.1	5	Metal	GEM	n/s
Pacific Telescope Skywatcher 1309	5.1	7	Metal	GEM	n/s
Pacific Telescope Skywatcher 1301	5.1	7.7	Metal	GEM	n/s
Stargazer Steve 4.25 Kit	4.25	10	Cardboard	Dob	16
Price Range: $300 to $500					
Bushnell Voyager 48 × 6"	6	8	Metal	Dob	n/s
Bushnell Voyager 48 × 8"	8	6	Metal	Dob	n/s
Celestron Firstscope 114 EQ	4.5	9	Metal	GEM	19
Discovery DHQ 6	6	8	Cardboard	Dob	41 (28)
Discovery DHQ 8	8	6	Cardboard	Dob	46 (28)
Edmund Astroscan	4.25	4	Plastic	Ball	13
Helios Explorer-114	4.5	8	Metal	GEM	n/s
Helios Explorer-150	6	5	Metal	GEM	n/s
Konus Konusuper 45	4.5	7.9	Metal	GEM	n/s
Konus Konuspace 114	4.5	8.7	Metal	GEM	n/s
Konus Konuspace 130	5.1	6.9	Metal	GEM	n/s
Meade DS114-EC	4.5	8	Metal	Alt-az	18
Meade DS2114-ATS	4.5	8.8	Metal	Alt-az	n/s
Meade DS127-EC	5	8	Metal	Alt-az	20
Murnaghan Astron 6	6	5	Metal	GEM	n/s
Novosibirsk Instruments TAL-1	4.3	7.3	Metal	GEM	42
Novosibirsk Instruments TAL-120	4.7	6.7	Metal	GEM	44
Orion Optics (UK) Europa 110	4.5	8	Metal	GEM	26
Orion Telescopes SpaceProbe 130ST	5.1	5	Metal	GEM	27
Orion Telescopes SkyQuest XT6	6	8	Metal	Dob	37 (24)
Pacific Telescope Skywatcher 150 Dob	6	6.7	Metal	Dob	n/s
Pacific Telescope Skywatcher 200 Dob	8	6	Metal	Dob	n/s
Stargazer Steve SGR-4	4.25	7.1	Cardboard	Alt-az	19
Stargazer Steve 6 Kit	6	8	Cardboard	Dob	24
Price Range: $500 to $1,000					
Celestron NexStar 114 GT	4.5	8.8	Metal	Alt-az	16
Celestron C150-HD	6	7	Metal	GEM	28
Celestron G-8N	8	5	Metal	GEM	43
Discovery DHQ 10	10	4.5 or 5.6	Cardboard	Dob	64 (36) or 66 (34)
Discovery Premium DHQ	10	6	Cardboard	Dob	63 (39)
Helios Apollo-150	6	6.7	Metal	GEM	n/s
Helios Explorer-200	8	5	Metal	GEM	n/s
Konus Konusuper 150	6	7.9	Metal	GEM	n/s

Brand/Model	Aperture	Focal Ratio	Tube Material[1]	Mount[2]	Total Weight (heaviest component)[3]
Meade Starfinder 12.5	12.5	4.8	Cardboard	Dob	96 (55)
Novosibirsk Instruments TAL-150	6	5	Metal	GEM	n/s
Novosibirsk Instruments TAL-2	6	8	Metal	GEM	88
Orion Optics (UK) Europa 150	6	5 or 8	Metal	GEM	33
Orion Optics (UK) Europa 200	8	4.4 or 6	Metal	GEM	47
Orion Telescopes SkyView Deluxe 6	6	5	Metal	GEM	39
Orion Telescopes SkyView Deluxe 8	8	4	Metal	Dob	45
Orion Telescopes SkyQuest XT8	8	6	Metal	Dob	42 (23)
Orion Telescopes SkyQuest XT10	10	5	Metal	Dob	58 (35)
Pacific Telescope Skywatcher 15075P	6	5	Metal	GEM	n/s
Pacific Telescope Skywatcher 2001P	8	5	Metal	GEM	n/s
Pacific Telescope Skywatcher 250 Dob	10	4.8	Metal	Dob	n/s
Parks Optical Companion Series	4.5	5	Fiberglass	Sold separately	n/s
Parks Optical Precision Series	6	6	Fiberglass	GEM	65
Parks Optical Precision Series	6	8	Fiberglass	GEM	67
Starsplitter Tube 8	8	6	Composite	Dob	45
Price Range: $1,000 to $2,000					
Discovery Premium DHQ	12.5	5	Cardboard	Dob	89 (61)
Discovery Premium DHQ	15	5	Cardboard	Dob	97 (68)
JMI NGT-6	6	5	Alum truss	SREM	26
Konus Konusky Evolution 200	8	5	Metal	GEM	n/s
LiteBox	12.5	5 to 6	Alum truss	Dob	55 plus optics
Mag One PortaBall 8	8	6	Alum truss	Ball	31 (24)
Meade Starfinder 16	16	4.5	Cardboard	Dob	170 (100)
Orion Optics (UK) GX150	6	5 or 8	Metal	GEM	46
Orion Optics (UK) GX200	8	6	Metal	GEM	53
Orion Optics (UK) Europa 250	10	4.8	Metal	GEM	57
Orion Optics (UK) GX250	10	4.8	Metal	GEM	59
Parks Optical Astrolight Nitelight Series	6	3.5	Fiberglass	GEM	35
Parks Optical Astrolight System	6	6 or 8	Fiberglass	GEM	42
Parks Optical Astrolight Nitelight Series	8	3.5	Fiberglass	GEM	35
Parks Optical Precision Series	8	6	Fiberglass	GEM	85
Starsplitter Compact 8	8	6	Two-pole	Dob	38 (30)
Starsplitter Compact II 8	8	6	Alum truss	Dob	41 (30)
Starsplitter Compact IV 8	8	6	Four-pole	Dob	35 (30)
Starsplitter Compact 10	10	6	Two-pole	Dob	43 (35)
Starsplitter Compact II 10	10	6	Alum truss	Dob	55 (45)
Starsplitter Compact IV 10	10	6	Four-pole	Dob	41 (35)
Starsplitter Tube 10	10	6	Fiberboard	Dob	70
Starsplitter Compact 12.5	12.5	4.8	Two-pole	Dob	60 (50)
Starsplitter Compact IV 12.5	12.5	4.8	Four-pole	Dob	58 (50)
Takahashi MT-130	5.2	6	Metal	Sold separately	12
Vixen GP-R130S	5.1	5.5	Metal	GEM	44

Brand/Model	Aperture	Focal Ratio	Tube Material[1]	Mount[2]	Total Weight (heaviest component)[3]
Vixen GP-R150S	5.9	5	Metal	GEM	52
Vixen GP-R200SS	7.9	4	Metal	GEM	53
Price Range: $2,000 to $3,000					
Discovery Truss Design	12.5	5	Alum truss	Dob	80 (49)
Discovery Truss Design	15	5	Alum truss	Dob	98 (57)
Discovery Premium DHQ	17.5	5	Cardboard	Dob	177 (115)
LiteBox	15	4.5 to 6	Alum truss	Dob	65 plus optics
LiteBox	18	4.5 to 6	Alum truss	Dob	75 plus optics
Mag One PortaBall 10	10	5	Alum truss	Ball	45 (35)
Meade Starfinder Equatorial 16	16	4.5	Cardboard	GEM	247 (62)
Orion Optics (UK) DX300	12	4 or 5.3	Metal	GEM	26
Parallax Instruments PI200	8	7.5	Metal	Sold separately	42
Parks Optical Superior Series	8	6	Fiberglass	GEM	185
StarMaster ELT 11	11	4.3 or 5.4	Four-pole	Dob	64 (38)
Starsplitter Compact II 12.5	12.5	4.8 or 6	Alum truss	Dob	67 (50)
Takahashi MT-160	6.3	6	Metal	Sold separately	18
Price Range: $3,000 to $5,000					
Discovery Truss Design	17.5	5	Alum truss	Dob	125 (75)
Excelsior Optics E-256	10	6	Metal	n/s	46
Excelsior Optics E-258	10	8	Metal	n/s	52
JMI NGT-12.5	12.5	4.5	Alum truss	SREM	120 (40)
Mag One PortaBall 12.5	12.5	4.8	Alum truss	Ball	62 (50)
Mag One PortaBall 14.5	14.5	4.3	Alum truss	Ball	72 (60)
Obsession 15	15	4.5	Alum truss	Dob	88 (60)
Obsession 18	18	4.5	Alum truss	Dob	109 (75)
Obsession 20	20	5	Alum truss	Dob	141 (90)
Parallax Instruments PI250	10	6.5	Metal	Sold separately	58
Parallax Instruments PI320	12.5	6	Metal	Sold separately	70
Parks Optical Superior Nitelight Series	8	3.5	Fiberglass	GEM	169
Parks Optical Superior Nitelight Series	10	3.5	Fiberglass	GEM	191
Parks Optical Superior Series	10	5	Fiberglass	GEM	205
Parks Optical Superior Series	12.5	5	Fiberglass	GEM	265
Sky Valley Ultra Light 12.5	12.5	≥ 4	Alum truss	Dob	60 (40)
Sky Valley Ultra Light 14.5	14.5	≥ 4	Alum truss	Dob	75 (53)
Sky Valley Ultra Light 16	16	≥ 4	Alum truss	Dob	100 (60)
Sky Valley Ultra Light 18	18	≥ 4	Alum truss	Dob	115 (75)
Sky Valley Rotating Tube 12.5	12.5	≥ 4	Alum truss	Dob	120 (65)
Sky Valley Rotating Tube 14.5	14.5	≥ 4	Alum truss	Dob	130 (75)
Sky Valley Rotating Tube 16	16	≥ 4	Alum truss	Dob	145 (105)
Sky Valley Rotating Tube 18	18	≥ 4	Alum truss	Dob	160 (120)
StarMaster ELT 12.5	12.5	5	Four-pole	Dob	72 (44)
StarMaster Truss 14.5	14.5	4.3	Alum truss	Dob	100 (36)

Brand/Model	Aperture	Focal Ratio	Tube Material[1]	Mount[2]	Total Weight (heaviest component)[3]
StarMaster Truss 16	16	4.3	Alum truss	Dob	116 (42)
Starsplitter Compact II 15	15	4.5	Alum truss	Dob	77 (60)
Starsplitter II 15	15	4.5	Alum truss	Dob	85 (70)
Starsplitter II 16	16	4.7	Alum truss	Dob	95 (80)
Starsplitter II 18	18	4.5	Alum truss	Dob	110 (90)
Starsplitter II 20	20	5	Alum truss	Dob	140 (120)
Takahashi MT-200	7.9	6	Metal	Sold separately	32
Price Range: Over $5,000					
Discovery Truss Design	20	4.5	Alum truss	Dob	150
Discovery Truss Design	22	4.5	Alum truss	Dob	165
Discovery Truss Design	24	4.5	Alum truss	Dob	185
JMI NGT-18	18	4.5	Alum truss	SREM	245 (75)
Obsession 25	25	5	Alum truss	Dob	240 (110)
Obsession 30	30	4.5	Alum truss	Dob	395 (175)
Parallax Instruments PI400	16	6	Metal	Sold separately	110
Parks Optical Superior Nitelight Series	12.5	3.5	Fiberglass	GEM	231
Parks Optical Observatory Series	12.5	5	Fiberglass	GEM	515 (375)
Parks Optical Superior Nitelight Series	16	3.5	Fiberglass	GEM	615
Parks Optical Observatory Series	16	5	Fiberglass	GEM	755 (375)
StarMaster Truss 18	18	4.3	Alum truss	Dob	132 (50)
StarMaster Truss 20	20	4.3	Alum truss	Dob	149 (59)
StarMaster Truss 24	24	4.2	Alum truss	Dob	186 (79)
Starsplitter II Light 20	20	4	Alum truss	Dob	135 (110)
Starsplitter II Light 22	22	4.5	Alum truss	Dob	155 (130)
Starsplitter II Light 24	24	4.5	Alum truss	Dob	170 (140)
Starsplitter II 25	25	5	Alum truss	Dob	250 (220)
Starsplitter II Light 28	28	4.5	Alum truss	Dob	250 (220)
Starsplitter II 30	30	4.5	Alum truss	Dob	400 (300)

E. Cassegrain Reflectors

Brand/Model	Aperture	Focal Ratio	Tube Material	Mount	Total Weight
Price Range: Under $2,000					
Novosibirsk Instruments TAL-200K	8	8.7	Metal	GEM	27 (OTA)
Vixen GP VC200L	7.9	9	Metal	GEM	15 (OTA)
Price Range: $2,000 to $5,000					
Parallax Instruments PI250R	10	9	Metal	Sold separately	40
Parallax Instruments PI250C	10	15	Metal	Sold separately	40
Parks Optical H.I.T. Astrolight Series	6	4 or 12	Fiberglass	GEM	39 (18)
Parks Optical H.I.T. Penta Series	6	4 or 12	Fiberglass	GEM	n/s
Parks Optical H.I.T. Penta Series	8	4 or 12	Fiberglass	GEM	n/s

Brand/Model	Aperture	Focal Ratio	Tube Material[1]	Mount[2]	Total Weight (heaviest component)[3]
Parks Optical H.I.T. Astrolight Series	8	4 or 12	Fiberglass	GEM	73 (18)
Parks Optical H.I.T. Superior Series	10	4 or 12	Fiberglass	GEM	203 (70)
Parks Optical H.I.T. Series	12.5	4 or 12	Fiberglass	GEM	249 (70)
Price Range: $5,000 to $10,000					
Takahashi Mewlon M-180	7	12	Metal	GEM	14 (OTA)
Takahashi CN-212	8.3	3.9 (Newt) 12.5 (Cas)	Metal	GEM	21 (OTA)
Takahashi Mewlon M-210	8.3	11.5	Metal	GEM	18 (OTA)
Takahashi Mewlon M-250	9.8	12	Metal	GEM	28 (OTA)
Price Range: $10,000 to $20,000					
OGS Ritchey-Chrétien-10	10	8.5	Metal or carbon fiber	Sold separately	30
OGS Classic Cassegrain-10	10	15	Metal or carbon fiber	Sold separately	30
OGS Ritchey-Chrétien -12.5	12.5	9	Metal or carbon fiber	Sold separately	40
OGS Classic Cassegrain-12.5	12.5	16	Metal or carbon fiber	Sold separately	40
Parallax Instruments PI320R	12.5	9	Metal	Sold separately	60
Parallax Instruments PI320C	12.5	16	Metal	Sold separately	60
Parallax Instruments PI400R	16	8.4	Metal	Sold separately	85
Parallax Instruments PI400C	16	14.25	Metal	Sold separately	85
RC Optical Systems	10	9	Carbon fiber	Sold separately	32
RC Optical Systems	12.5	9	Carbon fiber	Sold separately	45
RC Optical Systems	14	8 to 10	Carbon fiber	Sold separately	53
Takahashi BRC-250	10	5	Carbon fiber	GEM	34 (OTA)
Takahashi Mewlon M-300	11.8	11.9	Metal	GEM	55 (OTA)
Price Range: Over $20,000					
OGS Ritchey-Chrétien-14.5	14.5	7.9	Metal or carbon fiber	Sold separately	70
OGS Classic Cassegrain-14.5	14.5	16	Metal or carbon fiber	Sold separately	85
OGS Ritchey-Chrétien-16	16	8.4	Metal or carbon fiber	Sold separately	85
OGS Classic Cassegrain-16	16	14.25	Metal or carbon fiber	Sold separately	95
OGS Ritchey-Chrétien-20	20	8.1	Metal or carbon fiber	Sold separately	150
OGS Classic Cassegrain-20	20	16	Metal or carbon fiber	Sold separately	160
OGS Ritchey-Chrétien-24	24	8	Metal or carbon fiber	Sold separately	210

Brand/Model	Aperture	Focal Ratio	Tube Material[1]	Mount[2]	Total Weight (heaviest component)[3]
OGS Classic Cassegrain-24	24	16	Metal or carbon fiber	Sold separately	230
OGS Ritchey-Chrétien-32	32	7.6	Metal or carbon fiber	Sold separately	340
OGS Classic Cassegrain-32	32	10	Metal or carbon fiber	Sold separately	350
Parallax Instruments PI500R	20	8.1	Metal	Sold separately	130
Parallax Instruments PI500C	20	16	Metal	Sold separately	130
Parks Optical H.I.T. Series	16	4 or 12	Fiberglass	GEM	655 (306)
RC Optical Systems	16	8.4	Carbon fiber	Sold separately	75
RC Optical Systems	20	8.1	Carbon fiber	Sold separately	140

F. Exotic Reflectors

Price Range: Under $1,000

Brand/Model	Aperture	Focal Ratio	Tube Material[1]	Mount[2]	Total Weight (heaviest component)[3]
DGM Optics OA-4.0	4	10.4	PVC or Fiberglass	Dob	29 (18)

Price Range: $1,000 to $3,000

Brand/Model	Aperture	Focal Ratio	Tube Material[1]	Mount[2]	Total Weight (heaviest component)[3]
DGM Optics OA-5.1 ATS	5.1	10.8	Fiberglass	Dob	44 (28)
DGM Optics OA-5.5	5.5	10.1	Fiberglass	Dob	48 (30)
DGM Optics OA-6.5 ATS	6.5	10.4	Fiberglass	Dob	62 (40)
DGM Optics OA-7.0	7	9.7	Fiberglass	Dob	69 (45)

Price Range: Over $3,000

Brand/Model	Aperture	Focal Ratio	Tube Material[1]	Mount[2]	Total Weight (heaviest component)[3]
DGM Optics OA-9.0	9	9.6	Fiberglass	Dob	96 (60)
Takahashi Epsilon 160	6.3	3.3	Metal	GEM	17 (OTA)
Takahashi Epsilon 210	8.2	3	Metal	GEM	25 (OTA)
Takahashi Epsilon 250	9.8	3.4	Metal	Sold separately	42 (OTA)

G. Schmidt-Cassegrain Catadioptrics

Price Range: Under $1,000

Brand/Model	Aperture	Focal Ratio	Tube Material[1]	Mount[2]	Total Weight (heaviest component)[3]
Celestron C-5 Spotting Scope	5	10	Metal	Sold separately	6
Celestron G-5	5	10	Metal	GEM	17 (12)
Meade LXD55 (Schmidt-Newtonian)	6	5	Metal	GEM	n/s
Meade LXD55 (Schmidt-Newtonian)	8	4	Metal	GEM	n/s
Meade LXD55 (Schmidt-Newtonian)	10	4	Metal	GEM	n/s
Meade 203SC/300	8	10	Metal	GEM	45
Meade 203SC/500	8	10	Metal	GEM	55

Price Range: $1,000 to $2,000

Brand/Model	Aperture	Focal Ratio	Tube Material[1]	Mount[2]	Total Weight (heaviest component)[3]
Celestron NexStar 5	5	10	Metal	Single-arm Fork	18
Celestron Celestar 8	8	10	Metal	Fork	41

Brand/Model	Aperture	Focal Ratio	Tube Material[1]	Mount[2]	Total Weight (heaviest component)[3]
Celestron G-8	8	10	Metal	GEM	43
Celestron NexStar 8	8	10	Metal	Single-arm Fork	35
Celestron G-9.25	9.25	10	Metal	GEM	58
Meade LX10	8	10	Metal	Fork	49 (26)
Meade LX10 Deluxe	8	10	Metal	Fork	49 (26)
Meade LX90	8	10	Metal	Fork	51 (31)
Price Range: $2,000 to $3,000					
Celestron NexStar 8 GPS	8	10	Carbon fiber	Fork	68 (42)
Celestron Ultima 2000	8	10	Metal	Fork	49 (28)
Meade LX200GPS	8	6.3 or 10	Metal	Fork	69 (41)
Price Range: Over $3,000					
Celestron NexStar 11 GPS	11	10	Carbon fiber	Fork	91 (65)
Celestron CM-1100	11	10	Metal	GEM	118 (79)
Celestron CM-1400	14	10	Metal	GEM	164 (96)
Meade LX200GPS	10	6.3 or 10	Metal	Fork	86 (58)
Meade LX200GPS	12	10	Metal	Fork	120 (70)
Meade LX200	16	10	Metal	Fork	313 (120)

H. Maksutov Catadioptrics

(Note: MC: Maksutov-Cassegrain; MN: Maksutov-Newtonian)

Brand/Model	Aperture	Focal Ratio	Tube Material[1]	Mount[2]	Total Weight (heaviest component)[3]
Price Range: Under $1,000					
Celestron NexStar 4 (MC)	4	13	Metal	Single-arm Fork	11
Intes Micro MN56	5	6	Metal	Sold separately	11 (OTA)
Intes MK67 (MC)	6	12	Metal	Sold separately	10 (OTA)
LOMO Astele 70 (MC)	2.8	13.2	Metal	GEM	3 (OTA)
LOMO Astele 95 (MC)	3.7	12.2	Metal	GEM	19 (15)
Meade ETX-90 Spotting Scope (MC)	3.5	13.8	Metal	Sold separately	7
Meade ETX-90RA (MC)	3.5	13.8	Metal	Fork	9
Meade ETX-90EC (MC)	3.5	13.8	Metal	Fork	9
Meade ETX-105EC (MC)	4.1	14	Metal	Fork	12
Meade ETX-125 Spotting Scope (MC)	5	15	Metal	Fork	16
Meade ETX-125EC (MC)	5	15	Metal	Fork	19
Orion StarMax 90	3.5	12.9	Metal	GEM	16
Orion StarMax 102	4	12.7	Metal	GEM	20
Orion StarMax 127	5	12.1	Metal	GEM	37

Brand/Model	Aperture	Focal Ratio	Tube Material[1]	Mount[2]	Total Weight (heaviest component)[3]
Price Range: $1,000 to $2,000					
Intes MN61	6	6	Metal	Sold separately	20 (OTA)
Intes MK66 (MC)	6	12	Metal	Sold separately	12 (OTA)
Intes MK69 (MC)	6	6	Metal	Sold separately	10 (OTA)
Intes MK72 (MC)	7	10	Metal	Sold separately	n/s
Intes Micro MN66 (MN)	6	6	Metal	Sold separately	16 (OTA)
Intes Micro Alter 603 (MC)	6	10	Metal	Sold separately	13 (OTA)
LOMO Astele 102MN	4	5.5	Metal	GEM	38 (27)
LOMO Astele 133.5 (MC)	5.3	10.1	Metal	GEM or Fork	43 (27)
LOMO Astele 150 (MC)	5.9	14.2	Metal	GEM or Fork	45 (27)
LOMO Astele 152MN	6	4.4	Metal	GEM	45 (27)
Orion (UK) OMC-140 (MC)	5.5	14.3	Metal	GEM	8 (OTA)
TEC TEC-6 (MC)	6	12	Metal	Sold separately	11
Price Range: $2,000 to $4,000					
Intes MN71	7	6	Metal	Sold separately	n/s
Intes MK91 (MC)	9	13.5	Metal	Sold separately	30 (OTA)
Intes Micro MN76	7	5.9	Metal	Sold separately	30 (OTA)
Intes Micro Alter 809 (MC)	8	10	Metal	Sold separately	22 (OTA)
LOMO Astele 180 (MC)	7.1	10	Metal	GEM or Fork	50 (27)
LOMO Astele 203 (MC)	8	10	Metal	GEM or Fork	60 (27)
LOMO Astele 203MN	8	4.6	Metal	GEM	55 (27)
Meade LX200GPS (MC)	7	15	Metal	Fork	80 (52)
Orion (US) Argonaut 6 GP-DX (MN)	6	5.9	Metal	GEM	44 (21)
Questar Standard 3.5 (MC)	3.5	14.4	Metal	Fork	7
TEC MN7	7	6	Metal	Sold separately	20
TEC MN200/3.5	8	3.5	Metal	Sold separately	20
TEC MC200/11	8	11	Metal	Sold separately	20
TEC MC200/15	8	15.5	Metal	Sold separately	20
TEC MC200/20	8	20	Metal	Sold separately	20
Price Range: Over $4,000					
Astro-Physics 10F14MC	10	14.6	Metal	Sold separately	33
Intes Micro MN86	8	5.9	Metal	Sold separately	40
Intes Micro Alter 1008 (MC)	10	12.5	Metal	Sold separately	44
Intes Micro MN106	10	5.5	Carbon fiber	Sold separately	68
Intes Micro Alter 1208 (MC)	12	10	Metal	Sold separately	65
Intes Micro MN126	12	6	Carbon fiber	Sold separately	143
Intes Micro Alter 1408 (MC)	14	10	Metal	Sold separately	n/s
Intes Micro Alter 1608 (MC)	16	10 to 20	Metal	Sold separately	240
Intes Micro MN165	16	5	Carbon fiber	Sold separately	330
Questar Duplex 3.5 (MC)	3.5	14.4	Metal	Fork	7
Questar Astro 7 (MC)	7	13.4	Metal	Fork	19
Questar Classic 7 (MC)	7	14.3	Metal	Fork	26

Brand/Model	Aperture	Focal Ratio	Tube Material[1]	Mount[2]	Total Weight (heaviest component)[3]
TEC MC250/12	10	12	Metal	Sold separately	44
TEC MC250/20	10	20	Metal	Sold separately	44

All technical specifications were supplied by the manufacturers and have not been independently verified.

1. *Tube material:*
   Alum truss: *Open truss built from 6 or 8 aluminum poles.*
   Carbon fiber: *Solid-wall tube; material is very lightweight, yet very strong.*
   Cardboard: *Sonotube or similar concrete-form tubing commonly used in construction.*
   Composite: *Cardboardlike tube coated with a vinyl finish.*
   Fiberglass: *Solid-wall fiberglass tube.*
   Four-pole: *Open design that uses four poles to support eyepiece mount and diagonal mirror assembly.*
   Metal: *Solid-wall metal tube.*
   Plastic: *Solid-wall plastic tube.*
   Two-pole: *Open design that uses two poles to support eyepiece mount and diagonal mirror assembly.*
2. *Mount:*
   Alt-az: *Altitude-azimuth mount.*
   Ball: *Spherical end of telescope tube rides in a concave base.*
   Dob: *Dobsonian-style altitude-azimuth mount.*
   Fork: *Fork equatorial mount.*
   GEM: *German equatorial mount.*
   Single-arm fork: *Fork equatorial mount, but telescope is supported by only one arm.*
   SREM: *Split-ring equatorial mount.*
   n/s: *Not supplied; mounting must be purchased separately*
3. *"Total weight" includes both telescope and mounting. If the weight of the heaviest single part that a user is likely to carry out into the field is known, then it is listed immediately afterward in parentheses. In addition, the following abbreviations apply:*
   n/s: *Not supplied by manufacturer.*
   n/a: *Not applicable, usually denoting a telescope that is sold without a mounting; instead, a mounting must be purchased separately, which, depending on the mount used, may cause the total instrument weight to vary. In these cases, the weight quoted in "Heaviest Assembly" is that of the optical tube assembly (OTA). Bear in mind that the mounting will probably be heavier.*

   All weights are expressed in pounds (rounded to the nearest pound), except for binoculars, which are expressed in ounces. To convert, there are 16 ounces per pound and 2.2 pounds per kilogram.

Appendix B
Eyepiece Marketplace

Company	Series	Focal Length	Apparent Field of View	Eye Relief	Eyecup	Parfocal
0.965-Inch Barrel: Under $50						
Meade	MA	40	36	18	No	No
Meade	MA	25	40	16	No	No
Meade	MA	12	40	8	No	No
Meade	MA	9	40	6	No	No
Orion Telescopes	Explorer II	25	50	15	No	Yes
Orion Telescopes	Explorer II	17	50	11	No	Yes
Orion Telescopes	Explorer II	13	50	7	No	Yes
Orion Telescopes	Explorer II	10	50	4	No	Yes
Orion Telescopes	Explorer II	6	50	3	No	Yes
0.965-Inch Barrel: $50 and Above						
Edmund	Orthoscopic	25	47	22	No	No
Edmund	Orthoscopic	18	46	15	No	No
Edmund	Orthoscopic	12.5	44	10	No	No
Edmund	Orthoscopic	6	43	5	No	No
Edmund	Orthoscopic	4	41	3	No	No
Orion Telescopes	Plössl	32	50	25	Yes	Yes
Orion Telescopes	Plössl	26	50	16	Yes	Yes
Orion Telescopes	Plössl	20	50	13	Yes	Yes
Orion Telescopes	Plössl	17	50	10	Yes	Yes
Orion Telescopes	Plössl	12.5	50	8	Yes	Yes
Orion Telescopes	Plössl	10	50	5	Yes	Yes
Orion Telescopes	Plössl	7.5	50	4	Yes	Yes

Company	Series	Focal Length	Apparent Field of View	Eye Relief	Eyecup	Parfocal
1.25-Inch Barrels: Under $50						
Adorama	Plössl	40	52	31	No	No
Adorama	Plössl	32	46	22	No	No
Adorama	Plössl	25	50	22	No	No
Adorama	Plössl	20	52	20	No	No
Adorama	Plössl	17	52	13	No	No
Adorama	Plössl	12.5	52	8	No	No
Adorama	Plössl	10	52	7	No	No
Adorama	Plössl	7.5	52	5	No	No
Adorama	Plössl	6.3	52	5	No	No
Apogee	Plössl	30	40	n/s	No	No
Apogee	Plössl	25	52	20	Yes	No
Apogee	Plössl	20	52	18	Yes	No
Apogee	Plössl	15	52	14	Yes	No
Apogee	Plössl	10	52	11	Yes	No
Apogee	Plössl	6.5	52	6	Yes	No
Apogee	Plössl	5	52	5	Yes	No
Celestron	Plössl	25	50	22	Yes	Yes
Celestron	Plössl	17	52	13	Yes	Yes
Celestron	Plössl	13	52	8	Yes	Yes
Celestron	Plössl	10	52	7	Yes	Yes
Celestron	Plössl	8	52	5	Yes	Yes
Celestron	Plössl	6	52	5	Yes	Yes
Celestron	SMA	25	52	14	No	No
Celestron	SMA	17	52	13	No	No
Celestron	SMA	12	52	7	No	No
Celestron	SMA	10	52	6	No	No
Celestron	SMA	6	52	4	No	No
Discovery	Plössl	40	n/s	n/s	No	No
Discovery	Plössl	30	n/s	n/s	No	No
Discovery	Plössl	25	n/s	n/s	No	No
Discovery	Plössl	20	n/s	n/s	No	No
Discovery	Plössl	15	n/s	n/s	No	No
Discovery	Plössl	12.5	n/s	n/s	No	No
Discovery	Plössl	10	n/s	n/s	No	No
Discovery	Plössl	6.5	n/s	n/s	No	No
Discovery	Plössl	4	n/s	n/s	No	No
Edmund	RKE	28	45	25	No	No
Edmund	RKE	21	45	19	Yes	Yes
Edmund	RKE	15	45	13	Yes	Yes
Edmund	RKE	12	45	11	Yes	Yes
Edmund	RKE	8	45	8	Yes	Yes
Hands-on Optics	Plössl	40	46	31	Yes	No
Hands-on Optics	Plössl	30	49	22	Yes	No

Company	Series	Focal Length	Apparent Field of View	Eye Relief	Eyecup	Parfocal
Hands-on Optics	Plössl	25	52	22	Yes	No
Hands-on Optics	Plössl	20	52	15	Yes	No
Hands-on Optics	Plössl	17	52	13	Yes	No
Hands-on Optics	Plössl	12.5	52	8	Yes	No
Hands-on Optics	Plössl	10	52	7	Yes	No
Hands-on Optics	Plössl	9	50	7	Yes	No
Hands-on Optics	Plössl	7.5	52	5	Yes	No
Hands-on Optics	Plössl	6.3	52	5	Yes	No
Hands-on Optics	Plössl	4	52	4	Yes	No
Hands-on Optics	Super Plössl	40	44	32	Yes	Yes
Hands-on Optics	Super Plössl	32	47	24	Yes	Yes
Hands-on Optics	Super Plössl	26	52	20	Yes	Yes
Hands-on Optics	Super Plössl	20	52	16	Yes	Yes
Hands-on Optics	Super Plössl	17	53	13	Yes	Yes
Hands-on Optics	Super Plössl	12	52	10	Yes	Yes
Hands-on Optics	Super Plössl	10	52	8	Yes	Yes
Hands-on Optics	Super Plössl	7.5	52	6	Yes	Yes
Meade	MA	40	36	18	No	No
Meade	MA	25	40	16	No	No
Meade	MA	12	40	8	No	No
Meade	MA	9	40	6	No	No
Orion Telescopes	Explorer II	25	50	14.5	No	Yes
Orion Telescopes	Explorer II	17	50	11.5	No	Yes
Orion Telescopes	Explorer II	13	50	7	No	Yes
Orion Telescopes	Explorer II	10	50	5	No	Yes
Orion Telescopes	Explorer II	6	50	4	No	Yes
Rini	RKE	52	30	36	No	No
Rini	RKE	30	n/s	n/s	No	No
Rini	RKE	28	n/s	n/s	No	No
Rini	Modified Plössl	26	n/s	n/s	No	No
Rini	Modified Plössl	25	60	15	No	No
Rini	Modified Plössl	21	52	10	No	No
Rini	RKE	18	n/s	n/s	No	No
Rini	Modified Plössl	16	60	8	No	No
Siebert	Super Wide	45	42	29	No	No
Siebert	Super Wide	35	49	27	No	No
Siebert	Super Wide	32	52	27	No	No
Siebert	High Eye Relief	29	52	n/s	No	No
Siebert	Wide Angle	26	60	18	No	No
Siebert	Super Wide	21	80	7	No	No
Siebert	Wide Angle	15	65	7	No	No
Siebert	Wide Angle	12.5	65	7	No	No
Siebert	Wide Angle	10	65	7	No	No
Siebert	Wide Angle	7	65	6	No	No

Company	Series	Focal Length	Apparent Field of View	Eye Relief	Eyecup	Parfocal
1.25-Inch Barrels: $50 to $100						
Celestron	Plössl	40	46	31	Yes	No
Celestron	Plössl	32	52	22	Yes	No
Celestron	Ultima	12.5	51	9	Yes	Yes
Celestron	Ultima	7.5	51	5	Yes	Yes
Celestron	Ultima	5	50	4	Yes	Yes
Edmund	Erfle	20	65	n/s	No	No
Edmund	Orthoscopic	25	47	22	No	No
Edmund	Orthoscopic	18	46	15	No	No
Edmund	Orthoscopic	12.5	44	10	No	No
Edmund	Orthoscopic	6	43	5	No	No
Edmund	Orthoscopic	4	41	3	No	No
Meade	Plössl	40	44	29	Yes	No
Meade	Plössl	25	50	16	Yes	Yes
Meade	Plössl	16	50	10	Yes	Yes
Meade	Plössl	9.5	50	6	Yes	Yes
Meade	Plössl	6.7	50	4	Yes	Yes
Meade	Plössl	5	50	3	Yes	Yes
Meade	Super Plössl	26	52	18	Yes	Yes
Meade	Super Plössl	20	52	13	Yes	Yes
Meade	Super Plössl	15	52	9	Yes	Yes
Meade	Super Plössl	12.4	52	7	Yes	Yes
Meade	Super Plössl	9.7	52	5	Yes	Yes
Meade	Super Plössl	6.4	52	3	Yes	Yes
Orion Telescopes	Plössl	40	43	22	Yes	No
Orion Telescopes	Plössl	32	50	25	Yes	Yes
Orion Telescopes	Plössl	26	50	18	Yes	Yes
Orion Telescopes	Plössl	20	50	14	Yes	Yes
Orion Telescopes	Plössl	17	50	13	Yes	Yes
Orion Telescopes	Plössl	12.5	50	7	Yes	Yes
Orion Telescopes	Plössl	10	50	7	Yes	Yes
Orion Telescopes	Plössl	7.5	50	5	Yes	Yes
Orion Telescopes	Plössl	6.3	50	5	Yes	Yes
Orion Telescopes	Ultrascopic	25	52	17	Yes	Yes
Orion Telescopes	Ultrascopic	20	52	13	Yes	Yes
Orion Telescopes	Ultrascopic	15	52	10	Yes	Yes
Orion Telescopes	Ultrascopic	10	52	6	Yes	Yes
Orion Telescopes	Ultrascopic	7.5	52	5.3	Yes	Yes
Orion Telescopes	Ultrascopic	5	52	6	Yes	Yes
Orion Telescopes	Ultrascopic	3.8	52	5.3	Yes	Yes
Siebert	Wide Angle	4.9	65	7	Yes	No
Tele Vue	Plössl	25	50	17	Yes	Yes
Tele Vue	Plössl	20	50	14	Yes	Yes
Tele Vue	Plössl	15	50	10	Yes	Yes
Tele Vue	Plössl	11	50	8	Yes	Yes
Tele Vue	Plössl	8	50	6	Yes	Yes

Company	Series	Focal Length	Apparent Field of View	Eye Relief	Eyecup	Parfocal
1.25-Inch Barrels: $100 to $200						
Apogee	Super Easyview	6.5	52	18	Yes	No
Apogee	Super Easyview	5.5	50	12	Yes	No
Apogee	Super Easyview	3.6	48	8	Yes	No
Celestron	Axiom	23	70	10	Yes	Yes
Celestron	Axiom	19	70	10	Yes	Yes
Celestron	Axiom	15	70	7	Yes	Yes
Celestron	Ultima	42	36	32	Yes	No
Celestron	Ultima	35	49	25	Yes	No
Celestron	Ultima	30	50	21	Yes	Yes
Celestron	Ultima	24	51	18	Yes	Yes
Celestron	Ultima	18	51	13	Yes	Yes
Meade	Super Wide Angle	24.5	67	19	Yes	Yes
Meade	Super Wide Angle	18	67	14	Yes	Yes
Meade	Super Wide Angle	13.8	67	10	Yes	Yes
Meade	Ultra Wide Angle	6.7	84	11	Yes	Yes
Meade	Ultra Wide Angle	4.7	84	7	Yes	Yes
Meade	Super Plössl	40	44	30	Yes	No
Meade	Super Plössl	32	52	20	Yes	Yes
Orion Telescopes	Ultrascopic	35	49	25	Yes	No
Orion Telescopes	Ultrascopic	30	52	20.5	Yes	Yes
Sky Instruments	Speers-Waler	24	66	13	Yes	No
Sky Instruments	Speers-Waler	18	70	13	Yes	No
Sky Instruments	Speers-Waler	14	72	15	Yes	No
Sky Instruments	Speers-Waler	10	80	13	Yes	No
Takahashi	LE	24	52	17	Yes	Yes
Takahashi	LE	18	52	13	Yes	Yes
Takahashi	LE	12.5	52	9	Yes	Yes
Takahashi	LE	7.5	52	10	Yes	Yes
Takahashi	LE	5	52	10	Yes	Yes
Tele Vue	Nagler	4.8	82	7	Yes	Yes
Tele Vue	Plössl	40	43	28	Yes	No
Tele Vue	Plössl	32	50	22	Yes	Yes
Tele Vue	Zoom	8–24	55–40	15–20	Yes	No
Vixen (Orion)	Lanthanum LV	25	50	20	Yes	Yes
Vixen (Orion)	Lanthanum LV	21	50	20	Yes	Yes
Vixen (Orion)	Lanthanum LV	18	50	20	Yes	Yes
Vixen (Celestron)	Lanthanum LV	15	50	20	Yes	Yes
Vixen (Orion)	Lanthanum LV	14	50	20	Yes	Yes
Vixen (Orion)	Lanthanum LV	12.5	50	20	Yes	Yes
Vixen (Celestron)	Lanthanum LV	12	50	20	Yes	Yes
Vixen (Orion)	Lanthanum LV	10.5	50	20	Yes	Yes
Vixen (Celestron)	Lanthanum LV	10	50	20	Yes	Yes
Vixen (Orion)	Lanthanum LV	9.5	50	20	Yes	Yes

Company	Series	Focal Length	Apparent Field of View	Eye Relief	Eyecup	Parfocal
Vixen (Celestron)	Lanthanum LV	9	50	20	Yes	Yes
Vixen (Orion)	Lanthanum LV	7.5	50	20	Yes	Yes
Vixen (Celestron)	Lanthanum LV	6	45	20	Yes	Yes
Vixen (Orion)	Lanthanum LV	5.2	50	20	Yes	Yes
Vixen (Celestron)	Lanthanum LV	5	45	20	Yes	Yes
Vixen (Celestron)	Lanthanum LV	4	45	20	Yes	Yes
Vixen (Orion)	Lanthanum LV	3.8	50	20	Yes	Yes
Vixen (Celestron)	Lanthanum LV	2.5	45	20	Yes	Yes
Vixen (Orion)	Lanthanum LV	2.3	50	20	Yes	Yes
Vixen	Lanthanum LV Zoom	8–24	60–40	15–19	Yes	No

1.25-Inch Barrels: More than $200

Company	Series	Focal Length	Apparent Field of View	Eye Relief	Eyecup	Parfocal
Meade	Zoom	8–24	40–55	n/s	Yes	No
Pentax	XL	28	65	20	Yes	Yes
Pentax	XL	21	65	20	Yes	Yes
Pentax	XL	14	65	20	Yes	Yes
Pentax	XL	10.5	65	20	Yes	Yes
Pentax	XL	7	65	20	Yes	Yes
Pentax	XL	5.2	65	20	Yes	Yes
Sky Instruments	Speers-Waler	12	80	13	Yes	No
Sky Instruments	Speers-Waler	7	84	13	Yes	No
Sky Instruments	Speers-Waler Zoom	5–8	89–81	~13	Yes	No
Takahashi	LE	30	52	20	Yes	Yes
Tele Vue	Nagler Type 5	16	82	10	Yes	Yes
Tele Vue	Nagler Type 6	9	82	12	Yes	Yes
Tele Vue	Nagler	7	82	10	Yes	Yes
Tele Vue	Nagler Type 6	7	82	12	Yes	Yes
Tele Vue	Nagler Type 6	5	82	12	Yes	Yes
Tele Vue	Nagler Zoom	3–6	50	10	Yes	Yes
Tele Vue	Panoptic	19	68	13	Yes	Yes
Tele Vue	Panoptic	15	68	10	Yes	Yes
Tele Vue	Radian	18	60	20	Yes	Yes
Tele Vue	Radian	14	60	20	Yes	Yes
Tele Vue	Radian	12	60	20	Yes	Yes
Tele Vue	Radian	10	60	20	Yes	Yes
Tele Vue	Radian	8	60	20	Yes	Yes
Tele Vue	Radian	6	60	20	Yes	Yes
Tele Vue	Radian	5	60	20	Yes	Yes
Tele Vue	Radian	4	60	20	Yes	Yes
Tele Vue	Radian	3	60	20	Yes	Yes
VERNONscope	Brandon	32	50	26	Yes	Yes
VERNONscope	Brandon	24	50	19	Yes	Yes
VERNONscope	Brandon	16	50	13	Yes	Yes
VERNONscope	Brandon	12	50	10	Yes	Yes
VERNONscope	Brandon	8	50	6	Yes	Yes

Company	Series	Focal Length	Apparent Field of View	Eye Relief	Eyecup	Parfocal
1.25-Inch/2-Inch Barrel: $200 to $300						
Meade	Ultra Wide Angle	14	84	23	Yes	No
Meade	Ultra Wide Angle	8.8	84	16	Yes	Yes
Tele Vue	Panoptic	22	68	15	Yes	Yes
Vixen (Orion)	Lanthanum SW	22	65	20	Yes	Yes
Vixen (Orion)	Lanthanum SW	17	65	20	Yes	Yes
Vixen (Orion)	Lanthanum SW	13	65	20	Yes	Yes
Vixen (Orion)	Lanthanum SW	8	65	20	Yes	Yes
Vixen (Orion)	Lanthanum SW	5	65	20	Yes	Yes
Vixen (Orion)	Lanthanum SW	3.5	65	20	Yes	Yes
1.25-Inch/2-Inch Barrel: More than $300						
Collins	13 Piece	25	35	38	Yes	No
Tele Vue	Nagler Type 4	12	82	17	Yes	No
2-Inch Barrel: Under $100						
Rini	Erfle	45	52	30	No	No
Rini	Modified Plössl	38	62	20	No	No
Russell	Super Plössl	65	n/s	n/s	No	No
Russell	Plössl	32	52	n/s	Yes	No
Russell	SWA Konig	19	67	n/s	Yes	No
Russell	SWA Konig	17	65	n/s	Yes	No
Russell	SWA Konig	13	67	n/s	Yes	No
Russell	SWA Konig	11	67	n/s	Yes	No
Siebert	Wide Angle	51	52	27	No	No
Siebert	Wide Angle	45	57	27	No	No
Siebert	Wide Angle	38	60	26	No	No
Siebert	Wide Angle	35	62	26	No	No
Siebert	Wide Angle	32	67	24	No	No
Siebert	Wide Angle	19	75	7	No	No
2-Inch Barrel: $100 to $200						
Orion Telescopes	Optiluxe	50	45	40	No	No
Orion Telescopes	Optiluxe	40	62	27	No	No
Orion Telescopes	Optiluxe	32	58	22	No	No
Sky Instruments	Erfle	52	50	~25	Yes	No
Sky Instruments	Erfle	42	55	~20	Yes	No
Sky Instruments	Erfle	32	62	~13	Yes	No
Sky Instruments	Speers-Waler	30	67	16	No	No
Vixen	Lanthanum LV	30	60	20	Yes	Yes

Company	Series	Focal Length	Apparent Field of View	Eye Relief	Eyecup	Parfocal
2-Inch Barrel: $200 to $300						
Celestron	Axiom	40	70	21	Yes	Yes
Celestron	Axiom	34	70	16	Yes	Yes
Celestron	Axiom	19	70	10	Yes	Yes
Meade	Super Plössl	56	52	47	Yes	No
Meade	Super Wide Angle	40	67	27	Yes	No
Meade	Super Wide Angle	32	67	20	Yes	No
Tele Vue	Plössl	55	50	38	Yes	No
VERNONscope	Brandon	48	50	38	No	Yes
2-Inch Barrel: More than $300						
Celestron	Axiom	50	50	38	Yes	Yes
Pentax	XL	40	65	20	Yes	Yes
Takahashi	LE	50	50	40	Yes	Yes
Tele Vue	Nagler Type 5	31	82	19	Yes	No
Tele Vue	Nagler Type 4	22	82	19	Yes	No
Tele Vue	Nagler Type 4	17	82	17	Yes	No
Tele Vue	Panoptic	35	68	24	Yes	Yes
Tele Vue	Panoptic	27	68	19	Yes	Yes
Vixen (Orion)	Lanthanum SW	42	65	20	Yes	No

All technical specifications were supplied by the manufacturers and have not been independently verified.

Series: *The optical design or trade name of the particular eyepiece.*

Focal length: *The eyepiece's focal length, expressed in millimeters.*

Apparent field of view: *The eyepiece's apparent field of view, expressed in degrees. The abbreviation "n/s" indicates that the apparent field of view is not specified by the manufacturer.*

Eye relief: *The distance, in millimeters, that an observer's eye must be away from the eye lens in order to see the entire field. The abbreviation "n/s" indicates that the eye relief value is not specified by the manufacturer. The symbol "~" indicates an approximate value.*

Eyecup: *Useful for preventing stray light from entering through the corner of the observer's eye.*

Parfocal: *The eyepiece focuses at nearly the same distance as other eyepieces in the manufacturer's series. Occasionally, some eyepieces in a particular line may not be parfocal with all others in the same line, but instead, only with one or two others.*

Appendix C
The Astronomical Yellow Pages

Manufacturers

Here's a listing of all of the companies whose products are discussed throughout this book, along with a few other enterprises that, while not mentioned earlier, offer worthwhile services to the amateur astronomical community.

Absolute Astronomy
21 High Street, Noank, CT 06340
Phone: (860) 536-6002
Web site: http://www.optiguy.com
Product(s): Eyepiece and filter cases/holders/boxes

Adirondack Video Astronomy
35 Stephanie Lane, Queensbury, NY 12804
Phone: (888) 799-0107, (518) 793-9484
Web site: http://www.astrovid.com
Product(s): Video cameras

Adorama
42 West 18th Street, New York, NY 10011
Phone: (800) 223-2500, (212) 741-0466
Web site: http://www.adorama.com
Product(s): Binoculars, eyepieces

APM-Telescopes
Goebenstrasse 35, 66617 Saarbrücken, Germany
Phone: 49-681-954-1161
Web site: http://www.apm-telescopes.de
Product(s): Refractors

Apogee, Inc.
P.O. Box 136, Union, IL 60180
Phone: (877) 923-1602, (815) 568-2880
Web site: http://www.apogeeinc.com
Product(s): Refractors, eyepieces

Aries Instruments Co.
58 Ushakova st.kv.39, Kherson 325026 Ukraine
Phone: 380-55-227-9653
Product(s): Apochromatic refractors

Astro Haven
5778 Production Way, Suite 113, Langley, BC, V3A 4N4 Canada
Phone: (604) 539-2208
Web site: http://www.astrohaven.com
Product(s): Observatory domes

Astro-Lite
627 Ridge Road, Lewisberry, PA 17339
Phone: (717) 938-9890
Web site: http://www.astrolite-led.com
Product(s): LED flashlights

Astronomical Society of the Pacific
390 Ashton Avenue, San Francisco, CA 94112
Phone: (415) 337-1100
Web site: http://www.astrosociety.org
Product(s): *RealSky* software, *Mercury* magazine, posters, books

Astronomy Book Club
P.O. Box 6304, Indianapolis, IN 46209
Web site: http://abc.booksonline.com/abc
Product(s): Books

Astronomy Now
P.O. Box 175, Tonbridge, Kent TN10 4ZY United Kingdom
Phone: 44 (0) 1732-367-542
Web site: http://www.astronomynow.com
Product(s): *Astronomy Now* magazine

Astro-Physics, Inc.
11250 Forest Hills Road, Rockford, IL 61115
Phone: (815) 282-1513
Web site: http://www.astro-physics.com
Product(s): Apochromatic refractors, mountings

AstroSystems, Inc.
124 N. Second Street, LaSalle, CO 80645
Phone: (970) 284-9471
Web site: http://www.ezlink.com/~astrosys/
Product(s): Telescope kits, collimation tools, accessories

Astrozap
P.O. Box 502, Lakewood, OH 44107
Phone: (216) 631-2480 ext.230
Web site: http://www.astrozap.com
Product(s): Dew caps

Baader Planetarium GmbH
Zur Sternwarte, D-82291 Mammendorf
Phone: 49-8145-8802
Web site: http://www.baader-planetarium.de
Product(s): Solar filters

Bausch & Lomb/Bushnell Sports Optics Worldwide
9200 Cody Street, Overland Park, KS 66214
Phone: (800) 423-3537, (913) 752-3400
Web site: http://www.bushnell.com
Product(s): Binoculars, Newtonian reflectors

Bean, L. L.
Freeport, ME 04033
Phone: (800) 441-5713 (US and Canada), 0-800-891-297 (U.K.), (207) 552-3028 (elsewhere)
Web site: http://www.llbean.com
Product(s): Clothing, outdoor accessories

Beattie Systems, Inc.
2407 Guthrie Avenue, Cleveland, TN 37311
Phone: (423) 479-8566, (800) 251-6333
Web site: http://www.beattiesystems.com
Product(s): Intenscreen camera focusing screens

Bike Nashbar
4111 Simon Road, Youngstown, OH 44512
Phone: (800) 627-4227
Web site: http://www.nashbar.com
Product(s): Cold-weather clothing and accessories (bike stuff, too!)

Blaho Company
P.O. Box 419, Pleasant Hill, OH 45359
Phone: (937) 676-5891
Web site: http://www.blaho.com
Product(s): Binocular mounts

Bogen Photo Corporation (Manfrotto)
Via Livinallongo, 3-I-20139 Milano, Italy
565 East Crescent Avenue, P.O. Box 506, Ramsey, NJ 07446
Phone: (201) 818-9500, (212) 695-8166; in Italy: 39-02-566-0991
Web site: http://www.manfrotto.com, www.bogenphoto.com
Product(s): Tripods

Borg (Oasis Borg)
Hutech Corporation
23505 Crenshaw Boulevard, #212, Torrance, CA 90505
Phone: (310) 325-5511
Web site: http://www.sciencecenter.net/hutech
Product(s): Refractors, light-pollution reduction filters

Boyd Observatories
57607 Broughton Road, Ray Township, MI 48096
Phone: (810) 749-9525
Web site: http://www.boydobservatories.com
Product(s): Observatories

Brunton Company
620 East Monroe, Riverton, WY 82501
Phone: (307) 856-6559, (800) 443-4871
Web site: http://www.brunton.com
Product(s): Binoculars

Campmor
810 Route 17 North, Paramus, NJ 07653
Phone: (800) 226-7667, (201) 825-8300
Web site: http://www.campmor.com
Product(s): Clothing, outdoor accessories

Canon, Inc.
1 Canon Plaza, Lake Success, NY 11042
Phone: (800) 652-2666
6390 Dixie Road, Mississauga, ON L5T 1P7 Canada
Phone: (905) 795-1111 Woodhatch, Reigate, Surrey RH2 8BF United Kingdom
Phone: +44 (0)1737-220000
Web site: http://www.canon.com
Product(s): Binoculars

Carina Software
12919 Alcosta Boulevard, Suite 7, San Ramon, CA 94583
Phone: (925) 355-1266
Web site: http://www.carinasoft.com
Product(s): *Voyager II* software

Celestron International
2835 Columbia Street, Torrance, CA 90503
Phone: (310) 328-9560
Web site: http://www.celestron.com
Product(s): Binoculars, refractors, reflectors, catadioptric telescopes, eyepieces, accessories

Clear Sky Products
P. O. Box 721480, Berkley, MI 48072
Phone: (248) 547-2315
Web site: http://home.earthlink.net/~barrycnp
Product(s): TeleDome observatory

Collins Electro Optics
9025 E. Kenyon Avenue, Denver, CO 80237
Phone: (303) 889-5910
Web site: http://www.ceoptics.com
Product(s): Electronic eyepiece

Coronado Filters
PMB 320, Suite 105, 9121 East Tanque Verde Road, Tucson, AZ 85749
Phone: (520) 740-1561, (866) 786-9282
Web site: http://www.coronadofilters.com
Product(s): Hydrogen-alpha solar filters and telescopes

D&G Optical
2075 Creek Road, Manheim, PA 17545
Phone: (717) 665-2076
Web site: http://www.dgoptical.com
Product(s): Refractors, Cassegrain reflector

Damart
3 Front Street, Rollinsford, NH 03805
Phone: (800) 258-7300
Web site: http://www.damartusa.com
Product(s): Cold-weather clothing and accessories

David Chandler Company
P.O. Box 999, Springville, CA 93265
Phone: (559) 539-0900
Web site: http://www.davidchandler.com
Product(s): *Deep Space* computer software

DayStar
3857 Schaefer Avenue, Suite D, Chino, CA 91710
Phone: (909) 591-4673
Web site: http://www.daystarfilters.com
Product(s): Nebula filters, hydrogen-alpha solar filters

DGM Optics
P.O. Box 120, Westminster, MA 01473
Phone: (978) 874-2985
Web site: http://www.erols.com/dgmoptics
Product(s): Off-axis folded reflectors

Discovery Telescopes
615 South Tremont Street, Oceanside, CA 92054
Phone: (760) 967-6598, (877) 523-4400
Web site: http://www.discovery-telescopes.com
Product(s): Reflectors, short-tube refractor, eyepieces, accessories

Eagle Optics
2120 West Greenview Drive, Suite 4, Middleton, WI 53562
Phone: (800) 289-1132
http://www.eagleoptics.com
Product(s): Binoculars

Earth and Sky Products
1557 Lofty Perch Place, Santa Rosa, CA 95409
Phone: (707) 539-3398
Web site: http://www.astrosales.com
Product(s): Eyepieces, accessories

Eastman Kodak
Kodak Park, Rochester, NY 14650
Phone: (800) 242-2424, (800) 465-6325 (Canada)

Web site: http://www.kodak.com
Product(s): Film

Edmund Scientific Co.
60 Pearce Avenue, Tonawanda, NY 14150-6711
Phone: (800) 728-6999
Web site: http://www.scientificsonline.com
Product(s): Binoculars, reflectors, eyepieces

Equatorial Platforms
15736 McQuiston Lane, Grass Valley, CA 95945
Phone: (530) 274-9113
Web site: http://www.astronomy-mall.com
Product(s): Equatorial platforms for alt-azimuth mounts

ET Platforms
1176 Suzanne Court NW, Poulsbo, WA 98370
Phone: (360) 598-9035
Web site: http://www.etplatforms.com
Product(s): Equatorial platforms

Excelsior Optics
6341 Osprey Terrace, Coconut Creek, FL 33073
Phone: (954) 574-0153
Web site: http://www.excelsioroptics.com
Product(s): Newtonian reflectors (optical tube assemblies)

Fuji Photo Optical Co., Ltd. (Fujinon, Inc.)
1-324 Uetake, Omiya City, Saitama 330-8624 Japan
Phone: 048-668-2152
10 High Point Drive, Wayne, NJ 07470
Phone: (973) 633-5600
Web site: http://www.fujinon.co.jp, http://www.fujifilm.com
Product(s): Binoculars, film

Galaxy Optics
P.O. Box 2045, Buena Vista, CO 81211
Phone: (719) 395-8242
Web site: http://home.chaffee.net/~galaxy
Products: Astronomical mirrors

Glatter Laser Collimators
3850 Sedgwick Avenue, 14F, Bronx, NY 10463
Web site: http://www.collimator.com
Product(s): Laser collimators

Grabber Mycoal (Division of John Wagner Associates, Inc.)
4600 Danvers Drive Southeast, Grand Rapids, MI 49512
Phone: (800) 423-1233
Web site: http://www.grabberwarmers.com
Product(s): Hand and pocket warmers

Hands-On Optics
P.O. Box 412, Damascus, MD 20872
Phone: (301) 482-0000
Web site: http://www.handsonoptics.com
Product(s): Eyepieces, accessories

Helios Telescopes
Optical Vision Limited
Unit 2b, Woolpit Business Park, Woolpit, Bury St. Edmunds, Suffolk IP30 9RT, England
Phone: 44(0)1359-244-200
Web site: http://www.opticalvision.co.uk
Product(s): Refractors, reflectors

Helix Observing Accessories
P.O. Box 490, Gibsonia, PA 15044
Phone: (724) 316-0306
Web site: http://helix-mfg.com
Product(s): Laser collimator, LED flashlight

Intes Telescopes
33 Bolshaya Ochakovskaya Ul., Moscow, RF, 119361 Russia
Phone: 7-095-430-3625
Web site: http://freeyellow.com/members5/astrotelescopes
Product(s): Maksutov telescopes

Intes Micro Co., Ltd.
Shvernika 4, Moscow, 117036 Russia
Phone: 7-095-126-9903
Product(s): Maksutov telescopes

Jim Fly
7806 Wildcreek Trail SE, Huntsville, AL 35802
Phone: (256) 882-2523
Web site: http://home.earthlink.net/~flyj
Product(s): Collimation tools, observing chairs

Jim's Mobile, Inc. (JMI)
810 Quail Street, Unit E, Lakewood, CO 80215
Phone: (303) 233-5353, (800) 247-0304
Web site: http://www.jimsmobile.com
Product(s): Reflectors, accessories

JMB, Inc.
736 Oak Glen Circle, Fall Branch, TN 37656
Phone: (423) 348-8883
Product(s): Solar filters

Just-Cheney Enterprises
3613 Alderwood Avenue, Bellingham, WA 98225
Phone: (360) 738-3766

Web site: http://www.dewshields.com
Product(s): Dew shields

Kalmbach Publishing Co.
21027 Crossroads Circle, P.O. Box 1612, Waukesha, WI 53187
Phone: (800) 533-6644, (262) 796-8776
Web site: http://www.astronomy.com
Product(s): *Astronomy* magazine and related publications

Kendrick Astro Instruments
2920 Dundas Street West, Toronto, ON M6P 1Y8 Canada
Phone: (800) 393-5456, (416) 762-7946
Web site: http://www.kendrick-ai.com
Product(s): Dew prevention system, portable observatories

Konus
Via Mirandola, 45-37026 Settimo di Pescantina (VR), Italy
Phone: 39-045-676-7670
8359 NW 68th Street, Miami, FL 33166
Phone: (305) 592-5500
Web site: http://www.konus.com
Product(s): Refractors, reflectors

Kowa Optimed, Inc.
20001 South Vermont Avenue, Torrance, CA 90502
Phone: (310) 327-1913
Web site: http://www.kowascope.com
Product(s): Binoculars

Kufeld, Steve, Inc.
P.O. Box 6780, Pine Mountain Club, CA 93222
Phone: (661) 242-5421
Product(s): Telrad aiming device

LaserMax, Inc.
3495 Winton Place, Building B, Rochester, NY 14623
Phone: (716) 272-5420, (800) 527-3703
Web site: http://www.oemlasers.com
Product(s): Laser collimators

Leningrad Optical Manufacturing Organization (LOMO)
20 Chugunnay Ulitsa, St. Petersburg 194044 Russia
Phone: 7-812-248-5242
15 East Palatine Road, Unit 104, Prospect Heights, IL 60070
Phone: (847) 215-8800
Web site: http://www.lomoplc.com
Product(s): Maksutov telescopes, eyepieces, binocular eyepiece holders

Leupold and Stevens, Inc.
P.O. Box 688, Beaverton, OR 97075
Phone: (503) 526-5195
Web site: http://www.leupold.com
Product(s): Binoculars

LITEBOX Telescopes
1415 Kalakaua Avenue, Suite 204, Honolulu, HI 96826
Phone: (808) 524-2450
Web site: http://www.litebox-telescopes.com
Product(s): Newtonian reflectors

Losmandy (Hollywood General Machining)
1033 North Sycamore Avenue, Los Angeles, CA 90038
Phone: (323) 462-2855
Web site: http://www.losmandy.com
Product(s): Telescope mounts and accessories

Lumicon
6242 Preston Avenue, Livermore, CA 94550
Phone: (925) 447-9570, (800) 767-9576
Web site: http://www.lumicon.com
Product(s): Finderscopes, filters, coma corrector, photographic accessories

Mag One Instruments
16342 West Coachlight Drive, New Berlin, WI 53151
Phone: (262) 785-0926
Web site: http://www.mag1instruments.com
Product(s): Reflectors

Manfrotto Nord Srl.
Via Livinallongo, 3-I-20139 Milano, Italy
Phone: 39-02-566-0991
Web site: http://www.manfrotto.com
Product(s): Tripods (see also Bogen)

Meade Instruments Corporation
6001 Oak Canyon, Irvine, CA 92620
Phone: (800) 626-3233, (949) 451-1450
Web site: http://www.meade.com
Product(s): Binoculars, reflectors, refractors, catadioptric telescopes, eyepieces, filters, CCD cameras

Minolta Corporation
101 Williams Drive, Ramsey, NJ 07446

Phone: (201) 825-4000
Web site: http://www.minolta.com
Product(s): Binoculars

Miyauchi
177 Kanasaki, Minano-machi, Saitama 369-1621 Japan
Phone: 81-494-62-3371
Product(s): Binoculars

Mountain Instruments
1213 South Auburn Street, Colfax, CA 95713
Phone: (530) 346-9113
Web site: http://www.mountaininstruments.com
Product(s): Equatorial mounts

Murnaghan Instruments
1781 Primrose Lane, West Palm Beach, FL 33414
Phone: (561) 795-2201
Web site: http://www.e-scopes.cc/Murnaghan_Instruments_Corp56469.html
Product(s): Reflectors, refractors, eyepieces, CCD accessories

NatureWatch
P.O. Box 22062, 24 St. Peters Road, Charlottetown, Prince Edward Island, C1A 9J2 Canada
Phone: (902) 628-8095, (800) 693-8095
Web site: http://www.naturewatchshop.com
Product(s): Replacement wooden tripod legs

Nikon
1300 Walt Whitman Road, Melville, NY 11747
Phone: (800) 645-6687, (631) 547-4200
Web site: http://www.nikonusa.com
Product(s): Binoculars, cameras

North Peace International
Suite 810, 9707-110 Street, Capital Place, Edmonton, AB, Canada T5K 2L9
Phone: (877) 687-3223
Web site: http://www.telusplanet.net/public/northpc
Product(s): Portable observatories

Nova Astronomics
P.O. Box 31013, Halifax, NS B3K 5T9 Canada
Phone: (902) 499-6196
Web site: http://www.nova-astro.com
Product(s): *Earth-Centered Universe* computer software

Nova Optical Systems
14121 South Shaggy Mountain Road, Herriman, UT 84065
Phone: (801) 446-1802
Web site: http://www.astronomy-mall.com/regular/products/Nova
Product(s): Astronomical mirrors

Novak, Kenneth F., & Co.
P.O. Box 69, Ladysmith, WI 54848
Phone: (715) 532-5102
Product(s): Telescope components

Novosibirsk Instrument-Making Plant (TAL Instruments)
179/2 D. Kovalchuk St., Novosibirsk, 630049 Russia
Phone: 7-383-226-0765
Web site: http://www.telescopes.ru
North American Importer:
Talscopes
599 Norris Court, Kingston, ON, Canada
Phone: (866) 384-4144
Product(s): Reflectors, eyepieces, accessories

Oberwerk Corporation
2440 Wildwood, Xenia, OH 45385
Phone: (866) 244-2460
Web site: http://www.bigbinoculars.com
Products: Binoculars, binocular accessories

Obsession Telescopes
P.O. Box 804, Lake Mills, WI 53551
Phone: 920-648-2328
Web site: http://www.globaldialog.com/~obsessiontscp/OBHP.html
Product(s): Reflectors

Optical Guidance Systems (OGS)
2450 Huntingdon Pike, Huntingdon Valley, PA 19006
Phone: (215) 947-5571
Web site: http://www.opticalguidancesystems.com
Product(s): Cassegrain reflectors

Orion Optics (U.K.)
Unit 21 Third Avenue, Crewe, Cheshire CW1 6XU England
Phone: 44(0)1270-500-089
Web site: http://www.orionoptics.co.uk
Product(s): Reflectors, catadioptric telescopes

Orion Telescopes and Binoculars (U.S.)
P.O. Box 1815, Santa Cruz, CA 95062
Phone: (800) 447-1001, (831) 763-7000
Web site: http://www.telescope.com

Product(s): Binoculars, reflectors, refractors, catadioptric telescopes, eyepieces, accessories

Pacific Telescope Corporation
see Synta Optical Technology Corporation

Parallax Instruments
P.O. Box 303, Montgomery Center, VT 05471
Phone: (802) 326-3140
Web site: http://www.parallaxinstruments.com
Product(s): Reflectors, mounts

Parks Optical
P.O. Box 716, Simi Valley, CA 93062
Phone: (805) 522-6722
Web site: http://www.parksoptical.com
Product(s): Binoculars, refractors, reflectors, accessories

Pegasus Optics
P.O. Box 1389, Unit 27, Lot 30 Eagle Way, Brackettville, TX 78832
Phone: (830) 563-2578
Web site: http://www.icstars.com/pegasus
Product(s): Astronomical mirrors

Pentax Corporation
35 Inverness Drive East, P.O. Box 6509, Englewood, CO 80155
Phone: (800) 877-0155
3131 Universal Drive Mississauga, ON L4X 2E5 Canada
Phone: (905) 625-4930
Pentax House, Heron Drive, Langley, Slough, Berks SL3 8PN United Kingdom
Phone: 44(0)1753-792-792
Web site: http://www.pentax.com
Product(s): Binoculars, telescopes, eyepieces

Performance Bike Shops
1 Performance Way, Chapel Hill, NC 27514
Phone: (800) 727-2453, (919) 933-9113
Web site: http://www.performancebike.com
Product(s): Cold-weather clothing and accessories (bike stuff, too!)

Pfizer Consumer Health Care
235 East 42nd Street, New York, NY 10017
Phone: (212) 573-1000
Web site: http://www.pfizer.com/chc
Product(s): Anti-fogging spray for eyeglasses

Photon Instruments
122 East Main Street, Mesa, AZ 85201
Phone: (800) 574-2589, (480) 835-1767
Web site: http://www.photoninstrument.com
Product(s): Refractors

Project Pluto
168 Ridge Road, Bowdoinham, ME 04008
Phone: (207) 666 5750, (800) 777 5886
Web site: http://www.projectpluto.com
Product(s): *Guide* computer software

Quadrant Engineering
10138 Commercial Avenue, Penn Valley, CA 95946
Phone: (530) 432-5285
Web site: http://www.qei-motion.com
Product(s): Telescope mounts

Questar Corporation
6204 Ingham Road, New Hope, PA 18938
Phone: (215) 862-5277
Web site: http://www.questar-corp.com
Product(s): Maksutov-Cassegrain catadioptric telescopes and accessories

RC Optical Systems
3507 Kiltie Loop, Flagstaff, AZ 86001
Phone: (520) 773-7584
Web site: http://www.rcopticalsystems.com
Product(s): Cassegrain reflectors

Resource International
P.O. Box 134, Los Gatos, CA 95031-0134
Phone: (408) 356-1125
Web site: http://www.astronomy-mall.com
Product(s): DarkSide Baffles, DewGuard dew caps

Rigel Systems
26850 Basswood, Rancho Palos Verdes, CA 90275
Phone: (310) 375-4149
Web site: http://www.rigelsys.com
Product(s): LED flashlights, collimation tools, QuikFinder, accessories

Rini Eyepieces
c/o Surplus Shed, 407 Route 222, Balndon, PA 19510
Phone: (877) 778-7758, (610) 926-9226
Web site: http://www.surplusshack.com/lpm.html
Product(s): Eyepieces

Russell Optics
P.O. Box 263, Meadview, AZ 86444
Phone: (520) 564-2886
Web site: http://www.ctaz.com/~optics

Product(s): Eyepieces, binoviewers

Santa Barbara Instrument Group (SBIG)
147-A Castilian Drive, Santa Barbara, CA 93117
Phone: (805) 571-7244
Web site: http://www.sbig.com
Product(s): CCD cameras and accessories

Schmidling Productions, Inc.
18016 Church Road, Marengo, IL 60152
Phone: (815) 923 0031
Web site: http://user.mc.net/arf/ez-testr.htm
Product(s): Ronchi test eyepiece

ScopeTronix
1423 SE 10th Street Unit 1A, Cape Coral, FL 33990
Phone: (941) 945-6763
http://www.scopetronix.com
Product(s): Eyepieces, finderscopes, eyepiece caps, dew caps and anti-fogging heat strips

Siebert Optics
200 Short Johnson Road, Clayton, NC 27520
Phone: (919) 553-3980
Web site: http://www.siebertoptics.bizland.com
Product(s): Eyepieces, binoviewers

Sky Engineering
4630 N. University Drive, #329, Coral Springs, FL 33067
Phone: (954) 345-8726
Web site: http://www.skyeng.com
Product(s): Sky Commander digital setting circles

Sky Instruments
P.O. Box 3164, Vancouver, BC V6B 3Y6 Canada
Phone: (604) 270-2813
Product(s): Eyepieces, finderscopes, filters

Sky Publishing Corporation
P.O. Box 9111, Belmont, MA 02178
Phone: (800) 253-0245, (617) 864-7360
Web site: http:// www.skypub.com
Product(s): *Sky & Telescope* magazine, books

SkyTent International
1289 N. Fordham Boulevard, Chapel Hill, NC 27514
Phone: (877) 759-8368
Web site: http://www.skytent.com
Product(s): Portable observatories

Sky Valley Scopes
9215 Mero Road, Snohomish, WA 98290
Phone: (360) 794-7757
Web site: http://www.skyvalleyscopes.com
Product(s): Reflectors

Small Parts, Inc.
13980 N.W. 58th Court, P.O. Box 4650, Miami Lakes, FL 33014
Phone: (305) 557-7955, (800) 220-4242
Web site: http://www.smallparts.com
Product(s): Nuts, bolts, fasteners, knobs, and other assorted hardware

Software Bisque
912 12th Street, Golden, CO 80401
Phone: (800) 843-7599, (303) 278-4478
Web site: http://www.bisque.com
Product(s): *The Sky* computer software, Paramount equatorial mount

Sonfest
P.O. Box 360982, Melbourne, FL 23936
Phone: (321) 259-6498
Web site: http://www.sac-imaging.com
Product(s): CCD imagers

SPACE.com Canada, Inc.
284 Richmond Street East, Toronto, ON M5A 1P4 Canada
Phone: (416) 410-0259, (800) 252-5417
Web site: http://www.starrynight.com
Product(s): *Starry Night Deluxe* computer software

Star Instruments
555 Blackbird Roost, #5, Flagstaff, AZ 86001
Phone: (520) 774-9177
Web site: http://www.star-instruments.com
Product(s): Cassegrain mirrors and other telescope optics

Starbound
68 Klaum Avenue, North Tonawanda, NY 14120
Phone: (716) 692-3671
Product(s): Observing chair

Stargazer Steve
1752 Rutherglen Crescent, Sudbury, ON P3A 2K3 Canada
Phone: (705) 566-1314
Web site: http://stargazer.isys.ca
Product(s): Reflectors

Starlight Xpress, Ltd.
Foxley Green Farm, Ascot Road, Holyport, Berkshire, SL6 3LA United Kingdom
Phone: 44(0)1628-777-126

Web site: http://www.starlight-xpress.co.uk
Product(s): Starlight Xpress CCD cameras
StarMaster Telescopes
2160 Birch Road, Arcadia, KS 66711
Phone: (620) 638-4743
Web site: http://www.starmastertelescopes.com
Product(s): Reflectors, observing chair
Starsplitter Telescopes
3228 Rikkard Drive, Thousand Oaks, CA 91362
Phone: (805) 492-0489
Web site: http://www.starsplitter.com
Product(s): Reflectors
Steiner (in care of Pioneer Marketing)
97 Foster Road Suite 5, Moorestown, NJ 08057
Phone: (609) 854-2424, (800) 257-7742
Web site: http://www.pioneer-research.com
Product(s): Binoculars
Stellarvue
P.O. Box 7515, Auburn, CA 95604
Phone: (530) 878-0710
Web site: http://www.stellarvue.com
Product(s): Refractors
Stratton Video Brackets
103 West Marley Lane, Simpsonville, SC 29681
Phone: (864) 963-0282
Web site: http://www.strattonbrackets.com
Product(s): Brackets to fit for video and digital cameras onto SCTs
Swarovski
2 Slater Drive, Cranston, RI 02920
Phone: (401) 734-1800
Web site: http://www.swarovskioptik.com
Product(s): Binoculars
Swift Instruments, Inc.
952 Dorchester Avenue, Boston, MA 02125
Phone: (617) 436-2960, (800) 446-1116
Web site: http://www.swift-optics.com
Product(s): Binoculars
Synta Optical Technology Corporation
No. 89 Lane 4, Chia An W. Road, Long Tan Tao Yuan, Taiwan
No. 28 Huqiu Xi Road, Suzhou, China
Importer:
Pacific Telescope Corporation
160-11880 Hammersmith Way, Richmond, BC V7A 5C8 Canada
Phone: (604) 241-7027
Web site: http://www.skywatchertelescope.com
Product(s): Refractors, Newtonian reflectors, eyepieces, accessories
T&T Binoculars Mounts
18 Strong Street Extension, East Haven, CT 06512
Phone: (203) 469-2845, (203) 272-1915
Web site: http://www.benjaminsweb.com/TandT
Product(s): Binocular mounts
Takahashi Seisakusho Ltd.
41-7 Oharacho, Itabashiku, Tokyo 174-0061 Japan
U.S. Importer:
Land, Sea and Sky
3110 South Shepherd, Houston, TX 77098
Phone: (713) 529-3551
Web site: http://www.lsstnr.com
U.K. Importer:
True Technology, Ltd.
Red Lane Aldermaston Berks RG7 4PA England
Phone: 44(0)1189-700-777
Web site: http://www.truetech.dircon.co.uk
Product(s): Refractors, reflectors, mounts, eyepieces
Taurus Technologies
P.O. Box 14, Woodstown, NJ 08098
Phone: (856) 769-4509
Web site: http://www.taurus-tech.com
Product(s): Astrocameras, flip mirrors
Tech 2000
69 South Ridge Street, Monroeville, OH 44847
Phone: (419) 465-2997
Web site: http://www.accnorwalk.com/~tddi/tech2000
Product(s): Dual-axis Dobsonian tracking system
Technical Innovations, Inc.
22500 Old Hundred Road, Barnesville, MD 20838
Phone: (301) 972-8040
Web site: http://www.homedome.com
Product(s): Observatory domes
Tectron Telescopes
5450 NW 52 Court, Chiefland, FL 32626
Phone: (352) 490-9101
Web site: http://www.amateurastronomy.com/tools
Product(s): Collimation tools, *Amateur Astronomy* magazine

Telescope Engineering Company
15730 West 6th Avenue, Golden, CO 80401
Phone: (303) 273-9322
Web site: http://www.telescopengineering.com
Product(s): Maksutov telescopes

Tele Vue Optics
100 Route 59, Suffern, NY 10901
Phone: (854) 357-9522
Web site: http://www.televue.com
Product(s): Refractors, eyepieces

Thousand Oaks Optical
P.O. Box 4813, Thousand Oaks, CA 91359
Phone: (800) 996-9111, (805) 491-3642
Web site: http://www.thousandoaksoptical.com
Product(s): Glass solar filters, LPR filters

TL Systems
2184 Primrose Avenue, Vista, CA 92083
Phone: (760) 599-4219
Web site: http:// pw1.netcom.com/~tlsystem/
Product(s): Equatorial table kits, binoviewer kits

TMB Optical
P.O. Box 44331, Cleveland, OH 44144
Phone: (216) 524-1107
Web site: http://www.tmboptical.com
Main products: Apochromatic refractors

Tuma, Steven
1425 Greenwich Lane, Janesville, WI 53545
Phone: (608) 752-8366
Web site: http://www.deepsky2000.net
Product(s): *Deepsky 2000* computer software

Tuthill, Roger W., Inc.
11 Tanglewood Lane, Mountainside, NJ 07092
Phone: (800) 223-1063, (908) 232-1786
Web site: http://www.tuthillscopes.com
Product(s): Mylar solar filters, dew caps, finders, eyepieces, other astronomical accessories

Unitron, Inc.
170 Wilbur Place, P.O. Box 469, Bohemia, NY 11716
Phone: (631) 589-6666
Web site: http://www.unitronusa.com
Product(s): Binoculars, refractors

Universal Astronomics
27 Mendon Street, Worcester, MA 01604
Phone: (508) 797-4304
Web site: http://www.gis.net/~astronut
Product(s): Binocular mounts

Universal Workshop
Furman University, Greenville, SC 29613
Phone: (864) 294-2208, (888) 432-2264
Web site: http://www.kalend.com
Product(s): *Astronomical Calendar,* other books

Van Slyke Engineering
12815 Porcupine Lane, Colorado Springs, CO 80908
Phone: (719) 495-3828
Web site: http://www.observatory.org/vsengr.htm
Product(s): Accessories

VERNONscope & Co.
151 Court Street, Binghamton, NY 13901
Phone: (888) 303-3526, (607) 722-5697
Web site: http://www.vernonscope.com
Product(s): Brandon eyepieces, color filters, Barlow lenses, accessories

Virgo Astronomics
608 Falconbridge Drive, Suite 46, Joppa, MD 21085
Phone: (410) 679-7055
Web site: http://www.virgoastro.com
Product(s): Binocular mounts, observatory

Vixen Optical Industries, Ltd.
247 Hongo, Tokorozawa, Saitama 359 Japan
Phone: 81-042-944-4141
Web site: http://www.vixen.co.jp
Product(s): Refractors, eyepieces

William Optics
#5 Lane 23 Juian Street, Taipei, Taiwan
Phone: 011-886-2-2709-7196
Web site: http://www.optics.com.tw
Product(s): Refractors, mounts, accessories

Willmann-Bell, Inc.
P.O. Box 35025, Richmond, VA 23235
Phone: (800) 825-7827, (804) 320-7016
Web site: http://www.willbell.com
Product(s): Books, software

World Wide Software Publishing
P.O. Box 326, Elk River, MN 55330
Phone: (763) 274-1801
Web site: http://www.skymap.com
Product(s): *SkyMap Pro* software

Zeiss, Carl, Inc.
1 Zeiss Drive, Thornwood, NY 10594
Phone: (800) 338-2984
Web site: http://www.zeiss.com
Product(s): Binoculars

Dealers and Distributors

An important warning: If you are shopping by mail, ALWAYS ask about shipping charges BEFORE ordering. Some offer exceptionally low prices only to charge the consumer exorbitant (and unpublished) shipping and handling costs. Not only will these hidden costs offset any savings, these dealers may end up being more expensive!

United States

Arizona

Astronomy Shoppe
3202 East Greenway Road, Suite 1615, Phoenix, AZ 85032
Phone: (888) 999-6396, (602) 971-3170
Web site: http://www.darkskyinc.com
Product line(s): Celestron, Meade, Orion Telescopes, Tele Vue, more

Crazy Ed Optical
P.O. Box 446, Pearce, AZ 85625
Phone: (520) 826-1484
Web site: http://www.crazyedoptical.com
Product line(s): Astrosystems, HB2000 Publications, Rigel Systems, Telrad

Stellar Vision and Astronomy Shop
1835 South Alvernon, #206, Tucson, AZ 85711
Phone: (520) 571-0877
Web site: http://www.theriver.com/stellar_vision
Product line(s): Celestron, Meade, Takahashi, Tele Vue, more

Arkansas

Astro Stuff
63 Observatory Lane, Dover, AR 72837
Phone: (501) 331-3773
Web site: http://www.astrostuff.com
Product line(s): Bushnell, Celestron, Meade, Orion Telescopes, Tele Vue, Thousand Oaks Optical, more

California

AstronomyQuest.com
4790 Irvine Boulevard, #108, Irvine, CA 92620
Phone: (800) 807-8763, (714) 832-7180
Product line(s): Celestron, Discovery Telescopes, Meade

Earth and Sky Adventure Products
1557 Lofty Perch Place, Santa Rosa, CA 95409
Phone: (707) 539-3398
Web site: http://www.astrosales.com
Product line(s): Celestron, Intes

Lumicon
6242 Preston Avenue, Livermore, CA 94550
Phone: (925) 447-9570, (800) 767-9576
Web site: http://www.lumicon.com
Product line(s): Celestron, Lumicon, Meade, Tele Vue, more

Oceanside Photo and Telescope
1024 Mission Avenue, Oceanside, CA 92054
Phone: (800) 483-6287, (760) 722-3348
Web site: http://www.optcorp.com
Product line(s): Bausch & Lomb, Celestron, Discovery Telescopes, Meade, Tele Vue, more

Out of This World
45100 Main Street, Mendocino, CA 95460
Phone: (800) 485-6884, (707) 937-3324
Web site: http://www.discounttelescopes.com
Product line(s): Celestron, Meade, Nikon, Swarovski

Scope City
730 Easy Street, Simi Valley, CA 93065
Phone: (805) 522-6646
Web site: http://www.scopecity.com
Product line(s): Celestron, Edmund, JMI, Meade, Parks, Pentax, Questar, Tele Vue, JMI, more

Woodland Hills Camera & Telescopes
5348 Topanga Canyon Boulevard, Woodland Hills, CA 91364
Phone: (818) 347-2270, (888) 427-8766
Product line(s): Bushnell, Celestron, Fujinon, Losmandy, Meade, SBIG, Tele Vue, Zeiss, more

Colorado

Good Clear Skies Astronomical

2865 Halleys Court, Colorado Springs, CO 80906

Phone: (719) 440-2006

Web site: http://www.gcsastro.com

Product line(s): JMI, Intes, Intes Micro, Parallax Instruments, Takahashi

S & S Optika

5174 South Broadway, Englewood, CO 80110

Phone: (303) 789-1089

Web site: http://www.sandsoptika.com

Product line(s): Bausch & Lomb, Celestron, DayStar, Pentax, Takahashi, Zeiss, more

Florida

Colonial Photo and Hobby

634 North Mills Avenue, Orlando, FL 32803

Phone: (407) 841-1485, (800) 841-1485

Web site: http://www.colonialphotohobby.com

Product line(s): Celestron, Orion Telescopes, Meade, more

ScopeTronix

1423 SE 10th Street, Unit 1A, Cape Coral, FL 33990

Phone: (941) 945-6763

Web site: http://www.scopetronix.com

Product line(s): Celestron, LOMO, Meade

Georgia

Camera Bug, Ltd.

1799 Briarcliff Road, Atlanta, GA 30306

Phone: (800) 545-8509, (404) 873-4513

Web site: http://www.camerabug.com

Product line(s): Canon, Celestron, Meade, Minolta, Nikon, Pentax, Tele Vue, more

Illinois

Shutan Camera and Video

100 Fairway Drive, Vernon Hills, IL 60061

Phone: (800) 621-2248, (847) 367-4600

Web site: http://www.shutan.com

Product line(s): Celestron, JMI, Losmandy, Meade, Rigel, Tele Vue, Vernonscope, more

Maryland

Company Seven

Box 2587, Montpelier, MD 20708

Phone: (301) 953-2000

Web site: http://www.company7.com

Product line(s): Astro-Physics, AstroSystems, Celestron, Fujinon, Lumicon, Questar, Tele Vue, more

Massachusetts

Harvard Camera

1 Still River Road, Harvard, MA 01451

Phone: (978) 456-8100, (800) 287-6569

Product line(s): Bushnell, Celestron, Meade, Nikon, Pentax, Swift

Hunt's Photo and Video

100 Main Street, Melrose, MA 02176

Phone: (800) 924-8682, (781) 662-8822

Web site: http://www.huntsphotoandvideo.com

Product line(s): Bausch & Lomb, Celestron, Minolta, Nikon, Swift

Michigan

Rider's Hobby Shop

4035 Carpenter Road, Ypsilanti, MI 48197

Phone: (734) 971-6116

Web site: http://www.riders.com

Product line(s): Celestron, Meade, more

Montana

Astro Stuff

1550 Amsterdam Road, Belgrade, MT 59714

Phone: (406) 388-1205

Web site: http://www.astrostuff.com

Product line(s): Bushnell, Celestron, Meade, Orion Telescopes, Tele Vue, Thousand Oaks Optical, more

Internet Telescope Exchange

3555 Singing Pines Road, Darby, MT 59829

Phone: (406) 821-1980

Web site: http://www.burnettweb.com/ite/index.html

Product line(s): Borg, Intes, Intes Micro, JMI, Losmandy, Starlight Xpress, more

Nevada

Pocono West Optics

2580 S. Decatur, #3, Las Vegas, NV 89102

Phone: (800) 635-3607, (702) 252-7200

Product line(s): AstroSystems, Celestron, Edmund, Fujinon, JMI, Meade, Miyauchi, Questar, Pentax, Starbound, Tele Vue, Zeiss, more

New Hampshire
Rivers Camera Shop
454 Central Avenue, Dover, NH 03820
Phone: (800) 245-7963, (603) 742-4888
Web site: http://www.riverscamera.com
Product line(s): Celestron, Kendrick, Losmandy, Meade, Orion Telescopes, Questar, Rigel, Starbound, Tele Vue, more

New Jersey
Bergen County Camera
270 Westwood Avenue, Westwood, NJ 07675
Phone: (800) 841-4118, (201) 664-4411
Web site: http://www.bergencountycamera.com
Product line(s): Celestron, Meade, Tele Vue, more

Dover Photo Supply
25 East Blackwell Street, Dover, NJ 07801
Phone: (973) 366-0994
Product line(s): Celestron, Meade, Nikon, Swift, Tele Vue

Pisano Opticians
5 Church Street, Montclair, NJ 07042
Phone: (973) 744-6128
Web site: http://www.pisanooptics.com
Product line(s): Celestron, Meade, JMI, Orion Telescopes, Swift

New Mexico
New Mexico Astronomical
834 Gabaldon Road, Belen, NM 87002
Phone: (505) 864-2953
Web site: http://www.nmastronomical.com
Product line(s): Celestron, Discovery, Meade, Parks, Tele Vue, more

New York
Adorama
42 West 18th Street, New York, NY 10011
Phone: (800) 723-6726, (212) 647-9800
Web site: http://www.adorama.com
Product line(s): Celestron, Meade, Tele Vue, Thousand Oaks, Fujinon, Pentax, more

Astrotec
4132 Sunrise Highway, Oakdale, NY 11769
Phone: (631) 563-9009
Web site: http://www.astrotecs.com
Product line(s): Celestron, JMI, Kendrick, Meade, Orion Telescopes, Tele Vue, more

Berger Brothers Camera Exchange
209 Broadway, Amityville, NY 11701
Phone: (800) 262-4160, (631) 264-4160
Web site: http://www.planetelectronics.com
Product line(s): Celestron, Meade, Tele Vue

Camera Concepts
33 East Main Street, Patchogue, NY 11772
Phone: (631) 475-1118
Web site: www.cameraconcepts.com
Product line(s): Celestron, Meade, Tele Vue

Continental Camera
5795 Transit Road, Depew, NY 14043
Phone: (716) 681-8038
Web site: http://www.continentalcamera.com
Product line(s): Bushnell, Celestron, Meade

Focus Camera, Inc.
4419-21 13th Avenue, Brooklyn, NY 11219
Phone: (888) 221-0828, (718) 437-8810
Web site: http://www.focuscamera.com
Product line(s): Celestron, Fujinon, JMI, Meade, Nikon, Tele Vue, Thousand Oaks, more

Neptune Photo, Inc.
827-829 Franklin Avenue, Garden City, NY 11530
Phone: (800) 955-1110, (516) 741-4484
Web site: http://www.neptunephoto.com
Product line(s): Bushnell, Canon, Celestron, Meade, Minolta, Nikon, Zeiss

Science & Hobby
Route 299, New Paltz Plaza, New Paltz, NY 12561
Phone: (845) 255-3744, (800) 763-3212
Web site: http://www.scienceandhobby.com
Product line(s): Celestron, Meade, more

North Carolina
Camera Corner
2273 South Church Street, Burlington, NC 27215
Phone: (336) 228-0251, (800) 868-2462

Web site: http://www.cameracorner.com
Product line(s): Celestron, Meade, Tele Vue, more

Ohio

Eastern Hills Camera and Video
7875 Montgomery Road, Cincinnati, OH 45236
Phone: (800) 863-6138, (513) 791-2140
Web site: http://home.fuse.net/ehcamera
Product line(s): Celestron, Minolta, Pentax, Swift, Vixen

Pete's PhotoWorld, Inc.
7731 Beechmont Avenue, Cincinnati, OH 45255
Phone: (800) 736-7383
Web site: http://www.photoworld.com
Product line(s): Celestron, Meade, Apogee, Kendrick, Orion Telescopes, Nikon, Minolta, Pentax, more

Oklahoma

Astronomics
680 Southwest 24th Avenue, Norman, OK 73069
Phone: (800) 422-7876, (405) 364-0858
Web site: http://www.astronomics.com
Product line(s): Bushnell, Celestron, Coronado, Discovery, Meade, Takahashi, Tele Vue, TMB, more

Oregon

Hardin Optical
1450 Oregon Avenue, Bandon, OR 97411
Phone: (877) 447-4847
Web site: http://www.hardin-optical.com
Product line(s): Bushnell, Celestron, Meade, Nikon, Tele Vue, more

Pennsylvania

Cutler Camera
314 Horsham Road, Horsham, PA 19044
Phone: (215) 674-5727, (800) 220-5727
Product line(s): Canon, Celestron, Meade, Nikon, Tele Vue, more

Texas

Analytical Scientific
11049 Bandera Road, San Antonio, TX 78250
Phone: (800) 701-7827, (210) 684-7373
Web site: http://www.analyticalsci.com
Product line(s): Celestron, Meade, Tele Vue

Land, Sea and Sky (Texas Nautical Repair)
3110 South Shepherd, Houston, TX 77098
Phone: (713) 529-3551
Web site: http://www.lsstnr.com
Product line(s): Celestron, Meade, Miyauchi, Takahashi, Tele Vue

The Observatory
19009 Preston Road, Suite 114, Dallas, TX 75252
Phone: (972) 248-1450
Web site: http://www.theobservatoryinc.com
Product line(s): Canon, Celestron, Fujinon, Technical Innovations, LOMO, Meade, Orion Telescopes, Takahashi, Tele Vue, Zeiss, more

Skywatch Products
7600 Benbrook Parkway, #5, Benbrook, TX 76126
Phone: (817) 249-3767
Web site: http://members.aol.com/skywatchpr
Product line(s): Celestron, Coronado, Discovery, Parks, SBIG, Thousand Oaks, more

Washington

Anacortes Telescope & Wild Bird
9973 Padilla Heights Road, Anacortes, WA 98221
Phone: (800) 850-2001, (360) 588-9000
Web page: http://www.buytelescopes.com
Product line(s): Celestron, Fujinon, JMI, Meade, Miyauchi, Parallax, Questar, SBIG, StarMaster, Tele Vue, William Optics, Zeiss, more

Wisconsin

Eagle Optics
2120 West Greenview Drive, Suite 4, Middleton, WI 53562
Phone: (800) 289-1132, (608) 836-7172
Web site: http://www.eagleoptics.com
Product line(s): Bushnell, Celestron, Fujinon, Meade, Nikon, Questar, Steiner, Tele Vue, more

Canada

Alberta

The Science Shop
316 Southgate Centre, Edmonton, AB T6H 4M6
Phone: (780) 435-0519
Web site: http://www.thescienceshop.com
Product line(s): Celestron, Meade, Pacific Telescopes, Sky Instruments, Tele Vue

British Columbia

A & A Astronomical
1859 West 4th Avenue, Vancouver, BC V6J 1M8
Phone: (604) 737-4303
Web site: http://www.aa-astronomical.com
Product line(s): Celestron, Meade, Olympus, Pentax, Pacific Telescopes, Sky Instruments, Steiner, Zeiss

Island Eyepiece
Box 133, Mill Bay, BC V0R 2P0
Phone: (250) 743-6633, (250) 770-1477
Web site: http://www.islandeyepiece.com
Product line(s): Sky Instruments

Sirius Science and Nature
277 Main Street, Penticton, BC V2A 5B1
Phone: (250) 770-1477
Web site: http://www.siriusscience.com
Product line(s): Celestron, Meade, Pacific Telescopes, Sky Instruments, Tele Vue

Ontario

EfstonScience
3350 Dufferin Street, Toronto, ON M6A 3A4
Phone: (416) 787-4581, (888) 777-5255
Web site: http://www.e-sci.com
Product line(s): Celestron, Meade, Orion Telescopes, Pentax, Tele Vue, more

Focus Scientific, Ltd.
1489 Merivale Road, Nepean (Ottawa), ON K2E 5P3
Phone: (613) 730-7767
Web site: http://www.focusscientific.com
Product line(s): Celestron, Fujinon, Meade, Nikon, Orion Telescopes, Sky Instruments, more

Khan Scope Centre
3243 Dufferin Street, Toronto, ON M6A 2T2
Phone: (416) 783-4140, (800) 580-7160
Web site: http://www.khanscope.com
Product line(s): Bausch & Lomb, Celestron, JMI, Losmandy, Meade, SBIG, Pacific Telescopes, Tele Vue, Thousand Oaks, more

O'Neil Photo & Optical, Inc.
356 William Street, London, ON N6B 3C7
Phone: (519) 679-8840
Web site: http://www.oneilphoto.on.ca
Product line(s): Brunton, Celestron, Fujinon, Orion Telescopes, Pacific Telescopes, Sky Instruments, more

Perceptor
Brownsville Plaza, Box 38, Suite 201, Schomberg, ON L0G 1T0
Phone: (905) 939-2313
Product line(s): Celestron, Meade, Orion Telescopes, Questar, Tele Vue, Zeiss, more

Sky Optics
4031 Fairview Street, Unit 216B, Burlington, ON L7L 2A4
Phone: (905) 631-9944, (877) 631-9944
Web site: http://www.skyoptics.net
Product line(s): Celestron, Lumicon, Meade, Orion Telescopes, SBIG, Pacific Telescopes, Sky Instruments, Tele Vue, more

Prince Edward Island

NatureWatch
P.O. Box 22062, 24 St. Peters Road, Charlottetown, PE C1A 9J2
Phone: (902) 628-8095, (800) 693-8095
Web site: http://www.naturewatchshop.com
Product line(s): Meade, Sky Instruments

Quebec

Astronomie Plus (Lire La Nature, Inc.)
1198 Chambly Road, Longueuil, QC J4J 3W6
Phone: (450) 463-5072
Web site: http://www.broquet.qc.ca/lln
Product line(s): Apogee CCD instruments, Baader, Celestron, JMI, Meade, Nikon, Pacific Telescopes, Pentax, SBIG, Tele Vue, Thousand Oaks, Zeiss, more

La Maison de l'Astronomie P.L., Inc.
8074 rue St-Hubert, Montreal, QC H2R 2P3
Phone: (514) 279-0063
Web site: http://www.microid.com/maison
Product line(s): Celestron, Meade, Vixen, more

United Kingdom

SCS Astro
1 Tone Hill, Wellington, Somerset TA21 0AU
Phone: 44(0)182-366-5510
Web site: http://www.scsastro.co.uk
Product line(s): Apogee, Celestron, Intes, Kendrick, Lumicon, Meade, Orion Optics, Tele Vue, Vixen, more

Sherwood Photo, Ltd.
11-13 Great Western Arcade, Birmingham B2 5HU
Phone: 44(0)121-236-7211
Web site: http://www.sherwoods-photo.com
Product line(s): Celestron, Fujinon, Helios, Meade, Orion Optics, Vixen, more

Telescope House
63 Farringdon Road, London EC1M 3JB
Phone: 44(0)207-405-2156
Web site: http://www.telescopehouse.co.uk
Product line(s): Meade, Takahashi, Vixen, Tele Vue, more

True Technology, Ltd.
Woodpecker Cottage, Red Lane, Aldermaston, Berks RG7 4PA
Phone: 44(0)1189-700-777
Web page: http://www.truetech.dircon.co.uk
Product line(s): Apogee CCD cameras, JMI, Takahashi

Venturescope
11B Wren Centre, Westbourne Road, Emsworth, Hampshire PO10 7RL
Phone: 44(0)1243-379-322
Web site: http://www.telescopesales.co.uk
Product line(s): Celestron, Coronado, Fujinon, Helios, Intes, JMI, Kendrick, Miyauchi, Meade, Parks Optical, Vixen, more

Other Countries

Argentina

Laseroptics, S.A.
Lavalle 1634, Piso 3, 1048 Buenos Aires
Phone: 54-11-4374-6754
Product line(s): Meade

R. Jorge Saracco e Hijos S.A.
Buenos Aires
Phone: 54-11-4393-1000
Web site: http://www.saracco.com
Product line(s): Celestron

Australia

Adelaide Optical Centre
120 Grenfell Street, Adelaide, SA 5000
Phone: 61-08-8232-1050
Web site: http://www.adelaideoptical.com.au
Product line(s): Celestron, Kowa, Swarovski, Meade, Vixen, more

Advanced Telescope Supplies
P.O. Box 447, Engadine, NSW 2233
Phone: 61-02-9541-1676
Web site: http://www.ozemail.com.au/~atsscope
Product line(s): Apogee CCDs, Losmandy, SBIG, more

Astro Optical Supplies
9 Clarke Street, Crows Nest, NSW 2065
Phone: 61-02-9436-4360
13 Lower Plaza, 131 Exhibition Street, Melbourne, VIC 3000
Phone: 61-03-9650-8072
Product line(s): Celestron, Australian-made AOS reflectors

Astronomy & Electronics Centre
P.O. Box 45, Cleve, SA 5640
Phone: 61-08-8628-2435
Web site: http://www.users.centralonline.com.au/kagee/aec.htm
Product line(s): Astro-Physics, D&G Optical, Intes, Konus, Lumicon, Miyauchi, Parks, Takahashi, more

Tasco Sales (Australia) Pty Ltd.
19 Roger Street, Brookvale, NSW 2100
Phone: 61-02-9938-3244
Web site: http://www.tasco.com.au
Product line(s): Celestron, Vixen

Telescope & Binocular Shop
55 York Street, Sydney, NSW 2000
Phone: 61-02-9262-1344
Web site: http://www.bintel.com.au
Product line(s): Celestron, Discovery, Nikon, Orion Telescopes, Pentax, Tele Vue, Telrad, Thousand Oaks, more

York Optical and Scientific
92 Wentworth Avenue, Sydney, NSW 2010

Phone: 61-02-9211-1606
316 St. Paul's Terrace, Fortitude Valley, Brisbane, QLD 4006
Phone: 61-07-3252-2061
159A Scarborough Beach Road, Mount Hawthorn, WA 6016
Phone: 61-08-9201-1971
Shop 6-8, 256 Flinders Street, Melbourne, VIC 3000
Phone: 61-03-9654-3938
Web site: http://www.yorkoptical.com.au
Product line(s):AstroSystems, Bausch & Lomb, Carton, Celestron, Fujinon, Kendrick, SBIG, Starlight Xpress, York, Zeiss

Austria
Optikhaus Binder
Schottengasse 2, 1010 Wien
Phone: 43-1-533-6315
Web site: http://www.optikhaus-binder.at
Product line(s): Meade, Swarovski

Belgium
Lichtenknecker Optics
Kuringersteenweg 44, 3500 Hasselt
Phone: 32-11-253-052
Product line(s): Vixen
Pollux Products
Weiveldlaan 41 B33, B-1930 Zaventem
Phone: 32-2-725-5855
Product line(s): Meade

Brazil
Ominis Lux
Rua Berta, 97-Vila Mariana-CEP: 04120-040, São Paulo, SP
Phone: 55-11-5573-2083
Web site: http://www.omnislux.com.br
Product line(s): Meade
SOSECAL
Rua Brigadeiro Galvao, 894, Barra Funda, São Paulo, SP, CEP01151-000
Phone: 55-11-66-4455
Product line(s): Vixen

Denmark
Thomasco Optik Foto
Postboks 9, Djursvang 6, DK-2620 Albertslund
Phone: 45-4-364-1555
Web site: http://www.thomasco.dk
Product line(s): Kowa, Vixen

France
Medas S.A.
57, Avenue P. Doumer, B.P. 2658, 03206 Vichy Cedex
Phone: 33-4-70-301930
Web site: http://www.medas.fr
Product line(s): Astro-Physics, Celestron, Baader, SBIG, Swift, Vixen, more

Germany
Vehrenberg KG
Meerbuscher Strasse 64-78, 40670 Meerbusch-Osterath
Phone: 49-021-595-20321
Web site: http://www.vehrenberg.de
Product line(s): Celestron, Losmandy, Starlight Express, Tele Vue, Vixen , more

Ireland (Republic of)
Astronomy Ireland
P.O. Box 2888, Dublin 1
Phone: 353-1-847-0777
Web site: http://www.astronomy.ie
Product line(s): *Astronomy & Space* magazine, Astro-Physics, Canon, Celestron, JMI, Takahashi, Tele Vue, Vixen, more

Italy
Auriga S.R.L.
Via Quintiliano 30, 20138 Milano
Phone: 39-02-509-7780
Web site: http://www.auriga.it
Product line(s): Celestron, SBIG, Tele Vue, Vixen

Korea
Sun Du Science Corp.
216-16 Jayang-Dong, Kangjin-Ku, Seoul 133-190
Phone: 2-2-453-8178
Web site: http://www.sundu.co.kr
Product line(s): Vixen

Lebanon
Gulbenk Corporation
P.O. Box 11/3111, Beirut
Phone: 961-1-414-238
Product line(s): Vixen

Mexico
Casa Sauer
Liverpool 41, Col. Juarez, Mexico, DF 06600
Phone: 52-5566-5558
Web site: http://www.casa-sauer.com.mx
Product line(s): Vixen

Netherlands
Ganymedes
Midderdorpstraat 1, 1182 HX Amstelveen
Phone: 31-20-641-2083
Web site: http://www.ganymedes.nl
Product line(s): Celestron, Vixen

New Zealand
Skylab
163 St. Asaph Street, Christchurch
Phone: 64-3-366-2827
Web site: http://www.skylab.co.nz
Product line(s): Celestron, Sky Instruments, more

Norway
N.J. OPSAHL A/S
Holbergsgt. 4 3722 Skien
Phone: 47-35-522-531
Web site: http://www.njopsahl.no
Product line(s): Celestron, Vixen

Portugal
Emilio de Azevedo Campos & Ca., Ltd.
Rua da Senhora da Penha, 110-114 Apartado 4057, 4461 Senhora da Hora Codex
Phone: 351-2-957-8670
Product line(s): Vixen

Spain
Cottet Import S.A.
Avda. Puerta de L'Angel, No. 40, Barcelona 08002
Phone: 34-93-301-2232
Web site: http://www.cottet.es
Product line(s): Vixen

Switzerland
P. Wyss Photo Video en Gros
Dufourstrasse 124, CH8034 Zurich
Phone: 41-1-383-0108
Product line(s): Vixen

Taiwan (R.O.C.)
Nick Enterprise Co., Ltd.
12-3F, No. 198, Sec 2, Roosevelt Road, Taipei
Phone: 886-2-2365-5790
Web site: http://www.nick.com.tw
Product line(s): Kowa, Takahashi, Vixen

Thailand
T.M.K. Trading Co., Ltd.
45/2 Soi Soontornpimol, Jarumuang Road, Patumwan, Bangkok 10330
Phone: 66-2-214-0976
Product line(s): Vixen

United Arab Emirates
M.K. Trading Co., L.L.C.
P.O. Box 5418, Dubai
Phone: 971-4-225-745
Product line(s): Vixen

Appendix D
An Astronomer's Survival Guide

It's not always easy being an amateur astronomer. We are nocturnal creatures by nature, going outdoors when the rest of the world sleeps; braving cold, heat, bugs, and things that go bump in the night; always looking up when most everyone else is looking down (astronomers are the eternal optimists).

Here is a checklist of things that I like to bring along for a night under the stars. They make the experience much more pleasurable and the cold night a little warmer.

Astronomical

Telescope____ Binoculars____ Eyepieces____ LPR filter(s)____
Color filter(s)____ Star atlas____ List of things to look at____
Clipboard____ Pen/pencil____ Flashlight (red)____ Flashlight (white)____
Etc.____

____

Photographic

Camera(s)_____ Auxiliary lenses_____ Film_____ Tripod_____
Camera-to-telescope adapters____ Scotch mount_____
Drive corrector_____ CCD Camera/Computer_____
Cable releases (always bring two, in case one breaks)__________________
Etc.____

____

Miscellaneous

Sweatshirt_____ Long underwear_____ Heavy socks_____ Jacket_____
Winter coat_____ Gloves/mittens_____ Boots_____ Foot/hand warmers_____
Hat_____ Folding table_____ Chair_____ Insect repellent_____
Something warm to drink (nonalcoholic)_____ Food/snacks_____ Radio____
Etc.____

____

Appendix E
Astronomical Resources

For those looking to expand their involvement with the hobby and science of astronomy, the following organizations are excellent ways to delve deeper into the universe.

Special Interest Societies

Amateur Telescope Makers Association
17606 28th Avenue SE, Bothell, WA 98012
Web site: http://www.atmjournal.com

American Association of Variable Star Observers
25 Birch Street, Cambridge, MA 02138
Web site: http://www.aavso.org

American Meteor Society
Department of Physics-Astronomy,
SUNY-Geneseo, Geneseo, NY 14454
Web site: http://www.amsmeteors.org

Antique Telescope Society
1275 Poplar Grove Lane, Cumming, GA 30041
Web site: http://www1.tecs.com/oldscope

Association of Lunar and Planetary Observers
305 Surrey Road, Savannah, GA 31410
Web site: http://www.1p1.arizona.edu/alpo

Astronomical League
5675 Real del Norte, Las Cruces, NM 88012
Web site: http://www.astroleague.org

Astronomical Society of the Pacific
390 Ashton Avenue, San Francisco, CA 94112
Web site: http://www.astrosociety.org

International Dark-Sky Association
3545 North Stewart, Tucson, AZ 85716
Web site: http://www.darksky.org

International Meteor Organization
161 Vance Street, Chula Vista, CA 91910
Web site: http://www.imo.net

International Occultation Timing Association
2760 SW Jewell Avenue, Topeka, KS 66611–1614
Web site: http://www.occultations.org

International Supernovae Network
C.P. 7114, I-47100, Forli 39–543–72456 Italy
Web site: http://www.supernovae.net

National Deep Sky Observer's Society
1607 Washington Boulevard, Louisville, KY 40242
Web site: http://users.erols.com/njastro/orgs/ndsos.htm

World Wide Web Resources

The Internet holds a whole universe of resources for the amateur astronomer. Trying to come up with a brief list is impossible, but here are a few sites that might be of interest. I'll lead off the list with my own web site, http://www.philharrington.net, and its companion discussion group, "Talking Telescopes," at http://groups.yahoo.com/group/telescopes. At both sites you will find updates to this book as they become available, including new product reviews, as well as online discussions of telescope pros and cons. Why not join us?

Telescopes and Equipment

Astro-Mart (used equipment)
http://www.astromart.com

Astronomy Mall
http://www.astronomy-mall.com

Bill Arnett's Planetarium Software Review Page
http://www.seds.org/billa/astrosoftware.html

Ed Ting's Scope Reviews Home Page
http://www.scopereviews.com

Matt Marulla's ATM Page
http://www.atmpage.com

Richard Berry's Cookbook Camera Home Page
http://wvi.com/~rberry

Todd Gross's Weather and Astronomy Home Page
http://www.weatherman.com

Sky Information

Abrams Planetarium
http://www.pa.msu.edu/abrams

Astronomy Magazine
http://www.astronomy.com

Jet Propulsion Laboratory's Comet Observation Home Page
http://encke.jpl.nasa.gov

Sky & Telescope Magazine
http://www.skypub.com

Solar Eclipse Home Page
http://sunearth.gsfc.nasa.gov/eclipse/eclipse.html

Astronomy-Related Newsgroups and E-mail Lists

A number of free newsgroups and electronic-mail lists exist on several topics of interest to the amateur astronomer. Newsgroups are typically unmoderated and, as such, can often lead to some truly spirited debates. No need to go through a formal joining process. Instead, jump right in!

Newsgroup	Topic
alt.sci.planetary	Discussion of planetary research and space missions
sci.astro	General discussion group on anything astronomical
sci.astro.amateur	General discussion group with a slant toward amateur observations and equipment
sci.astro.hubble	Discussions of the latest findings from the Hubble Space Telescope
sci.astro.planetarium	Planetarium professionals compare notes on shows and equipment
sci.astro.research	Researchers post questions and answers about various research-related topics

E-mail lists, on the other hand, are more specialized and are usually moderated by the list owner, who may or may not rule with a light touch. They often require you to register with the group, though the step is often more a formality than anything. The table below lists several active mail lists and gives details on how to join them. (Digest versions are often available, as well; consult the instructions that you will receive via e-mail after you subscribe initially.)

Mail Lists

List name	Subject	To Join, Send an E-mail Note to	Include the Following Message (without the quotes)
Astro	General astronomy	majordomo@mindspring .com	Put "subscribe astro" in body of message
AstroCCD	Digital astronomy (Italian)	astroccd-subscribe @ yahoogroups.com	(None required)
Astro_España	Astronomy and astro-computing (Spanish)	listserv@listserv.rediris.es	Put "subscribe ASTRO_ESPANA firstname lastname" in body of message
Astrolist	General astronomy (Danish)	astrolist-subscribe @sunsite.auc.dk	(None)
AstroMart	Classified ads for amateurs	astromart-request@lists .best.com	Put "subscribe" in body of message
Astronomer	Amateur astronomy	astronomer-subscribe @yahoogroups.com	(None)
Astronomia	Amateur astronomy (Spanish)	servidor@correo.dis .ulpgc.es	Put "subscribe astronomia" in body of message
Astrophotography	Film-based astronomical imaging	majordomo@seds.org	Put "subscribe astro-photo" in body of message
ATM	Amateur telescope making	majordomo@shore.net	Put "subscribe atm" in body of message
Big Dob	Building and using large-aperture telescopes	listserver@ucsd.edu	Put "subscribe YOUR E-MAIL ADDRESS bigdob-l" in body of message
CCD	Homemade CCD cameras	listserv@listserv.wwa.com	Put "subscribe" in the message subject line
Celestron Users	Celestron telescopes and accessories	celestronuser-request @lists.best.com	Put "subscribe" in the body of the message
Dark-Sky	Light pollution	DarkSky-list-subscribe @yahoogroups.com	(None)
Dynascope	Criterion Dynascope telescopes	dynascope-subscribe @yahoogroups.com	(None)
Eclipse	Eclipse enthusiasts	listproc@hydra.carleton.ca	Put "subscribe eclipse FIRSTNAME LASTNAME" in body of message
Netastrocatalog	Shared deep-sky observations	majordomo@latrade.com	Put "subscribe netastrocatalog YOUR E-MAIL ADDRESS" in body of message
Meade Advanced Product User	Meade products	mapug-request@shore.net	Put "subscribe" in body of message

List name	Subject	To Join, Send an E-mail Note to	Include the Following Message (without the quotes)
Questar Users	Questar telescopes	majordomo@shore.net	Put "subscribe alt-telescopes-questar" in body of message
SBIG User	Santa Barbara Instrument Group products	sbiguser-request@lists.best.com	Put "subscribe" in body of message CCD cameras
Sct-User	Schmidt-Cassegrain and Maksutov-Cassegrain telescopes	sct-user-subscribe@yahoogroups.com	(None)
Shallow Sky	Lunar and planetary observation	majordomo@shallowsky.com	Put "subscribe shallow-sky" in body of message
Starry Nights	Observational astronomy	telescopes-subscribe@yahoogroups.com	(None)
Talking Telescopes	Astronomical equipment	telescopes-subscribe@yahoogroups.com	(None)

Appendix F
English/Metric Conversion

Most amateur astronomers in the United States will speak of a telescope's aperture in terms of inches, while the rest of the world uses centimeters. The table below acts as a translation table to help convert telescope apertures from one system to another. Recall that there are 2.54 centimeters per inch.

English (in.)	Metric (cm)
2	5
3.1	8
4	10
6	15
8	20
10	25
12	30
14	35
16	40
18	45
20	50
24	60
30	75
32	80
36	90

Star Ware Reader Survey

To help prepare for the next edition of *Star Ware*, I'd like you to take a moment and complete the following survey. Photocopy it, fill it out, and send it to the address shown at the bottom. An online version is also available at the Star Ware Home Page (http://www.philharrington.net) or via e-mail by writing to phil@philharrington.net. Feel free to reproduce it in your club newsletter as well.

Please answer all (or as many) of the questions below as possible. If you own more than one telescope, I'd like to hear about each. *Use as much room as you want!* I'll save each response and reference them when it comes time to write the reviews. Please include your address should I need to get hold of you for questions or clarifications. Rest assured that your address will not be given to anyone, nor will any company ever see your survey.

Thanks in advance!

Your name: ______________________________

Address: ______________________________

City: ______________ State/Country: ________ Zip: ___________

E-mail: ______________________________

How long have you been involved in astronomy? ______________

Do you consider yourself a: Beginner Intermediate Advanced

TELESCOPE

How many telescopes do you own? ______________

Telescope model: ______________________________

How old is it? ____________ Are you the original owner? ____________

What do you like about it?

____

____

What don't you like about it?

____

____

Lived up to your expectations? ____________ Buy it again? ____________

Have you ever had to contact the company about a problem?

____

Was it resolved? Explain.

____

____

EYEPIECES

What eyepieces do you own?

____

How do they work? Any particular likes or dislikes? (Please list your impressions separately for each.)

____

____

ACCESSORIES

What accessories do you own? (Anything ... binoculars, books, software, filters, finderscope, etc.)

____

____

____

Any particular likes or dislikes? (Please list your impressions separately for each.)

____

____

____

____

Your vote for the best telescope of yesteryear (only models that are no longer made). ____

Anything you think should be included in the next edition?

____

____

Please mail this survey to Phil Harrington, c/o John Wiley & Sons, 111 River Street, Hoboken, NJ 07030, or respond by e-mail to phil@philharrington.net.

Index